Springer
Proceedings in Physics 22

Springer Proceedings in Physics

Managing Editor: H. K. V. Lotsch

Volume 1 *Fluctuations and Sensitivity in Nonequilibrium Systems*
Editors: W. Horsthemke and D. K. Kondepudi

Volume 2 *EXAFS and Near Edge Structure III*
Editors: K. O. Hodgson, B. Hedman, and J. E. Penner-Hahn

Volume 3 *Nonlinear Phenomena in Physics* Editor: F. Claro

Volume 4 *Time-Resolved Vibrational Spectroscopy*
Editors: A. Laubereau and M. Stockburger

Volume 5 *Physics of Finely Divided Matter* Editors: N. Boccara and M. Daoud

Volume 6 *Aerogels* Editor: J. Fricke

Volume 7 *Nonlinear Optics: Materials and Devices* Editors: C. Flytzanis and J. L. Oudar

Volume 8 *Optical Bistability III* Editors: H. M. Gibbs, P. Mandel,
N. Peyghambarian, and S. D. Smith

Volume 9 *Ion Formation from Organic Solids (IFOS III)* Editor: A. Benninghoven

Volume 10 *Atomic Transport and Defects in Metals by Neutron Scattering*
Editors: C. Janot, W. Petry, D. Richter, and T. Springer

Volume 11 *Biophysical Effects of Steady Magnetic Fields*
Editors: G. Maret, J. Kiepenheuer, and N. Boccara

Volume 12 *Quantum Optics IV* Editors: J. D. Harvey and D. F. Walls

Volume 13 *The Physics and Fabrication of Microstructures and Microdevices*
Editors: M. J. Kelly and C. Weisbuch

Volume 14 *Magnetic Properties of Low-Dimensional Systems*
Editors: L. M. Falicov and J. L. Morán-López

Volume 15 *Gas Flow and Chemical Lasers* Editor: S. Rosenwaks

Volume 16 *Photons and Continuum States of Atoms and Molecules*
Editors: N. K. Rahman, C. Guidotti, and M. Allegrini

Volume 17 *Quantum Aspects of Molecular Motions in Solids*
Editors: A. Heidemann, A. Magerl, M. Prager, D. Richter, and T. Springer

Volume 18 *Electro-optic and Photorefractive Materials* Editor: P. Günter

Volume 19 *Lasers and Synergetics* Editors: R. Graham and A. Wunderlin

Volume 20 *Primary Processes in Photobiology* Editor: T. Kobayashi

Volume 21 *Physics of Amphiphilic Layers* Editors: J. Meunier, D. Langevin,
and N. Boccara

Volume 22 *Semiconductor Interfaces: Formation and Properties*
Editors: G. Le Lay, J. Derrien, and N. Boccara

Springer Proceedings in Physics is a new series dedicated to the publication of conference proceedings. Each volume is produced on the basis of camera-ready manuscripts prepared by conference contributors. In this way, publication can be achieved very soon after the conference and costs are kept low; the quality of visual presentation is, nevertheless, very high. We believe that such a series is preferable to the method of publishing conference proceedings in journals, where the typesetting requires time and considerable expense, and results in a longer publication period. Springer Proceedings in Physics can be considered as a journal in every other way: it should be cited in publications of research papers as *Springer Proc. Phys.*, followed by the respective volume number, page number and year.

Semiconductor Interfaces: Formation and Properties

Proceedings of the Workshop,
Les Houches, France
February 24 – March 6, 1987

Editors: G. Le Lay,
J. Derrien, and N. Boccara

With 243 Figures

Springer-Verlag Berlin Heidelberg New York
London Paris Tokyo

Professor Dr. Guy Le Lay
C.R.M.C.2, CNRS, Campus de Luminy, Case 913,
F-13288 Marseille, Cedex 09, France

Professor Dr. Jacques Derrien
L.E.P.E.S., CNRS, BP 166, F-38042 Grenoble, France

Professor Dr. Nino Boccara
Centre de Physique Théorique, F-74310 Les Houches, France

ISBN-13: 978-3-642-72969-0 e-ISBN-13: 978-3-642-72967-6
DOI: 10.1007/978-3-642-72967-6

Library of Congress Cataloging-in-Publication Data. Semiconductor interfaces: formation and properties: proceedings of the workshop, Les Houches, France, February 24—March 6, 1987/ editors, G. Le Lay, J. Derrien, and N. Boccara. – (Springer proceedings in physics; v. 22) "A collection of lectures that were given at the International Winter School on Semiconductor Interfaces: Formation and Properties held at the Centre de Physique des Houches from 24 February to 6 March, 1987" – Pref. Includes index. 1. Semiconductors – Surfaces – Congresses. 2. Surface chemistry – Congresses. I. Le Lay, G. (Guy) II. Derrien, J. (Jacques) III. Boccara, Nino. IV. International Winter School on Semiconductor Interfaces: Formation and Properties (1987: Centre de Physique des Houches) V. Series. QC611.6.S9S46 1987 537.6'22—dc19 87-28658

Softcover reprint of the hardcover 1st edition 1987

Printing: Weihert-Druck GmbH, D-6100 Darmstadt
Binding: J. Schäffer GmbH & Co. KG., D-6718 Grünstadt
2153/3150-543210

Preface

The trend towards miniaturisation of microelectronic devices and the search for exotic new optoelectronic devices based on multilayers confer a crucial role on semiconductor interfaces. Great advances have recently been achieved in the elaboration of new thin film materials and in the characterization of their interfacial properties, down to the atomic scale, thanks to the development of sophisticated new techniques.

This book is a collection of lectures that were given at the International Winter School on *Semiconductor Interfaces: Formation and Properties* held at the Centre de Physique des Houches from 24 February to 6 March, 1987. The aim of this Winter School was to present a comprehensive review of this field, in particular of the materials and methods, and to formulate recommendations for future research.

The following topics are treated:

(i) Interface formation. The key aspects of molecular beam epitaxy are emphasized, as well as the fabrication of artificially layered structures, strained layer superlattices and the tailoring of abrupt doping profiles.

(ii) Fine characterization down to the atomic scale using recently developed, powerful techniques such as scanning tunneling microscopy, high resolution transmission electron microscopy, glancing incidence x-ray diffraction, x-ray standing waves, surface extended x-ray absorption fine structure and surface extended energy-loss fine structure.

(iii) Specific physical properties of the interfaces and their prospective applications in devices.

We wish to thank warmly all the lecturers and participants, as well as the organizing committee, who made this Winter School a success.

Finally, we gratefully acknowledge financial support from the Centre National de la Recherche Scientifique, the Université de Provence and IBM-France.

Marseille, Grenoble
Gif-sur-Yvette, May 1987

G. Le Lay
J. Derrien
N. Boccara

1. N. Leblanc
2. Vu Thien Binh
3. Y. Borensztein
4. J. Derrien
5. J. Geurts
6. R.S. Bauer
7. P. Chiaradia
8. D. Paget
9. A. Cricenti
10. G. Le Lay
11. A. Kiejna
12. H.J. Drouhin
13. J. Bonnet
14. R. Strumpler
15. L. Voisin
16. P. Galtier
17. C. Priester
18. C. Gaonach
19. L. Stauffer
20. G. Shaofang
21. G. Mathieu
22. Y. Chabal
23. M. Sauvage
24. G. Gillmann
25. J.Y. Veuillen
26. P. Boher
27. J.M. Moison
28. F. Grey
29. K. Hricovini
30. M. Prietsch
31. F. Rochet
32. L. Johansson
33. S. Doriot-Ghestin
34. S. Rousset
35. J. Mercier

Contents

Part I Introduction

An Introduction to the Formation and Properties of Semiconductor Interfaces. By C.A. Sébenne ... 2

Part II Experimental Study of the Formation of Semiconductor Interfaces

Formation of Semiconductor Interfaces During Molecular Beam Epitaxy. By K. Ploog (With 15 Figures) ... 10

Build-up and Characterization of "Artificial" Surfaces for III–V Compound Semiconductors. By J.M. Moison ... 43

Part III Structural Characterization of the Interfaces

Atomic Structure of Semiconductor Surfaces By G. Le Lay (With 5 Figures) ... 48

Monolayer Sensitive X-Ray Diffraction Techniques: A Short Guided Tour Through the Literature. By M. Sauvage-Simkin ... 66

SEXAFS for Semiconductor Interface Studies By G. Rossi (With 8 Figures) ... 69

XANES and XARS for Semiconductor Interface Studies By G. Rossi (With 10 Figures) ... 88

Recent Progress in Electron Spectroscopy: Application to the Local Geometry Determination at Surfaces and Interfaces By J. Derrien (With 7 Figures) ... 102

On the Use of Electron Microscopy in the Study of Semiconductor Interfaces. By J.-P. Chevalier (With 4 Figures) ... 109

Analytical Scanning Electron Microscopy Under Ultra High Vacuum By P. Morin ... 114

Scanning Tunneling Microscopy and Spectroscopy
By F. Salvan (With 7 Figures) 119

Field Emission Microscopy for Analysis of Semiconductor Surfaces
By V.T. Binh (With 9 Figures) 126

Surface and Interface Studies with MeV Ion Beams
By C. Cohen (With 15 Figures) 134

Surface Characterization by Low-Energy Ion Scattering
By E. Taglauer (With 11 Figures) 150

Part IV **Electronic Properties of Interfaces**

Band Structure Theory of Semiconductor Surfaces and Interfaces
By C. Calandra and F. Manghi (With 13 Figures) 162

Electronic Properties of Semiconductors: Fermi Level Pinning in Schottky Barriers and Band Line-up in Semiconductors
By F. Flores (With 10 Figures) 182

Photoemission and Inverse Photoemission from Semiconductor Interfaces. By F.J. Himpsel (With 11 Figures) 196

Photoelectron Spectroscopies: Probes of Chemical Bonding and Electronic Properties at Semiconductor Interfaces
By G. Hollinger (With 18 Figures) 210

Two-Photon Photoemission in Semiconductors
By J.M. Moison (With 2 Figures) 232

Formation and Electrical Properties of Metal-Semiconductor Contacts.
By L. Lassabatère (With 14 Figures) 239

Deep Level Transient Spectroscopy for Semiconductor Surface and Interface Analysis. By E. Rosencher (With 10 Figures) 273

Admittance Spectroscopy of Interface States in Metal/Semiconductor Contacts. By P. Muret (With 3 Figures) 282

Part V **Optical and Vibrational Properties of Interfaces**

Optical Properties of Surfaces and Interfaces
By P. Chiaradia (With 8 Figures) 290

Vibrational Properties at Semiconductor Surfaces and Interfaces
By Y.J. Chabal (With 11 Figures) 301

Raman Scattering from Interface Regions: Structure, Composition and Electronic Properties
By J. Geurts and W. Richter (With 9 Figures) 328

Part VI **Interfaces: Present Status and Perspectives**

Role of Interfaces in Semiconductor Heterostructures
By C. Delalande and G. Bastard (With 17 Figures) 340

The Physics of Metal Base Transistors
By E. Rosencher, F. Arnaud d'Avitaya, P.A. Badoz, C. d'Anterroches, G. Glastre, G. Vincent, and J.C. Pfister (With 13 Figures) 360

Perspectives on Formation and Properties of Semiconductor Interfaces
By R.S. Bauer, R.H. Miles, and T.C. McGill (With 13 Figures) 372

Index of Contributors 389

Part I

Introduction

An Introduction to the Formation and Properties of Semiconductor Interfaces

C.A. Sébenne

Laboratoire de Physique des Solides, associé au CNRS,
Université Pierre et Marie Curie, F-75252 Paris Cedex 05, France

1. IMPORTANCE OF INTERFACES IN SEMICONDUCTORS.

The role of interfaces is multiple and essential : any semiconductor-based device has interfaces between its heart, which is a piece of homogeneous single crystal, and the outside world. This outside world can be as friendly as the same crystal, differing only through its doping, in the case of a homojunction. It can be as hostile and unpredictable as the deleterious atmosphere of a big city. Besides giving a finite size to a device, interfaces act in building finite and sometimes driven potential walls for charge carriers, in letting free carriers in and out, or other excitations such as photons in and out, etc. The degree of control needed for the electronic properties of an interface is dictated by the bulk properties which characterize the semiconductor : essentially, the band gap, the extrinsic properties, the dynamical response to excitations. Mastering the bulk doping down to 10^{15} cm^{-3} means, with regard to the surface, a mastering of the gap interface states below the 10^{10} cm^{-2} range, compared to, typically, 10^{15} atoms per cm^2 along a single atomic plane. There are not that many semiconductors for which the intrisic bulk properties are that well controlled. Therefore huge difficulties are expected when trying to approach an equivalent control over interfaces. Then, the game in many device technologies consists in trying to circumvent the interface problem through various tricks which remain unavoidable in the present state of the art. In the following sections, the main driving forces which govern the physical properties of semiconductor interfaces are analysed and the specificities of their experimental approach are underlined before making a brief statement of the challenges for the future.

2. DRIVING FORCES AND EXAMPLES.

Dealing with the bulk properties of a semiconductor crystal implies, first, the knowledge of its crystallographic structure, then, the knowledge of its electronic band structure. This has to be transposed to interfaces where the periodic character of the structure is obviously lost along at least one direction (the case of homojunctions is kept aside throughout the following) : the atomic geometry and its local variations are the corner stone of the problem, and the electronic structure is linked to them. In order to discuss the various situations, the semiconductor interfaces will be divided into two families : a family where a perfect interface can be defined, for example by continuity of the crystal structure or, at least, of the nature and the number of the bonds, and a family where the notion of perfect interface is lost, for example in metal-semiconductor interfaces.

2.1 Near perfect semiconductor interfaces.

Considering a semiconductor crystal, an ideally perfect interface would be one where all the semiconductor atoms along the ending plane are similarly and uniformly bound, without distortion, to the outside atoms with an atomically abrupt compositional change. Then, the atomic arrangement and the corresponding electronic structure are intrinsic properties of the interface. Consequently any departure from the perfect structure, being a step, a vacancy (on either side), an impurity or a more complex local distortion, is a defect to which specific electron states are associated, giving the extrinsic properties of the interface. In practice, the accommodation of the various driving forces involved at such abrupt interfaces, which are essentially the strengths, the lengths and the angles of the bonds, is never made without the presence of some defects. Their origins and consequences will be illustrated by various examples.

- $Ga_xAl_{1-x}As$ / GaAs

One of the systems closest to perfection is the AlAs - GaAs heterojunction. Taking for example the (100) plane, which is alternately made of As atoms and cations (Ga or Al), a perfect interface would occur if, from one side of a given As plane to the other, the composition of the cation planes were switching from Ga to Al. However, the lattice parameters of AlAs (5.61 Å) and GaAs (5.6534 Å) differ by 0.8 % and one side of the heterojunction stresses the other. In the case of a thin layer, for example AlAs on a GaAs substrate, the latter imposes its own parameter, and either the layer is elastically deformed (the hydrostatic pressure needed to induce a 0.8 percent decrease of the AlAs lattice parameter is about 18 kbars ; then the uniaxial pressure normal to the AlAs layer equivalent to the two-dimensional extension stress exerted by the GaAs substrate is of the order of 6 kbars) or dislocations are generated to reduce or release the strains. Actually, in practical systems, one focusses on direct gap compounds and $Ga_xAl_{1-x}As$ layers with x in the range below 0.5 are usually grown. The compression or the strain-induced defects are therefore less important, while still there. However, the use of an alloyed compound introduces a statistical distribution effect along the interface. Of course, specific defects such as steps or kinks, besides point defects, can also be defined along these almost perfect interfaces.

Considering the electronic properties, the interface is a simple switch from mostly covalent As - Ga bonds to As - Al similar bonds with practically no local distortion : except in the case of point defects, which are not specific to the interface itself, no localized electron states are expected there. At most, depending on the preparation conditions, some compositional gradient may occur over a few layers normal to the interface, making the band discontinuity slightly softer.

- Ge / Si interfaces

Like in the first example, a noticeable lattice mismatch of 4 percent exists between the two covalent elemental semiconductors. However, it is too large to be accommodated through an epitaxial strained layer with an atomically abrupt interface. Starting, for example, with a clean Si substrate, either an amorphous Ge layer is formed when the temperature is too low, or some Ge - Si alloying occurs, not necessarily homogeneous along the interface, when the temperature is raised, making in any case the atomic arrangement a complicated problem.

The electronic structure is expected to display localized states all over the gaps since the situation leads to Ge - Ge, Si - Si and probably some Ge and Si dangling bonds in addition to the expected Ge - Si bonds at the interface, all these being variously deformed in accordance with the local atomic geometry. From a fundamental point of view, it seems an interesting system where several parameters can be studied.

- Ge - GaAs interfaces

Considering the lattice parameters, this is one of the most satisfactory situation,since the relative difference at room temperature is below 7×10^{-4}.However, the interface has to accommodate three elements, each with a different valency. It leads either to amorphous layers when the formation temperature is too low, or to some mixing if the temperature is raised. Even if a satisfactory compromise is found, the electronic structure is hard to keep under control since specific bonds occur at the interface, Ge - Ga and Ga - As, where the charge transfer is not obvious. Moreover, the bulk electronic properties on both sides may be changed since Ge acts either as a donor or an acceptor when substituting to Ga or As in GaAs, while As and Ga act respectively as donor and acceptor when substituting in Ge.

- $Ga_xIn_{1-x}As$ / InP interfaces

The proper adjustment of x allows a perfect matching of crystal parameters, at a given temperature. However, since one side of the interface is an alloy, the matching is only an average and at the interface some statistical distortion occurs due to the various covalent radii involved. Moreover Ga - P bonds exist only at the interface. These first two points concern the perfect interface. However, in practice, the formation of such an interface cannot be obtained within a stationary thermodynamical regime : it means that the existence of a compositional gradient, at least normal to the interface, is unavoidable, making the description of the interface in terms of defects very difficult. As a consequence, the prediction of the actual electronic structure is a perilous game, especially when the experimental determination is far from accurate,to say the least.

- InAs - GaSb

In spite of a good matching of the lattice parameters, a complicated interface can be expected since stronger bonds involving smaller atoms, Ga - As, must coexist there with weaker bonds involving bigger atoms, In - Sb. Therefore compositional fluctuations both normal and along the interface are expected.

- GaAs - ZnSe, GaSb - ZnTe, InSb - CdTe , GaP - CuCl, InAs - CuI ...

The valence changes from one side of the junction to the other bring a chemically more confusing situation at the interface where the notion of perfection is now very far away, whatever the lattice parameter matching.

- Si - silicides

Besides heterojunctions, another family of heterostructures displays high-quality interfaces obtained through the growth of several transition metal silicides on silicon. These MSi_2 compounds (with M : Ni, Co, Cr, W, Mo, Pt, Pd...) are metallic and have a well-adapted crystal structure, usually within 1 or 2 percent. This is the only situation where the influence of interface defects on Schottky barrier properties can be checked experimentally.

- Si - SiO_2

The silicon-silicon dioxide is the best, and unique, demonstration that crystal structure continuity is not a necessary condition to get an interface with a very low density of electrically active defects. In fact, it is the angular adaptability of the bonds in the glassy material which allows a full saturation of the available Si bonds. Actually, two phenomena contribute to such a perfect achievement : the interface smoothing which is a consequence of the thermal oxidation microscopic process, and the presence of some hydrogen to remove residual silicon dangling bonds. As a result, the residual density of electron interface states in the gap can get below 10^9 cm^{-2} along the (100) surface.

2.2 Unmatched semiconductor interfaces

When the guiding effect of some local structure continuity is removed, the tendency to energy minimization at a semiconductor interface will be a compromise between several conflicting forces. Without trying to cover all cases, a few significant examples will be given simply to illustrate the kind of driving forces one has to deal with.

- Semiconductor - Vacuum interfaces

Along such interfaces, the semiconductor must move its own atoms in order to insure the proper charge redistribution normal to the surface so as to screen the existence of the surface to the bulk material. This is at the expense of the tetrahedral symmetry, which however tries to save itself at the best. The partly ionic character of the bonds in compound semiconductors helps to the compromise with a limited distortion : for example the GaAs bond, usually along the (110) plane, makes an angle of about 27° with this plane upon reconstruction of the cleaved surfaces, acting then on the surface state position. In purely covalent semiconductors such as Si, the reconstruction has to make unequivalent at least two surface atoms. This is not always possible, and for example there is a generation of surface vacancies in the 7 x 7 reconstruction of Si(111) in order to save the tetrahedral bonding, to minimize the distortion and to bring the proper charge transfer.

- Metal - Semiconductor interfaces

This is certainly the most difficult case, since from one side of the interface to the other, one has to switch from a mostly covalent material with localized electrons and four first neighbours forming a tetrahedron around each atom, to a metal with delocalized electrons and twelve first neighbours in a close-packed structure for each atom. The main driving forces which are involved in the establishment of atomic geometry at the interface, either locally regular or not, and macroscopically homogeneous or not, can be tentatively listed as follows :

* a certain number of metal atoms may be used to ensure the necessary charge transfer to end the semiconductor (Ag or Ga on Si).

* the size of the metal atoms may let survive a certain density of semiconductor dangling bonds (In on GaAs).
* the formation of stronger bonds may induce a chemical exchange between the metal and the semiconductor (example : Al substituting In in the last layer at Al / InP interfaces) or the formation of a mixed layer (Cu or Au / GaAs).

In general the interface acts on the semiconductor as a buffer layer upon which the metal atoms are more or less free to arrange themselves as they wish. It may give rise to a pseudo epitaxy when metal atoms pack as dense plane along the surface and grow as oriented crystals, without true parameter matching with the substrate.

- Other interfaces

The main driving forces expressed in the case of metal-semiconductor interfaces are still valid, and the knowledge in depth of chemistry is fundamental in the foresight of what may occur (phosphate growth by oxidizing InP, ohmic contact formation using metal alloys on Si or GaAs, etc...).

3. EXPERIMENTAL AND THEORETICAL APPROACHES

A semiconductor interface is a two-dimensional system, more or less uniform and homogeneous, with a third dimension of a few angstroems. The determination of its physical properties requires the use of various specific techniques, both experimental and theoretical. It is the goal of the present winter school to give an insight into these approaches. As a general presentation, experiments follow three complementary ways :

- Surface physics approaches

Mostly used under vacuum, these techniques are able to produce important sets of information, on atomic composition and geometry as well as on the main electronic properties of a semiconductor surface at most covered by a foreign layer no more than a few angstroems thick. The recent availability of scanning tunnel microscopy brings a brighter future to these approaches. The stage of surface defect physics is now within reach.

- Differential bulk physics approaches

The improvement in sensitivity of many bulk spectroscopy techniques renders their use fruitful to study completed interfaces, mostly through differential methods. Simple atomic geometries are attainable.

- Electric and photoelectric approaches

These properties are directly related to devices and very accurate measurements have been made. They characterize in detail the static and dynamic features of the electronic response of various interfaces through essentially interface charge density variations. However, these effects do not allow one to trace the causes, and explanations proceed mostly through phenomenological models.

Most theoretical approaches had to limit themselves to some fundamental concepts concerning clean surfaces, band offset at perfect heterojunctions, the origin of a Schottky barrier. This is mainly due to the lack of interfaces where a detailed knowledge of the actual atomic geometry has been reached.

The present winter school intends to show that

- The determination of physical properties of semiconductor interfaces is a young and interesting problem : very few systems are effectively known and it involves most solid state physics concepts.
- The availability of a wider and wider variety of experimental approaches will make realistic as a goal the determination of the physical properties of some simple interfaces.
- Linking the microscopic properties of an interface with the electrical properties of a device would seem to be the ultimate goal and it is not out of reach : in that respect, the continuous progress, over 25 years, on the detailed understanding of the Si - SiO_2 interface is very encouraging.

Part II

Experimental Study of the Formation of Semiconductor Interfaces

Formation of Semiconductor Interfaces During Molecular Beam Epitaxy

K. Ploog

Max-Planck-Institut für Festkörperforschung, D-7000 Stuttgart 80, Fed. Rep. of Germany

1. INTRODUCTION

Molecular beam epitaxy (MBE) is a sophisticated, yet flexible, technique for growing epitaxial thin films of semiconductors, metals and dielectrics by impinging thermal-energy beams of atoms or molecules onto a heated substrate under ultrahigh vacuum (UHV) conditions / 1 / . The method provides atomic abruptness between layers of different lattice-matched and also lattice-mismatched crystalline materials at their interfaces or heterojunctions. The interfaces between epitaxial semiconductor layers of different composition or doping are used to confine electrons and holes to two (or even one)-dimensional motion. The challenge for the design and growth of artificially layered materials is to minimize scattering from impurities, alloy clusters or interface irregularities so that carriers can move freely along the interfaces.

In this article we discuss in detail the significant factors to attain high-quality MBE growth of monocrystalline III-V semiconductors, including growth system, substrate preparation, growth mechanism and growth conditions, in-situ analysis and growth control, and dopant incorporation, with a tutorial emphasis. In addition, we describe a few examples for (a) the control of the interface quality by monitoring the oscillations in the intensity of the reflection high energy electron diffraction (RHEED) and (b) the investigation of structural interface disorder effects in superlattices (SL) and multi quantum well heterostructures (MQWH) by high-resolution double-crystal X-ray diffraction. Finally, in a separate chapter we present selected examples for the application of interface formation during MBE to modify the bulk properties of semiconductors through bandgap (or wavefunction) engineering by grading the composition both abruptly (on an atomic scale) or gradually. In the final chapter only keynotes are given and figures are not included, as they appear elsewhere, but original references are cited. We focus on III-V compound semiconductors because they have been dominating the worldwide MBE activities.

2. TECHNOLOGY AND FILM GROWTH PROCESS

2.1 History of MBE and General Considerations

The basic ideas of the MBE growth process were developed in 1958 by Günther / 2 / for the deposition of stoichiometric films of compound semiconductors formed from elements of group III and V of the periodic table (e.g. GaAs, $Al_xGa_{1-x}As$, InAs, InSb etc.). The crucial point for achieving stoichiometry is that most of the constituent elements of III-V semiconductors have largely different vapour pressures so that these materials exhibit considerable decomposition at the respective evaporation temperature. In his so-called three-temperature method, Günther used the group-V-element source oven, kept at temperature T_1 (e.g. 300 °C for arsenic), to maintain a steady vapour pressure in his static vacuum chamber. The source oven with the group-III-element, kept at a much higher temperature T_3 (e.g. 950 °C for gallium), was providing a flux of atoms incident on the substrate that was critical for the condensation rate. The choice of the substrate temperature was most crucial. This intermediate temperature T_2 had to be increased to a value that allowed the condensation of the III-V compound but ensured that the excess group-V-component which had not reacted was re-evaporated from the substrate surface. Subsequently, in the middle 1960's, Arthur / 3, 4 / performed the first fundamental studies of the kinetic behaviour of Ga and As_2 species evaporating from or impinging onto GaAs surfaces using pulsed molecular beam techniques under UHV conditions that led him to a rough understanding of the growth mechanism. These investigations were soon followed by the application of the same technology, now called molecular beam epitaxy (MBE), by Cho / 5, 6 / to the growth of thin films for practical device fabrication purposes. A detailed bibliography spanning the first three decades of molecular beam epitaxy has been compiled for silicon by Bean and McAfee / 7 / and for III-V compounds by Ploog and Graf / 8 / .

The characteristic features of MBE are (a) the slow growth rate of one monolayer per second that permits the control of layer thickness within nanometers, by compiling atomic layer upon atomic layer, (b) the reduced growth temperature (e.g. 550 °C for GaAs) which causes only negligible bulk diffusion at abrupt material interfaces, (c) the ability to produce extremely abrupt material interfaces by the application of mechanical shutters positioned in the path of each molecular beam, (d) the progressive smoothing of the growing surface for most substrate orientations due to the specific non-equilibrium growth mode, and (e) the facility for in-situ analysis to ensure that the desired surface reaction conditions are reached before growth is started and are maintained while the crystal is growing. The resulting unique capability of molecular beam epitaxy for deposition control at the monolayer level opened the possibility for the growth of a new generation of exceptionally complex multilayer structures which include artificial superlattices, crystalline semiconductor - metal - semiconductor and semiconductor - dielectric - semiconductor composites, as well as metastable structures.

The schematic illustration in Fig. 1 shows the basic evaporation process for the growth of III-V semiconductors to be carried out under UHV conditions. The molecular beams reacting on the surface of the heated substrate crystal (usually a (001)oriented GaAs, InP, or GaSb slice) are generated by thermal evaporation of the constituent elements, or their compounds, contained in temperature-controlled effusion cells. These cells are arranged such that the central portion of their flux distribution intersects the substrate at an orifice-substrate distance rang-

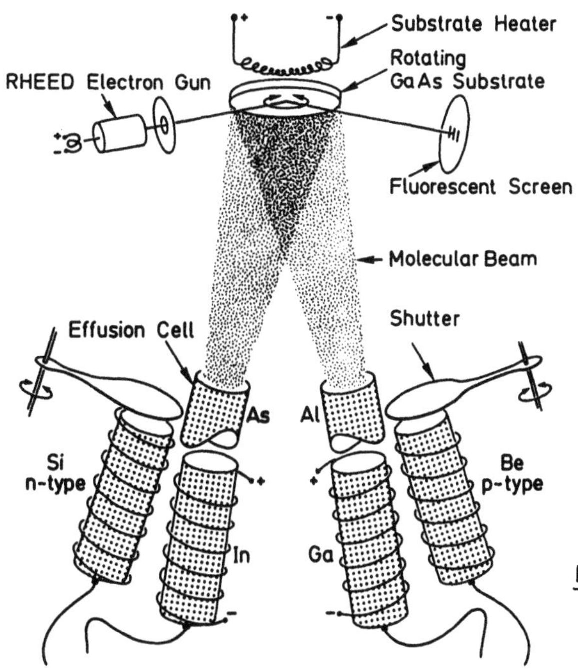

Fig. 1 Schematic illustration of MBE growth process for III-V semiconductors

ing from 100 to 180 mm in various MBE systems. Each source is provided with a separate, externally controlled, mechanical shutter. Operation of these shutters allows the rapid change of the beam species arriving at the substrate in order to alter abruptly the composition of the growing film. At the usual, slow, MBE growth rates of 0.1 to 1.5 μm/hr the shuttering time is much shorter than the time needed for the growing of a monolayer. Therefore, abrupt material interfaces can be achieved which are not washed out by any bulk diffusion because of the low MBE growth temperatures / 9 /. Uniformity in thickness and composition of the grown film depends on the uniformity of the molecular beams across the substrate and on the uniformity of substrate heating, in particular when growth of III-III-V and III-V-V ternary material is carried out at increased substrate temperature. The uniformity of the molecular beams depends not only on the beam fluxes at the substrate center but also on the geometrical relation between the effusion cells and the substrate.

The group-III-component for the growth of III-V semiconductors is obtained by evaporation of the respective element which produces only atomic species. The evaporation of group-V-elements, however, produces tetrameric molecules (P_4, As_4, Sb_4), while incongruent evaporation of the III-V compound itself yields dimeric molecules (P_2, As_2) / 10 /. It is not possible to produce monomeric beams of the group-V-elements. The film growth rate and the composition of $Al_xGa_{1-x}As$ layers is essentially determined by the flux of the group-III-elements, because of their unity sticking coefficient, whereas the stoichiometry is ensured as long as an excess flux of group-V-molecules is supplied (for details see Sect. 2.3). In practice, the III-V compounds are grown with a 2- to 10-fold excess of the group-V-element in order to keep the elemental group-V/III ratio > 1 on the growing surface, also at higher substrate temperatures. The excess arsenic reevaporates from the growth surface. The incorporation of controlled amounts of electrically active impurities into the growing film, for the purpose of doping, is achieved by using separate effusion cells which contain the desired doping element.

The intensities of the molecular beams incident on the growing surface are controlled by the temperature of the respective effusion cells. For typical growth rates of 1 to 2 monolayers per second, the fluxes required at the substrate are approximately 10^{14} to 10^{15} atoms $cm^{-2}s^{-1}$ for the group-III-elements, 10^{15} to 10^{16} atoms $cm^{-2}s^{-1}$ for the gorup-V-elements, and 10^7 to 10^{11} atoms $cm^{-2}s^{-1}$ for the dopants. These fluxes are produced with cell pressures of 10^{-3} to 10^{-2} Torr for the main constituent elements and correspondingly lower pressures for the dopant elements. Continuous changes in the chemical composition of the growing film are achieved by programmed variation of the cell temperatures. Abrupt changes are obtained by operating the respective mechanical shutters.

2.2 MBE Growth Apparatus

The slow growth rate achievable with MBE and the obvious film purity requirements necessitate extremely low impurity levels in the growth chamber. The whole operation is thus carried out in a bakeable stainless steel UHV system with a base pressure of 2×10^{-11} Torr which is usually pumped by ion pumps and Ti sublimation pumps. Additional LN_2 cryopanels keep, in the vicinity of the growing crystal, the partial pressure of gases with high sticking coefficient, e.g. oxygen containing species, even below 10^{-14} Torr. In practice, the total pressure in the growth chamber may rise, during deposition, above 10^{-9} Torr due to scattered beam species and, in the case of III-V compounds, to the excess of group-V-element used. However, the UHV condition is always maintained with respect to impurity species. This UHV environment is ideal for surface analytical studies, such as Auger electron spectroscopy (AES), quadrupole mass spectroscopy (QMS), and reflection high energy electron diffraction (RHEED), that examine the substrate surface prior to epitaxial growth and provide a high degree of in-situ growth control. In the initial development period, surface analysis performed during deposition played a major role for the understanding of the general growth mechanisms of MBE / 11 - 14 /.

The majority of the UHV equipment used for MBE growth until 1977 was home designed, i.e. it was built by commercial UHV manufacturers according to the instruction of the user, and it consisted basically of a single deposition-analysis chamber without sample load-lock device. Since 1978 more standardized MBE systems composed of several basic UHV building blocks, such as growth chamber, substrate loading and preparation unit, and optional surface analysis chamber, each with a separate pumping system, have become commercially available. This more sophisticated but modular design allows the whole instrumentation to be matched easily to most individual requirements. In Fig. 2 we show schematically the top view of an advanced multi-chamber UHV system designed primarily for material growth. The assembly of the effusion cells is positioned such that the center of the beams is directed quasi-horizontally towards the substrate. The part of the system essential for crystal growing consists of the growth chamber, the sample preparation, and the load-lock chamber, which are separately pumped and interconnected via large diameter channels and isolation valves. The analytical chamber is an option for performing detailed surface analytical studies on MBE grown material without exposing the sample to outside environments.

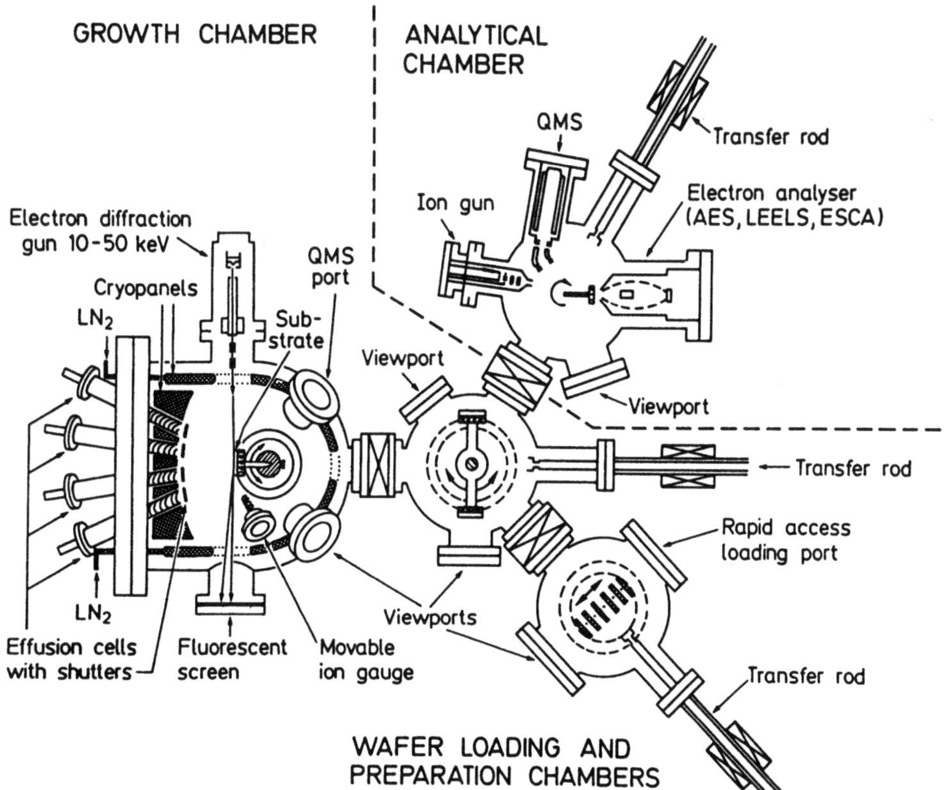

Fig. 2 Schematic diagram (top view) of a multichamber MBE system. The analytical chamber is an option for performing detailed surface analytical studies.

The growth chamber, usually of 450 mm inner diameter (ID), holds a large-diameter removable source flange which provides support to eight effusion cells, their shutters, and the LN_2 shroud. The cells are fixed to individual 38 mm ID flanges for easy refilling and exchanging, and they are spaced and oriented such that their beams converge on the substrate in the growth position. The beam sources are thermally isolated from each other by a LN_2 cooled radial vane baffle, which prevents chemical cross-contamination. For the growth of reactive Al-containing material ($Al_xGa_{1-x}As$, $Al_xIn_{1-x}As$ etc.) a second cryopanel surrounds the substrate such that the stainless steel bell-jar is internally lined by this cryopanel. This arrangement also reduces contamination due to outgassing of the chamber walls.

In MBE systems used predominantly for continuous film growth, only a few monitoring techniques are required. A RHEED facility with primary energies between 10 and 50 keV, operated in the small glancing angle reflection mode, is used to monitor, on a fluorescent screen, the structure of the outermost layers of the substrate before and during epitaxial growth. Deviations from established growth patterns can be instantly identified so that adjustments of the operating parameters can be readily made. Originally the RHEED technique was used during MBE growth for the following monitoring purposes: (a) cleaning and annealing process of the substrate surface, (b) initial stages of epitaxial growth, and (c) changes of the surface structure during heterostructure growth or when variations of the elemental arrival rate or of the substrate temperature are intentionally performed during growth /12/. Recently, temporal intensity oscillations in the features of the RHEED pattern have been employed to study MBE growth dynamics and the formation of semiconductor heterojunctions / 14 - 17 / . To observe RHEED intensity oscillations, the intensity of particular features (usually the specularly reflected beam) is measured via optical fibre with a well collimated entrance aperture using a photomultiplier. The oscillation period of the specular beam intensity provides a continuous growth rate monitor with atomic layer precision, which is now used for absolute beam flux calibration.

A movable ion gauge is used to measure molecular beam fluxes before commencement of growth by turning it into the substrate position to measure the direct beam density and then out of the beams to measure the background density / 18 / . Although not element specific, the ion gauge beam flux monitor reaches sufficient accuracy for most requirements during MBE growth, provided it is calibrated by weighing a deposit or by measuring the film thickness after growth. In addition to the inexpensive ion gauge, a quadrupole mass spectrometer (QMS) may be used as a true element-specific detector to monitor the species emerging from the molecular beam sources and to control the background-gas composition. The ionizer of the spectrometer, which must be encircled by a LN_2 cooled shroud, should be placed within line of sight of the effusion cells and in close proximity of the substrate position during operation. The QMS is a valuable aid in setting up the initial conditions in a newly installed growth system, because a close correlation of important operating parameters with process results can be readily made. However, it is difficult to use the mass spectrometer as a quantitative element-specific sensor in closed-loop computer controlled MBE systems, because of the uncertain long-term stability of the instrument due to contamination inside the growth chamber.

2.3 Effusion Cells for Molecular Beam Generation

The intensity of the molecular beam is determined by the vapour pressure of the starting materials (elements) and by the evaporation geometry. The design of high-quality molecular beam sources fabricated from non-reactive refractory materials is the most important requirement for successful MBE growth. The evaporation sources must provide stable high-purity molecular beams of the required intensity and uniformity, and they must withstand operating temperatures up to 1400 °C without themselves contributing to the molecular beams.

For accurate kinetic studies relevant to the growth process, Knudsen effusion cells with small orifice should be used to produce collision-free thermal-energy beams of the constituent elements which are directed to the substrate. When the vapour pressure of the material heated in the cell is such that the mean free path is larger than the orifice diameter, molecular flow of the emerging species is ensured. The beam flux $J(\phi)$ on a substrate at distance L(cm) from the source with angle ϕ between cell axis and the normal of the substrate surface (see Fig. 3a) is then given by

$$J(\phi) = 1.11*10^{22}\ [(AP/L^2)(MT)^{-1/2}]\ \cos\phi \quad \text{molecules cm}^{-2}\text{s}^{-1} \qquad (1)$$

where A (cm^2) is the orifice area, P (Torr) the equilibrium vapour pressure in the cell at temperature T (K), and M is the molecular weight. The vapour pressure versus temperature relationship for elemental source materials can be expressed by

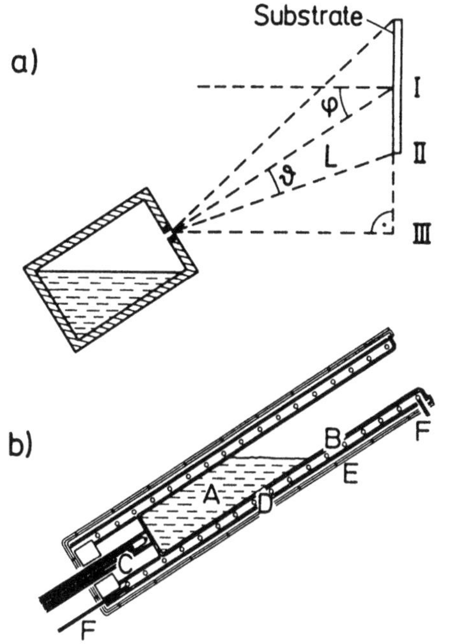

Fig. 3 Effusion cell for molecular beam epitaxy. (a) Definition of coordinates for the calculation of the flux distribution from a true (equilibrium) Knudsen cell across a non-axially mounted substrate. (b) Schematic diagram of a non-equilibrium MBE effusion cell without taper made from pyrolytic boron nitride (PBN) and refractory metals; (A) evaporant, (B) PBN tubular crucible, (C) PBN thermocouple sheath, (D) outer PBM tube for tantalum heater wire support and isolation, (E) crimped tantalum foil heat shield (F) tantalum wire.

$$\log P(T) \simeq \frac{A}{T} + B \log T - C \qquad (2)$$

where A, B, and C are element-specific empirical constants given in Ref. / 19 / . If we take Ga as a typical example, the vapour pressure at T = 1283 K (1010 C) is 3×10^{-3} Torr. Inserting M = 70, A = 1 cm^2 and   = 10 cm, the flux on the substrate is 8×10^{14} atoms $cm^{-2}s^{-1}$.

It is important to note that Eq. (1) is strictly valid only for a true Knudsen effusion cell, i.e. a cell containing condensed phase and vapour at equilibrium and with an ideal orifice of area A in a diaphragm of negligible thickness as effusion aperture. In practice, holes drilled in a lid of finite thickness have to be used as non-ideal orifices with a finite orifice wall. The orifice has thus a somewhat collimating effect on the molecular beam. Herman / 20 / has summarized analytical formulae for the angular intensity distribution of molecular beams from single and multi channel non-ideal orifices. Unfortunately, only very limited experimental data for examination of the accuracy of the calculations are available.

For practical film growth, nonequilibrium (Langmuir-type) effusion cells of sufficient material capacity for reduced filling cycles are used. In general, the cells are of either cylindrical or conical (tapered) shape, and they have a length of about 5 - 10 times greater than their largest diameter. They have a large aperture, i.e. the crucibles have no lid, in order to achieve large epitaxial areas with uniform thickness at the required growth rate of 1 - 2 monolayers/s. Dayton / 21 / has first pointed out that a crucible of straight cylindrical shape acts as a collimator, particularly in the vertical position, so that the flux exhibits a more sharply peaked angular distribution when the charge level falls. A computer model for the effusion process from cylindrically or conically shaped sources was developed by Curless / 22 / and applied to the growth of GaAs. The observed variation of the Ga flux over a 4 cm^2 substrate was in good agreement with the model. The important result is that most of the typical MBE effusion cells yield progressively greater nonuniformity in the films as the charge is depleted. In some MBE systems an effort is made to balance this variation in uniformity from cell depletion by inclining the rotating substrate with respect to the common central axis of the entire effusion cell assembly. Saito and Shibatomi / 23 / performed a systematic investigation of the dependence of the uniformity of the molecular beams across the substrate on the geometrical relationship between the Langmuir-type effusion cells and the substrate. The authors used conical crucibles having a taper ϕ_0 and a diamter D of the cell aperture, which are inclined at an angle ϕ to the normal of the continuously rotating substrate (see Fig. 4 for illustration). The diameter of the uniform area on the substrate was found to depend primarily on the distance between the substrate and the effusion cell, the taper of the cell wall, and the diameter of the cell aperture. The optimization of these parameters resulted in a reduction of the thickness variation of GaAs and $Al_xGa_{1-x}As$ layers to less than $\pm$ 1% over 3-inch wafers / 23 / .

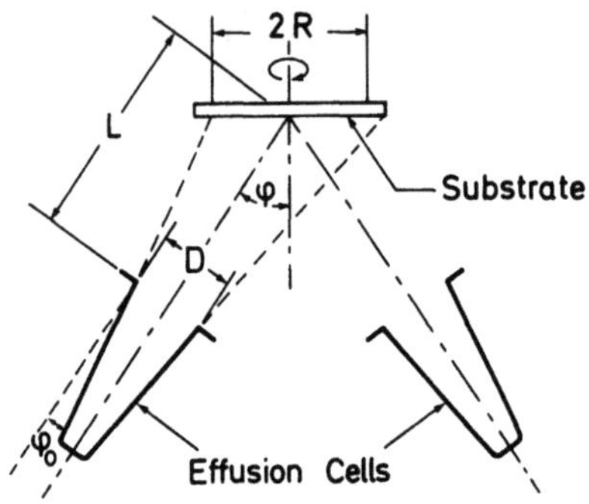

Fig. 4 Schematic configuration of effusion cells versus substrate used by Saito and Shibatomi / 23 / for MBE growth of highly uniform layers.

In Fig. 3b we show schematically a resistance-heated MBE effusion cell of cylindrical shape made from pyrolytic boron nitride (PBN) which provides sufficiently uniform and extremely stable beam fluxes in a quasi-horizontal evaporation system. The Ta heater wire is supported by two concentric PBN tubes, which are formed by the cylindrical PBN crucible extended downwards by an additional tube for thermocouple housing, and by an outer PBN tube of larger diameter. To prevent distortion of the RHEED pattern by stray magnetic fields, the heater windings of the cells should be noninductively wound. The enclosure of the chemically stable butt-welded W-Re thermocouple forms a thermally isolated "black" body. Using this configuration, the temperature reading at the thermocouple location differs by less than ± 5 C from the actual temperature of the evaporant. The outer PBN cylinder is surrounded by several turns of thin crimped Ta foil for an effective thermal isolation of the cell. This design fulfills the requirements of rapid thermal response, uniform heating, and low radiant power loss. Each effusion cell with electrical power and thermocouple feedthroughs is mounted on a separate flange with provisions for accurate cell adjustment and rapid substitution. Ta and Mo are used exclusively for all support and electrical connections to the cell. All other specially shaped parts of the design shown in Fig. 3 are made from PBN, which is chemically very inert. Crucibles with a volume of 20 - 50 cm^3 for group-III-elements and 40 - 100 cm^3 for group-V-elements as well as small capacity crucibles with a volume of 2.5 cm^3 for dopant elements have been built with this configuration in the laboratory of the author.

When using the nonequilibrium (Langmuir-type) effusion cells described before, the formula (1) should be multiplied by a correction factor F which, however, cannot be calculated directly. Therefore, the beam flux emerging from the cell must be monitored intermittently using the movable ion gauge placed in the substrate growth position. To convert the ion gauge measurements into (approximate) absolute flux densities, Eq. (1) has to be modified by including the ionization efficiency η relative to nitrogen (for which the nominal pressure reading of the ion gauge is calibrated) given by the empirical relation / 24 /

$$\eta \simeq 0.6\ (Z/14) + 0.4 \qquad (3)$$

where Z is the number of electrons in the atoms or molecules. The calibration of the ion gauge mounted in the beam path can be made absolute (to $\pm$ 5%) for the group-III-elements by relating the measured ion current to the weight of each element separately deposited on a plate in the beam path. For the group-V-elements, however, only relative calibrations are possible because of non-unity sticking coefficients (an improved calibration procedure for As_4 and P_4 has been described by Foxon et al. / 25 /). During practical MBE growth, the ion gauge is mainly used to ensure reproducibility of fluxes, not to measure absolute magnitudes.

Precise temperature stability and reproducibility of the effusion cells are essential. Temperature fluctuations of $\pm$ 1 C result in beam flux variations ranging from $\pm$ 2 to 4%. Since the film growth rate is proportional to the arrival rate of the group-III-element, these temperature fluctuations cause a fluctuation in the growth rate of the same order of magnitude. In the case of ternary alloys, such as $Al_xGa_{1-x}As$, $Ga_xIn_{1-x}As$, $Al_xIn_{1-x}As$ etc., these fluctuations produce large variations in alloy composition, e.g. by about $\pm$ 4% if the beam intensity of both group-III-elements fluctuates by $\pm$ 2% each. These fluctuations are deleterious for ternary alloys which require precise control of composition in order to match their lattice to that of the binary substrate. While for the growth of $Al_xGa_{1-x}As$ on GaAs substrates the ternary is closely lattice-matched to the binary substrate for all values of x, growth of $Ga_xIn_{1-x}As$ on InP is only lattice matched if x = 0.47. A variation of 1% in the InAs mole fraction from x = 0.47 causes a lattice mismatch, expressed by the ratio of the difference in lattice constant between the ternary layer and the InP substrate, of $\Delta a/a = 7 \times 10^{-4}$. For high-quality material $\Delta a/a$ should be less than 10^{-3} / 26 / . Therefore, it is necessary to control the alloy composition to better than 1%, which requires an accuracy in temperature control of better than 0.02% for the effusion cells.

In the growth of III-V compounds and alloys, the choice of the group-V-element species (dimeric or tetrameric) may have a significant effect on the structural, electrical and optical properties of the films, including concentration of majority carrier traps /27/, minority carrier lifetime / 28 / , and occupancy of donor and acceptor sites by amphoteric dopants / 29 / . These differences in film properties are related to differences in the surface chemistry of the growth process between dimer and tetramer species, which are discussed in Sect. 2.6. There are three methods of producing P_2 and As_2. First, incongruent evaporation of the binary III-V compounds yields dimeric molecules / 10 / . Second, the dimeric species are obtained by thermal decomposition of gaseous PH_3 or AsH_3 / 30, 31 / which, however, produces a large amount of hydrogen in the UHV system. Third, dimers can be produced from the elements by using a two-zone effusion cell, in which a flux of tetramers is formed conventionally and passed through an optically baffled high-temperature stage where complete conversion to P_2 or As_2 occurs above 900 $^{\circ}$C. The design of two-zone effusion cells has been described by several authors / 27, 29, 32, 33 /.

Severe problems in the growth of heterojunctions of exact stoichiometry may arise from the existence of transients in the beam flux intensity when the shutter is opened or closed. The temperature of the melt in the crucible is affected by the radiation shielding provided by a closely spaced mechanical beam shutter. Therefore, a flux transient lasting typically 1 - 3 min. occurs when the shutter is opened and the cell is establishing a new equilibrium temperature. Several approaches have been proposed to reduce or eliminate these flux transients, including the increase of the distance between cell aperture and shutter to more than 3 cm / 34 / or the application of a two-crucible configuration / 35 / . Elimination of the flux transients is of particular importance for the growth of $Al_xIn_{1-x}As$ / $Ga_xIn_{1-x}As$ heterojunctions lattice-matched to InP substrates.

The PBN effusion cells described here operate at temperatures that range from 250 to 1400 $^{\circ}$C. In cases where the source material has too low a vapour pressure to be cleanly evaporated by these resistively heated crucibles, electron beam evaporators have to be used. Details of this type of evaporator applied to silicon MBE have been described by Ota / 36 / and by Bean / 37 / .

2.4 Substrate Processing

In general MBE growth of III-V semiconductors is performed on (001) oriented ($\pm 0.1^{o}$) substrate slices 300 - 600 µm thick. The preparation of the growth face of the substrate from the polishing stage to the in-situ cleaning stage in the MBE system is of crucial importance for epitaxial growth of ultrathin layers and heterostructures with high purity and crystal perfection and with accurately controlled interfaces on an atomic scale. We shall describe in detail the cleaning methods for GaAs and InP, which are the most important substrate materials for deposition of III-V compounds. In any case, the substrate surface should be free of crystallographic defects and clean on an atomic scale, i.e. less than 0.01 monolayer of impurities. The first step always involves chemical etching, which leaves the surface covered with some kind of volatile protective oxide. After insertion in the UHV system this oxide is removed by heating. To avoid surface dissociation of III-V semiconductors, this heating must be carried out in a beam of the group-V-element (P_4/P_2 or As_4/As_2).

Various cleaning methods have been described for (001) oriented GaAs, which mainly use chemical etching based on a H_2O_2 / H_2SO_2 / H_2O mixture / 1, 11, 38 / . However, the formation of oval defects on MBE grown GaAs and $Al_xGa_{1-x}As$ surfaces elongated in the < $01\bar{1}$ > direction and with densities ranging from 10^3 to 10^5 cm^{-2}, has long been a severe problem for practical application. We have recently developed a new method for GaAs substrate preparation which effectively reduces the oval-defect density to less than 10^2 cm^{-2} and allows storage of prepared substrates in air under dust-free conditions for several weeks without any degradation. The reproducible preparation of a contamination-free substrate surface was improved as follows / 39 / : The GaAs wafer is first polished with diamond paste to remove saw-cut damage followed by etch-polishing on an abrasive-free lens paper soaked with NaOCl solution leaving a mirror-like finish. Then the wafer is simply placed twice in concentrated H_2SO_4 kept at 300 K and stirred ultrasonically. The slice is then carefully rinsed in water to remove all SO_4^{2-}ions (careful ultrasonic stirring may accelerate SO_4^{2-} removal from the surface) and finally blown dry with filtered N_2 gas. The important final step is heating of the wafer to 300 ^oC in air under dust-free conditions for about 3 min. During this heating process a stable Ga-rich surface oxide on the (001) GaAs substrate is reproducibly generated to protect the surface from carbon contamination. The passivated wafer is then either soldered with liquid In to a conventional Mo substrate plate or it is fixed to a special Mo substrate holder designed for direct-radiation substrate heating. Both methods give similar results for the subsequent growth of selectively doped n-$Al_xGa_{1-x}As$ / GaAs heterostructures with high-mobility two-dimensional electron gas (2DEG). After transfer to the MBE growth chamber the surface oxide is thermally desorbed by heating the substrate wafer to 550 ^oC in a flux of arsenic. When the desorption process is finished a clear (2 x 4) surface reconstruction is observed in the RHEED pattern.

The preparation of (001)InP substrates requires extra care during polishing and handling, because InP is much softer than GaAs. In addition, the thermal cleaning procedure in UHV has to be modified substantially, since incongruent evaporation of InP starts already at T_s = 365 ^oC while native oxides on InP are more stable than GaAs oxides / 26, 40 / . After degreasing the wafer successively in trichloroethylene, acetone, and water, the saw-cut damage is removed by polishing the substrate on lens paper soaked with 0.5% bromine-methanol solution to a mirror-like finish. The substrate is then etched for 10 min. in a (3:1:1) solution of sulfuric acid, hydrogen peroxide, and water kept at 48 ^oC. This etch may be followed by an additional 3 min. etch in 0.3% bromine-methanol solution. Finally, the substrate is passivated in water for a very short time (about one minute) and blown dry with filtered nitrogen. The optional oxidizing bromine-methanol etch provides only minor carbon contamination while preserving a protective oxide coverage. In order to remove the surface oxides, the InP substrate must be heated to above 500 ^oC in UHV, about 140 ^oC above its congruent temperature. This would give rise to the formation of In droplets on the InP surface. Therefore, an impinging flux of arsenic or phosphorus is required to replace the phosphorus of the InP which is lost by preferential thermal desorption. Heating of the substrate in an As_4 flux of $J(As_4) = 10^{15}$-10^{16} cm^{-2} s^{-1} (or P_4 flux, respectively) for a few minutes at 505 ^oC is sufficient to completely remove the surface oxides and to provide a cleaned InP surface / 26, 40 / .

The actual cleaning methods required for a specific substrate depend on the material and contaminants involved. For materials like Si or Ge reactive etching or sputter etching with low-energy inert-gas ions (0.5 - 1.0 keV) are employed followed by careful annealing cycles / 36, 37 / .

Until recently, most of substrate wafers were soldered with liquid In (at 160 ^{o}C) to a Mo mounting plate / 11 / . This practice provides good temperature uniformity due to the excellent thermal sinking , and it is advantageous for irregularly shaped substrate slices. However, the increasing demand for production oriented post-growth processing of large-area GaAs wafers and the availability of large-diameter (> 2 in.) GaAs substrates has fostered the development of In-free mounting techniques. Technical details of direct-radiation substrate heaters have been described by several authors / 41 - 44 / .

Immediately after mounting the etched substrate wafer, the Mo plate with the freshly prepared substrate is remotely exchanged with a plate holding the processed substrate from the previous growth run using a substrate exchange load-lock system (Fig. 2). The specimen is transferred between the chambers by trolley and/ or magnetically coupled transfer mechanisms. In advanced MBE systems, an additional intermediate UHV chamber for storage of several substrate wafers is added which may contain facilities for further surface treatment prior to epitaxy (ion beam sputtering, preheating etc.). The Mo mounting plate holding the substrate is fixed by a bayonet joint to the internally heated Mo heater block which remains always inside the growth chamber. During sample exchange, the cryopanels in the growth chamber are held at LN_2 temperature, and the effusion cells are kept at their usual operating temperature. The sample exchange procedure is completed in a few minutes, and a new growth run can be started without distortion of the growth conditions. The application of interlock systems, about 5 years ago, has drastically increased film purity and growth reproducibility in MBE technology.

The internally heated Mo block, used for controlled substrate heating during epitaxial growth, is attached to a special manipulator mounted on a separate flange which also contains power and thermocouple feedthroughs. This manipulator correctly positions the wafer relative to the sources, heats it to the required temperature, and rotates it azimuthally for optimum film uniformity. The distance between cell orifices and substrate ranges from 100 to 250 mm in various systems. The temperature of the substrate mounted on the remotely exchanged Mo plate is measured with a butt-welded W-Re thermocouple which passes through the Mo heater block and is rigidly fixed at a certain distance (e.g. 2 mm) below the surface to which the substrate is fixed. This arrangement does not allow to measure the absolute temperature of the substrate surface accurately. However, it is possible to obtain and maintain a given temperature in a reproducible manner, particularly when the sample exchange device is used, and this is all that is required for successful film growth. A sophisticated method of calibrating the surface temperature of substrates attached to the heater block uses infrared pyrometry with the correct emissivity setting. For the case of GaAs, e.g., the temperature measured within the heating block and the actual temperature of the substrate surface can be correlated in the range of interest using the well-established temperature of 550 ^{o}C / 11, 39 /, at which the passivating oxide film on GaAs evaporates, and the well-known temperature of 630 ^{o}C / 3, 45 / which indicates the limit of congruent evaporation of GaAs. Both events can be monitored by using the RHEED or - if available - the AES facility.

In modern MBE systems continuously azimuthally rotating substrate holders with a maximum speed of 125 rpm are now used, because even the most sophisticated geometrical arrangement of the effusion cells cannot avoid considerable (> 10%) deviations from layer uniformity across wafers of more than 25 mm diameter / 46 / . When the substrate is rotating about its azimuthal axis during growth with 4 - 10 rpm at 1 µm/hr growth rate, epitaxial films with thickness and compositional uniformity to better than $\pm$ 2% over 50 mm diameter substrates can be achieved routinely. The rotating substrate holder also facilitates growth of high-quality ternary and quaternary III-V compounds which require close lattice match to the substrate ($Al_xIn_{1-x}As$ or $Ga_xIn_{1-x}As$ on InP, $Ga_xIn_{1-x}P$ on GaAs etc.).

2.5 In-Situ Growth Monitoring by RHEED

In this section we concentrate on the more conventional application of RHEED to monitor the surface symmetry of the various reconstructed surfaces of III-V semiconductors. A more detailed analysis of RHEED patterns provides additional information on surface morphology, disorder and topography. The relation between temporal intensity oscillations in the RHEED pattern and MBE growth dynamics and interface control will be discussed in Sect. 4.1. RHEED provides important information about surface cleaning and proper growth conditions, as its forward scattering geometry makes it fully compatible with the growth process (see Fig. 2 for illustration), and it is therefore used in the first few minutes of growth for every run. The availability of this facility in the growth chamber is thus essential.

A semiconductor surface of the same fundamental crystallographic orientation can have different structures because of the phenomenon of surface reconstruction. The surface would have the (1 x 1) structure if the bulk atom position would be maintained at the surface. However, in general the symmetry at the surface is reduced by reconstruction which results from the rehybridization of bound orbitals of surface atoms in order to lower the free energy of the surface. This reordering of the outermost layers often results in a surface symmetry which is modified with respect to that of the bulk lattice, i.e. the size of the unit cell has a larger periodicity. The notation of Wood / 47 / describes the new primitive cell in terms of the dimensions of the unreconstructed unit cell. In the case of the (001) surface of GaAs there are several reconstructions. The most stable is the (2 x 4) structure (Fig. 5) where a twofold and fourfold increase in the periodicity occurs along two orthogonal [110] directions. The existence of surface reconstruction leads to additional features in the RHEED pattern at fractional intervals between the bulk diffraction streaks. Therefore, information contained in the RHEED patterns comprises (a) the symmetry and periodicity of ordered layers near the surface and (b) the position of atoms within the unit mesh. The form of the reconstruction of the (001) surface of III-V semiconductors can qualitatively be correlated to surface stoichiometry, which is an important growth parameter / 48 - 50 / .

An important aspect of RHEED patterns obtained from atomically flat surfaces is the appearance of streaks normal to the shadow edge instead of spots, as shown in Fig. 5. For explanation the thermal diffuse scattering mechanism was proposed by Holloway and Beeby / 51 / , but so far a true satisfactory interpretation does not exist. The simplest treatment of this phenomenon is the assumption that the penetration of the electron beam under conditions of grazing incidence is restricted to the outermost layer of the crystal surface. In this case only this first layer contributes to the diffracted intensities, i.e. the usual picture of the reciprocal lattice point is drawn out into a one-dimensional rod perpendicular to the surface (in practice, however, the penetration depth of the electron beam is not totally restricted to the outermost layer, and a model using a modulated reciprocal lattice rod is probably more appropriate). Using this relaxation of the third Laue condition, the diffraction pattern can then be visualized as an intersection of the Ewald sphere, whose radius is large at electron energies > 5 keV, with a set of reciprocal lattice rods, as shown in Fig. 6. In this way the spacing between reciprocal lattice rods can be related to the spacing between rows of atoms in the surface layer. For practical MBE growth it is further important that, as the topography changes from a flat to a rough surface, the streaks change from intense, short, and well defined to longer, diffused, and spotty, and finally to a pure spotted pattern arising from transmission of electrons through surface asperities. We have to realise, however, that the length and width of streaks are not necessarily related to the topography / 50 / .

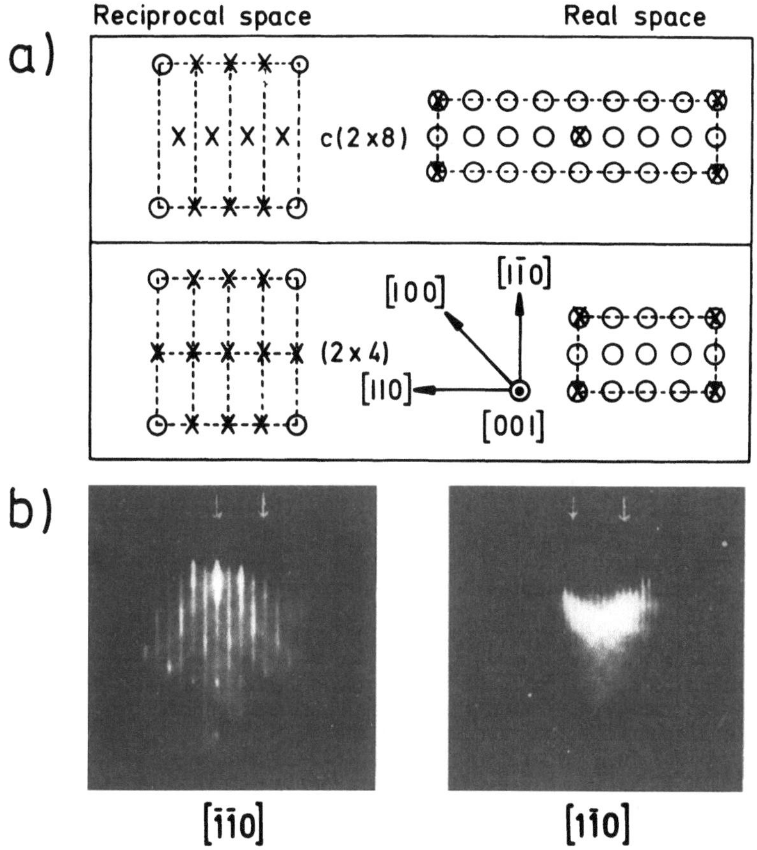

Fig. 5 (a) Real-space and reciprocal-space representations for the c(2 x8) and (2 x 4) surface reconstruction on growing (001) GaAs. (b) RHEED patterns of the c(2 x 8) [or (2 x 4)] surface structure in different azimuths.

Even when the kinematic treatment based on the reciprocal lattice rod concept of the Ewald sphere is used, the detailed interpretation of RHEED data from reconstructed III-V semiconductor surfaces during MBE growth has remained controversial. In general, the surface periodicity normal to the beam azimuth can be evaluated from the Ewald sphere construction taking the distance between the streaks on the fluorescent screen, the camera constant, and the energy (wavelength) of the incident electrons. Measurements along two or three different azimuths are often required to determine the surface symmetry unambiguously.For illustration we show in Fig. 5 the real and reciprocal space representation together with the corresponding zero-order Laue zone RHEED patterns at two different azimuths for the most important (001) GaAs-(2 x 4) surface reconstruction. Multiazimuthal measurements become particularly important if a two-dimensional surface disorder (or domain) exists which is separated by one-dimensional (or antiphase) boundaries.

The analysis also of intermediate Laue zones of a RHEED pattern from measurements at a single azimuth is possible in the absence of a pronounced surface disorder. An elegant method given by Hernandez-Calderon and Höchst / 52, 53 / will be described in some detail by means of Fig. 6. The continuous rods of the reciprocal lattice are normal to the surface plane and possess a two-dimensional translational symmetry in that plane. The Ewald sphere intersects all rods contained in the projection of this sphere onto the surface. Diffracted beams are thus observed in the angular

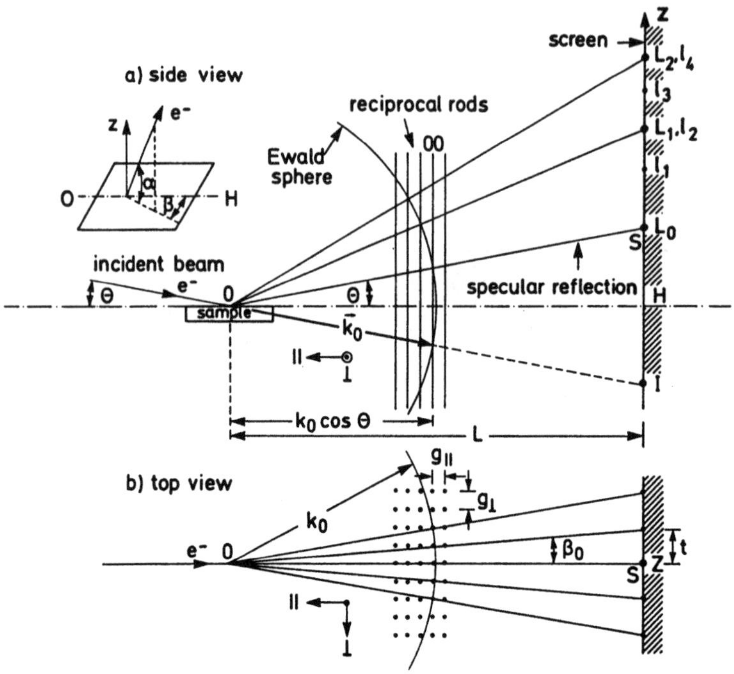

Fig. 6 Ewald construction to depict the origin of a typical RHEED pattern during MBE growth (not to scale). (a) Side view where L_n corresponds to the Laue zones of the ideal surface. The intermediate spots l_n that are present in the case of $g_{||} = \frac{1}{2} g^{i}_{||}$ are shown for illustration. (b) Top view with a projection of the Ewald sphere on a plane parallel to the sample surface (after Hernandez-Calderon and Höchst / 52 /).

directions α and β.. The electron wave vector k_0 in a typical RHEED experiment is about 10^2 Å^{-1}. Using a reciprocal surface vector g, we obtain a characteristic value of g/k_0 of about 10^{-2}. Thus, the Ewald sphere touches only a few rods on both sides of the (00) rod. The angle between these reflections is given by / 52 /

$$\beta_0 = \tan^{-1}\,[g_{\perp}/(k_0^2 - g_{\perp}^2)^{1/2}] = \tan^{-1}(t/L) \tag{4}$$

where t is the distance between streaks and L is the sample-to-screen distance. From Eq. (4) we obtain the relation

$$d_{||} = t = L\omega_0 \tag{5}$$

as $g_{\perp} << k_0$ and $g_{\perp} = 2\,\pi/d_{||}$, where $d_{||}$ is the distance between equivalent rows of atoms parallel to the incident beam, and where λ_0 is the electron wavelength according to

$$\lambda_0 = 12.27/[V_0^{*}(1 + 0.978{*}10^{-6}V_0)]^{1/2} \tag{6}$$

where V_0 is the electron accelerating voltage. With a fixed energy and geometry we can thus determine surface lattice constants with a precision of 10^{-3}.

The intersections of each row of rods normal to the incident beam are projected on a screen over arcs of circumference. The distance between these arcs is determined by the periodicity in the $g_{||}$ direction. The separation between spots on the same arc provides information on the periodicity in the $g_{||}$ direction [Eq. (4)]. Hernandez-Calderon and Höchst / 52 / have described the determination of the magnitude of $g_{||}$ which allows a detailed analysis of intermediate Laue zones of a

RHEED pattern (Fig. 6b). From this analysis the authors were able to positively determine the surface reconstruction of (100) α-Sn. A more empirical method to obtain the reconstruction of the surface, which is also used in practical MBE growth, is via the measurement of the streak separation under different azimuths, as those shown in Fig. 5.

We have already mentioned that in some cases the length and width of streaks in the RHEED pattern may be related to the topography of the growth surface. The simplest case is surface roughness or aspherities. They usually appear during heat treatment of the chemically cleaned substrate on such a scale that the glancing incidence beam produces a transmission diffraction pattern consisting of diffraction spots instead of streaks. At typical electron energies of 20 - 30 keV, the aspherities must have a thickness of less than 100 nm in beam direction to allow the formation of transmission patterns. If GaAs is grown on a slightly C-contaminated (001) GaAs substrate, {511} facets are formed during the onset of growth / 54 / . The presence of these facets produces additional streaks in the RHEED pattern which are not normal to the shadow edge due to diffraction from the facet planes. As a result, arrowhead-like features show up in the RHEED pattern with an angle of about 19° between these streaks and the surface normal, which are characteristic for {511} facet planes / 54 / .

In addition to providing information on topography, the streak shape of RHEED patterns can be used to evaluate the effects of two-dimensional (2D) surface disorder / 53 / . The average size of ordered surface regions may be restricted due to the lack of perfect ordering in a particular direction (azimuth). In this case the ideal one-dimensional (1D) reciprocal lattice rod becomes two-dimensional, i.e. it forms a solid ellipsoidal cylinder in the Ewald sphere construction. The existence of domains having a strong ordering direction on the surface is indicated by the lengthening and broadening of the fractional and integral order beams in that direction for two orthogonal [110] azimuths, where the domain extension is restricted. While the streaks are thus long and broad with the beam parallel to the short domain side, short and narrow streaks are produced when the beam is parallel to the long domain side. Joyce et al. / 53 / pointed out that the domains on the (2x4) reconstructed (001) GaAs surface are separated by one-dimensional (or antiphase) boundaries which give rise to curved streaks in the [010] azimuth. A detailed analysis of the curved streaks in intermediate azimuths allowed the authors to relate the (2x4) and c(2x8) reconstruction occurring on many (001) oriented III-V semiconductor surfaces simply by the surface disorder effect.

We have mentioned in Sect. 2.2 that most III-V semiconductors are grown with a substantial excess of the more volatile group-V-element. For GaAs these As-stabilized growth conditions yield a (2x4) [or c (2x8)] surface reconstruction,as shown in the RHEED pattern of Fig. 5. The As-stabilized (2x4) structure is stable over a wide range of substrate temperatures / 55 / , and smooth (001) surfaces with steps down to atomic dimensions can be routinely achieved with these conditions, provided the starting substrate surface is sufficiently clean. If MBE growth is performed with an elemental III/V flux ratio of 1 or at substrate temperatures close to the limit of congruent evaporation, the Ga-stabilized (4x2) [or c(8x2)] surface reconstruction is obtained. Depending on the incident arsenic and gallium fluxes and on the substrate temperature, reversible transitions between the two principal surface structures on (001) GaAs are possible; these involve a change of the surface composition. Several intermediate structures, e.g. (3x1), (1x6), (4x6), (3x6), and mixtures, can be observed within very narrow ranges of growth conditions. Whereas the transitions between these structures are not sharp, the final change to the Ga-stabilized (4x2) structure is very abrupt. Furthermore, only a very small increase in the Ga flux above the minimum value required to produce the Ga-stabilized structure results in the formation of free gallium on the surface. This means that the practical use of conditions with a Ga (4x2) structure is limited, because it yields samples with dull surfaces and in some cases with evidence of a build-up of Ga droplets. The close relationship between reconstruction effects and surface chemistry,as well as the correlation with gain or loss of arsenic from the surface,has been determined by several authors / 11, 48 - 50 / .

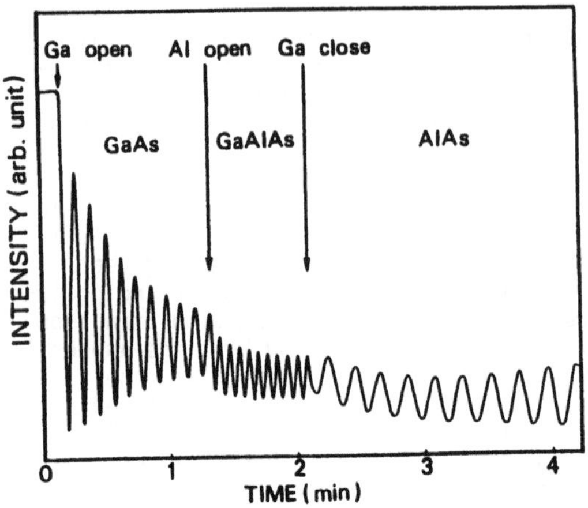

Fig. 7 RHEED intensity oscillation of the specular beam obtained in [100] azimuth on (001) GaAs substrate during continuous growth of a GaAs / $Al_xGa_{1-x}As$ / AlAs heterostructure / 16 / .

Wood / 56 / , Neave et al. / 14 / , van Hove / 15 / , and Sakamoto / 16 / have discovered pronounced periodic intensity oscillations in the specularly reflected and diffracted beams in the RHEED pattern during growth of GaAs, AlAs, $Al_xGa_{1-x}As$ etc. (Fig. 7). The period of the oscillations corresponds exactly to the time required to grow a monolayer of GaAs (or AlAs or $Al_xGa_{1-x}As$) on the (001) substrate surface. A monolayer GaAs is defined as one complete layer of Ga plus one complete layer of As having a thickness equivalent of $a_o/2$. These RHEED intensity oscillations provide direct evidence that MBE growth occurs predominantly in a 2D layer-by-layer growth mode. The changing intensity reflects variations in the step density of each layer when growth proceeds. To a first approximation we can assume that the oscillation amplitude reaches its maximum when the monolayer is completed (maximum reflection). Although the fundamental principles underlying the occurrence of these oscillations and the damping of their amplitude are not completely understood, the method is now widely used to monitor and to calibrate absolute growth rates in real time with monolayer resolution (see Sect. 4.1.).

2.6 Kinetics and Mechanism of Growth Processes

The growth of binary and ternary III-V semiconductors, especially GaAs, by MBE has been studied by two methods. Modulated molecular beam spectroscopy has been used to investigate the surface chemical processes / 57 / and dynamic RHEED measurements have been used to determine the involved growth mechanisms / 58 / . Additional Monte Carlo simulations of growth / 59 / have improved our understanding of the kinetic processes. From these results detailed kinetic models were established for the growth of GaAs from beams of Ga and As_2 and Ga and As_4 (Fig. 8) / 10, 60 - 62 /. Gallium has a unity sticking coefficient on (001) GaAs at 500 °C. Condensation of As_2 and As_4 occurs only when Ga adatoms are present on the GaAs surface. The As_2 molecules are first adsorbed into a mobile weakly bound precursor state. Dissociation of adsorbed As_2 in a simple first-order process occurs as the molecules encounter single Ga lattice atoms while migrating on the surface. In the absence of free Ga adatoms on the surface, no permanent condensation of As_2 occurs although the species exhibit a measurable surface lifetime. The sticking coefficient of As_2 is therefore a function of the arrival rate of Ga, and is unity only for monolayer coverage of Ga atoms. Consequently, stoichiometric GaAs can be grown from As_2 with a relative arsenic-to-gallium flux ratio larger than or equal to unity.

A similar model is valid for the interaction of Ga and As_4 molecular beams with a heated (001) GaAs surface (Fig. 8b). The As_4 molecules, generated from elemental

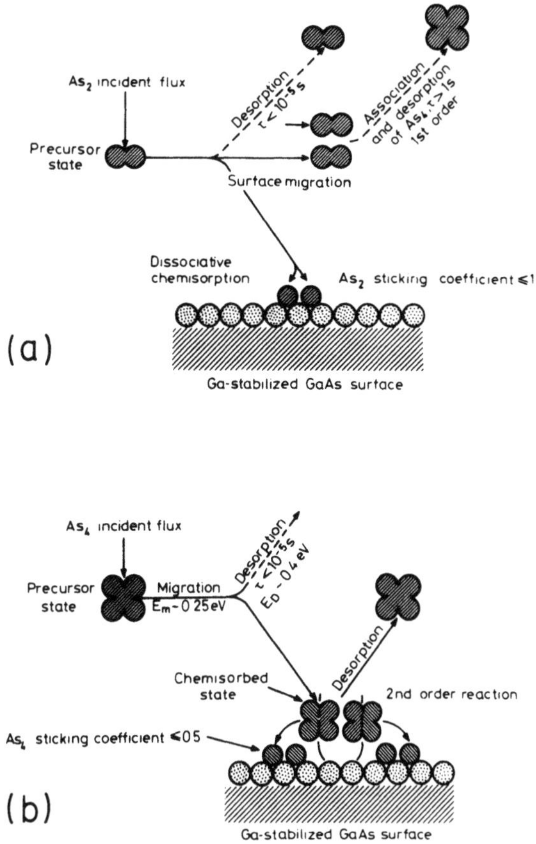

Fig. 8 Models for MBE growth of GaAs (a) from Ga and As_2 species and (b) from Ga and As_4 species, developed by Foxon and Joyce / 61, 62 / .

arsenic sources, are first adsorbed into a mobile precursor state while the Ga adatom population controls the condensation and reaction of As_4. The sticking coefficient of As_4, however, can never exceed 0.5 and become unity, even when a monolayer of Ga is present on the surface. This crucial behaviour is caused by the complex pairwise interaction process of As_4 molecules chemisorbed on adjacent Ga lattice atoms (second-order process), as indicated in Fig. 8b. Consequently, because of the smaller sticking coefficient, a larger amount of excess As_4 as compared to As_2 is required for growth of stoichiometric GaAs with an As-stabilized (2x4) surface reconstruction. The differences in growth mechanism between the two arsenic species have pronounced effects on the non-equilibrium concentration of native defects incorporated during MBE growth.

The growth models depicted in Fig. 8 imply that almost all Ga atoms incident on the surface are incorporated into the growing epitaxial layer and the growth rate is thus controlled by the Ga flux / 11 / . These models are also valid for other binary III-V semiconductors and to a good approximation also for ternary III-III-V alloys / 63 / . A good compositional control of the growing III-III-V alloy films can be achieved by supplying excess group-V-species and adjusting the flux densities of the impinging group-III-beams, as long as the substrate temperature is kept below the congruent evaporation limit of the less stable of the constituent binary III-V compounds (e.g. GaAs in the case of $Al_xGa_{1-x}As$). At higher growth temperatures, however, preferential desorption of the more volatile group-III-element (i.e. Ga from $Al_xGa_{1-x}As$) occurs, so that the final film composition is not only determined by the added flux ratios but also by the differences in the desorption rates. To a first approximation we can estimate the loss rate of the group-III-elements from their vapour pressure data / 19 / . This assumption is reasonable because the vapour pressure of the element over the compounds, i.e. Ga over GaAs, is similar to the vapour pressure of the element over itself / 45 / . The results are summarized in Table 1. The surface of alloys grown at high temperatures will thus be enriched in the less volatile group-III-element. As a consequence, we expect a significant loss of In in $Ga_xIn_{1-x}As$ films grown above 550 ^oC

Table 1 Approximate loss rate of group III elements in monolayers per second estimated from vapour pressure data

Temperature (^oC)	Al	Ga	In
550	-	-	0.03
600	-	-	0.3
650	-	0.06	1.4
700	-	0.4	8
750	0.05	2	30

and a loss of Ga in $Al_xGa_{1-x}As$ films grown above 650 ^oC, and an intermittent detailed calibration basedon measured film composition is recommended for accurate adjustments of the effusion cell temperatures.

The growth of ternary III-V-V alloy films (e.g. GaP_yAs_{1-y}) by MBE is more complicated, because even at moderate substrate temperatures, the relative amounts of the group-V-element incorporated into the growing film are not simply proportional to their relative arrival rate / 25 / . The factors controlling this incorporation behaviour are at present not well understood. It is therefore extremely difficult to obtain a reproducible compositional control during MBE growth of III-V-V alloys when solid phosphorus and arsenic sources are employed.

2.7 Dopant Incorporation

Application of MBE grown films in device structures requires the control of the electrical and optical material properties by the incorporation of small amounts of impurity (dopant) elements from additional effusion cells. The electronic properties of the grown films depend on the flux of the dopant atoms, on their sticking coefficient and on their incorporation behaviour and electrical activity. The required dopant concentration is typically in the range 10^{16} - 10^{18} atoms cm^{-3} or 10^{-6} - 10^{-4} atomic fraction. The most common dopants used during MBE growth have unity sticking coefficients over a large range of growth conditions / 10 / . This behaviour is indicated by a linear behaviour of carrier concentration versus reciprocal dopant cell temperature (Clausius-Clapeyron-type plots), which are shown in Fig. 9 for four representative dopants of MBE GaAs. In Table 2 we have summarized a few important properties of the most promising dopant elements used for GaAs grown by MBE. Many of these dopants are also suitable for other III-V compounds. Hereafter we will discuss in some detail the characteristics of the four dopants displayed in Fig. 9.

For p-type doping Be is used, while for n-type it is either Si, Ge or Sn. For doping levels below 1 x 10^{19} cm^{-3} Be behaves as an almost ideal shallow acceptor in MBE grown III-V compounds / 64 / . Each incident Be atom produces one ionized impurity species providing, in GaAs, an acceptor level 29 meV above the valence band edge. The observed doping level is thus simply proportional to the arrival rate. The very low diffusion coefficient of Be in MBE GaAs ensures excellent control of the depth distribution of this p-type dopant, and abrupt doping profiles can be achieved / 65 / . At doping levels above 5 x 10^{19} cm^{-3}, however, the surface morphology and the luminescence properties degrade / 66 / and the diffusion of Be is enhanced / 67, 68 / if the samples are grown at a substrate temperature above

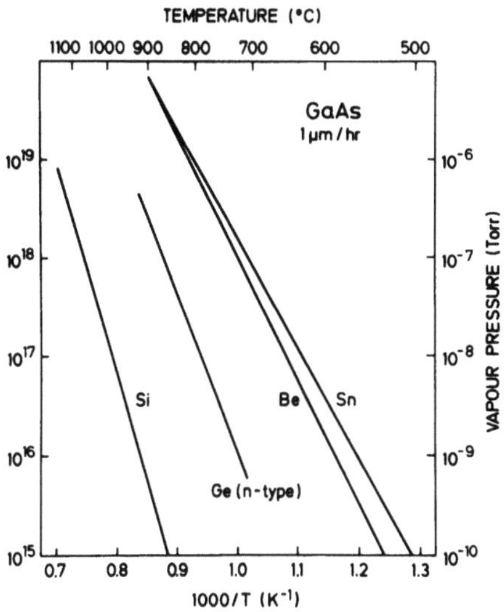

Fig. 9 Room-temperature carrier concentration in MBE grown GaAs as a function of four dopant effusion cell temperatures for constant growth rate, substrate temperature and As_4-to-Ga flux ratio (Clausius-Clapeyron-type plots). The data were obtained from Hall effect and capacitance-voltage measurements. They are also valid for other III-V semiconductors.

Table 2 Properties of widely used dopant elements in GaAs grown by MBE

Element	Incorporation behaviour	Degree of compensation	Maximum achievable carrier concentration (cm^{-3})	Remarks
Be	acceptor only	low	6×10^{19}	vapour is highly toxic
Si	predominantly donor	fairly low with As-stabilized conditions	1×10^{19}	high source temperature required
Ge	depends strongly on growth conditions	can be high	4×10^{18}	acceptor with Ga-stabilized and donor with As-stabilized conditions
Sn	donor only	low	6×10^{19}	tends to accumulate at the growth surface
S Se Te	donors only	low	3×10^{19}	molecular sources like PbS, PbSe, PbTe or electrochemical cells for controllable incorporation required

550 °C. Lowering of the substrate temperature to 500 °C makes feasible Be acceptor levels up to 2×10^{20} cm^{-3} with perfect surface morphology and reduced Be diffusion / 69 / . It is important to note that Be is highly toxic. In addition, the commercially available Be source material is at best 99.99% pure and contains a certain amount of metallic impurities. Unfortunately, no other suitable p-type dopant is available for MBE growth of GaAs and other III-V semiconductors.

The group-IV-element Si is primarily incorporated on Ga sites during MBE growth under As-stabilized conditions, yielding n-type material of fairly low compensation / 65 / . The observed doping level is simply proportional to the dopant arrival rate provided care is taken to reduce the water and carbon monoxide level during growth. The upper limit of $n = 1 \times 10^{19}$ cm^{-3} for the free-electron concentration in GaAs was originally attributed to the enhanced autocompensation, i.e. the incorporation of Si on As sites and on interstitials / 70 , 71 / . Recent investigations / 72, 73 / indicate, however, that more probably nitrogen evolving from the PBN crucible containing Si and heated to about 1300 ^{o}C causes the compensating effect. This effect can be overcome by the use of a very low growth rate of about 0.1 µm/hr / 72 / . The possibility of Si migration during MBE growth of $Al_xGa_{1-x}As$ films at high substrate temperatures and/or with high donor concentration has been the subject of controverse discussions, because of its deleterious effects on the properties of selectively doped $Al_xGa_{1-x}As$ / GaAs heterostructures (see / 74 / and references therein). Gonzales et al. / 74 / provided some evidence that only at high doping concentrations ($> 2 \times 10^{18}$ cm^{-3}) Si migration might occur in $Al_xGa_{1-x}As$ films as the result of a concentration-dependent diffusion process which is enhanced at high substrate temperatures. Finally, it is important to note that the incorporation of Si atoms on either Ga or As sites during MBE growth depends on the orientation of the GaAs substrate. Wang et al. / 75 / found that in GaAs grown on (111)A, (211)A and (311)A orientations the Si atoms predominantly occupy As sites and act as acceptor, while they occupy Ga sites and act as donors on (001), (111)B, (211)B, (311)B, (511)A, (511)B and higher-index orientations. Based on these results, Miller / 76 / and Nobuhara et al. / 77 / could grow a series of lateral p-n junctions on graded steps of a (001) GaAs substrate surface.

The group-IV-element Ge is an amphoteric dopant in MBE GaAs, and it can be used to prepare either p- or n-type films depending on the growth conditions / 78 - 80 /. On As sites Ge acts as an acceptor, while on Ga sites Ge acts as a donor. The site incorporation depends thus critically on the arsenic-to-gallium flux ratio and on the substrate temperature. When the surface of the growing GaAs film exhibits a Ga-stabilized (4 x 2) reconstruction, the Ga atom surface population is increased. The codeposited Ge atoms are incorporated predominantly on As sites, and thus p-doped films result. The major problem with MBE growth of p-type GaAs doped with Ge is the small stability range of the required Ga-stabilized growth conditions. With As-stabilized growth conditions Ge is predominantly incorporated on Ga sites and acts as donor in GaAs. The observed free-carrier concentration versus reciprocal Ge effusion cell temperature plots (see Fig. 9) implies that the degree of autocompensation of Ge-doped n-GaAs does not depend on the doping level. Owing to the amphoteric nature of the dopant, compensation of highly Ge-doped GaAs films is considerable and the maximum achievable free-carrier concentration is lowered to 4×10^{18} cm^{-3}.

A widely used n-type dopant for many III-V semiconductors in any epitaxial growth technique is Sn which is not amphoteric and acts as a shallow donor only. The measured free-electron concentration is directly proportional to the Sn flux in the molecular beams, and high concentrations up to 5×10^{19} cm^{-3} can be achieved. The major disadvantage of using Sn is its tendency to accumulate at the GaAs surface during growth / 81, 82 / . Wood and Joyce / 83 / observed that the Sn incorporation is surface rate-limited. Before a steady-state donor concentration in the growing GaAs film can be achieved, it is necessary to build up a steady-state surface population of Sn, which may be as large as 0.1 monolayer. This produces dips in the carrier-concentration profile at the film-substrate interface, which can be overcome only by predeposition of Sn prior to GaAs growth. Consequently, no sharp doping profiles can be realized with Sn doping.

The group-VI-elements S, Se, Te cannot be incorporated during MBE growth by simply evaporating the elements in separate dopant cells. A method to use these elements for MBE GaAs involves synergic reactions using the lead chalcogenides PbS, PbSe, or PbTe / 84, 85 / . A second method for the incorporation of S and Se into GaAs uses the electrolysis of a solid state silver electrolyte / 86 / . At elevated growth temperature (> 600 ^{o}C), however, the incorporation efficiency of the group-VI-elements in GaAs as well as in $Al_xGa_{1-x}As$ is reduced. To a certain extent this loss of the dopant may be suppressed by increasing the As_4 overpressure during growth.

The discussion of this section has shown that as yet no one element of Table 2 can be considered truly as the best choice for MBE growth. The nature of the required doping profile and the application of the wafer determine to a large extent which dopant element is used.

3. INITIATION OF GROWTH

For III-V compounds, the oxide passivation layer serves as a protection for the chemically etched substrate from atmospheric contamination before epitaxial growth. After mounting the Mo transfer plate with the fixed substrate, the introduction chamber is pumped down to 10^{-8} Torr, and the substrate is heated (350 °C for GaAs) to desorb the water from the substrate mounting plate. In the meantime, the LN_2 shrouds of the growth chamber are cooled and the effusion cells are brought up to the desired temperatures. When the pressure in the introduction (or intermediate preparation) chamber has reached the low 10^{-9} Torr range, the substrate is transferred to the growth chamber. There it is heated to a higher temperature (< 630 °C for GaAs) under a flux of the group-V-element to evaporate the oxides from its surface (see Sect. 2.4). At this stage the substrate, provided proper preparation procedures were followed, is nearly atomically clean and ready for epitaxial growth. This thermal cleaning of a chemically etched surface produces a surface which is sometimes rough on a microscopic scale as indicated by a spotty RHEED pattern. After the deposition of several tens of nm (of GaAs, e.g.) by MBE the surface has become atomically flat and produces a streaked diffraction pattern with additional features due to reconstruction (see Fig. 5)

Using the ability to produce abrupt interfaces and doping variations, multilayered single-crystal structures with dimensions of only a few atomic layers can be obtained by actuating the shutters in front of the effusion cells in a predetermined manner. This potential for excellent dimensional control of MBE is indicated in the transmission electron micrographs (TEM) of Fig. 10. The growth rate and the layer thickness are calibrated by the following procedure. First, estimated layer thicknesses of 1 to 5 µm are grown on the substrate using different temperatures of the group-III-element effusion cells in sequential growth runs. The cleaved cross-

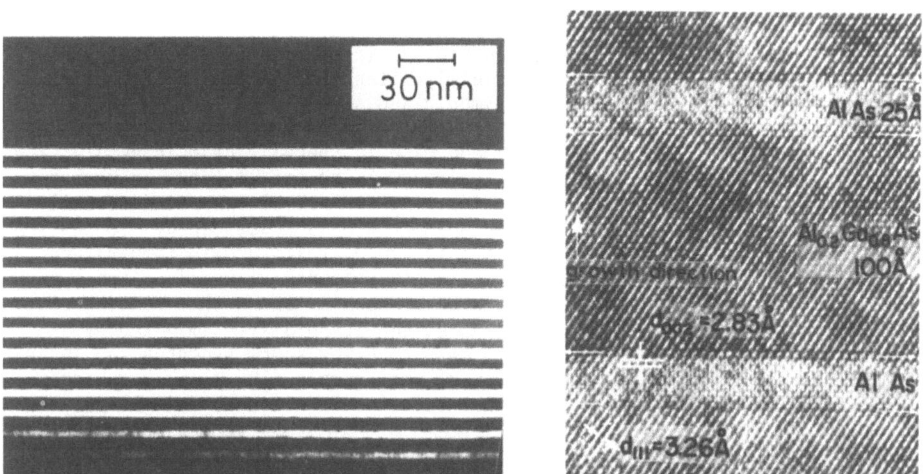

Fig. 10 (a) (110) cross sectional TEM of an AlAs / GaAs superlattice;
(b) High-resolution lattice image of a 10 nm $Al_{0.2}Ga_{0.8}As$ / 2.5 nm AlAs superlattice using (110) electron beam incidence / 88 / .

sections of the layers are then stained or etched to delineate the interface and then examined with a scanning electron microscope. The actual growth rate is then determined by the measured layer thickness divided by the growth time. To account for possible flux transients in periodic structures, the period and the layer thickness in properly designed superlattice configurations are determined by high-resolution double-crystal X-ray diffraction (see Sect. 4.2 for details).

The unique capability of MBE to create a variety of mathematically complex compositional and doping profiles in semiconductors arises from the conceptual simplicity of this process. Accurately controlled temperatures have a direct, calculable effect upon the growth process. This simplicity allows composition control from x = 0 to x = 1 in $Al_xGa_{1-x}As$ with a precision of ± 0.001 and doping control, both n- and p-type, from 10^{14} cm^{-3} to the 10^{19} cm^{-3} range with a precision of +1%. The accuracy is largely determined by the care with which the growth rate and doping level were previously calibrated in test layers.

4. EXAMINATION OF INTERFACE QUALITY

The heterointerfaces between epitaxial layers of different composition ($Al_xGa_{1-x}As$ / GaAs, $Al_xIn_{1-x}As$ / $Ga_xIn_{1-x}As$, $Al_xGa_{1-x}Sb$ / GaSb, Si / Si_xGe_{1-x}, etc.) are used to confine electrons or holes to two-dimensional (2D) motion. Recent developments in materials characterization have helped significantly to elucidate complex processes and phenomena connected with the microstructure of materials and interactions at abrupt heterointerfaces. However, the experimentally realisable interface perfection is so high that conventional methods of profiling, which involve sputtering to section the material followed by some method of composition determination, such as AES or secondary ion mass spectroscopy (SIMS), do not have the required resolution. The available methods to obtain structural and compositional information with the needed level of spatial resolution are limited to transmission electron microscopy (TEM) of cross sections, double-crystal X-ray diffraction, and to a certain extent photoluminescence combined with absorption measurements. In this chapter we present a few examples for monitoring the interface quality by RHEED intensity oscillations and for studying interface disorder effects by X-ray diffraction.

4.1 RHEED Intensity Oscillations

For closely lattice-matched systems it is possible to prepare interfaces by MBE so that compositional changes occur over no more than one monolayer, because MBE growth occurs predominantly in a 2D layer-by-layer growth mode. Direct evidence for this 2D growth mode is provided by the observation of oscillations in the intensity of the RHEED pattern / 14 - 16 / , as shown in Fig. 7. Inspection of this figure reveals that damped oscillations occur immediately after the initiation of growth. We assume that the equilibrium surface is smooth and strongly reflective to electrons so that the specular beam intensity is high. As growth commences, 2D clusters form randomly on the surface, leading to a decrease in specular beam intensity. The minimum occurs at half-layer coverage. The damping of the oscillations is probably due to deviations from this ideal model in that the surface becomes statistically distributed over several incomplete layers. When growth is suspended by closing the group-III-element shutters and the layer is maintained at its growth temperature in a beam of arsenic, the reflectivity recovers to its original value, i.e. the surface becomes smoother. Briones et al. / 89 / have recently shown that damping of the RHEED oscillations during MBE growth of GaAs can be totally eliminated by a periodic phase-locked perturbation of the growth front by short synchronized interruptions of the arsenic flux. This cessation of the arsenic flux greatly enhances the surface migration of Ga atoms and thus promotes the formation of single-phase surface domains.

The oscillatory nature of the RHEED intensities provides direct real-time evidence of compositional effects and growth modes during the formation of heterointerfaces. As for the widely used $Al_xGa_{1-x}As$ / GaAs heterojunctions, the sequence of layer growth is critical for compositional gradients and crystal perfection, which in turn is important for optimizing 2D transport properties. When the Al flux is switched at the maximum of the intensity oscillations, the first period for the growth sequence from ternary alloy to binary compound corresponds neither to the $Al_xGa_{1-x}As$ growth rate nor to the steady-state GaAs rate, but shows some intermediate value (Fig. 7). For the growth sequence from binary compound to ternary alloy or between the two binaries an intermediate period does not exist. A possible explanation for this phenomenon can be found in the relative surface diffusion lengths of the group-III-elements Al and Ga, which were estimated to be $\lambda_{Al} \cong 3.5$ nm and $\lambda_{Ga} \cong 20$ nm on (100) surfaces under typical MBE growth conditions / 17 / . These differences in cation diffusion rates have striking consequences on the nature of the interface. While a GaAs layer should be covered by smooth terraces of 20 nm mean length between monolayer steps, those on an $Al_xGa_{1-x}As$ layer would only be 3.5 nm apart. The important result of this qualitative estimate is that the GaAs/ $Al_xGa_{1-x}As$ interface is much smoother on an atomic scale than the inverted structure. Direct experimental evidence for this distinct difference in binary-to-ternary layer growth sequence is obtained from inspection of the high-resolution TEM micrograph depicted in Fig. 10b / 88 / . This lattice image of an $Al_{0.2}Ga_{0.8}As$ / AlAs superlattice shows clearly that the heterointerface is abrupt to within one atomic layer only when the ternary alloy is grown on the binary compound, but not for the inverse growth sequence.

Since the nature of the heterointerface is critical for optimizing excitonic as well as transport properties in quantum wells, various attempts have been made to minimize the interface roughness (or disorder) by modified MBE growth conditions. The most successful modification is probably the method of growth interruption at each interface. Growth interruption allows the small terraces to relax into larger terraces via diffusion of the surface atoms. This reduces the step density and thus simultaneously enhances the RHEED specular beam intensity which can be used for real-time monitoring. The time of closing both the Al and the Ga shutter (while the As shutter is left open) apparently depends on the actual growth condition. Values ranging from a few seconds to several minutes have been reported by different authors / 90, 91 / , in particular for the $Al_xGa_{1-x}As$ / GaAs interface and if the full recovery of the specular beam intensity has been allowed for. We have found that in most cases (i.e. growth rate < 1 µm/hr and growth temperature < $650^{\circ}C$) growth interruption of less than 100 s is sufficient to minimize the interface roughness of the Al-Ga-As system.

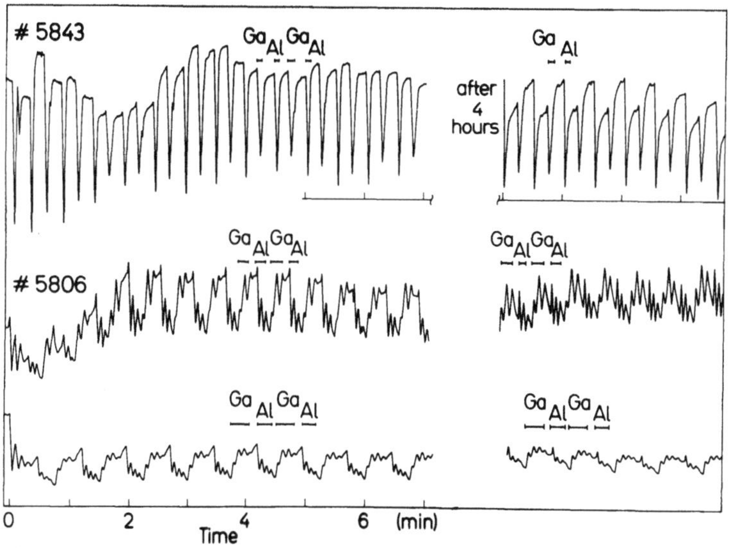

Fig. 11 RHEED intensity oscillations of specularly reflected electron beam from (001) surface in [100] azimuth during growth of $(GaAs)_m(AlAs)_m$ superlattices with m = 1, 2, and 3 (from top to bottom).

In Fig. 11 we show the RHEED intensity sequence for growth of ultrathin-layer $(GaAs)_m(AlAs)_m$ superlattices with m = 1, 2, and 3. The growth rates of GaAs and AlAs monolayers were accurately controlled by the period of the intensity oscillations of the specularly reflected beam in the RHEED pattern, and actuation of the effusion cell shutters was synchronized to these oscillations. The synthesis of each superlattice period requires four steps. At first a monolayer (or bi- or trilayer) of GaAs is deposited. In the second step the crystal growth is stopped for 5 s and the surface is only exposed to the arsenic flux. The third step involves the deposition of a monolayer (or bi- or trilayer) of AlAs. Finally, the fourth step is the same as the second one. The shutters in front of Ga and Al effusion cells, respectively, are opened at the time when the intensity of the specularly reflected beam has reached its maximum value and closed after one (or two or three) period(s). The maximum intensity indicates that the growth surface has become extremely smooth. A low growth rate of 1 monolayer per 5 s ensures that the crystal growth occurs indeed in a 2D layer-by-layer mode. The purpose of the second and the fourth step of this growth sequence is that each layer is completed before the next one starts. The formation of high-quality layered crystals has been demonstrated by the appearance of distinct satellite peaks around the Bragg reflections in the X-ray diffraction pattern (see next Sect. 4.2). Our detailed luminescence investigations / 92 / revealed that the ultrathin-layer $(GaAs)_m(AlAs)_m$ superlattices with m < 4 indeed represent a new artificial semiconductor material with novel electronic properties.

4.2 Double Crystal X-Ray Diffraction

High-angle X-ray diffraction is a powerful non-destructive technique for investigation of interface disorder effects in superlattices and multi quantum well heterostructures, if a detailed analysis of the diffraction curves is performed. We have recently developed a semi-kinematical approach of the dynamical theory of X-ray diffraction to determine the strain profile, the composition (periodicity), and the interface quality of $Al_xGa_{1-x}As$/GaAs and of $Al_{0.48}In_{0.52}As/Ga_{0.47}In_{0.53}As$ heterostructures and superlattices / 93 / . In the following we briefly outline those parts of the theory that are relevant to the experiments.

The scattering of X-rays from strained single crystals is described by Taupin's formalism of the dynamical theory of X-rays / 94 / . The semikinematical approximation of Petrashen / 95 / uses the first iteration of Taupin's equation for the amplitude ratio of the diffracted and incident waves and is valid if the thickness of the deformed layer is small compared with the X-ray extinction length. The reflectivity is then given in integral form which reduces the calculation time for a diffraction curve. For our calculation the epitaxial layer is devided into n lamellae of equal thickness $\Delta = z_0/n$, where z_0 is the total thickness of the layer measured in units of the extinction length divided by π. The reflectivity R(y) is given by the product of the diffraction curve for an undeformed crystal $R_p(y) = |g - y|^2$ and a deformation-dependent factor

$$R(y) = R_p(y) \left| 1 - 2iy \sum_{j=1}^{n} \exp(i\phi_j) \frac{\sin(y-s_j)\Delta}{y-s_j} \right| . \tag{7}$$

For a crystal without a centre of inversion we have

$$y = \frac{-2b \sin(2\Theta_B)\Delta\Theta - (1-b)(\chi_0^r + i\chi_0^i)}{2C\,(|b|(\chi_h^r\chi_{\bar{h}}^r - \chi_h^i\chi_{\bar{h}}^i))^{1/2}} , \tag{8}$$

$$g = \text{sign}(y) \left(y^2 - 1 - 2i \frac{\chi_{\bar{h}}^r\chi_h^i + \chi_h^r\chi_{\bar{h}}^i}{\chi_h^r\chi_{\bar{h}}^r + \chi_h^i\chi_{\bar{h}}^i}\right)^{1/2} \tag{9}$$

and

$$\phi_j = -y\Delta \quad (2(n-j)+1) + \Delta s_j + 2\Delta + \sum_{j=j+1}^{n} s_j \qquad (10)$$

where $\chi_h^{r,i} = r_e \frac{\lambda^2}{\pi V} F_h^{r,i}$ are the h-th Fourier coefficients of the polarizability multiplied by 4π, λ is the X-ray wavelength, r_e is the classical electron radius, V is the unit cell volume and F^r, F^i are the real and imaginary part of the structure factor. $b=\gamma_o/\gamma_h$ is the asymmetry factor, and $\gamma_o = \sin(\theta_B-\alpha)$ and $\gamma_h = \sin(\theta+\alpha)$ are the direction cosines of the incident and diffracted waves respectively, θ_B is the kinematical Bragg angle and α the angle between crystal surface and reflecting lattice plane. Δ is the deviation from the Bragg angle. C is the polarization factor, which is equal to unity for σ-polarization and cos $2\theta_B$ for π-polarization.
s_j is the normalized strain for the j-th lamella. This constant value within each lamella is given by the equation

$$s = \frac{\lambda \, |b|^{1/2}}{2\pi C \, (\chi_h^r\chi_{\bar{h}}^r - \chi_h^i\chi_{\bar{h}}^i)^{1/2}} \; \frac{\partial}{\partial s_h} \; (\vec{h}\vec{u}) \qquad (11)$$

where $\bar{h}$ is the diffraction vector, $\bar{u}$ is the displacement field, and s_h is the direction of the diffracted wave. If there is no shear strain, as in epitaxial layers grown in (001) and (111) orientation, and if the crystal is bound by the xy plane and the diffraction plane is the xz plane, Eq. (11) takes the form

$$s(z) = \frac{2|b|^{1/2}\sin^2\theta_B}{(\chi_h^r\chi_{\bar{h}}^r - \chi_h^i\chi_{\bar{h}}^i)^{1/2}} (\varepsilon_{zz}(z)\cos^2\alpha + \varepsilon_{xx}(z)\sin^2\alpha + (\varepsilon_{zz}(z) - \varepsilon_{xx}(z))\frac{\sin 2\alpha}{2\tan\theta_B}). \qquad (12)$$

Here the depth dependence of the normalized strain is taken into account, and $\varepsilon_{zz}(z)$ and $\varepsilon_{xx}(z)$ are the strains at depth z perpendicular and parallel to the crystal surface, respectively.

The sinusoidal term in Eq. (7) describes the amplitude of a single lamella, whereas ϕ_j in the exponential term describes the phase of a single lamella. One of the interesting consequences of Eq. (7) is the possibility of observing oscillations on the X-ray diffraction curve, i.e., Pendellösung fringes, as a result of interferences of waves scattered at different depths in the epitaxial layer (see $\Delta\omega$ in Fig.12).

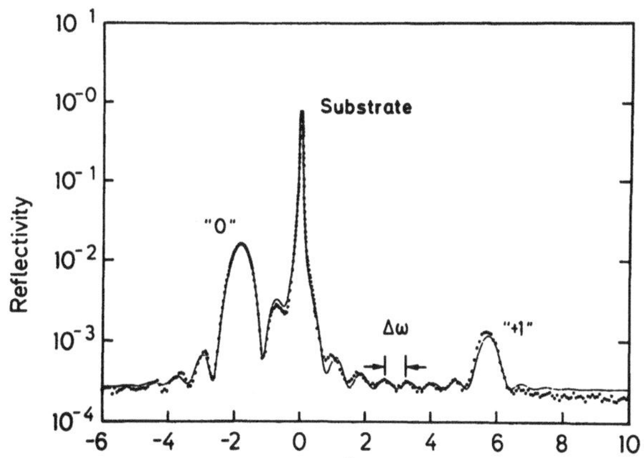

Fig. 12 Experimental (dotted curve) and theoretical (solid line) X-ray diffraction curve of an $(AlAs)_{44}$ $(GaAs)_{46}$ superlattice in the vicinity of the (004) reflection using $CuK\alpha_1$ radiation.

If we assume that s(z) is constant in the whole epilayer of the thickness D, the sinusoidal term of the amplitude in Eq. (7) oscillates with the period π. For the argument of the sinusoidal term we then obtain the relation

$$yD = \pi. \tag{13}$$

Using Eq. (8) and considering the fact that D must be multiplied by l_{ex} we find the known relation between the angular spacing of the Pendellösung fringes $\Delta\omega$ and the thickness D of the strained surface layer, i.e.

$$D = \frac{\lambda\,|\gamma_h|}{\Delta\omega \sin(2\theta_B)}. \tag{14}$$

If we apply the same evaluation to a superlattice, i.e., a one-dimensional periodic structure with a period length T, we find the identical relation as given in Eq. (14). The whole epilayer thickness is replaced by the superlattice period T, and the oscillation spacing is replaced by the angular distance between the satellite peaks which occur in X-ray diffraction curves from superlattices. The amplitude of the deformation factor in Eq. (7), which describes the scattering of the deformed layer, has a maximum if $y = \bar{s}$, where $\bar{s}$ is the average value of s(z) in the strained layer. This consideration leads to the relation between (i) the angular spacing of the main diffraction peaks of the strained layer and of the unstrained crystal, $\Delta\theta^{o}$, and (ii) the average strain value $\bar{\varepsilon}_{zz}$, which exists in the strained layer.

Semiconductor superlattices are one-dimensional periodic structures consisting of thin layers of alternating composition. Therefore, the depth variation of the strain in Eq. (11) and the depth variation of the structure factor in the deformation-dependent term of Eq. (7) must be included. The depth distribution of the strain and of the chemical composition are determined by a comparison of the calculated X-ray diffraction curve with the experimental data. A certain strain and an approximate composition distribution is initially presumed from the employed growth conditions. The best coincidence of the theoretical and experimental X-ray diffraction curve is then found by varying successively the values for the structural parameters, i.e. the strain and composition profiles as well as the epilayer thickness.

The X-ray diffraction measurements are performed with a computer-controlled double-crystal X-ray diffractometer in non-dispersive (+, -) Bragg geometry. An asymmetrically cut (100) Ge crystal was used for monochromizing and collimating the X-rays / 93 / . As the lattice parameter and the scattering factors of superlattices are subject to a one-dimensional modulation in growth direction, the diffraction patterns consist of satellite reflections located symmetrically around the Bragg reflections, as shown in Fig. 12 for an AlAs/GaAs superlattice. From the position of the satellite peaks the superlattice periodicity can be deduced. Detailed information about thickness fluctuations of the constituent layers, inhomogeneity of composition, and interface quality can be extracted from the halfwidths and intensities of the satellite peaks. The excellent agreement between experimental and theoretical diffraction curve in Fig. 12 indicates extremely abrupt AlAs/GaAs interfaces to within one monolayer.

Since MBE growth occurs predominantly in a two-dimensional layer-by-layer growth mode, the compositional changes at heterointerfaces of closely lattice-matched materials, like AlAs/GaAs, should occur over no more than one monolayer. However, for the widely used $Al_xGa_{1-x}As$/GaAs heterojunction it is well established that the sequence of layer growth is critical for compositional gradients and crystal perfection, which in turn strongly affect the excitonic and the transport properties of quantum wells. While the GaAs/$Al_xGa_{1-x}As$ heterointerface is abrupt to within one monolayer when the ternary alloy is grown on the binary compound, this is not the case for the inverse growth sequence under typical MBE growth conditions. In X-ray diffraction the existence of interface disorder manifests itself in an increase of

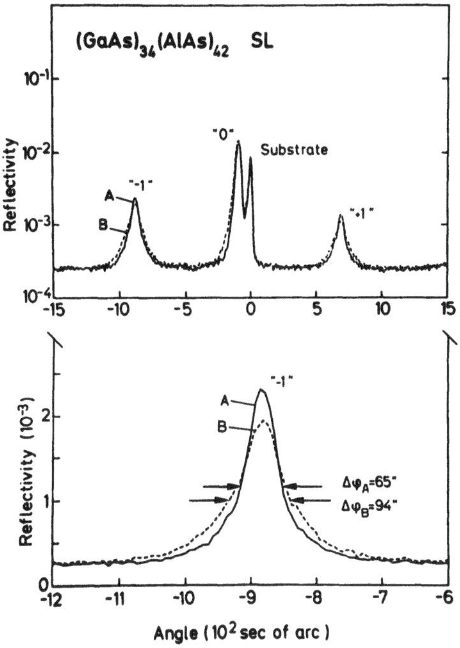

Fig. 13 X-ray diffraction curves of two AlAs/GaAs superlattices grown (A) with and (B) without growth interruption at the heterointerfaces, recorded with CuKα1 radiation in the vicinity of the quasi-forbidden (002) reflection.

the halfwidths and a decrease of the intensities of the satellite peaks, as shown in Fig. 13. During MBE growth of these two $(AlAs)_{42}(GaAs)_{34}$ superlattices, the adjustment of the shutter motion at the transition from AlAs to GaAs and vice versa was changed in the two growth runs. Sample A was grown with growth interruption at each AlAs/GaAs and GaAs/AlAs interface, whereas sample B was grown continuously. While the positions of all the diffraction peaks of sample A coincide with those of sample B, the halfwidths of the satellite peaks from sample A are narrower and their reflected intensities are higher. A growth interruption of 10 s was sufficient to smooth the growing surface which then provides sharp heterointerfaces. When the heterojunctions are grown continuously, the monolayer roughness of the growth surface leads to a disorder and thus broadening of the interface. In X-ray diffraction this broadening manifests itself as a random variation of the superlattice period of about one lattice constant (∿ 5.6 Å) for sample B.

The quantitative evaluation of the interface quality by X-ray diffraction becomes even more important if the lattice parameters of the epilayer have to be matched to those of the substrate by appropriate choice of the layer composition, as for $Al_{0.48}In_{0.52}As/Ga_{0.47}In_{0.53}As$ superlattices lattice-matched to InP substrates. In Fig. 14 we show the X-ray diffraction pattern of such a superlattice with a periodicity of 20.4 nm. For the investigation of this all-ternary material system the X-ray diffraction patterns were recorded in the vicinity of the symmetric (002) and (004) reflections and of the asymmetric (224) and (044) reflections using $CuK\alpha_1$ radiation. Both symmetric and asymmetric diffraction data yield the average lattice strain perpendicular, $\bar{\varepsilon}_{zz}$, and parallel, $\bar{\varepsilon}_{xx}$, to the (001) substrate surface. The lattice strains $\bar{\varepsilon}_{zz}$ and $\bar{\varepsilon}_{xx}$ are correlated with the angular distance $\Delta\theta_I$ and $\Delta\theta_{II}$ between the substrate diffraction maximum and the main epitaxial layer peak ("0"-peak) by the equation

$$\begin{pmatrix} \bar{\varepsilon}_{zz} \\ \bar{\varepsilon}_{xx} \end{pmatrix} = \begin{pmatrix} A_I & B_I \\ A_{II} & B_{II} \end{pmatrix}^{-1} \begin{pmatrix} \Delta\theta_I \\ \Delta\theta_{II} \end{pmatrix} \quad \text{with} \tag{15}$$

$$A_{I,II} = \cos\alpha_{I,II} *[\cos\alpha_{I,II} *\tan\theta_{I,II} + \sin\alpha_{I,II}],$$

$$B_{I,II} = \sin\alpha_{I,II} *[\sin\alpha_{I,II} *\tan\theta_{I,II} - \cos\alpha_{I,II}] \tag{16}$$

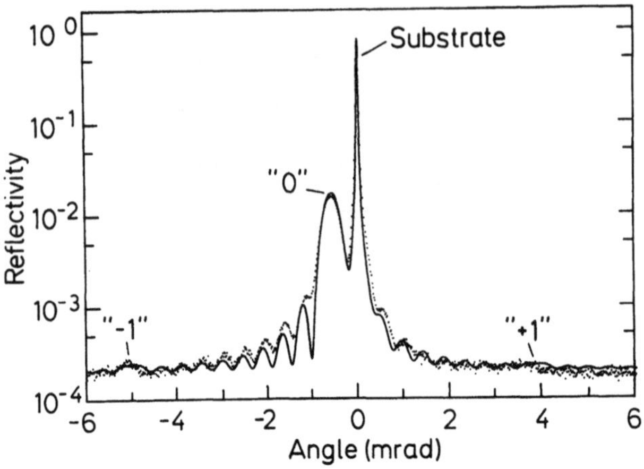

Fig. 14 $CuK\alpha_1$ (004) diffraction pattern of $Al_xIn_{1-x}As/Ga_xIn_{1-x}As$ superlattice on (001) InP with $L_Z = L_B$ = 10.2 nm (.... experiment, ——— theory).

and

$$\bar{\varepsilon}_{zz} = (\bar{d}_e^{\perp} - d_s) / d_s ,$$
$$\bar{\varepsilon}_{xx} = (\bar{d}_e^{\parallel} - d_s) / d_s . \qquad (17)$$

The indices I and II hold for symmetric and asymmetric reflections, respectively. Here, $\bar{d}_e^{\perp}$ and $\bar{d}_e^{\parallel}$ are the average interplanar spacings of the epitaxial layer perpendicular and parallel to the crystal surface, while d_s is the interplanar spacing of the substrate crystal, respectively. θ_I and θ_{II} are the kinematic Bragg angles, and α_I α_{II} the angles between crystal surface and reflection planes I and II. The evaluation of our experimental diffraction data reveals that the superlattice is not misoriented with respect to the substrate crystal and the $\bar{\varepsilon}_{xx}$ = 0 for all samples, i.e. the lattice spacing parallel to the crystal surface in the constituent $Ga_xIn_{1-x}As$ and $Al_xIn_{1-x}As$ layers and in the InP substrate crystal are the same. Hence it follows that the lattice strain in the $Ga_xIn_{1-x}As$ and $Al_xIn_{1-x}As$ epilayers perpendicular to the crystal surface is a measure of their mole fraction x. The relation between elastic strain and chemical composition in the $Ga_xIn_{1-x}As$ layers is given by

$$x = \frac{1}{a_{GaAs} - a_{InAs}} * [a_{InP}(1 + \frac{c_{11}}{c_{11} + 2c_{12}} \varepsilon_{zz}) - a_{InAs}] \qquad (18)$$

where c_{11} and c_{12} are the elastic stiffness constants of the epilayer material. For the $Al_xIn_{1-x}As$ layers the lattice constant a_{GaAs} in Eq. (18) must be replaced by that of a_{AlAs}.

In Fig. 14 we show the experimental (dotted line) and the theoretically fitted (solid line) diffraction patterns obtained from a representative $Al_xIn_{1-x}As$ / $Ga_xIn_{1-x}As$ superlattice in the vicinity of the symmetrical (004) $CuK\alpha_1$ reflection. From the theoretical diffraction pattern we obtain the thickness as well as the lattice strain of the individual layers. The chemical compositions are determined by using Eq. (18). In Table 3 we summarize the measured and the calculated structural data from four $Al_xIn_{1-x}As/Ga_xIn_{1-x}As$ superlattices grown under different conditions. The growth conditions were adjusted such that for all samples the

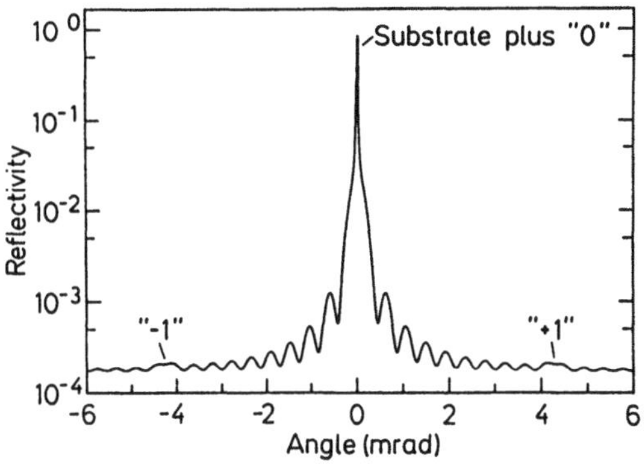

Fig. 15 Calculated diffraction pattern of a perfectly lattice-matched $Ga_xIn_{1-x}As/Al_xIn_{1-x}As$ superlattice on (001) InP [$CuK\alpha_1$ (004) reflection]

Table 3 Structural parameters of four 10-period $Al_xIn_{1-x}As/Ga_xIn_{1-x}As$ superlattices grown by MBE with different temperature settings of the group-III-element effusion cells, as determined by X-ray diffraction.

Sample No.	Thickness of $Ga_xIn_{1-x}As$ layers (nm)	Thickness of $Al_xIn_{1-x}As$ layers (nm)	Average lattice mismatch of superlattice $\bar{\epsilon}_{zz}$ ($\times 10^{-4}$)	Lattice strain in $Ga_xIn_{1-x}As$ layers $\bar{\epsilon}_{zz}$ ($\times 10^{-4}$)	Lattice strain in $Al_xIn_{1-x}As$ layers $\bar{\epsilon}_{zz}$ ($\times 10^{-4}$)	Mole fraction x of $Ga_xIn_{1-x}As$ layers	Mole fraction x of $Al_xIn_{1-x}As$ layers
5649	10.6	10.6	- 4.5	+ 6.0	- 15.0	0.464	0.487
5653	10.2	10.2	+ 9.4	9.5	9.3	0.462	0.470
5654	10.2	10.2	- 4.0	0.0	- 8.0	0.468	0.482
5655	10.2	10.2	- 3.5	7.0	- 14.0	0.464	0.486

barrier width L_B equals the well width L_Z. The lattice mismatch perpendicular to the (001) growth face for the tetragonally distorted epilayers on InP is less than / 2.8×10^{-3} / . i.e. less than the lattice mismatch in the AlAs/GaAs system. The Pendellösung fringes observed between the main diffraction peaks and the satellite peaks "-1" and "+1" demonstrate the excellent thickness and composition homogeneity perpendicular and parallel to the crystal surface.

In Fig. 15 we show the theoretical diffraction pattern for a perfectly lattice-matched $Ga_{0.468}In_{0.532}As/Al_{0.477}In_{0.523}As$ superlattice with $L_B = L_Z = 10.6$ nm. It should be noted that the satellite peaks "-1" and "+1" have almost disappeared. This finding is in contrast to that observed in strained layer superlattices, where a strain periodicity produces strong satellite peaks. The low intensity and the Pendellösung fringes in Fig. 15 are caused only by the periodicity of the structure factors in the superlattice. A weak intensity of the satellite peaks is also observed if the lattice strains of the $Ga_xIn_{1-x}As$ layers are of the same magnitude, which occurs in the sample displayed in Fig. 14 (see also Table 3).

These few examples indicate that important details of the interface structure can be extracted from high-angle X-ray diffraction if a semi-kinematical treatment of the dynamical theory is applied to analyse the experimental data.

5. CURRENT RESEARCH ACTIVITIES

In this section we will briefly define some selected areas of current research activities on the application of interface formation during MBE growth to modify the bulk properties of semiconductors through bandgap engineering. An extensive survey on these research activities can be found in Refs. / 96 - 98 / . In the following we classify the various heterojunctions and superlattice systems and give references for the most active areas of research.

5.1 GaAs / $Al_xGa_{1-x}As$

The GaAs / $Al_xGa_{1-x}As$ system has a lattice mismatch of only 0.16% for x = 1 and represents the prototype materials system for research on artificially layered semiconductor structures.

Dynamical Properties of Two-Dimensional Excitons (excitons in an electric field, optical dephasing of excitons)

E.O. Göbel, J. Kuhl, and R. Höger, J. Lumin. 30, 541 (1985)
Y. Horikoshi, A. Fischer, and K. Ploog, Jpn. J. Appl. Phys. 24, 955 (1985)
H.J. Polland, L. Schultheis, J. Kuhl, E.O. Göbel, and C.W. Tu, Phys. Rev. Lett. 55, 2610 (1985)
R.T. Collins, K. von Klitzing, and K. Ploog, Phys. Rev. B33, 4378 (1986); Appl. Phys. Lett. 49, 406 (1986)
L. Vina, R.T. Collins, E.E. Mendez, and W.I. Wang, Phys. Rev. B 33, 5939 (1986)
L. Schultheis, A. Honold, J. Kuhl, and C.W. Tu, Phys. Rev. B 34, 9027 (1986); Superlattices and Microstructures 2, 441 (1986)

Quantum Hall Effect

K. von Klitzing, Rev. Mod. Phys. 58, 519 (1986)
R.E. Prange and S.M. Girvin, The Quantum Hall Effect, Springer Ser. Contemp. Phys. 2 (1986)
H.L. Störmer, J.P. Eisenstein, and A.C. Gossard, Phys. Rev. Lett. 56, 85 (1986)
Z. Schlesinger, W.I. Wang, and A.H. McDonald, Phys. Rev. Lett. 58, 73 (1987)

(Resonant) Tunneling and Ballistic Transport

T.C.L.G. Sollner, W.D. Goodhue, P.E. Tannenwald, C.D. Parker, and D.D. Peck, Appl. Phys. Lett. 43, 588 (1983)
M. Tsuchiya, H. Sakaki, and J. Yoshino, Jpn. J. Appl. Phys. 24, L 466 (1985)
N. Yokoyama, K. Imamura, S. Muto, S. Hiyamizu, and H. Nishi, Jpn. J. Appl. Phys. 24, L 853 (1985)
T. Nakagawa, H. Imamoto, T. Kojima, and K. Ohta, Appl. Phys. Lett. 49, 73 (1986)
R.A. Davies, M.J. Kelly, and T.M. Kerr, Phys. Rev. Lett. 55, 1114 (1985)
E.E. Mendez, E. Calleja, and W.I. Wang, Phys. Rev. B 34, 6026 (1986)
T.W. Hickmott, Phys. Rev. B 32, 6531 (1985)
M. Heiblum, M.I. Nathan, D.C. Thomas, and C.M. Knoedler, Phys. Rev. Lett. 55, 2200 (1985)

Folded Phonons and Interface Modes in Superlattices

A.K. Sood, J. Menendez, M. Cardona, and K. Ploog, Phys. Rev. Lett. 54, 2111 (1985); Phys. Rev. Lett. 54, 2115 (1985); Phys. Rev. Lett. 56, 1751 (1986)

Plasmon Resonances and Single Particle Excitations

G. Fasol, N. Mestres, H.P. Hughes, A. Fischer, and K. Ploog, Phys. Rev. Lett. 56, 2517 (1986)

Electron-Phonon Interactions

M.A. Brummel, R.J. Nicholas, M.A Hopkins, J.J. Harris, and C.T. Foxon, Phys. Rev. Lett. 58, 77 (1987)

GaAs on Si Substrates (lattice mismatch is about 4%)

The following problems have to be solved: (i) antiphase disorder, (ii) misfit dislocations, and (iii) interface charge and cross doping.
R. Fischer, H. Morkoc, D.A. Neumann, H. Zabel, C. Choi, N. Otsuka, M. Longerbone, and L.P. Erickson, J. Appl. Phys. 60, 1640 (1986)
H. Kroemer, Mater. Res. Soc. Symp. Proc. 67, 3 (1986)
H. Morkoc, Mater. Res. Soc. Symp. Proc. 67, 149 (1986)

5.2 GaSb / $Al_xGa_{1-x}Sb$

The lattice mismatch is 0.66% for x = 1. The accomodation of the lattice mismatch by elastic deformations (strain) may affect the subband levels.

Y. Ohmori, Y. Suzuki, H. Okamoto, Jpn. J. Appl. Phys. 24, L 657 (1985); J. Appl. Phys. 59, 3760 (1986)
K. Ploog. Y. Ohmori, H. Okamoto, J. Wagner, and W. Stolz, Appl. Phys. Lett. 47, 384 (1985)
A. Forchel, U. Cebulla, G. Tränkle, U. Ziem, H. Kroemer, S. Subbana, and G.Griffiths, Appl. Phys. Lett. 50, 182 (1987)
A. Forchel, U. Cebulla, G. Tränkle, E. Lach, T.L. Reinecke, H. Kroemer, S. Subbana, and G. Griffiths, Phys. Rev. Lett. 57, 3217 (1986)

5.3 InAs / GaSb

The lattice mismatch is 0.61% for this system. The unique broken-gap band line-up at the interface leads to a spatial separation of carriers with electrons in the InAs layers and holes in the GaSb layers. In particular, the two-dimensional gas with coexisting electrons and holes has been studied.

E.E. Mendez, L. Esaki, and L.L. Chang, Phys. Rev. Lett. 55, 2216 (1985)
S. Washburn, R.A. Webb, E.E. Mendez, L.L. Chang, and L. Esaki, Phys. Rev. B 33, 8848 (1986)
L.M. Claessen, J.C. Maan, M. Altarelli, P. Wyder, L.L. Chang, and L. Esaki, Phys. Rev. Lett. 57, 2556 (1986)

5.4 $Ga_xIn_{1-x}As$/InP and $Ga_xIn_{1-x}As$ / $Al_xIn_{1-x}As$ Lattice Matched to InP

These two materials systems are of significant importance for advanced photonic devices.
Y. Kawamura, K. Wakita, and H. Asahi, Electron. Lett. 21, 1168 (1985)
Y. Kawamura, K. Wakita, H. Asahi, and K. Kurumada, Jpn. J. Appl. Phys. 25, L928 (1986)
J. Wagner, W. Stolz, J. Knecht, and K. Ploog, Solid State Commun. 57, 781 (1986)
W.T. Tsang and E.F. Schubert, Appl. Phys. Lett. 49, 220 (1986)
T. Fujii, Y. Nakata, S. Muto, and S. Hiyamizu, Jpn. J. Appl. Phys. 25, L598 (1986)

5.5 GaAs / $Ga_{0.8}In_{0.2}As$ on GaAs substrate

The lattice mismatch of $Ga_{0.8}In_{0.2}As$ is 1.43% with respect to GaAs, and below a certain critical thickness the $Ga_{0.8}In_{0.2}As$ layers are heavily strained ("strained layer superlattices"). The built-in strain in $Ga_{0.8}In_{0.2}As$ / GaAs superlattices was used to remove the degeneracy of the valence band at the Brillouin zone center so that the light-hole band can be preferentially populated.

J.E. Schirber, I.J. Fritz, and L.R. Dawson, Appl. Phys. Lett. 46, 187 (1985)

5.6 Si / Si_xGe_{1-x} Strained-Layer Superlattices

The lattice mismatch between Si and Ge is about 4%. Modification of the band line-up by strain distribution was used to form 2D electron or hole gas. Observation of folded phonons.

G. Abstreiter, H. Brugger, T. Wolf, H.J. Jorke, and H.J. Herzog, Phys. Rev. Lett. 54, 2441 (1985)
H. Brugger, G. Abstreiter, H. Jorke, H.J. Herzog, and E. Kasper, Phys. Rev. B 33, 5928 (1986)
D.J. Lockwood, M.W.C. Dharma-Wardana, J.M. Baribean, and D.C. Houghton, Phys. Rev. B 35, 2243 (1987)

5.7 CdTe / HgTe

Transition from finite to zero-gap semiconductor

J.P. Faurie, IEEE J. Quantum Electron. QE-22, 1656 (1986)

5.8 ZnSe / $Zn_xMn_{1-x}Se$

2D magnetic semiconductor superlattices

L.A. Kolodzieski, R.L. Gunshor, and N. Otsuka, IEEE J. Quantum Electron. QE-22, 1666 (1986)

5.9 PbTe / $Pb_{1-x}SnTe$ and PbTe / $Pb_xEu_{1-x}Se_yTe_{1-y}$

Transport and magneto-optical properties

G. Bauer, Surf. Sci. 168, 462 (1986)
D.L. Partin, J.P. Heremans, and C.M. Thrush, Superlattices and Microstructures 2, 459 (1986)

5.10 Doping Superlattices in GaAs

Space-charge induced modulation of real-space energy bands

K. Ploog and G.H. Döhler, Adv. Phys. 32, 285 (1983)
K. Ploog, A. Fischer, and E.F. Schubert, Surf. Sci. 174, 120 (1986)

Acknowledgment

This work was sponsored by the Bundesministerium für Forschung und Technologie of the Federal Republic of Germany.

References

1. For a recent extensive review on MBE see: The Technology and Physics of Molecular Beam Epitaxy, ed. by E.H.C. Parker (Plenum, New York 1985)
2. K.G. Günther: Z. Naturforschg. A 13, 1081 (1958)
3. J.R. Arthur: J. Phys. Chem. Solids 28, 2257 (1967)
4. J.R. Arthur: J. Appl. Phys. 39, 4032 (1968)
5. A.Y. Cho: J. Appl. Phys. 41, 782 (1970)
6. A.Y. Cho: J. Vac. Sci. Technol. 8, S 31 (1971)
7. J.C. Bean and S.R. McAfee: J. Physique 43, Colloque C5, 153 (1982)
8. K. Ploog and K. Graf: Molecular Beam Epitaxy of III-V Compounds (Springer, Berlin, Heidelberg 1984)
9. A.C. Gossard: Thin Solid Films 57, 3 (1979)
10. C.T. Foxon and B.A. Joyce: In Current Physics in Materials Science, Vol. 7, ed. by E. Kaldis (North-Holland, Amsterdam 1981) 1
11. A.Y. Cho and J.R. Arthur: Progr. Solid State Chem. 10, 157 (1975)
12. K. Ploog: J. Vac. Sci. Technol. 16, 838 (1979)
13. R. Ludeke: J. Vac. Sci. Technol. 17, 1241 (1980)
14. J.H. Neave, B.A. Joyce, P.J. Dobson, and N. Norton: Appl. Phys. A 31 1 (1983)
15. J.M. van Hove, C.S. Lent, P.R. Pukite, and P.F. Cohen: J. Vac. Sci. Technol. B 1 , 741 (1983)
16. T. Sakamoto, H. Funabashi, K. Ohta, T. Nakagawa, N.J. Kawai, T. Kojima, and K. Bando: Superlattices and Microstructures 1, 347 (1985)
17. B.A. Joyce, P.J. Dobson, J.H. Neave, K. Woodbridge, J. Zhang, P.K. Larsen, and B. Bölger: Surf. Sci. 168, 423 (1986)

18. C.E.C. Wood: In Physics of Thin Films, Vol. 11, ed. by W.R. Hunter and G. Haas (Academic, New York 1980) 35
19. R.E. Honig and D.A. Kramer: RCA Review 30, 285 (1969)
20. M.A. Herman: Vacuum 32, 555 (1982)
21. B.B. Dayton: Trans. 2nd American Vacuum Soc. Symp. 5, 5 (1961)
22. J.A. Curless: J. Vac. Sci. Technol. B 3, 531 (1985)
23. J. Saito and A. Shibatomi: Fujitsu Tech. J. 21, 190 (1985)
24. T.A. Flaim and P.D. Ownby, J. Vac. Sci. Technol. 8, 661 (1971)
25. C.T. Foxon, B.A. Joyce, and M.T. Norris: J. Cryst. Growth 49, 132 (1980)
26. K.Y. Chen, A.Y. Cho, W.R. Wagner, and W.A. Bonner, J. Appl. Phys. 52, 1015 (1981)
27. J.H. Neave, P. Blood, and B.A. Joyce: Appl. Phys. Lett. 36, 311 (1980)
28. G. Duggan, P. Dawson, C.T. Foxon, and G.W. L' Hooft: J. Physique 43, Colloque C5, 129 (1982)
29. H. Künzel, J. Knecht, H. Jung, K. Wünstel, and K. Ploog: Appl. Phys. A 28, 167 (1982)
30. M.B. Panish: J. Electrochem. Soc. 127, 2729 (1980)
31. A.R. Calawa: Appl. Phys. Lett. 38, 701 (1981)
32. T. Henderson, W. Kopp, R. Fischer, J. Klem, H. Morkoc, L.P. Erickson, and P.W. Palmberg: Rev. Sci. Instrum. 55, 1763 (1984)
33. D. Huet, M. Lambert, D. Bonnevie, and D. Dufresne: J. Vac. Sci. Technol. B 3, 823 (1985)
34. T. Mizutani and K. Hirose: Jpn. J. Appl. Phys. 24, L119 (1985)
35. P.A. Maki, S.C. Palmateer, A.R. Calawa, and B.R. Lee: J. Electrochem. Soc. 132, 2813 (1985)
36. Y. Ota: Thin Solid Films 106, 3 (1983)
37. J.C. Bean: J. Vac. Sci. Technol. A 1, 540 (1983)
38. L.L. Chang and K. Ploog: Molecular Beam Epitaxy and Heterostructures (Martinus Nijhoff, Dordrecht 1985) NATO Adv. Sci. Inst. Ser. E 87
39. H.J. Fronius, A. Fischer, and K. Ploog: Jpn. J. Appl. Phys. 25, L137 (1986)
40. G.J. Davies, R. Heckingbottom, H. Ohno, C.E.C. Wood, and A.R. Calawa: Appl. Phys. Lett. 37, 290 (1980)
41. K. Oe and Y. Imamura: Jpn. J. Appl. Phys. 24, 779 (1985)
42. L P. Erickson, G.L. Carpenter, D.D. Seibel, P.W. Palmberg, P. Pearah, W. Kopp, and H. Morkoc: J. Vac. Sci. Technol. B 3, 536 (1985)
43. D.E. Mars and J.N. Miller: J. Vac. Sci. Technol. B 4, 571 (1986)
44. A.J. Springthorpe and P. Mandeville: J. Vac. Sci. Technol. B 4, 853 (1986)
45. C.T. Foxon, J.A. Harvey, and B.A. Joyce: J. Phys. Chem. Solids 34, 1693 (1973)
46. A.Y. Cho and K.Y. Cheng: Appl. Phys. Lett. 38, 360 (1981)
47. E.A. Wood, J. Appl. Phys. 35, 1306 (1964)
48. A.Y. Cho: J. Appl. Phys. 47, 2841 (1976)
49. K. Ploog and A. Fischer: Appl. Phys. 13, 111 (1977)
50. J.H. Neave and B.A. Joyce: J. Cryst. Growth 44, 387 (1978)
51. S. Holloway and J.L. Beeby: J. Phys. C 11, L247 (1978)
52. I. Hernandez-Calderon and H. Höchst: Phys. Rev. B 27, 4961 (1983)
53. B.A. Joyce, J.H. Neave, P.J. Dobson, and P.K. Larsen: Phys. Rev. B 29, 814 (1984)
54. G. Laurence, F. Simondet, and P. Saget: Appl. Phys. 19, 63 (1979)
55. A.Y. Cho, J. Appl. Phys. 42, 2074 (1971)
56. C.E.C. Wood: Surf. Sci. 108, L441 (1981)
57. C.T. Foxon, M.R. Boundry, and B.A. Joyce: Surf. Sci. 44, 69 (1974)
58. B.A. Joyce, P.J. Dobson, J.H. Neave, and J. Zhang: Surf. Sci. 174 (1986)
59. A. Madhukar: Surf. Sci. 132, 344 (1983)
60. J.R. Arthur: Surf. Sci. 43, 449 (1974)
61. C.T. Foxon and B.A. Joyce: Surf. Sci. 50, 434 (1975)
62. C.T. Foxon and B.A. Joyce: Surf. Sci. 64, 293 (1977)
63. C.T. Foxon and B.A. Joyce: J. Cryst. Growth 44, 75 (1978)
64. M. Ilegems: J. Appl. Phys. 48, 1278 (1977)
65. K. Ploog, A. Fischer, and H. Künzel: J. Electrochem. Soc. 128, 400 (1981)
66. N. Duhamel, P. Henoc, F. Alexandre, and E.V.K. Rao: Appl. Phys. Lett. 39, 49 (1981)
67. D.L. Miller and P.M. Asbeck: J. Appl. Phys. 57, 1816 (1985)

68. Y. C. Pao, T. Hierl, and T. Cooper: J. Appl. Phys. 60, 201 (1986)
69. J.L. Lievin and F. Alexandre: Electron. Lett. 21, 413 (1985)
70. E. Nottenburg, H.J. Bühlmann, M. Frei, and M. Ilegems: Appl. Phys. Lett. 44, 71 (1984)
71. J.M. Ballingall, B.J. Morris, D.J. Leopold, and D.L. Rode: J. Appl. Phys. 59, 3571 (1986)
72. M. Heiblum, W.I. Wang, L.E. Osterling, and V. Deline: J. Appl. Phys. 54, 6751 (1983)
73. R. Sacks and H. Shen: Appl. Phys. Lett. 47, 374 (1985)
74. L. Gonzales, J.B. Clegg, D. Hilton, J.P. Gowers, C.T. Foxon, and B.A. Joyce: Appl. Phys. A 41, 237 (1986)
75. W.I. Wang, E.E. Mendez, T.S. Kuan, and L. Esaki: Appl. Phys. Lett. 47, 826 (1985)
76. D.L. Miller. Appl. Phys. Lett. 47, 1309 (1985)
77. H. Nobuhara, O. Wada, and T. Fujii: electron. Lett. 23, 35 (1987)
78. A.Y. Cho and I. Hayashi: J. Appl. Phys. 42, 4422 (1971)
79. C.E.C. Wood, J. Woodcock, and J.J. Harris: Inst. Phys. Conf. Ser. 45, 28 (1979)
80. K. Ploog, A. Fischer, and H. Künzel: Appl. Phys. 22, 23 (1980)
81. A.Y. Cho: J. Appl. Phys. 46, 1733 (1975)
82. K. Ploog and A. Fischer: J. Vac. Sci. Technol. 16, 838 (1978)
83. C.E. Wood, and B.A. Joyce: J. Appl. Phys. 49, 4854 (1978)
84. C.E.C. Wood: Appl. Phys. Lett. 33, 770 (1978)
85. D.S. Jiang, Y. Makita, K. Ploog, and H.J. Queisser: J. Appl. Phys. 53, 999 (1982)
86. D.A. Andrews, R. Weckingbottom, and G.J. Davies: J. Appl. Phys. 54, 4421 (1983)
87. D.A. Andrews, R. Heckingbottom, and Davies: J. Appl. Phys. 60, 1009 (1986)
88. Y. Suzuki and H. Okamoto: J. Appl. Phys. 58, 3456 (1985)
89. F. Briones, D. Golmayo, L. Gonzales, and A. Ruiz: J. Cryst. Growth 81, 387 (1987)
90. M. Tanaka, H. Sakaki, and J. Yoshino: Jpn. J. Appl. Phys. 25, L155 (1986)
91. F. Voillot, A. Madhukar, J.Y. Kim. P. Chen. N.M. Cho, W.C. Tang, and P.G.Newman: Appl. Phys. Lett. 48, 1009 (1986)
92. T. Isu, D.S. Jiang, and K. Ploog: Appl. Phys. A (1987)
93. L. Tapfer and K. Ploog: Phys. Rev. B 33, 5565 (1986)
94. D. Taupin: B ull. Soc. Fr. Minéral. Cristallogr. 87, 496 (1964)
95. P.V. Petrashen: Fiz. Tverd. Tela (Leningrad) 16, 2168 (1974); ibd. 17, 2814 (1975)
96. Special issue of IEEE J. Quantum Electron. QE-22, No. 9 (1986)
97. Proc. 6th Int. Conf. Electron. Prop. Two-Dimens. Systems (EP2DS-VI), Surf. Sci. 170 (1986)
98. Proc. 2nd Int. Conf. Modulated Semicond. Struct. (MSS-2), Surf. Sci. 174 (1986)

Build-up and Characterization of "Artificial" Surfaces for III–V Compound Semiconductors

J.M. Moison

Laboratoire de Bagneux, Centre National d'Etudes des Télécommunications, 196, Avenue Henri Ravéra, F-92220 Bagneux, France

With molecular-beam epitaxy, we can now build on a given material monolayer-thin overlayers, which give to its surface structural and electronic properties different from its natural ones. The performances and limitations of this approach to surface "engineering" are presented.

1. Introduction

In the last ten years, progress in the growth techniques, especially molecular beam epitaxy (MBE), for group-IV or III-V semiconductor materials has been impressive [1]. High-quality materials can now be built into complex structures, such as superlattices or graded heterojunctions. The quality of the structures themselves, as defined by the accuracy with which the planned geometry is obtained, has been clearly demonstrated in the case of GaAs/GaAlAs superlattices [2] where the interface between well and barrier is atomically flat over distances up to 1000 Å, and the well width is the one planned plus or minus one monolayer. New structures are already appearing, such as superlattice alloys or ultra-thin quantum wells with non-lattice-matched compounds [3] whose pattern size is on the monolayer scale.

These performances, which are due to the flexibility of III-V compounds and to their stubborn tendency to layer-after-layer MBE growth, are obtained in an environment which is compatible with surface physics experiments, with respect to crystalline order and cleanliness. All this gives us the opportunity of using MBE as a tool for changing the structure or composition of semiconductor surfaces, and even for building on a given bulk various surfaces with properties different from those of the "native" surface. We present here some recent experiments performed in that spirit.

2. Geometry, composition and crystal structure of overlayers

The first prerequisite to such an approach is clearly that the desired structure, i.e. a monolayer-sized overlayer of a material A on another material B, will actually be such. Especially, intermixing of atoms or roughening of the surface should not occur, even on the monolayer scale. This is in a sense a harder challenge than building up the more conventional heterostructures described above, since the geometry must then be controlled down to the submonolayer level, in spite most often of enormously steep composition and lattice parameter gradients. At this level, which represents a kind of quantum limit since sizes involved are atomic ones, we may encounter intrinsic mechanisms preventing us from reaching such an accuracy [4].

Assuming that the starting substrate is perfectly flat, the two main such mechanisms are 1) the switch from layer-after-layer growth to island growth leading to roughening and 2) surface segregation leading to intermixing. Problem 1) is associated with the lattice mismatch between materials A and B. The A overlayer must be all the thinner to avoid strain relaxation by island growth as the mismatch is greater; roughly speaking, 1000 Å for 1%, but 10 Å for 5% [5]. Most mismatches between group-IV or III-V semiconductors lie in the percent range (4% for Si/GaAs, 7% for InAs/GaAs,...) except for AlAs/GaAs which is nearly lattice-matched. This means that the growth of one or two monolayers is possible in the laminar regime, but usually not much further, so that care must be exercised.

Problem 2) results from segregation mechanisms which bring to the surface substrate B atoms, while overlayer A atoms are driven under this overlayer inside the substrate. This process, which takes place only during the growth by purely surface-located exchanges, may occur at temperatures where genuine bulk diffusion is negligible. It leads to various effects such as segregation in ternary III-V compounds, exchange reaction in metal/semiconductor contacts, or, in the case we consider, intermixing. For instance, during the growth of AlAs on GaAs, of Si on GaAs and GaP, of Ge on GaAs, of GaSb on AlSb,... [6], a sizeable fraction of a monolayer of the substrate is driven on top of the growing overlayer. By Auger spectroscopy and photoemission measurements, we find that this fraction is quite high for common systems, 0.8 for GaAs/InAs, 0.9 for AlAs/GaAs, and 0.95 for AlAs/InAs [7]. On the other hand, the reverse systems reveal no significant segregation, which shows that the tendency to surface segregation in III-As systems follows the order In > Ga > Al. This makes the systems InAs/GaAs, GaAs/AlAs, and InAs/AlAs, with others like InAs/InP [8], GaAs/GaP [9], Ge/Si [10] or As/Si [11], and possibly GaAs/Si good candidates for our purpose.

Finally, it is clear that this aim is all the more clearly achieved as the growth of the overlayer is a registered growth and that an epitaxial overlayer is obtained. If this is rather natural for a lattice-matched system like GaAs/AlAs, it is not the case for non-lattice matched systems, even though they involve a common tetrahedral coordination in the same zinc-blende or diamond lattice. However, with an overlayer thickness limited to a similar range as the one mentioned above, growth most often proceeds by building an overlayer strained in the growth plane to fit the substrate lattice constant, and therefore also strained in the perpendicular direction. This phenomenon, known as pseudomorphism, allows the build-up of structures like strained-layer superlattices. For monolayer-sized thicknesses, epitaxial growth is then most often obtained. However, it must be emphasized that this leads to large deformations in the overlayer, whose properties could then be altered significantly and in a rather unpredictable way.

3. Electronic properties of overlayers

Only a few studies, experimental or theoretical, have been devoted to the study of the electronic properties of the structures we discuss here, i.e. the bulk of a semiconductor covered by an ultra-thin layer of another one. Such a structure, because of the thinness of the top layer, cannot be described as an heterojunction, but must be studied on its own. The few results of the literature involve the density of surface states, as obtained from UV-photoemission measurements on Ge/Si [10], InAs/InP [8], InAs/GaAs and GaAs/GaP [9], and compared to those of the free materials involved. Surprisingly enough, it is found on these four examples that the density of surface states of the structure - and also the pinned Fermi level - is similar to that of the surface of the overlayer material, the vacuum level being taken as the common reference. In other words, it looks like the structure has the bulk states of the bulk material and the surface states that the overlayer material has when fitted on its own bulk: Ge/Si has Ge-like surface states.

This is indeed surprising, especially in view of what we said about the distortions caused by pseudomorphism in non-lattice-matched structures, which all four are. However, surface states are associated with surface bonds and may plausibly not be influenced much by the underlying substrate, in spite of the pseudomorphism deformations. However, at present, it is not clear whether this rather crude description of our limited data has a physical basis or is only empirical, and both experimental and theoretical studies are still needed. Nevertheless, we may have here a very simple way to tailor the surface of a covalent material by "grafting" onto its bulk a surface with the properties it had when still on its bulk parent.

4. Towards surface engineering ?

In many electronic or optronic devices, problems arise from defective surface or interface electronic properties leading to Fermi level pinning or interface recombination. These problems are usually solved by empirical treatments, and no method for engineering (mostly curing !) at will surface properties is available at present. The problem is actually that of replacing the native surface of the semiconductor with another one having the desired properties. The standard solution is oxidation in the gas or liquid phase, plus often hydrogenation to saturate the few remaining dangling bonds. The use of the structures discussed above is another way. For instance, building an InAs surface on InP moves the surface states away from the band gap, thus decreasing their density within it. This leads to a tenfold decrease of the surface recombination velocity as seen in the threefold increase of the luminescence yield of InP [12] and to an increase of the MIS performances [13]. Grafting an InAs surface on GaAs has the inverse effect, and we observe concurrently a decrease by half of the luminescence yield of GaAs. Finally, the surface of GaAs/GaP has roughly the same states as the intrinsic GaP surface. Those three cases then display what can be called from an applied point of view a positive, a negative and a neutral effect.

Artificial III-V surfaces (pseudomorphic overlayers of monolayer thickness) can now be built on III-V compounds by MBE techniques in spite of the large lattice mismatches involved, except in cases (AlAs/InAs, GaAs/InAs,...) where exchanges between substrate and overlayer atoms alter the desired structure. Because of the overlayer thinness, the final structure cannot be described as a genuine heterojunction, but rather as the surface of the overlayer material fitted on top of the bulk substrate. The properties of this artificial surface are similar to those the overlayer has when fitted on its own bulk and may then be foreseen. This may open the way to genuine surface - and interface - engineering.

Acknowledgments

Results presented here are the fruit of teamwork with F.Houzay, F.Barthe, M.VanRompay, and C.Guille, whose participation is gratefully acknowledged. I also benefited from many dicussions with C.Sébenne.

References

[1] See for instance in "The technology and physics of molecular-beam epitaxy", edited by E.H.C.Parker, Plenum Press, New York 1985

[2] B.Deveaux, J.Y.Emery, A.Chomette, B.Lambert, and M.Baudet, Appl.Phys.Lett. 45,1078(1984)

[3] L.Goldstein, F.Glas, J.Y.Marzin, M.N.Charasse, and G.LeRoux, Appl.Phys.Lett. 47,1099(1985)

[4] J.M.Moison, in "Advanced Materials for Telecommunications", edited by P.A.Glassow, Y.I.Nissim, J.P.Noblanc and J.Speight, les Editions de Physique, les Ulis 1986

[5] Theories are derived from: J.H.Van der Merwe, J.Appl.Phys. 34,123(1962), J.W.Matthews, J.Vac.Sci.Technol. 12,126(1975); experimental data are reviewed in reference [4]

[6] R.A.Stall, J.Zilko, V.Swaminathan, and N.Schumaker, J.Vac.Sci.Technol. B3,524(1985) C.Raisin, H.Tegmousse, and L.Lassabatère, Le Vide et les Couches Minces, 41,241(1986); P.C.Zalm, P.M.J.Maree, and R.I.J.Olthof, Appl.Phys.Lett. 46,597(1985); T.deJong, W.A.S.Douma, J.F.van der Veen, F.W.Saris, and J.Haisma, Appl.Phys.Lett. 42,1037(1983); W.Monch, R.S.Bauer, H.Gant, and R.Murschall, J.Vac.Sci.Technol. 21,498(1982); Ping Chen, D.Bolmont, and C.A.Sébenne, Surf.Sci. 132,505(1983)

[7] J.M.Moison, C.Guille, F.Houzay, and M.Bensoussan, to be published

[8] J.M.Moison, M.Bensoussan, and F.Houzay, Phys.Rev. B34,2018(1986)

[9] J.M.Moison, M.VanRompay, C.Guille, F.Houzay, F.Barthe, and M.Bensoussan, Proceedings of the 18th Int. Conf. on the Physics of Semiconductors, Stockholm 1986

[10] Ping Cheng, D.Bolmont, and C.A.Sébenne, Solid State Commun. 46,6689(1983)

[11] M.A.Olmstead, R.D.Bringans, R.I.G.Uhrberg, and R.Z.Bachrach, Phys.Rev. B34,6041(1986)

[12] J.M.Moison, M.VanRompay, and M.Bensoussan, Appl.Phys.Lett. 48,1362(1986)

[13] R.Blanchet, P.Viktorovitch, J.Chave, and C.Santinelli, Appl.Phys.Lett. 46,761(1985)

Part III

Structural Characterization of the Interfaces

Atomic Structure of Semiconductor Surfaces

G. Le Lay

CRMC2-CNRS, Campus de Luminy, Case 913,
F-13288 Marseille Cedex 09, France

Many new experimental results and theoretical advances support a definite model for each peculiar surface structure of a given low index surface plane of silicon, of germanium and of a large variety of III-V covalent semiconductors. Even the most elusive reconstruction of silicon : Si(111)7x7 seems now to be solved, thereby bringing a journey of a quarter century close to its end.

1. INTRODUCTION

As was highlighted by the Panel on Condensed Matter Physics of the Third National (US) Research Council Survey of Physics "Physics through the 1990s" [1] semiconductor interfaces are of basic interest both from a fundamental and a practical point of view. The control of their formation using such techniques as molecular beam epitaxy (MBE), with which materials may be constructed one molecule at a time, and organo-metallic chemical vapor deposition has permitted the fabrication of novel materials that do not occur in nature, such as artificial periodic superlattices consisting of alternating layers of different semiconductors, different metals or semiconductors and metals. The ability to tailor their structures and properties has led to the discovery and elucidation of many new phenomena such as the quantized Hall effect appearing in purely two-dimensional electron gases. K. von KLITZING, the discoverer of the quantum Hall effect was awarded the 1985 Nobel Prize in Physics. However, the formation of an interface starts from the creation of a surface, before growing new layers. The substrate surfaces will play a very important role in the initial stages of the epitaxy and further will condition the quality of the interfaces. Thus, in the past decade, there has been an ever increasing stimulus to study their structural, electronic and vibronic properties on a microscopic level [2].

Immense progress has been achieved with the development and improvement of quite a number of surface-sensitive experimental techniques, especially using bright, polarized and pulsed sources of synchrotron radiation. Major advances come from techniques that yield real-space images and permit one to "see" individual atoms : scanning tunneling microscopy (STM), which can resolve single atoms and has a vertical resolution of less than 0.1 A, and transmission electron microscopy (TEM), which can resolve structures as small as 1.5 Å. The inventors of these techniques -G. BINNIG and H. ROHRER in 1982- on the one hand, -E. RUSKA in 1932- on the other hand, were awarded the 1986 Nobel Prize in Physics.

Correlatively with these experimental developments great advances in the theoretical description of electronic structures of semiconductor surfaces and in the calculation of optimum atomic configurations using total energy minimization techniques have permitted the prediction of the surface relaxation and reconstruction of a large variety of covalent semiconductors.

As a matter of fact, the surfaces of semiconductors are generally reconstructed (or at least relaxed) ; in such a case the size of the crystallographic unit cell at the surface is larger than in the bulk because the surface atoms rearrange themselves into a geometric structure that may be different from that of the underlying bulk lattice. As underlined by TROMP et al.[3] this phenomenon of surface reconstruction is a consequence of the fact that in a bulk-like terminated surface the atoms are not fourfold coordinated because some of their neighbors are missing. The broken bonds of such an "ideal" surface give rise to a high surface energy that drives the surface atoms to rearrange themselves in such a way as, for example, to decrease the number of broken bonds. Concomitantly, these unusual surface bonds also give rise to new localized electronic quantum states that do not exist in the bulk of the crystal and that could be atomically resolved in space and energy by these authors [4] both on Si(111) and Si(100) surfaces. Hence, with the new technique of current imaging tunneling spectroscopy (CITS) the scanning tunneling microscope has "bridged the gap between surface electronic structure and surface geometric structure and has fulfilled the theorist's dream (or nightmare) in which calculation and experiment can be compared on an atomic level".

However, although electronic and geometric structures of surfaces are closely related to each other, traditionally different experimental probes, several of which will be presented in these proceedings, have been used to study surface geometry and surface electronic structure ; the reason being that most techniques that are sensitive to the geometric structure (e.g. grazing incidence X-ray diffraction, X-ray standing waves, surface extended X-ray absorption fine structure, electron diffraction , ion and atom scattering ...) cannot address the electronic structure, whereas techniques that are well suited for studying surface electron states (e.g. photoemission, inverse photoemission, electron energy loss spectroscopy...) have no direct access to the geometric structure [3].

In this contribution I will give a brief status report of our present knowledge of the atomic structure of low-index semiconductor surfaces. However, we must keep in mind that the subject is rapidly developing, especially thanks to the applications of the new surface probes just mentioned. Therefore, if a general consensus seems to be reached on quite a number of reconstructions, it is not absolutely impossible, especially because of the large number of independent structural parameters to be determined, that our present representation might be questioned in the near future. In this respect, we should remember that "it is the devil that has invented the surfaces" as W. PAULI said, a penetrating vision that was recently recalled by G. BINNIG and H. ROHRER [5].

This article will be organized as follows. First I will describe the geometry of cleaved Si(111) and Ge(111) surfaces, then address the reconstructions of (111) surfaces after annealing before discussing some interesting phase transitions. Secondly, I will present the crystallographic structure of Si(100) and Ge(100) surfaces. Finally I will describe the relaxation of cleaved GaAs(110) and summarize our current knowledge of the different reconstructions observed on other polar surfaces.

2. CLEAVED Si AND Ge(111)-2x1 RECONSTRUCTIONS

2.1 Si(111)2x1

In early low energy electron diffraction (LEED) work by FARNSWORTH [6] and LANDER et al.[7] it was reported that (111) surfaces of Si and Ge cleaved in vacuum exhibit a 2x1 reconstruction (if care is taken in orienting the samples with respect to the cleaver blade in order to have the wedge parallel to the ‹$\bar{2}$11› crystal direction, single domain 2x1 reconstructed surfaces may be prepared). For an ideal surface i.e. a truncated bulk structure (see Fig. 1) one has isolated broken bonds and expects half-filled orbitals around the Fermi level [8]. The reconstruction is the result of rebonding after cleavage. For 20 years it was believed that this reconstruction could be explained by the so-called buckling model proposed by HANEMAN in 1961 [9] in which alternate ‹1$\bar{1}$0› rows of atoms in the topmost surface layer are raised (dehybridization of sp^3 bonding towards a p^3 coordination, s character of the dangling bond) and lowered (sp^2 coordination, p dangling bond).

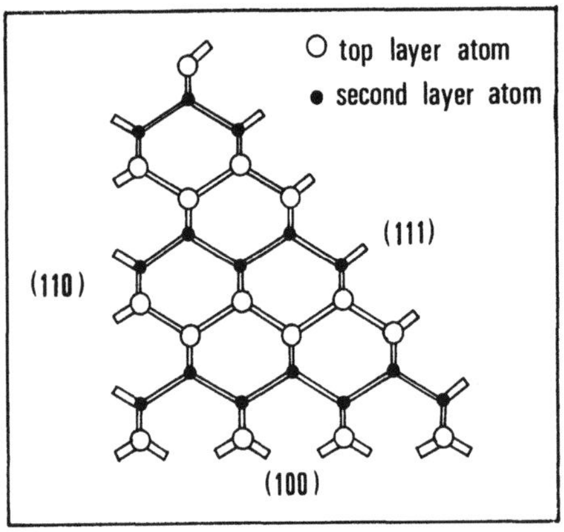

Fig.1- Ideal model of a silicon or germanium crystal, viewed along a ‹110› direction, with the three low index surface planes.

Stimulated by an angle-resolved photoemission (AR-UPS) study by HIMPSEL et al.[10] that discarded the buckling model, PANDEY [11] proposed, purely on theoretical considerations, a completely different geometric model : the π-bonded chain model. In this model (Fig. 2) the top two layers of the (111) surface reconstruct to form fivefold and sevenfold rings, the resulting atomic positions being similar to those of an ideal (110) Si surface. Correlatively the surface atoms rearrange to form parallel, polyacetylene like, zig-zag chains, with dangling bonds on nearest-neighbor sites, instead of next-nearest-neighbor sites for the ideal surface, such that π-bonding between adjacent chain atoms becomes possible.

Unlike the buckling model, which is an ionic one, with charge transferred from the p_z to the s dangling orbitals, the π-bonded chain model is a covalent one : the broken bond orbitals split into bonding/antibonding π/π^* pairs respectively below and above the Fermi level.

Although the formation of the π-bonded chain reconstruction involves large displacements and extensive rebonding of surface atoms NORTHRUP and COHEN [12], on the basis of total energy calculations, could show that the barrier for the transition to this structure is smaller than ~ 0.03 eV per surface atom, so that it can be realized by

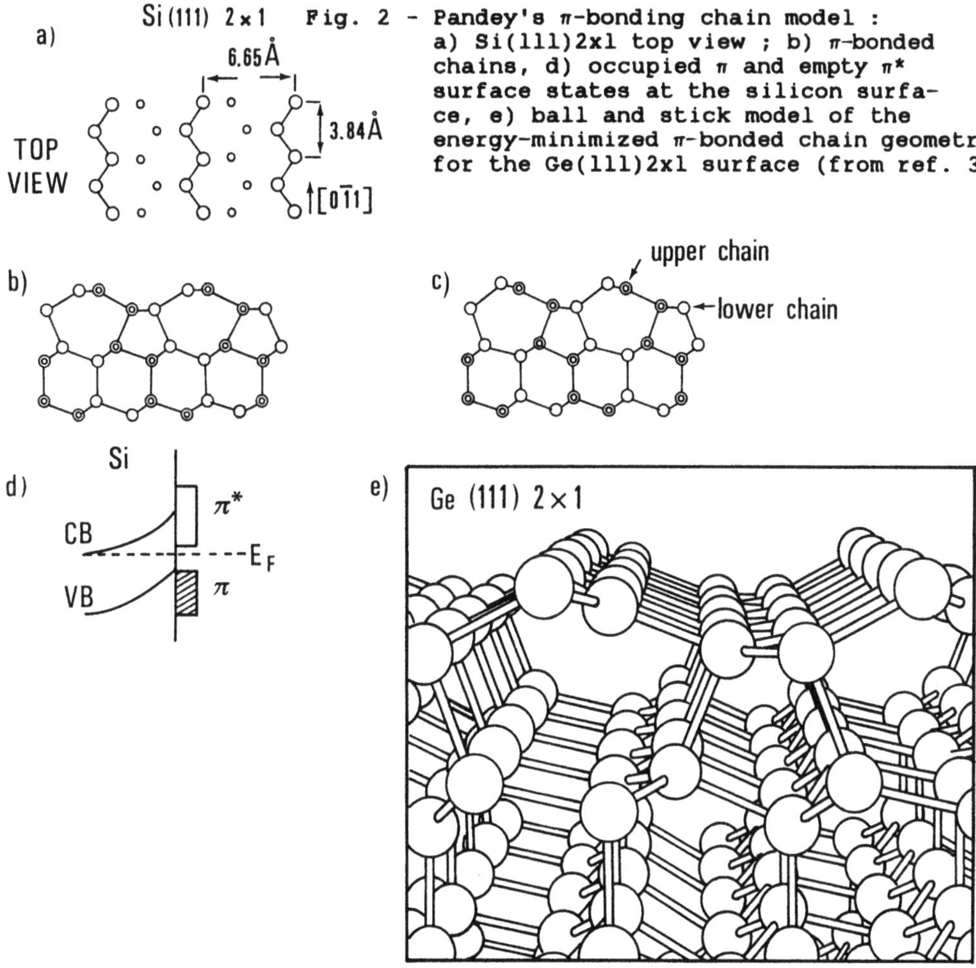

Fig. 2 - Pandey's π-bonding chain model : a) Si(111)2x1 top view ; b) π-bonded chains, d) occupied π and empty π^* surface states at the silicon surface, e) ball and stick model of the energy-minimized π-bonded chain geometry for the Ge(111)2x1 surface (from ref. 36).

the cleavage process. -In this respect it is noteworthy that just after the crystal cleavage in ultra-high vacuum (UHV) ~10^{-10} torr at a cleavage temperature ranging from 4.2 to 300 K, low energy electron diffraction (LEED) measurements show that the Si(111) cleaved surfaces always possess a 2x1 superstructure [13].- The energy-minimized geometry [12] (the energy gain per surface atom is rather large ~ 0.3 eV) includes a tilt of the chain by buckling of the atoms along the chain as indicated in Fig. 2c.

Several experimental studies have provided strong support for the π-bonded chain model. Early optical absorption measurements revealed the presence of two surface bands [14] separated by an energy gap of about 0.45 eV. With AR-UPS [10,15,16,17] the dispersion of the lower band, as well as with inverse photoemission [18] that of part of the upper band, has been observed ; it is in fairly good agreement with the results from calculations for the energy-minimized geometry. Recently, optical techniques making use of polarized light : photothermal displacement spectroscopy [18], polarization-dependent differential reflectivity [19,20] as well as non-linear, optical second harmonic generation (SHG), which reveals the existence of one single (1$\bar{1}$0) mirror [21] (orthogonal to the π-bonded chains), have reinforced this model.

Moreover, medium-energy ion scattering experiments by TROMP et al.[22] have yielded optimum atomic positions in general agreement with the π-bonded chain model also introducing a moderate tilt in the top chain and a small tilt in the second layer chain. In addition, while early LEED results were not compatible with Pandey's original geometry, recent studies including buckling of the chains and subsurface distortions do support the model [23,24].

Beautiful STM images, with exceptionally high resolution (as a result of very sharp probe tips) have been obtained by FEENSTRA et al.[25,26]. They observe a relatively large $\langle\bar{2}11\rangle$ corrugation amplitude along with no observable $\langle 0\bar{1}1\rangle$ corrugation amplitude which is consistent with and provides support for the π-bonded chain model. Moreover, an asymmetry observed in the $\langle\bar{2}11\rangle$ corrugation may be assigned to a tilt (buckling) of the chains. In addition, various types of structural defects in the form of protrusions along the chains and crossing over between chains are imaged.

The electronic properties of the Si(111) 2x1 surface were also determined with spectroscopic measurements by STM [27]. Measurements of the tunneling current versus voltage identify the 0.45 eV gap separating the occupied (π-bonding) and unoccupied (π-antibonding) bands of surface states. Real space images of these surface states reveal a striking phase reversal of the corrugation when the polarity of the bias is reversed (see Fig. 6 of ref. [28]), illustrating that the STM maps charge density rather than atomic positions.

Finally, the dynamical properties of the Si(111)2x1 surface have also been investigated. The calculations of ALERLAND et al.[29] predict chain localized modes displaying strong dispersion as expected for a one dimensional chain. While two recent studies [30,31] of the azimuthal dependence of high resolution electron energy loss spectroscopy (HR-EELS) measurements on Si(111)2x1 were interpreted in the context of the π-bonded chain model, although no strong anisotropy of the 56 meV vibrational excitation studied was found, two new investigations, one of the temperature dependence of transitions across the surface state-band gap (optical absorption) [32], the other one of high-resolution inelastic helium scattering [33], reveal a non-dispersive mode at 10.5 meV in contradiction to the lattice-dynamical calculations mentioned above. This favors the notion of localized oscillators that could be realized by the dimers of the π-bonded molecule model proposed by CHADI [34]. However, as pointed out by HEINZ et al.[21] this model violates the mirror-plane symmetry evidenced by SHG, and according to a calculation by NIELSEN et al.[35] it has much higher total energy than the chain model.

Hence, it could be that those dynamical results are like the black sheep in a well-behaved family ! The main question to be answered is whether the vibronic structure of the silicon (111)2x1 reconstruction can be reconciled with the π-bonded chain model or whether structural models allowing for a quasi-localized vibration have to be introduced. In this respect the role played by the numerous defects imaged by STM could be crucial ...

2.2 Ge(111)2x1

Although not as widely studied as the silicon one, the Ge(111)2x1 cleaved reconstruction is believed to be very similar. The geometry of the energy-minimized π-bonded chain model calculated by NORTHRUP and COHEN [36] is shown in Fig. 2e. AR-UPS data strongly support this model, especially measurements of a heavily n-doped crystal, which permitted direct study of the empty surface state band [37].

Nevertheless, germanium presents its own characteristics. LEED studies have revealed that while the cleaved silicon (111) surface always possesses a 2x1 superstructure at a cleavage temperature T_{Cl} ranging from 4.2 to 300 K, the cleaved Ge(111) surface structure depends on T_{Cl} : beyond 60K the Ge(111)2x1 superstructure is always observable by LEED, however, at $T_{Cl} < 40K$ the LEED pattern may show only a 1x1 atomic structure [13]. Also recent EELS measurements of the temperature (30-395 K) dependence of the surface-state gap show a larger effect than for Si but little evidence for the defect induced states usually observed even on excellent single-domain 2x1 cleaves of Si(111) [38]. Unfortunately, no STM work on cleaved germanium has been published until now ; indeed comparison with the study of FEENSTRA et al.[26] revealing orientational disorder of the π-bonded chains on Si(111)2x1 and associated defect states in or near the band gap of the 2x1 surface-state energies would be highly desirable and instructive !

2.3 Thermal treatments

The 2x1 surface is metastable and converts to a much larger superstructure after annealing : 7x7 for silicon, c-(2x8) for germanium. The Si(111)2x1→7x7 phase transition has been monitored by SHG measurements during thermal annealing and observed to occur on a time scale of seconds at ~ 275° C [21]; besides, a range of transformation temperatures from ~ 200° to 400° C has been well correlated with the step density on the surface [39]. The Ge(111)2x1→c(2x8) transition was first observed by ellipsometry at 632 nm [40] ; it has been reexamined by photothermal displacement investigations on the temperature dependence of the surface optical gap in the range 300-450 K [41]. Here again, upon heating rapidly at ~ 120°C, the same time scale of seconds for the conversion was obtained.

3. Si(111)7x7 AND Ge(111)c-2x8

3.1 Si(111)7x7

The Si(111)7x7 is a very complicated surface with a unit cell 49 times larger than that of the ideal Si(111) surface : it has created a very strong interest since its discovery nearly three decades ago [42] and a large number of models have been proposed to explain the available experimental data, some of them (up to 1983) have been classified by TAKAYANAGI [43]. As underlined by HIMPSEL [8], to fully characterize this surface is a formidable task : one needs to determine the 735 coordinates of all atoms that are significantly displaced from their ideal bulk lattice sites (3 coordinates per atom, 49 atoms per cell, 5 layers deep reconstruction). This explains why until recently none of the proposed models could be considered as reliable. Nevertheless a breakthrough came with the historical (1983) STM study of BINNIG et al.[44] providing the first real space image of this surface, which revealed 12 hills, valleys and deep holes in the corners of the unit cell (see Fig. 3a). One year later TAKAYANAGI et al.[43,45] proposed on the basis of a transmission electron diffraction (TED) intensity analysis a "dimer, adatom, stacking-fault (DAS)" structure model. This model also accounts for the ion-scattering experiments [46,47,48] which have provided evidence for a significant rearrangement of the atoms beneath the surface, interpreted by BENNETT et al.[49] in terms of a surface stacking fault, i.e., stacking sequences of atom layers in the selvedge which are different from the cubic structure.

Figure 3b shows TAKAYANAGI's original model. The 7x7 unit cell is divided into two antiphase equilateral triangles. In one of the triangles the surface contains a stacking fault that is due to a

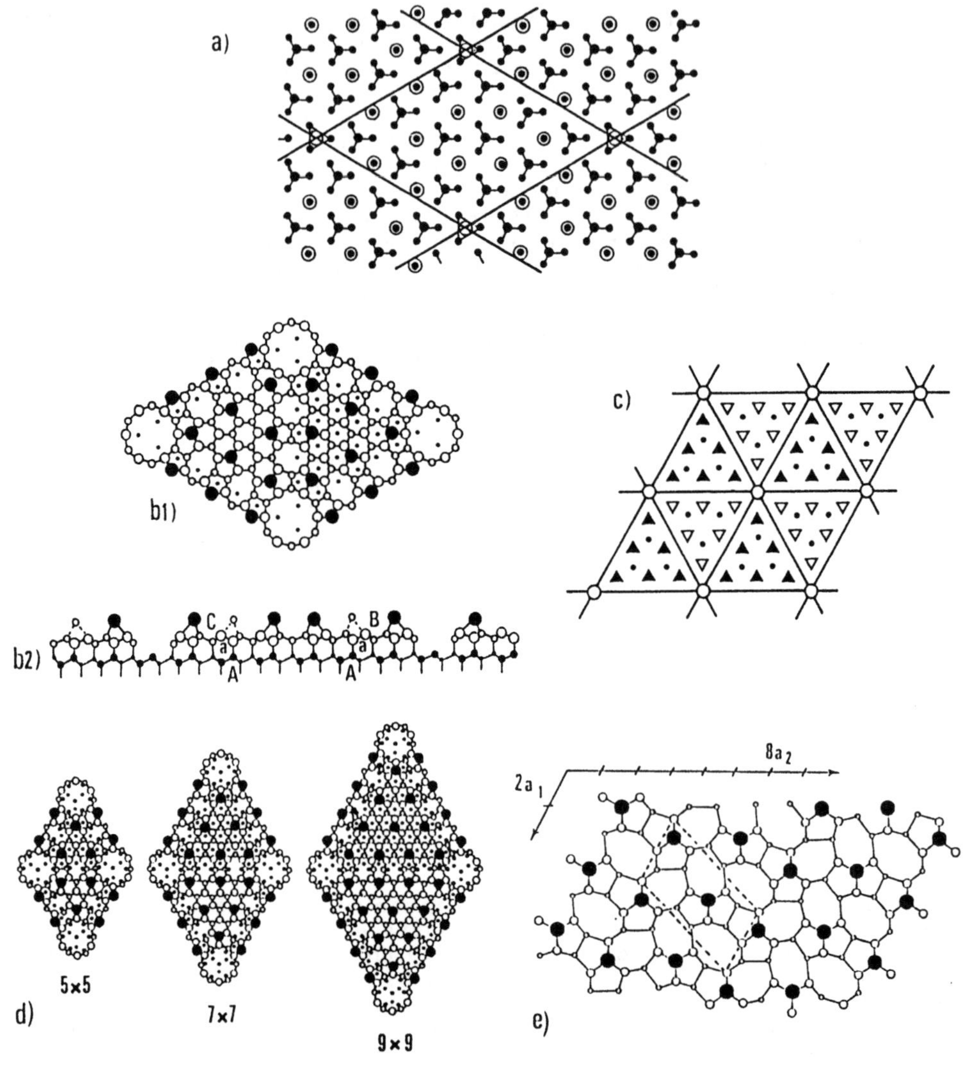

Fig. 3- a) Schematic view of the STM images of the Si(111)7x7 surfaces : large filled circles are adatoms, large empty circles are corner vacancies and double circles are restatoms.
b) DAS model of Si(111)7x7.
1/ Top view : the filled large circles are adatoms, the medium open circles are in a stacking fault in the left triangular subunit, the small open circles are dimers.
2/ Side view : atoms on the lattice plane along the long diagonal of the 7x7 unit cell are shown with larger circles than those behind them. The stacking sequence in the right half ... aB is the same as in bulk siilicon, in the left half the stacking sequence is faulted ... aC.
c) Wall (solid line) and domain configuration of the 7x7 structure (from Ref. 50).
d) DAS model of 5x5, 7x7 and 9x9 reconstructions.
e) Top view of Ge(111)-c(2x8) DAS model. The two-dimensional symmetry is Cm. The primitive mesh is shown by dashed lines.

rotation of the bond directions in the outer layer by 60°. The faulted (stacking sequence (bCcA)a/C) and unfaulted (normal (bCcA)a/B stacking sequence) regions are tied together by dimer bonds (9 dimers) along the edges of the two triangular subunits. On top of this structure 12 adatoms are located on threefold atop (T_4) sites directly above second layer atoms. The adatoms bond to three atoms in the surface plane (42 atoms) replacing three dangling bonds by a single one. Thus thirty-six of the 42 surface atoms are bonded to adatoms and their broken bonds are saturated. The remaining six surface layer atoms, the so-called "restatoms", maintain their threefold coordination. At the corner of the cell a 12 bonded ring of atoms surrounds a large "hole" (a vacancy) consistent with the STM images of the surface. This DAS model has been further described by its authors [50] as a network of walls (a wall consists of a chain of dimers at the sides of a triangular subcell) and domains (a domain corresponds to a triangular subcell of the 7x7 mesh) as shown in Fig. 3c. In such domains six adatom clusters (open or solid triangles in Fig. 3c) and three adatoms (dots in Fig. 3c) arranged locally in 2x2 are considered.

In addition to the experimental results already mentioned, other powerful methods have given strong experimental evidence for the presence of the main features of the DAS model. We mention especially the glancing incidence X-ray diffraction study of ROBINSON et al.[51] and the very last X-ray-standing-waves studies (PATEL et al.[52], DEV et al.[53]), which in addition to the distribution of matter parallel to the surface yield reliable structural information on the distribution of surface atoms normal to the surface. At this point it is noteworthy that previous conclusions, regarding XSW data, that were given as conflicting with the DAS model [54,55] have now been denied. In this respect the new variant version proposed by YANG et al.[56], to circumvent this early (now erased) discrepancy, consisting of modified-milkstool building blocks instead of adatoms and having three stacking faults in both halves of the unit cell, probably relies on mined ground.

Thus until now the DAS model has successfully passed the most stringent structural tests. However, as pointed out by GUO-XIN QIAN and CHADI [57] its merit is not a priori clear because of the delicate balance between the driving mechanism which stabilizes the 7x7 reconstruction : the energy reduction arising from a low dangling bond number (19 instead of 49) and the competing energy increase resulting from large angular strains. But enough confidence in its validity has stimulated several sophisticated calculations of the total energy of the surface [58,59,60] and of its electronic structure [57,61]. It has thus become clear that it is more favorable for adatoms to reside on threefold atop T_4 sites rather than on threefold hollow H_3 sites (energy reduction of 0.64 eV per adatom) [61]. Moreover, using a semiempirical tight-binding based energy minimization calculation GUO-XIN QIAN and CHADI have obtained for the Si(111)7x7 DAS model the lowest surface energy ever calculated for a (111) surface : relative to the ideal unrelaxed Si(111) surface the energy is lowered by 0.403 eV per 1x1 unit cell ; it is also slightly lower (0.04 eV per 1x1 cell) than the π-bonded chain structure of the cleaved surface (this is corroborated by the calculations of TERSOFF [59] who uses an empirical interatomic Morse-type potential). The calculated atomic displacements, both transverse and normal, appear to be in fairly good agreement with the last XSW results [52,53].

The theoretical study by NORTHRUP [61] of the origin of surface states as well as the calculated surface electronic density of states and the atomic origin of the electronic structure by GUO-XIN QIAN and CHADI [57] is in good agreement with the experimental data. It is generally accepted now that the Si(111)7x7 surface exhibits three surface states [62] : a dispersing state S_3 (at ~ -1.7 eV with respect to Fermi level E_F) and two non-dispersing states S_1 (at ~ -0.2 eV) and S_2 (at ~ -0.8 eV). Recently UHRBERG et al.[63] have shown that the S_3 state, for off normal emission has a p_z character, which is in agreement with the theoretical studies mentioned above. Using the novel technique of current-imaging-tunneling spectroscopy (CITS), which uses a real space image of the tunneling current to measure directly the spatial distribution of surface states HAMERS et al.[3,4] could localize the Si state on the adatoms, the S_2 state on the restatoms and the S_3 state on the back bonds of the adatoms. In addition, the mirror symmetry of the broken-bond states on the restatoms with respect to the short diagonal of the unit cell indicates a mirror symmetry of the atomic positions in the two triangular subunits, which directly proves the presence of a stacking fault in one half of the unit cell.

Finally it is noteworthy to highlight the possibilities offered by the comparison between the bias-dependent STM images and the ones obtained from simple atomic charge superposition (ACS) calculations [3,64] as suggested by TERSOFF and co-workers [65]. A comparison of images calculated with an exponentially decaying, spherically symmetric charge density assigned to each atom in the crystal, for the models proposed by BINNIG et al.[44] (12 adatoms and a corner vacancy), by SNYDER [66] (milkstool model instead of adatoms : pyramidal clusters of four atoms are located on the surface in the same geometric arrangement as the adatoms in BINNIG's model), by CHADI [67] (12 adatoms and five and sevenfold rings along the unit cell edges in a manner similar to that occurring in the π-bonded chain structure for the Si(111) 2x1 surface), by McRAE and PETROFF [68] (the outer two double layers contain two different stacking faults in the two halves of the unit cell), and finally by TAKAYANAGI (DAS model) shows a striking agreement with this last model.

Last, but not least, an essential feature of the DAS model is its transferability to other nxn (of odd periodicity) and 2xm (of even periodicity) reconstructed surfaces as shown in Figs. 3d,3e. Indeed, such new reconstructions i.e. 5x5, 9x9, c-(4x2), 2x2 plus a $\sqrt{3}x\sqrt{3}$-R(30°) have been observed locally by STM [69] on silicon (111) surfaces that display an apparent 1x1 LEED structure, prepared by a combination of laser and thermal annealing. These new reconstructions probably stem from defects, strain fields and local energy barriers to structural transformation. This essential role played by strain is confirmed by the beautiful high-resolution transmission-electron-microscope (HRTEM) profile images of a Si(111)5x5 structure which is induced and stabilized by a tensile (epitaxial) strain [70] and is to be contrasted with the observation that the clean germanium c(2x8) structure (to be discussed below) is changed to a (7x7) one by means of a small compressive (epitaxial) strain [71], both Ge and Si 7x7 surfaces having a close structural similarity according to LEED and high-energy ion scattering results. Besides it is noteworthy that the adsorption of germanium atoms on the silicon (111) surface has been used as a "decoration technique" in the XSW measurements mentioned above [52,53] to study the silicon (111) reconstructions. All these studies, and other ones, confirm the similarities of these different structures easily interpreted in the framework of the DAS model.

3.2 Ge(111)-c(2x8)

The c(2x8) reconstruction of clean annealed Ge(111) surfaces which corresponded until recently to an unsolved structure has been also analysed by TAKAYANAGI and TANISHIRO [50] in the same context. YANG and

JONA [72] and CHADI and CHIANG [73] deduced from the extinction rule observed in LEED and RHEED patterns that the Ge(111)-2x8 structure has a centered lattice ; the dimer-chain structure presented in Fig. 3e constructed by alternating two types of walls, with 4 adatoms on "top sites" and 12 surface layer atoms per c(2x8) unit cell (thus markedly reducing the total number of dangling bonds from 16 to 4) yields calculated diffraction intensities (giving rise to weak (0,1/4) and (0,3/4) reflections as compared with (0,1/2) and (0,1/8) ones, clearly observed in RHEED patterns [74] but only weakly in LEED patterns [75]) which agree fairly well with the diffraction data. This centered 2x8 structure with adatom clusters arranged in a 2x2 structure is also consistent with the STM images obtained by BECKER et al [76] who also observed 2x2 and c(4x2) cells which appear to serve as the building blocks from which the c(2x8) structure is built. This arrangement also partly reproduces the distribution of protrusions seen in the tunneling images (Fig. 3 in ref. 76). Finally, as postulated by TAKAYANAGI and TANISHIRO [50] hair-pin-like and striped-wall arrangements of dimer chains might correspond to the floating of 2x8 walls associated with a floating 2x8 phase, as pointed out first by KANAMORI and OKAMOTO [77], for the higher temperature phase of the Ge(111) surface. Indeed, this demands further investigations especially in view of the fact that a tight-binding energy-minimization calculation by GUO-XIN QIAN and CHADI [57] indicates that the energy reduction of the completely relaxed c(2x8) dimer-chain model from an ideal Si(111) surface (not a Ge(111) one ! but comparison should be possible) is only 0.18 eV per 1x1 surface area instead of 0.40 eV for the 7x7 DAS structure.

3.3 High Temperature phase transitions

The low temperature c(2x8) reconstruction of Ge(111) disappears at 500K while the 7x7 structure of Si(111) converts to a 1x1 one at 1100K. These phase transitions are reversible and have stimulated several investigations.

The disappearance of the c(2x8) superstructure of Ge(111) was attributed by PHANEUF and WEBB [75], after a detailed LEED study, to a first-order transition to an incommensurate reconstruction. Above the transition oblong-shaped intensity features at half-order positions split, broaden and become fainter with increasing temperature. Two possible models have been proposed to account for this diffraction : first an incommensurate 2x1 striped phase (3 orientational domains) containing quasi-periodic walls perpendicular to the undoubled direction of the 2x1 unit mesh or secondly a quasi-periodic arrangement of domains of a 2x2 reconstruction. In the latter case the wall network can breathe and distort to make room for wall meandering, yielding, as just mentioned, a floating phase. In fact, it is this last structure which seems the more plausible in view of the tunneling images [76] and in the light of the dimer-chain model [50] discussed for the c(2x8) phase.

The 7x7 $\rightleftarrows$ 1x1 phase transition on Si(111) was carefully reexamined by BAUER et al [78,79] using a new technique : low energy electron reflection microscopy (LEERM). The last interpretation of the data assign this transition to a 2nd order phase transition but with a very small critical exponent ($\beta \sim 0.07$). Beautiful micrographs show that upon cooling below the transition temperature, the 7x7 domains preferentially nucleate at steps of monoatomic height. The domains expand on the terraces in three easy directions of growth reflecting the threefold symmetry of the substrate.

4. Si(100) AND Ge(100) SURFACE STRUCTURES

4.1 Si(100)

The determination of the detailed atomic structure of the Si(100) surface, one of the backbones of silicon technology, has been greatly complicated by the experimental difficulty in preparing a well-ordered surface. The most frequently reported structure is the two-domain 2x1 reconstruction first observed by SCHLIER and FARNSWORTH [80] and recently imaged in profile by HRTEM [81], but LEED studies have revealed higher-order reconstructions such as c(4x2) domains [82,83] and also 2xn ($6<n<10$) superstructures obtained by quenching from high temperatures [84], while He-atom diffraction experiments [85] have indicated the presence of p(2x2) and also c(4x2) domains in addition to the 2x1 domains.

There is now almost universal agreement that the building block of the observed reconstructions is a surface dimer formed by dangling bond pairing of adjacent atoms in the top-most layer (Figure 4). As summarized by POLLMAN et al.[2] the former vacancy, symmetric pairing and conjugated chain models of Si(100)2x1 which predicted a metallic surface had to be discarded on the basis of the ARUPS data of HIMPSEL and EASTMAN [86], which revealed a semiconducting surface. This led CHADI [87] to propose the asymmetric dimer model shown in Fig. 4c. Since then a wealth of experimental evidence and theoretical calculations (see reference 2 for a very well documented review from last year) appear to be consistent with the asymmetric dimer reconstruction of Si(100)2x1, although some recent conflicting results concerning its electronic structure [88,89] and especially the assignment of the metallic state observed at the $\bar{\Gamma}$ and $\bar{J}$ points in the surface Brillouin zone, for highly n doped samples [62,89], partially challenge the model in its simple form.

The conflict will probably be resolved thanks to the latest STM observations [90] which reveal rows of oblong protrusions attributed to dimers but show a rather complex surface topography : both buckled and non-buckled dimers are observed as well as a high density of vacancies (missing dimers). In defect-free areas only symmetric (non-buckled) dimers are observed while buckled dimers appear to be stabilized near surface defects. However, these surprising observations do not rule out the asymmetric dimer model : the room temperature symmetric configuration probably corresponds to the time-average equilibrium configuration of the dimers which may flip dynamically on a time scale shorter than the STM measurement time.

Indeed, there are two fully equivalent asymmetric dimers at the surface (left atom up and right atom down and vice versa, see Fig. 4c) with two equivalent orientations. Such a system can be directly mapped onto a two-dimensional Ising model ; in this respect it is noteworthy that a phase transition at about 200 K from a low temperature c(4x2) phase to a high-temperature 2x1 phase recently observed by LEED [91] has been naturally explained with the buckled-dimer model as a second order order-disorder transition induced by a short range driving force from an ordered antiferromagnetic ground state to a disordered paramagnetic state above the critical temperature T_c = 200K, in accordance with a renormalization-group study by IHM et al.[92] and a Monte Carlo simulation by SAXENA et al.[93]. In fact, the zig-zag structures, frequently observed in the STM images [90], are attributed to rows of asymmetric dimers in which the direction of buckling alternates from dimer to dimer along the row ; correlation between the phase of buckling in adjacent rows of these alternating buckled dimers gives rise to local c(4x2) and p(2x2) reconstructions, when adjacent rows are out of phase or in phase, respectively.

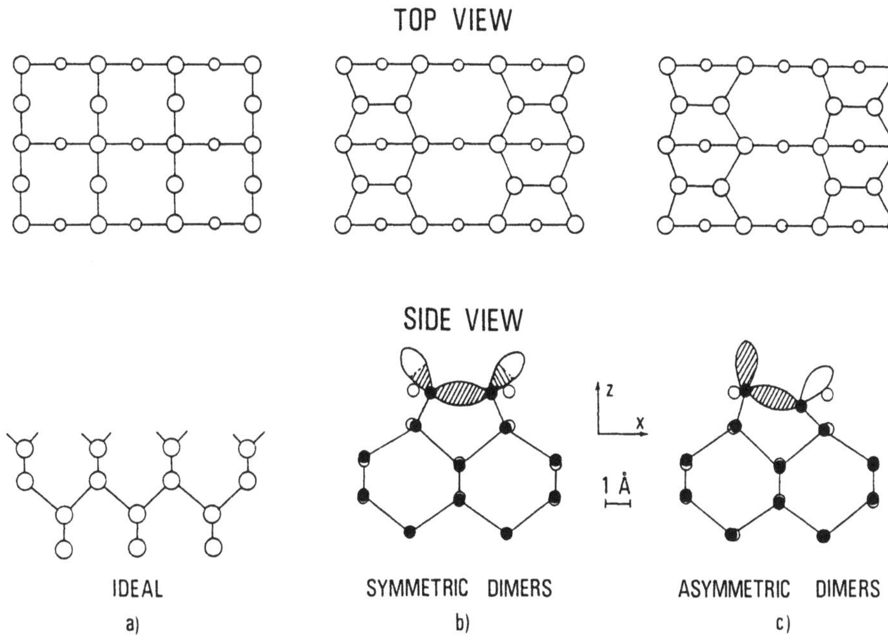

Fig. 4 - a) Top and side view of the ideal Si(100) surface.
b) Top and side view of the symmetric-dimer model.
c) Top and side view of the asymmetric-dimer model.

The open circles in the lower half of the figure give bulk positions. Surface states originating from dangling and bridge bonds are shown schematically, the hatching indicates the filling of the associated surface bonds.

As mentioned above, the (100) 2x1 surfaces contain a relatively large number of defects (density ~ 10%) visualized in the STM micrographs. These defects bear some similarity with the individual missing dimers, as proposed by PANDEY [94], which should lower the surface energy because of the reduction of broken bonds present at the surface and a π-bonding between them (albeit at the expense of elastic strain). Indeed, such a high defect density should affect in some way typically the LEED I-V profiles and the electronic band structure measurements.

Finally we should mention that after detailed LEED studies, the high-order (2xn) reconstructions of Si(100) [95,96] have been interpreted as due to ordering of such missing-dimer defects.

4.2 Ge(100)

Most of the points discussed above apply equally well to the germanium (100) surface structures. At RT a 2x1 reconstruction was determined by LEED [97], although hints of quarter order spots are sometimes reported. At low temperatures (100 K) a sharp c(4x2) is observed [98]. The pioneering X-ray diffraction study of EISENBERGER and MARRA [99] indicated that the Ge(100)-2x1 reconstruction involves substantial surface and subsurface displacements. This has been confirmed by recent Rutherford back scattering (RBS) experiments [98,100] which indicate a strong correlation in the nature of the reconstructions between Si and Ge(100) in agreement with the asymmetric dimer model, a point further supported by theoretical calculations of the electronic structure [101]. Interestingly, however, the transition to the c(4x2) structure was reported to proceed in two stages [102], first along a dimer row and then perpendicular to it (only this last case occurs for Si(100) [91]) ; that is, the dimers in a particular row parallel to the ‹100› and ‹010› azimuths first align at ~ 250 K in a one-dimensional antiferromagnetic fashion, while the orientation of these dimer rows with respect to one another remains random, then at T ~ 220K the rows align into the c(4x2) structure. This structural transition is accompanied by a metal-insulator transition which has been analyzed in detail by KEVAN [102] with ARUPS, and shown to be driven by a short-range coupling of dimers by direct interactions between dangling bonds on neighboring dimers. The elementary excitation corresponding to the metallic state observed above the disordering transition temperature being related to the flipping of a single dimer.

As a final remark I will emphasize the fact that here again the dynamics involved is an essential ingredient of semiconductor surface reconstructions. This demands further investigations to clarify the overall situation as in the case of the (111) surface, both on the experimental side with e.g. HREELS or optical absorption or atom scattering (see section 2) and on the theoretical side as was undertaken by the group of FLORES [103].

5. COMPOUND SEMICONDUCTOR SURFACE STRUCTURES

5.1 Clean relaxed GaAs (110)

The structure of the cleaved (110) surface of GaAs is the prototype of the structures of the (110) faces of III-V compound semiconductors [2,104] ; it is worth mentioning that it claims the title to be the first semiconductor surface structure to be determined by LEED (in 1976) [105]. The gross features of the relaxation (1x1 surface mesh) are given in Fig. 5. The relaxation involves a bond-length-conserving rotation with $\omega = 29° \pm 3°$ of the As atom outwards from the surface plane and of the Ga atoms inwards, a downwards relaxation of the (rotated) top layer by 0.05 ± 0.1 Å and a counter-rotation of the subsurface layer yielding a perpendicular displacement of 0.06 ± 0.1 Å. Since 1976 refinements of the structure have been reported over the years ; an up-to-date history of the determination of this relaxation may be found in [106].

New interest was added to the problem of the structure of GaAs(110) with the controversial exchange of letters [106,107] of a new conceivable bond relaxation model with $\omega \approx 7°$. However, as discussed in [2], the question now seems to be settled, the new structure being rejected on the basis of recent medium energy ion scattering data [108], as well as fully dynamical LEED calculations [105] and detailed energy minimization studies [2,109].

Supporting this view is the comparative study of the (110) surface of InSb by COWELL et al.[110], which shows strong similarities with GaAs. Even more instructive are the comments of these authors on the methodology used in LEED analyses and on the utilization of (reliability) R-factors ... ; it seems to me that they very pertinently stress the fact that -i- a potential structure established by an indirect surface structure method like LEED (or any other method) should be regarded as only provisional as long as it has not been confirmed by (at least!) one other, physically different, technique. -ii- in some cases LEED may not be able to distinguish with a reasonable degree of certainty between some structures.

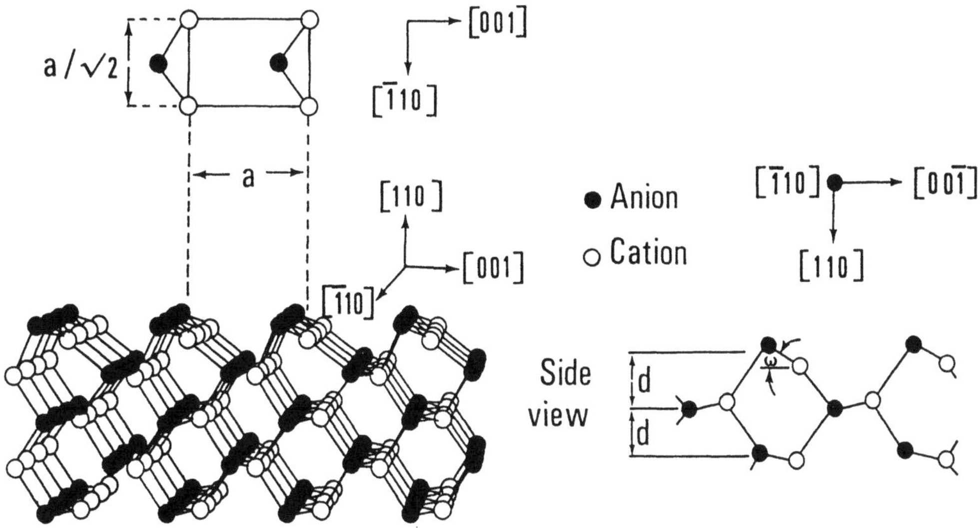

Fig. 5
Bond rotation model of the relaxed (110) surface of a III-V compound, the relaxation is characterized by a rotation of the plane of the anion-cation bonds in the top layer by an angle ω with respect to the surface plane (from Ref. 100)

As a matter of fact, STM images of the surface morphology of GaAs(110) have been obtained by FEENSTRA and FEIN [111], if the topographic images are consistent with the relaxation model described above, nevertheless detailed geometrical quantities, like the Ga-As tilt angle ω, could not be directly extracted.

5.2 Other reconstructions of III-V compounds

Finally, I will just indicate that in the last few years, surface reconstructions on other polar surfaces of GaAs and III-V compounds have attracted much attention both experimentally and theoretically ; for the large variety of reconstructions displayed, the surface stoichiometry appears to be a crucial factor (i.e. typically the As coverage for GaAs). For the (111)2x2 surfaces strong evidence for a vacancy reconstruction mechanism has come from the accurate glancing-incidence X-ray diffraction work of BOHR et al [112] on InSb.

6. SYNOPSIS

The atomic structures of clean semiconductor surfaces are a challenging most important problem in surface science and a subject of debate since the discovery of the 7x7 reconstruction of the (111) surface of silicon by LEED in 1959 [42]. However, due to strong multiple scattering the analysis of LEED intensities for large unit cells is hardly amenable. However, LEED was very successful in the elucidation of the geometry of the relaxed 1x1 surface of cleaved GaAs(110).

In the 1980's a great impetus in the determination of the superstructures of semiconductor surfaces (e. g. Si and Ge(111)2x1 cleaved surfaces, Si(111)7x7, Ge(111)-c(2x8), Si and Ge(100)2x1 and c(4x2) clean prepared surfaces) has been achieved thanks to their observation in real space with the spatially resolved STM probe and the use of UHV transmission electron microscopes. Developments in existing techniques : transmission electron diffraction, grazing incidence X-ray diffraction, X-ray standing waves, ion beam crystallography, SEXAFS, photoemission, etc... have also strongly contributed to the recent progress as have correlatively large advances in the theoretical description of the geometric and electronic structures of semiconductor surfaces.

Finally, it is the comparison of several structure-sensitive experimental data with the theoretical calculations that allow different models to be tested. Presently a consensus has been reached for Pandey's π-bonded chain model for the Si and Ge(111) 2x1 superstructures, while best agreement is obtained for the adatom stacking-fault model of TAKAYANAGI for Si(111)7x7 and Ge(111)-c(2x8) and the dimer model for Si and Ge(100) ; yet, should we consider that the correct structures of these surfaces are established "beyond all reasonable doubt" ?

REFERENCES

1. Physics Today, April 1986
2. J. Pollmann, R. Kalla, P. Kruger, A. Mazur, G. Wolfgarten : Appl. Phys. A41, 21 (1986)
3. R.M. Tromp, R.J. Hamers, J.E. Demuth : Science 234, 304 (1986)
4. R.J. Hamers, R.M. Tromp, J.E. Demuth : Phys. Rev. Lett. 56 1972 (1986)
5. G. Binnig, H. Rohrer : Pour la Science, Oct. 1985, p. 25
6. H.E. Farnsworth : Conf. N.Y. Acad. on Clean Surfaces (1962)
7. J.J. Lander, G.W. Gobeli, J. Morrison : J. Appl. Phys. 34 2298 (1963)
8. F.J. Himpsel : Appl. Phys. A38, 205 (1985)
9. D. Haneman : Phys. Rev. 127, 1093 (1961)
10. F.J. Himpsel, P. Heinemann, D.E. Eastman : Phys. Rev. B24, 2003 (1981)
11. K.C. Pandey : Phys. Rev. Lett. 47, 1913 (1981) -ibid- 49, 223 (1982)
12. J.E. Northrup, M.L. Cohen :J. Vac. Sci. Technol.21,333 (1982)
13. V.A. Grazhulis, V.F. Kuleshov : Appl. Surface Sci. 22/23, 14 (1985)
14. G. Chiarotti, S. Nannarone, R. Pastore, P. Chiaradia : Phys.Rev. B4, 3398 (1971)
15. R.I.G. Uhrberg, G.V. Hansson, J.M. Nicholls, S.A. Flodstrom : Phys. Rev. Lett. 48, 1032 (1982)
16. F. Houzay, G. Guichar, R. Pinchaux, G. Jezequel, F. Solal, A. Barsky, P. Steiner, Y. Petroff : Surface Sci. 132, 40 (1983)
17. P. Martensson, A. Cricenti, G.W. Hansson : Phys. Rev. B32, 6959 (1985)
18. M.A. Olmstead, N.M. Amer : Phys. Rev. Lett. 52, 1148 (1984). -ibid- Phys. Rev. B29, 1048 (1984)
19. P. Chiaradia, A. Cricenti, S. Selci, G. Chiarotti : Phys. Rev. Lett. 52, 1148 (1984)
20. S. Selci, P. Chiaradia, F. Ciccaci, A. Cricenti, N. Sparvieri, G. Chiarotti : Phys. Rev. B31, 4096 (1985)
21. T.F. Heinz, M.M.T. Loy, W.A. Thompson : Phys. Rev. Lett. 54, 63 (1985)
22. R.M. Tromp, L. Smit, J.F. Van der Veen : Phys. Rev. B30, 6235 (1984)
23. F.J. Himpsel, P.M. Marcus, R. Tromp, I.P. Batra, M.R. Cook, F. Jona, H. Liu : Phys. Rev. B30, 2257 (1984)
24. H. Sakama, A. Kawasu, K. Ueda : Phys. Rev. B34, 1367 (1986)
25. R.M. Feenstra, W.A. Thompson, A.P. Fein : Phys. Rev. Lett. 56, 608 (1986)

26. R.M. Feenstra, W.A. Thompson, A.P. Fein : J. Vac. Sci. Technol. A4, 1315 (1986)
27. J.A. Stroscio, R.M. Feenstra, A.P. Fein : Phys. Rev. Lett. 57, 2579 (1986)
28. C.F. Quate : Physics Today 26 (August 1986)
29. D.C. Alerland, D.C.Allan, E.J. Mele : Phys. Rev. Lett. 55, 2700 (1985)
30. W.A. Thompson, A.J. Schell-Sorokin, J.E. Demuth : Phys. Rev. B34 3007 (1986)
31. V. del Pennino, M.G. Betti, C. Mariani, C.M. Bertoni : Sol.State Commun. 60, 337 (1986)
32. F. Ciccacci, S. Selci, G. Chiarotti, P. Chiaradia : Phys. Rev. Lett. 56, 2411 (1986)
33. U. Harten, J.P. Toennies, Ch. Woll : Phys. Rev. Lett. 57 2947 (1986)
34. D.J. Chadi : Phys. Rev. B26, 4762 (1982)
35. O.H. Nielsen, R. Martin, D.J. Chadi, K. Kunc : J. Vac. Sci. Technol. B1, 714 (1983)
36. J.E. Northrup, M.L. Cohen : Phys. Rev. B27, 6553 (1983)
37. J.M. Nicholls, P. Martensson, G.V. Hansson : Phys. Rev. Lett. 54, 2363 (1985)
38. J.E. Demuth, R. Imbihl, W.A. Thompson : Phys. Rev. B34, 1330 (1986)
39. A. Frendenhammer, W. Monch, P.P. Amer : J. Phys. C13, 4407 (1980)
40. G. Quentel, R. Kern : Surface Sci. 135, 325 (1983)
41. M.A. Olmstead, N.M. Amer : Phys. Rev. B33, 2564 (1986)
42. R.E. Schlier, H.E. Farnsworth : J. Chem. Phys. 30, 917 (1951)
43. K. Takayanagi : J. Micros. 136, 287 (1984)
44. G. Binnig, H. Rohrer, Ch. Gerber, E. Weibel : Phys. Rev. Lett. 50, 120 (1983)
45. K. Takayanagi, Y. Tanishiro, M. Takahashi, S. Takahashi : J. Vac. Sci. Technol. A3, 1502 (1985) ; Surface Sci. 164, 367 (1985)
46. R.J. Culbertson, L.C. Feldman, P.J. Silverman : Phys. Rev. Lett. 45, 2043 (1980)
47. R.M. Tromp, E.J. Van Loenen, M. Iwami, F.W. Saris : Solid State Commun. 44, 971 (1982)
48. R.M. Tromp, E.J. Van Loenen : Surface Sci. 155, 441 (1985)
49. P.A. Bennett, L.C. Feldman,; Y. Kuk, E.G. McRae, J.E. Rowe : Phys. Rev. B28, 3656 (1983)
50. K. Takayanagi, Y. Tanishiro : Phys. Rev. B34, 1034 (1986)
51. I.K. Robinson, W.K. Waskiewicz, P.H. Fuoss, J.B. Stark, P.A. Bennett : Phys. Rev. B33, 7013 (1986)
I.K. Robinson : J. Vac. Sci. Technol. A4, 1309 (1986)
52. J.R. Patel, P.E. Freeland, J.A. Golovchenko, A.R. Kortan, D.J. Chadi, G. Guo-Xin Qian : Phys. Rev. Lett. 57, 3077 (1986)
53. B.N. Dev, G. Materlik, F. Grey, R.L. Johnson, M. Clausnitzer : Phys. Rev. Lett. 57, 3058 (1986)
54. S.M. Durbin, L.E. Berman, B.W. Batterman, J.M. Blakely : Phys. Rev. Lett. 56, 236 (1986)
55. J.R. Patel, J.A. Golovchenko, J.C. Bean, R.J. Morris : Phys. Rev. B31, 6884 (1985)
56. W.S. Yang, R.G. Zhao (preprint)
57. Guo-Xin Qian, D.J. Chadi : Phys. Rev. (in press)
58. Guo-Xin Qian, D.J. Chadi : J. Vac. Sci. Technol. B4, 1079 (1986)
59. J. Tersoff : Phys. Rev. Lett. 56, 632 (1986)
60. T. Yamaguchi : Phys. Rev. B34, 1085 (1986)
61. J.E. Northrup : Phys. Rev. Lett. 57, 154 (1986)
62. P. Martensson : thesis, Linkoping University, 1986
63. R.I.G. Uhrberg, G.V. Hansson, J.M. Nicholls, P.E.S. Persson : Phys. Rev. B31, 3805 (1985)

64. R.M. Tromp, R.J. Hamers, J.E. Demuth : Phys. Rev. B34, 1388 (1986)
65. J. Tersoff, D.R. Hamann : Phys. Rev. B32, 5044 (1985)
66. L.C. Snyder : Surface Sci. 140, 101 (1984)
67. D.J. Chadi : Phys. Rev. B30, 4470 (1984)
68. E.C. McRae, P.M. Petroff : Surface Sci. 147, 385 (1984)
69. R.S. Becker, J.A. Golovchenko, G.S. Higushi,B.S. Swartzentruber: Phys. Rev. Lett. 57, 1020 (1986)
70. A. Ourmard, D.W. Taylor, J. Berk, B.A. Davidson, L.C. Feldman, J.P. Marmaerts : Phys. Rev. Lett. 57, 1332 (1985)
71. H.J. Gossmann, J.C. Bean, L.C. Feldman, E.G. McRae, I.K. Robinson : Phys. Rev. Lett. 55, 1106 (1985)
72. W.S. Yang, F. Jona : Phys. Rev. B29, 899 (1984)
73. D.J. Chadi, C. Chiang : Phys. Rev. b23, 1843 (1981)
74. T. Ichikawa, S. Ino : Surface Sci. 221 (1979)
75. R.J. Phaneuf, M.B. Webb : Surface Sci. 164, 167 (1985)
76. R.S. Becker, J.A. Golovchenko, B. S. Swartzentruber :Phys. Rev. Lett. 54, 2678 (1985)
77. J. Kanamori, M. Okamoto : J. Phys. Soc. Jpn. 54, 4636 (1985)
78. E. Bauer : Proceedings of 10th Int. Seminar on Surf. Physics, Piechowice, May 1986
79. W. Telieps, E. Bauer : Surface Sci. 162, 163 (1985)
80. R.E. Schlier, H.E. Farnsworth : J. Chem. Phys. 30, 918 (1959)
81. J.M. Gibson, M.L. McDonald, F.C. Unterwald : Phys. Rev. Lett. 55, 1765 (1985)
82. J.J. Lander, J. Morrison : J. Chem. Phys. 37, 729 (1962)
83. T.D. Poppendieck, T.C. Ngoc, M.B. Webb : Surface Sci. 75, 285 (1978)
84. K. Muller, E. Lang, L. Hammer, W. Grim, P. Heilman, K. Heinz : in Determination of Surface Structure by LEED, edited by P.M. Marcus and F. Jona (Plenum, New York, 1984), p. 483
85. M.J. Cardillo, G.E. Becker : Phys. Rev. B21, 1497 (1980)
86. F.J. Himpsel, D.E. Eastman : J. Vac. Sci. Technol. 16, 1297 (1979)
87. D.J. Chadi : J. Vac. Sci. Technol. 16, 1290 (1979)
88. A. Goldmann, P. Koke, W. Monch, G. Wolfgarten, J. Pollmann : Surface Sci. 169, 438 (1986)
89. P. Martensson, A. Cricenti, G. Hansson : Phys. Rev. B33, 8855 (1986)
90. R.J. Hamers, R.M. Tromp, J.E. Demuth :Phys. Rev.B34,5343 (1986)
91. T. Tabata, T. Aruga, Y. Murata : Surface Sci. 179, L63 (1987)
92. J. Ihm, D.H. Lee, J.D. Joannopoulos, J.J. Xiong : Phys. Rev. Lett. 51, 1872 (1983)
93. A. Saxena, E.T. Gaulinski, J.D. Gunton : Surface Sci. 160, 618 (1985)
94. K.C. Pandey : in Proceedings of the Seventeenth International Conference on the Physics of Semiconductors, edited by D.J. Chadi and W.A. Harrison (Springer-Verlag, New York, 1985) p. 55
95. J.A. Martin, D.E. Savage, W. Moritz, M.G. Lagally : Phys. Rev. Lett. 56, 1936 (1986)
96. T. Aruga, Y. Murata : Phys. Rev. B34, 5654 (1986)
97. J.C. Fernandez, W.S. Yang, H.D. Shih, F. Jona, D.W. Jepsen, P.M. Marcus : J. Phys. C14, L55 (1981)
98. R.L. Culbertson, Y. Kuk, L.C. Feldman : Surface Sci. 167, 127 (1986)
99. P. Eisenberger, W.C. Marra : Phys. Rev. Lett. 46, 1081 (1981)
100. For a review on ion beam crystallography,see J.F. Van der Veen: Surface Sci. Rep. 5/6, 199 (1985)
101. P. Kruger, A. Mazur, J. Pollmann, G. Wolfgarten : Phys. Rev. Lett. 57, 1468 (1986)
102. S.D. Kevan : Phys. Rev. B32, 2344 (1985)
103. C. Tejedor, F. Flores, E. Louis : J. Phys. C19, 543 (1986)

104. A. Kahn : Surface Sci. Rep. 3, 193 (1983)
105. A.Q. Lubinsky, C.B. Duke, B.W. Lee, P. Mark : Phys. Rev. Lett. 36, 1058 (1976)
106. C.B. Duke, A. Paton : Surface Sci. 164, L797 (1985)
107. M.W. Puga, G. Xu, S.Y. Tong : Surface Sci. 164, L789 (1985)
108. L. Smit, T.E. Derry, J.F. Van der Veen : Surface Sci. 150, 245 (1985)
109. J.C. Mailhoit, C.B. Duke, D.J. Chadi : Surface Sci. 149, 366 (1985)
110. P.G. Cowell, M. Prutton, S.P. Tear : Surface Sci. 177, L915 (1986)
111. R.M. Feenstra, A.P. Fein : Phys. Rev. B32, 1394 (1985)
112. J. Bohr, R. Feidenhans's, M. Nielsen, M. Toney, R.L. Johnson, I.K. Robinson : Phys. Rev. Lett. 54, 1275 (1985)

Monolayer Sensitive X-Ray Diffraction Techniques: A Short Guided Tour Through the Literature

M. Sauvage-Simkin

LURE, CNRS-MEN-CEA, Bât. 209D, F-91405 Orsay, France and Laboratoire de Minéralogie-Cristallographie, Universités Pierre et Marie Curie et Paris VII, Associé au CNRS, 4, Place Jussieu, F-75252 Paris Cedex, France

The advent of powerful X-ray sources, rotating anode X-ray generators and Synchrotron Radiation sources, has made possible the detection of X-ray diffraction signals or of secondary effects excited by diffracted X-rays, produced in atom monolayers.

Such methods have found numerous applications in the characterization of semiconductor surfaces and interfaces and can be classified in two main groups : surface or superficial layer scattering collected in a grazing incidence geometry, with application to the determination of the atomic structure in surface unit cell, the second group is concerned with the excitation of surface or interface atoms by Bragg diffraction-induced X-ray standing waves.

1. SURFACE SCATTERING

A surface diffraction experiment consists in collecting Bragg reflexions corresponding to diffraction vectors parallel or nearly parallel to the crystal surface, representative of the surface translational symmetry. For reconstructed surfaces, the surface periodicity is usually a multiple m x n of the bulk projected unit cell, which implies that surface Bragg peaks may have fractional indices when referred to the bulk reciprocal lattice. As a consequence, the diffracted intensities which convey the specific information on the atomic displacements involved in the reconstruction are collected at positions where the bulk contribution is symmetry forbidden. For a purely bidimensional scattering object, the Bragg reflexion intensity distribution normal to the object plane is flat, however, several layers may participate in the reconstruction process which results in a modulation of the out-of-plane scattering.

The atomic structure of the surface unit cell is then derived from the intensity measurements through a classical structure factor analysis, based on the X-ray kinematical diffraction theory, and using standard crystallographic methods developed for the 3-dimensional case : Patterson function, difference map, a major simplification when compared to LEED structure analysis.

For a detailed description of the method, the reader is referred to two recent review articles by ROBINSON /1/ and FUOSS, LIANG and EISENBERGER /2/.

Among successful semiconductor surface structure determination or confirmation when models had already been produced by other approaches, one can quote the Si(111) 7x7 /3/ with specific information on the 6-fold symmetry of the surface layers /4/ and on the stability of the reconstruction when the crystal surface is buried under an amorphous Si-a layer /5/, the Ge(100) 2x1 /6/ and the III-V compounds InSb(111) 2x2 /7,8,9/ and GaSb(111) 2x2 /9,10/.

As mentioned above, a detailed study of the out-of-plane scattering enables to estimate the actual thickness of the reconstructed region /5,6/. Information on the surface roughness may also be derived through such measurements /11/ and indeed, an intensive effort is presently taking place in the field of roughening transitions.

Surface phase transitions of order-disorder type (Si(111) 7x7 → 1x1 at 1100 K /12/) or induced by stoichiometry changes in the surface layer (α → β transition in the Pb/Ge(111) √3 x √3 R30 phase /13/, can be followed.

Leaving the true monolayer regime one can obtain useful information on the relaxation of epitactic strains in heterojunctions : the strain-induced 7x7 reconstruction of Ge/Si(111) has been studied /14/ and the transition from elastic to plastic relaxation in the $Ge_{1-x}Si_x$/Si(111) systems has been followed as a function of the epilayer thickness and composition /15/, a pioneering experiment in the field being the case of the Al/GaAs(100) heterojunction /16/.

Elaborate set-ups fulfilling the requirements of both the ultra-high vacuum and single crystal diffractometry (multiple circle instruments) are installed in the various Synchrotron Radiation facilities /9,17,18/, with possible in-situ MBE growth of the sample /19,20/.

Most intensity measurements are performed slightly above the critical incidence for total reflexion to avoid complex refraction effects while still taking advantage of a small penetration depth which minimizes the bulk background. Nevertheless, the theory of X-ray diffraction below the critical incidence has been worked out in both the kinematical /21/ and dynamical /22,23/ approaches and predicts specific interference effects which are observed when an accurate definition of the incidence angle is achieved /24,25/.

2. X-RAY STANDING WAVE EXPERIMENTS

This second class of measurements deals with the detection of signals related to the deexcitation processes (X-ray fluorescence, Auger electrons, photoelectrons...) after excitation of the atoms by X-ray standing waves.

Such waves, beating with the lattice periodicity of the host single crystal under Bragg reflexion, can be made to sweep through the actual atomic network as the crystal rocking curve is described. The response of a given atom to X-ray excitation, as a function of the angular setting (required accuracy in the sub-arc second range) enables then to locate the atom with respect to the single crystal atomic planes.

Detailed dynamical theory calculations taking into account absorption and possible anomalous dispersion effects may be found in papers by BEDZYK and MATERLIK /26/ and AUTHIER /27/. The method has been applied to the assignment of implanted atom sites (As /28/ and Be /29/ in silicon) or to the registry of adsorbed layers (Br/Si(111) /30/ and Br/Ge(111) /31/). Information on surface reconstruction can be derived with the help of chemically equivalent adsorbate such as Ge/Si(111) 7x7 /32/ and buried heterointerfaces may be studied : $NiSi_2$(111) /32/.

Contrary to the surface Bragg scattering method, the location of atoms in a standing wave experiment does not require long-range order but only a limited distribution of site occupancy.

3. ACKNOWLEDGEMENTS

The friendly cooperation of all the authors quoted in this review who let me know about their most recent results is gratefully acknowledged.

4. REFERENCES

1. I.K. Robinson "Surface Crystallography" in Handbook on Synchrotron Radiation. Ed. D.E. Moncton and G. Brown, North-Holland 1987

2. P.A. Fuoss, K.S. Liang and P. Eisenberger. "Advances in Surfaces and Low-dimensional Systems" in Synchrotron Radiation Research. Ed. R.Z. Bachrach - Plenum (1987)
3. I.K. Robinson, W.K. Waskiewiz, P.A. Fuoss, J.B. Stark, P.A. Bennet. Phys. Rev., B33 (1986) 7013
4. I.K. Robinson, Phys. Rev., B35 (1987) in press
5. I.K. Robinson, W.K. Waskiewicz, R.T. Tung and J. Bohr, Phys. Rev. Lett., 57 (1986) 2714
6. F. Grey, Internal Report-Aarhus University (1987)
7. J. Bohr, R. Feidenhans'l, M. Nielsen, M. Toney, R.L. Johnson and I.K. Robinson, Phys. Rev. Lett., 54 (1985) 1275
8. R. Feidenhans'l, J. Bohr, M. Nielsen M. Toney, R.L. Johnson, F. Grey and I.K. Robinson, Festkörperprobleme XXV 545 (1985)
9. R. Feidenhans'l, P.H.D. Thesis "Solving Surface Structure with X-ray Diffraction" Aarhus University, Denmark 1986
10. R. Feidenhans'l, M. Nielsen, F. Grey, R.L. Johnson and I.K. Robinson (1987), to be published
11. I.K. Robinson, Phys. Rev., B33 (1986) 3830
12. I.K. Robinson in "Structure of Surfaces" Ed. M.A. Van Hove and S.Y. Tong, Springer-Verlag (1985)
13. R. Feidenhans'l, J.S. Pedersen, M. Nielsen, F. Grey and R.L. Johnson, Surf. Science, 178 (1986) 927
14. H.J. Gossmann, J.C. Bean, L.C. Feldman, E.G. Macrae and I.K. Robinson, Phys. Rev. Lett. (1985)
15. J.C. Bean, L.C. Feldman, A.T. Fiovy, S. Nakahara and J.K. Robinson, J. Vac. Sc. Technol., A2 (1984) 436
16. W.C. Marra, P. Eisenberger and A.Y. Cho, J. Appl. Phys., 50 (1979) 6927
17. P.H. Fuoss and I.K. Robinson, Nucl. Inst. Meth., 222 (1984) 171
18. S. Brennan and P. Eisenberger, Nucl. Inst. Meth., 222 (1984) 164
19. P. Claverie, Thèse de Docteur-ingénieur, Université de Clermont (1986)
20. E. Vlieg, A. Van't Ent, A.P. De Jongh and J.F. Van der Veen (1987) to be published
21. G.H. Vineyard, Phys. Rev., B26 (1982) 4146
22. P.L. Cowan, Phys. Rev., B32 (1985) 5437
23. A.M. Afanas'ev and M.K. Melkonyan, Acta Cryst., A29 (1983) 207
24. P.L. Cowan S. Brennan, T. Jach, M.J. Bedzyk and G. Materlik, Phys. Rev. Lett., 57 (1986) 2399
25. A.L. Golovin, R.M. Imamov and S.A. Stepanov, Acta. Cryst., A40 (1984) 225
26. M.J. Bedzyk and G. Materlik, Phys. Rev., B32 (1985) 6456
27. A. Authier, Acta Cryst., A42 (1986) 414
28. S. Kjaer Andersen, J.A. Golovchenko and G. Mair, Phys. Rev. Lett., 37 (1976) 1141
29. N. Hertel, G. Materlik and J. Zegenhagen, Z. Phys., B58 (1985)
30. J.A. Golovchenko, J.R. Patel, D.R. Kaplan, P.L. Cowan and M.J. Bedzyk, Phys. Rev. Lett., 49 (1982) 560
31. M.J. Bedzyk and G. Materlik, Phys. Rev., B31 (1985) 4110
32. E. Vlieg, A.E.M.J. Fischer, J.F. Van der Veen, B.N. Dev and G. Materlik, Surface Science, 178 (1986)
33. B.N. Dev, G. Materlik, F. Grey, R.L. Johnson and M. Clausnitzer, Phys. Rev. Lett., 57 (1986) 3058

SEXAFS for Semiconductor Interface Studies

G. Rossi

Laboratoire pour l'Utilisation du Rayonnement Electromagnetique, CNRS, CEA, MEN; Université de Paris-Sud, F-91405 Orsay, France

DEFINITIONS

SEXAFS: Surface (sensitive) Extended X-ray Absorption Fine Structure

THE PHENOMENON: Oscillatory modulations of the X-ray absorption cross section above characteristic edges in condensed matter

THE MEASURE: X-ray absorption as a function of the photon energy hν;
Particle impact ionisation as a function of E_p

THE DETECTION: Fluorescence Yield,
electron yield
Auger elastic electron yield
stimulated desorption of ions, neutrals

THE INTERPRETATION: in the range from 50-100 eV above threshold up to 500-1000 eV:
SCATTERING THEORY: interference between the photoelectron that leaves the excited atom and the back-diffused amplitude due to a single scattering event with the core electrons of neighbouring atoms
GENERAL: measure of the density of final electron states accessible via dipole transitions.

ABSTRACT

Structural information on the atomic arrangements at the early stage of formation of semiconductor interfaces and heterojunctions is needed along with the determination of the structure of the electron states in order to put on a complete experimental ground the discussion of the formation of solid junctions. Amongst the structural tools that have been applied to the interface formation problem Surface-EXAFS [1-4] is probably the best suited since the local configuration of the interface atoms can be directly measured independent from the presence of long - range order or of well-known bulk phases. We propose some comments on the applicability of SEXAFS to the interface formation problem, a brief review of the merits of the technique, an overview of selected studies from the recent literature, and a status report of the SEXAFS interface studies underway.

SEXAFS of interface systems:

The problems that one can address with the SEXAFS tool while studying the growth and formation of metal/semiconductor and/or semiconductor/semiconductor interfaces [5,6] are:

1) the bond lengths between the metal (semiconductor) adsorbed atoms and the surface of a semiconductor (metal) single crystal substrate at the chemisorption stage;

2) the adsorption site, i.e. the chemisorption position of the adatoms on (within, below) the substrate surface, thanks to the polarisation dependence of SEXAFS;

3) the changes in local coordination of the majority specie (the substrate) at the surface (e.g. adsorbate induced reconstruction);

4) the local coordination of the minority specie (i.e. the adsorbate specie) in the reactive interdiffusion stage (precursor compound, unreacted clusters);

5) the local coordination of the majority specie (substrate material) in the new interface phases;

6) the local coordination of the minority specie in the nucleated interface compound at the stage of compound formation;

7) the local coordination of the majority species in the interface compounds;

8) the presence of more than one specie in metastable equilibrium at the interface;

9) tangled informations on the vibrational and electron mean free path properties.

Before detailing the data acquisition and analysis procedures that allow to address the listed problems, it is important to evaluate accuracy of the information that can be retrieved from SEXAFS data on each of the listed physical quantities, i.e. to define the merit of the technique in the particular application of interface science.
With reference to the list above, the accuracy of the SEXAFS method is:

1) very good; this is the classical employment of SEXAFS and has been well established in studies of surface/gas adsorbate systems. The typical accuracy for data with a good energy range (300-700 eV) and good signal-to-noise ratio is of 0.02 Å, [3] and is always better than 0.03 Å on bond lenghts; ±15% on coordination numbers in case of a single chemisorption environment. The accuracy can be just as good in the presence of two or more chemisorption phases if they differ in bond distance by more than 1Å, or if the first neighbour specie is different and the relative phase shifts are much different. In case of multishell contributions to the same EXAFS oscillations the accuracy decreases somewhat, or even a lot in the worst cases.

2) very good; the adsorption positions on single crystal surfaces are often uniquely assigned from the analysis of both polarisation dependent bond lengths and relative coordination numbers.[2,7] The relative, polarisation-dependent, amplitudes of the EXAFS oscillations indicate without ambiguity the chemisorption position if such position is the same for all adsorbed atoms. More than one chemisorption site could be present at a time (surface defect sites or just several of the ideal surface sites). If the relative population of the chemisorption sites is of the same order of magnitude, then the analysis of the data becomes difficult, or just impossible.

3) to be established; it all depends on the detection mode and on the absolute surface sensitivity of the most surface-sensitive detection mode. The way to proceed is to measure differential SEXAFS spectra on one adsorption edge of the substrate material, where the difference is made between the clean substrate and the exposed substrate, or between a bulk sensitive detection of the EXAFS, (TY, FY,

high-energy AEY) and a surface-sensitive measure (low - energy AEY). [8] Attempts have been made on this ground, but final reports have not yet appeared in the literature.

4) very good if the interdiffusion sites/local coordination is one dominant configuration. If more configurations are present the accuracy decreases. Identical consideration as in 1, 2, apply.

5) to be established; analogous considerations as to point 3 apply, with less strict conditions on the surface sensitivity since the reacted interdiffused region can be several layers thick. Yet the interdiffusion region is more likely heterogeneous from the point of view of the substrate material (different first nearest neighbour environments) than for the adsorbate, therefore considerations on the resolution of multiple phases also apply.

6) very good if one well-defined compound is nucleated, second and higher neighbour distances can be measured in favourable cases. The problem of multishells analyses becomes identical to the bulk EXAFS case.

7) to be established; see point 5.

8) it depends on the case; it can be very good if the two phases have first distances sufficiently different (>1 Å). This is not the case if two silicide phases, for example, are formed. In this case the presence of more than one phase can only be inferred from the inconsistency of the coordination numbers with any single one of the silicide phases, but the relative amounts or any better details are lost. It can be quite easy instead if one silicide phase coexists with metal clusters, or with a metal-rich phase where the nearest neighbours are metal atoms instead of silicon atoms. In this case, even if the absolute distances are similar, the phase shift difference can be strong enough to separate the experimental frequencies and allow an accurate analysis.

9) such analysis is possible only if the interface phase is very well defined, and if temperature-dependent measurements are done and compared.[9,10] Due to the multiple parameters in the SEXAFS amplitude term care must be taken not to overinterpret the data. Debye Waller effects could be in fact simulated by ordering transformation of the interface phase as a function of temperature and so on. If a single phase interface with order at least to the second nearest neighbour is recognised, then a temperature-dependent Debye Waller, and mean free path analysis can be attempted.

Why do we need to know bond lengths and coordination numbers with so high accuracy on interface systems ?
The interface formation problem is a difficult one. The best proof is represented by the wide experimental and theoretical effort of the last twenty years which has set the frame of interface science, but has left still wide open the key questions on the Schottky barrier versus ohmic phenomenology of the junctions, the band gap discontinuity distribution on the heterojunctions, the kinetic thresholds for compound nucleation at a solid/solid interface and so on. The quest for accuracy is generalised to all spectroscopies and other physical probes (transport measurements, quantitative analysis), since it is believed that the intimate nature of the macroscopic behaviour of the interfaces is to be grasped only by looking at the microscopic scale to the electrical and crystallographic properties of the interfaces. The differences in bond length and coordination number for different silicide compounds of the same

metal-Si binary system are small. Metal-rich silicides have first Me-Si distances which differ by a few 10^{-2} A. Recognition of the exact silicide phase formed at the interface, or accurate definition of an interface-specific reacted phase are results of paramount importance for the overall understanding of the interface phenomenology, and for the meaningful comparison with spectroscopic data [6] and with theoretical calculations of the band structure [5] and of the thermodynamical properties. It is definitively clear that the accuracy of SEXAFS is needed in the interface studies, and that the next advancements of interface science are to be expected to come from high resolution and very accurate, microscopic scale studies.

RUDIMENTS OF THE TECHNIQUE

The SEXAFS technique is a particular surface-sensitive detection mode applied to an EXAFS measurement, performed in a particular environment, the Ultra High Vacuum, which is needed in order to prepare and protect the atomic-scale cleanliness of surface and interface systems for the whole time length of the experiment. As such the technique borrows equipment and procedures from the surface science spectroscopies,[2] and exploitation procedures of the synchrotron radiation source from the conventional, non vacuum, EXAFS method.[1] The physics of EXAFS is described, in scattering theory, as the modulations of the X-ray absorption coefficient as a function of excitation energy which are due to the interference between the photoexcited electron and the backscattered amplitude due to the presence of scattering centres (the neighbouring atoms) in condensed matter. The scattering of the photoelectron wave is assumed to be due to the point scattering centres represented by the very localised charge stored in the core levels of the neighbouring atoms. In the SEXAFS (i.e. in the EXAFS) regime the photoelectron kinetic energy is so high (from 50-100 to up to 1000 eV) that the scattering power of the atomic charge is rather low, so that only single scattering events are of importance, and therefore the EXAFS signal contains intrinsically informations on the two - body correlation function (i.e one backscattering event and interference on the excited atom site).[11] In general terms SEXAFS is the measure of the empty density of states for extended energy ranges starting at 50-100 eV above the Fermi level.

The surface sensitivity is ensured by detecting the decay products of the photoabsorption process instead of the direct optical response of the medium (transmission, reflection). In particular one can measure the photoelectrons, Auger electrons, secondary electrons, fluorescence photons, photodesorbed ions and neutrals which are ejected as a consequence of the relaxation of the system after the photoionization event. No matter which detection mode is chosen, the observable of the experiment is the interference processes of the primary photoelectron with the backscattered amplitude.

Detailed descriptions of all the SEXAFS detection techniques are abundant in the literature.[2,3] The most popular insofar are the electron techniques: the measure of the secondary (or total) electron yield which is ejected from the sample as a consequence of the excitation and decay of the primary core hole ensures rather intense signals with constant escape depth set by the minimum kinetic energy detected (typically in the secondary tail) but low adsorbed/substrate contrast; the Auger electron yield allow the tunability of the surface sensitivity as a function of the final state energy of the particular Auger transition which is chosen. The relaxation of the primary core hole causes a cascade of Auger (and fluorescence) characteristic transitions. The probability of each secondary decay process is constant, since the Auger (and fluorescence) decays are transitions between the atomic discrete energy levels, it follows that a direct proportionality (with a possibly unknown proportionality factor, but a constant one!) holds between the primary photoionization event, and the relative EXAFS modulation, and any of the Auger (and fluorescence) decays of the excited atom. This fact allows one to attempt to analyse the depth variations of the local environments of a diffused atomic specie for example. AEY also gives a better adsorbate to substrate contrast since the secondary signal from the substrate at the same energy of the adsorbate Auger peak may be relatively small, but a worse signal to-noise ratio. A table, inspired by the recent one published by P. Citrin,[3] is proposed, which summarises the merit of each SEXAFS detection technique.

	SURFACE SENSITIVITY	ABSOLUTE SENSITIVITY	COST/ DIFFICULTY
BEST	AEY	FY, PY	TY
	PY	AEY	PY
	TY		FY
WORST	FY	TY	AEY

The interface studies in so far have been done in electron detection mode, AEY and TY, whilst fluorescence seems to be a promising complement, with applications in the special case of multilayer arrays.

WHAT THE SEXAFS DATA LOOK LIKE

SEXAFS scans are X-ray absorption versus photon energy curves starting a few tens of eV lower energy than a characteristic absorption edge (50-100 eV below threshold) and ending at as high energy as meaningful (and possible) above the edge, possibly 500-1000 eV above the threshold, typically 300-700 eV above the threshold.(fig.1) The spectrum contains the photoabsorption from all the electron states of lower binding energy than the excitation energy, which contribute a sloping, continuous background, since, in general, all EXAFS oscillations from these states are reduced to less than the statistical noise by the very large photoelectron kinetic energies for such shallower states and the correspondent very low scattering power of the atoms, and by the absorption jump and EXAFS modulations due to the characteristic absorption edge,which is met in the chosen energy interval. For a homogeneous, ordered, system the EXAFS oscillations could still be seen at 1000 eV above the edge if the only damping factor were the mean square relative displacements of the atoms (thermal vibrations, Debye-Waller-like term); practically, a number of phenomena limit the data range, which can be exploited for the analysis. The limitations are of two kinds: intrinsic to the optical behaviour of the sample, i.e. intrinsic to the SEXAFS technique, and extrinsic due to the presence of partially avoidable artefacts, and to the always improvable data quality (signal-to-noise ratio, signal to-background ratio).

Intrinsic limitations are: 1) photoelectron elastic peaks from the adsorbate or the substrate, sweeping through the electron analyser window; 2) Bragg glitches represented by multiple spikes and dips in the spectrum; 3) Higher energy absorption edges from the adsorbed or substrate material.

For metal/semiconductor interfaces the limitations coming from 1) are not very severe,since the cross sections for photoemission and the lifetime broadening of such deep state photopeaks along with the

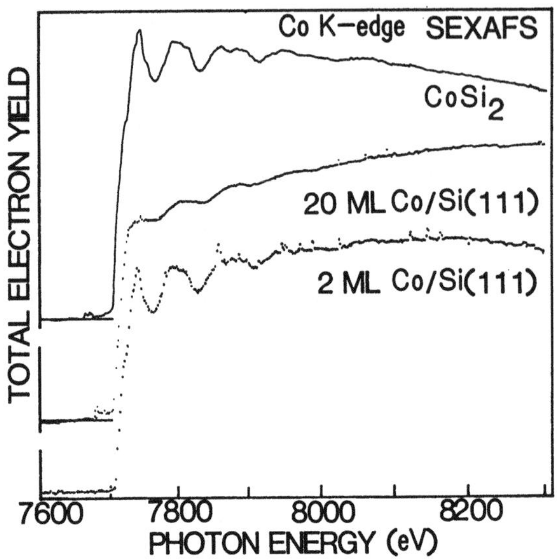

figure 1: SEXAFS raw data on the K edge of Co overlayers on Si(111)7x7 surfaces, and $CoSi_2$. The bottom curve is for 2 monolayers (ML) of Co on Si(111) as deposited at room temperature (RT), the reaction of Co with the Si substrate is seen from the similarity of the EXAFS oscillations with the $CoSi_2$ standard (top curve). The central curve represents the SEXAFS spectrum for 30 ML Co/Si(111), which is dominated by unreacted Co.[25]

poor energy resolution of the photon source (typically 3-10 eV for hard X-rays) contribute to smear the photoelectron elastic peak. Furthermore , this problem can be drastically reduced or eliminated by adopting the TY or FY detection. The limitation coming from 2) are, on the other hand,severe, in particular for high quality semiconductor single crystal substrates. The glitches, sharp peaks and dips in the SEXAFS spectrum are due to abrupt changes in the effective excitation of the adsorbate and/or substrate atoms which arise from the occurrence of Bragg diffraction of the incoming photon beam on the crystallographic planes of the substrate crystal. Such diffraction events are unavoidable,and depend on the wavelength of the X-rays and the relative orientation of the x-ray electric field vector and the crystallographic planes. One way to describe the phenomenon is to consider the interference between the incident beam (which in non Bragg conditions has a given optical path in the sample substrate) and the Bragg-diffracted beam.[12] The standing-wave field which results has peaks and nodes that travel within the crystal when the incident photon energy is varied (or the incidence angle of the light on the sample is varied), changing the effective excitation, i.e. the effective field intensity, in the region of the sample where the measured signal comes from, producing peaks and dips. The phenomenon is quite annoying for semiconductor, defect-free, single crystals since the Bragg diffractions are coherent. Polycrystals or crystals with a high density of dislocations (like most metal single crystals) are less of a problem, since the Bragg conditions are met over a broadened interval of X-ray energy and angular incidence; (of course this does not change the nature of the problem,since instead of peaks one can have broad modulations of the photoexcitation, which could be misinterpreted as EXAFS oscillations). In our own experience the Bragg glitches are one major problem in the exploitation of the SEXAFS data from interfaces having the semiconductor as the substrate. The time consuming research of angular (both polar and azymuthal) tuning of the sample in order to move the glitches out of a reasonable energy window (≥ 300 eV) above the edge is a leit-motive of the SEXAFS experimental runs.

The limitations due to the presence of a higher absorption edge from the adsorbate or from the substrate are rather a guideline to the choice of the proper way to do the experiment than actual limitations, unless the energy range of the synchrotron radiation outlet available is particularly short. For SEXAFS in the hard X-ray regime (hv > 5 KeV) the choice is often easy; in the soft X-ray range (100-4000 eV), on the contrary, if high angular momentum edges are sought, the total absorption spectra may become crowded with peaks and some edges cannot be exploited.

The extrinsic limitations are common to all spectroscopies: 1) signal-to-noise ratio. This limits the precision (reproducibility of the measures) and range of the SEXAFS data. Although at current levels of synchrotron radiation output intensities data precision can be as good as ±0.01 Å [3] it takes long acquisition times, which may introduce systematic errors like sample evolution and detection instabilities. Higher electron orbit stability in the storage rings and more intense photon fluxes are quickly becoming available, and are expected to continuously improve over the next decade. A better source will certainly prompt more applications and better quality data in SEXAFS on interfaces. Furthermore, the strong bonds which are typically formed at semiconductor interfaces should not be significantly perturbed by more intense X-ray beams (one should think differently for the future of SEXAFS on organic or biological adsorbate systems...). At the current state of the art SEXAFS of submonolayer interfaces still requires acquisition time of hours (1-6 hours) which, for unstable systems, or easily contaminated systems, means data averaging over more than one sample preparation.

2) The surface sensitivity is determined by the escape depth of the detected signal (elastic Auger detection) and/or by the penetration depth of the excitation (total reflection geometry).[13] Practically, the escape depth of the measure in Auger mode is determined by the electron analyser resolution. This is an important point since one is generally tempted to lower the analyser resolution in order to increase the count rate around a characteristic Auger peak. In doing so the actual measure is a sum of the elastic Auger peak (the only surface-sensitive one), the partial electron yield tails on both sides of the peak (if the peak is 10 eV wide and one sets the analyser at 20-30 eV effective resolution the partial yield part of the count rate may be significant) and the partial yield background from the adsorbate and substrate underneath the elastic peak. Probably the best way of doing the measure is to normalise the AEY by a partial yield PY measure done with the same analyser at little different kinetic energy, like 20 or 30 eV shifted from the elastic Auger peak. This should assure that the background measured at those particular kinetic energies is properly normalised, and that the truly surface-sensitive signal is isolated. Such approach has not been done yet due to the relatively low Auger yield signals. The actual surface sensitivity of the state of the art AEY SEXAFS data is therefore worse than the ideal one, but significant improvements will come from higher photon fluxes and better analysers.

RUDIMENTS OF ANALYSIS

The analysis of the SEXAFS data is basically identical to the analysis of conventional EXAFS data (fig.2).[1] We will simply recall the basic ideas that sustain the conventionally used Fourier

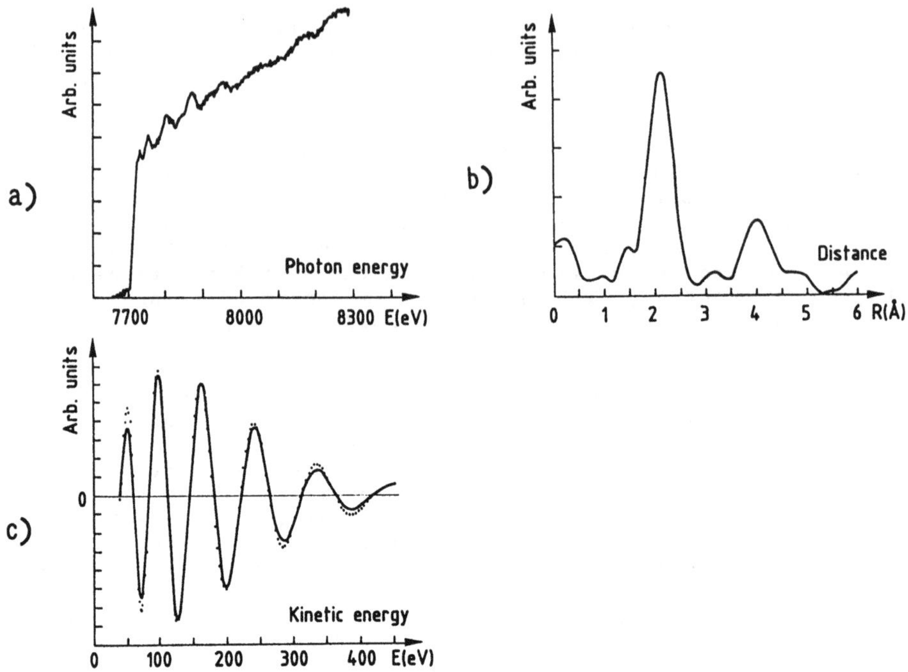

figure 2: Steps of SEXAFS analysis applied to 0.3 ML Co/Cu(111). The curve a) is the raw absorption spectrum over the Co K-edge. The curve b) is the Fourier transformation of the background subtracted EXAFS oscillations. The main peak represents the frequency corresponding to the first nearest neighbour scattering. The curve c) is the inverse Fourier transformation of the first nearest neighbour contribution. This curve represents the first neighbour EXAFS contribution to the absorption coefficient, and contains the distance and coordination number (polarization-dependent) information. This can be retrieved by fitting the curve with simulated EXAFS data for a proper geometry. The result in this case is Co-Cu distance = 2.51 Å, and coordination number 3.[33,34]

analysis of the SEXAFS data, once more the reason for being essential is that excellent reviews are available in widely diffused journals that were written by the promoters of the technique and warrant exhaustivity on the subject.[1-5] The (S)EXAFS signal is defined as:

$$X(h\nu) = \frac{\mu\,(h\nu) - \mu_{atomic}(h\nu)}{\mu_{atomic}(h\nu)}$$

i.e. the oscillatory part of the X-ray absorption coefficient measured in the condensed matter system that deviates from the monotonic (in the hν range starting at ≈50 eV after the threshold energy) atomic absorption coefficient. The origin of the X-ray absorption spectrum is taken at the energy of the threshold for the core level excitation, at this energy the photoelectron has zero kinetic energy final state, above this energy the photoelectron wavevector is defined by:

$$k = \frac{2m\,(h\nu - E_{binding})}{\hbar^2} \quad .$$

The EXAFS formula in the approximation of single scattering is the following:

$$X(h\nu) = -\sum_j \frac{N_j}{kR_j^2} e^{-\frac{2R_j}{\tau(k)}} e^{-2\sigma_j^2 k^2} A_j(k) \sin(2kR_j + \phi_j(k))$$

which is the sum over all the neighbour shells j containing N_j atoms at the distance R_j from the absorber of the sinusoidal EXAFS contributions whose amplitude A_j is damped by the two exponential factors, and whose frequency is determined by the distance R_j and by the total phase shift undergone by the photoelectron wave during the scattering events.[1] Amplitude and phase shift are characteristic functions of the scattering pair (the excited atom, which we know, and the backscattering atom at the distance R_j which is the unknown). Since in the adopted approximation the atoms of the solid are considered as point-scatterer, only the localised charge, i.e. the core electrons, determines the A(hν) and ϕ(hν) functions. In other words the delocalised charge between the atoms does not contribute significantly to the scattering and therefore the A(hν) and ϕ(hν) functions are, so to say, chemistry independent. This is a very fundamental hypothesis (very well verified indeed within the accuracy limits ±0.02 A) of EXAFS, and allows the analysis of an unknown system, say an interface between a transition metal and silicon, by using the amplitudes and phase shifts from a model compound of known crystallography, say a silicide.

The damping factors take into account: 1) the mean free path τ(k) of the photoelectron; the exponential factor selects the contributions due to those photoelectron waves which make the round trip from the central atom to the scatterer and back without energy losses; 2) the mean square value of the relative displacements of the central atom and of the scatterer. This is called Debye-Waller-like term since it is not referred to the laboratory frame, but is a relative value, and it is temperature dependent, of course.[14] It is important to remember the peculiar way of probing the matter that EXAFS does: the source of the probe is the excited atom which sends off a photoelectron spherical wave, the detector of the distribution of the scattering centres in the environment is again the same central atom that receives the backdiffused photoelectron amplitude. This is a unique feature since all other crystallographic probes are totally (source and detector) or partially (source or detector) "external probes", i.e. the measured quantities are referred to the laboratory reference system.

One SEXAFS specific feature is the polarisation dependence of the amplitude. This derives from the high anisotropy of the surface and of ultrathin interfaces, that we may consider as quasi two-dimensional systems. The relative orientation of the X-ray electric vector with respect to the surface (interface) normal does represent a preferential excitation for those atom pairs aligned along the electric vector; e.g. with the electric vector perpendicular to the surface (interface) plane the EXAFS amplitude will be maximum for the atom pairs aligned normal, or almost normal to the surface (interface). The electric vector can be also aligned, within the surface plane, along different crystallographic directions. The advantage of the polarisation properties of SEXAFS is that a test for the chemisorption site can be done on the basis of relative polarisation-dependent amplitudes, independently from the SEXAFS distance analysis.[2,7]

CASE HISTORIES

Some of the interface systems that have been more extensively studied by SEXAFS and whose results seem sound are shortly reviewed and commented in this section of the paper. An updated list of the interface studies which have been attempted or are underway is added below.

Ni/Si(111) and (100) interface growth at room temperature:
Bulk-sensitive and surface-sensitive Auger yield SEXAFS on the Ni K-edge (8333 eV) provided the first complete local information on a reactive near noble metal/Si(111) interface (fig.3).[15] The relevance of such system, along with the isoelectronic systems Pd/Si(111), Pt/Si(111), is great, since these are amongst the best silicide-forming interfaces. A great deal of spectroscopy and thermodynamical measurements existed on the Ni/Si junction, but the atomic scale interface crystallographic information lacked. The lowest coverage investigated was 0.5 ML and at this submonolayer stage the SEXAFS signal is almost identical to that from bulk $NiSi_2$ [Ni-(8)Si = 2.34 Å]: the Ni-Si distance is 2.37±0.03 Å and 6-7 Si nearest neighbours coordinated with each Ni. This result, obtained from the analysis of the absolute amplitude of the SEXAFS as measured with normal incidence of the SR on the sample surface, supports the assignment of a sixfold interstitial site between the top and second Si(111) layer. To accommodate the Ni interstitial, it is suggested a vertical expansion of the top three Si atoms by ~0.8 Å. This stage is interpreted as the formation of a template for the epitaxial growth of $NiSi_2$ on the Si(111) surface. At higher coverages the amplitude of the SEXAFS oscillations decreases monotonously. The explanation given to this phenomenon is the creation of Ni antisite defects within the cubic structure of $NiSi_2$: the phase opposition of the Ni-Ni backscattering and the Ni-Si backscattering determines the damping of the total backscattering amplitude function. The amplitude analysis was based on the comparison of the experimental Fourier transform peaks with simulated data of various environments and coordination numbers.(fig.3) Only the population of Ni antisites within the $NiSi_2$, corresponding therefore to an interface average stoichiometry higher than $NiSi_2$, compare favourably with the experiment. Such procedure is a truly empirical procedure, since no further assumptions are made in the construction of the simulated data than in the Fourier analysis of the original experimental data.

Submonolayers of Ni on Si(100) were also studied in similar conditions by the same scientists.[3] The SEXAFS spectrum for 0.5 ML Ni-Si(100) is very similar to that for Ni-Si(111). At higher coverages the evolution of the Ni-Si(100) interface differs from the Si(111) case with indications of extended interdiffusion rather than silicide nucleation. The results have not been given a final report yet.

The chemisorption site of Ni on Si(111) is well established from this study. The epitaxial growth mode on the Si(111) substrate is strongly supported by the SEXAFS amplitude analysis. These explanations are at odds with those derived from ion backscattering data, which propose clustering of Ni_2Si covered by a more diluted layer.[16] A part the difficulty in establishing the exact nature of the crystallographic information derived from different techniques, we suggest that a major problem in silicide growth is the exact kinetical conditions of the interface growth.[6] Different substrate conditions, evaporation rates and availability of extra thermal energy (irradiation, heating) are very likely responsible for different growth modes. Although the local coordination of the adsorbate should be unchanged, the actual growth mode, homogeneity or heterogeneity of the interface (reacted layer vs. islanding or iceberging), the average concentration of the intermixed regions may vary a lot. This point calls for systematic SEXAFS (for example, but other spectroscopies too, of course) studies of the interface growth following several controlled, kinetic paths.

The data on the Si(100) are somewhat preliminary. In spite of the different substrate geometry and density, the chemisorption coordina-

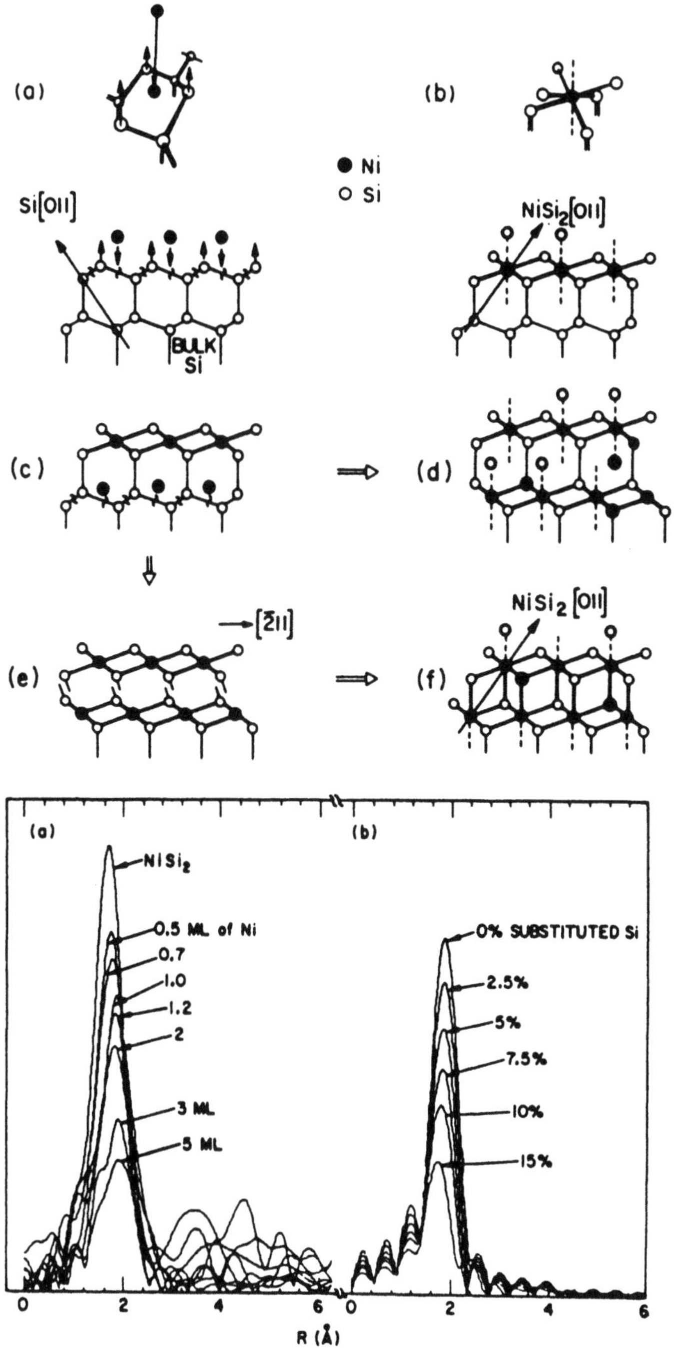

figure 3: SEXAFS data and analysis for the Ni/Si(111) interface prepared at RT. The left curves are the Fourier transformations of the data, the curves in the right panel are SEXAFS simulations for the sixfold interstitial site of the Si(111) double layer, and for the $NiSi_2$ epitaxial structure with Ni antisite defects. (see the text, and ref. 15) The top sketches are for the adsorption site and for the interface growth mechanism suggested by the authors.[15]

tion is basically the same, which, in our opinion, is quite consistent with the stability of the silicide-like bonding structure observed at all near noble metal/Si interfaces.

The poor resolution of these data prevents the analysis of the lineshape of the edge region, which is expected to change along with the chemical changes at the interface, as it is shown in the case of Co/Si(111).[17] Another consequence of the poor resolution is the little sensitivity to higher shell contributions. In fact the second nearest neighbours are not seen in the interface spectra, but they are seen with a much reduced amplitude even in the $NiSi_2$ standard pointing to a low overall sensitivity of the data to higher than first shell scattering, and therefore not necessarily to disorder beyond the first coordination shell at the interface. One can safely remark that SEXAFS has prompted a significant advance in the knowledge of the Ni/Si system, showing the similarity of the local coordination of Ni at the submonolayer chemisorption stage with $NiSi_2$.

The Pt/Si(111) interface growth at room temperature.

Total Electron Yield SEXAFS (and XARS) measurements were obtained on the L_3 edge of Pt submonolayers and monolayers deposited onto Si(111)7x7 at nominal room-temperature conditions (fig.4).[18] The analysis of the normal incidence spectra for 0.8±0.2 ML Pt/Si(111) gives an unambiguous assignment of the sixfold interstitial chemisorption site between the top and second Si(111) layer. The Pt-(6±1)Si bond length derived is 2.48±0.03 Å, which is a typical Pt-silicide bond length (Pt_2Si=2.46Å; PtSi= average 2.5Å). The deformation of the Si sixfold interlayer cage due to the Pt interstitial is discussed on the basis of the comparison of the 0.8 ML SEXAFS data and of simulated data which have been constructed using as inputs the experimental phase shift and backscattering amplitude and first neighbour distance (the one derived from the SEXAFS analysis) and higher neighbour distances corresponding to the various hypothesis of environment.(fig.5) The sixfold interlayer site with vertical and lateral displacement of the six Si nearest neighbours, and smaller displacements also in the second coordination shell, reproduce very

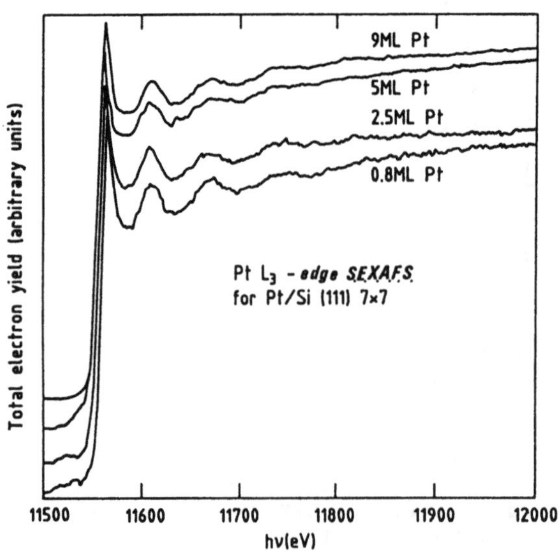

figure 4: Pt L_3 edge SEXAFS raw data for Pt monolayers onto Si(111)7x7 at room temperature.[18]

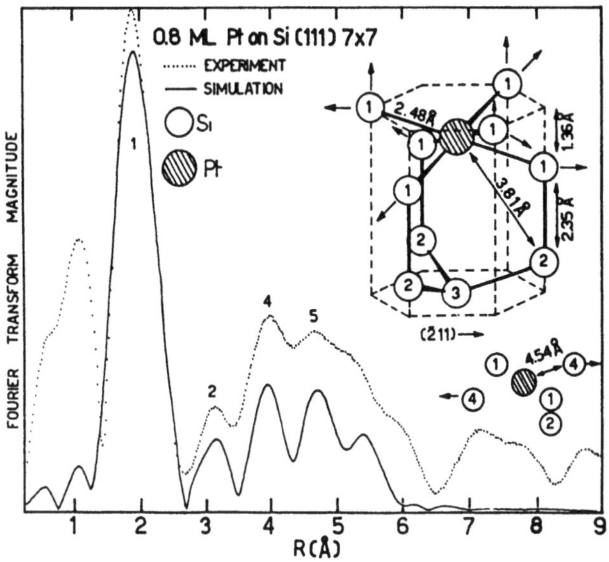

figure 5: Fourier transform of Pt L_3 edge SEXAFS data for 0.8 ML Pt on Si(111)7x7 (dots), sixfold interstitial site SEXAFS simulation (solid line), and derived chemisorption model. The peaks beyond the first Pt-Si distance (2.48±0.03 Å) are fitted by the higher distances up to the 5th Pt-Si distance, and show the deformation of the Si cage, as indicated in the model, hosting the interstitial Pt atom.[18]

well the experimental peaks, although a rather large error bar must be allowed for higher than first shell Fourier components (also due to the noise level seen at low K in the spectra). This is an important point: the changes in the substrate crystallography due to the presence of the adsorbate can be measured with SEXAFS either by directly measuring in surface sensitive mode the substrate EXAFS, or by measuring with high resolution and a large K range the higher shell contributions in the adsorbate SEXAFS spectrum. The local coordination of Pt at the submonolayer stage is therefore basically identical to PtSi [Pt-(6)Si], apart the bond angles, but there is a very important difference: the SEXAFS spectrum is fully explained by all Si neighbours, i.e. no evidence of Pt second nearest neighbours is found, which makes the interface at this stage non silicide-like. This is an important point: the molecular orbital bonding structure between Pt and Si, with typical silicide bond length, is determined within the first cage, whilst the silicide-like electronic structure, which is also determined by the d-d interaction between Pt second neighbours, is lacking at the chemisorption stage, i.e. the ultrathin interface is not silicide-like.[6] For higher coverages the SEXAFS shows a dominant configuration for the diffused Pt in the substrate: Pt-(4±1)Si at 2.46 Å.(fig.6) At 5 ML a second peak in the experimental Fourier transformation is only explained, at least qualitatively, by a second neighbour shell of Pt atoms. This corresponds to a local Pt_2Si configuration, and may represent the nucleation stage of the silicide within the Pt-Si solid solution produced by the intermixing. The growth mode derived from SEXAFS for the Pt/Si system is quite different from the one above summarised for Ni/Si. One difference is that the Pt silicides do not grow epitaxially on Si(111), and that both the PtSi (orthorhombic, fig.7) and Pt_2Si (tetragonal) have crystal structures of lower symmetry than Si and cannot nucleate starting

Fig. 6

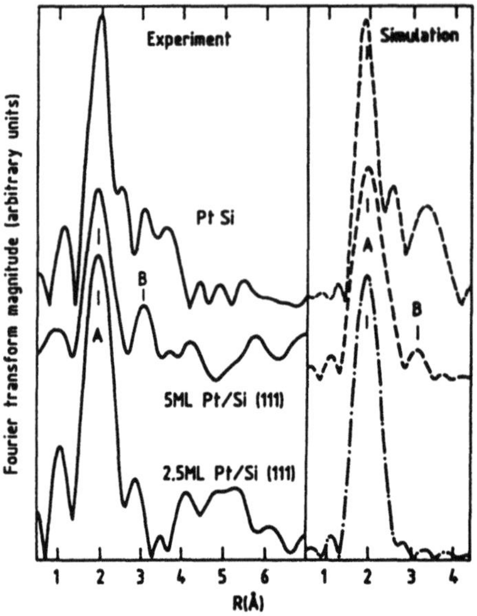

Fig. 7

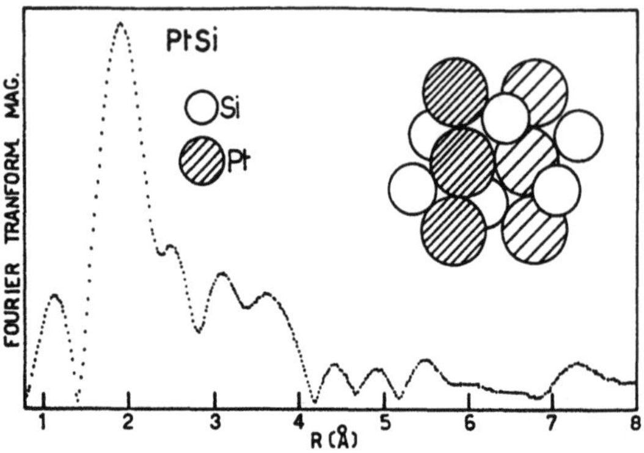

figure 6: Fourier transformations of Pt L_3 egde SEXAFS data and simulated data for 2.5 ML and 5 ML of Pt deposited onto Si(111), and PtSi standard. The peak B is fitted by assuming a local Pt_2Si coordination.[18]

figure 7: Fourier transform and model of the PtSi standard.

from the diamond structure of the Si substrate. The nucleation within the disordered solid solution suggests a reaction phenomenology similar to a phase separation process. The evidence of such mechanism must be strengthened by other experiments, and by a careful kinetic study (in the described experiment the heating of the substrate during Pt evaporation was not controlled).

The high resolution (better than 3 eV over the whole range) allows the discussion of the lineshape of the L_2 and L_3 edge resonances, which is discussed in the following paper.

One can safely remark that SEXAFS has made possible a direct picture of the local environment at the Pt-Si(111) interface. The sixfold interstitial site is the same as the Ni/Si(111) [as well as Ag/Si(111)$\sqrt{3}$x$\sqrt{3}$,(fig.8) and Co/Si(111)]. Pt-Pt coordination starts to be seen only at 5 ML where the local coordination resembles that of Pt_2Si suggesting a silicide cluster nucleation stage at the Pt/Si interface which follows the intermixing stage.

Si/Ge(111) and Ge/Si(111) preliminary results.

Evaporated Si on Ge(111) is amorphous at room temperature with the bulk Si-Si bond length of 2.35±0.02 Å.[19] After mild annealing the Si K-edge SEXAFS spectra show both Si-Si and Si-Ge distances of 2.44±0.02 Å. Both the distances are well resolved in the spectra thanks to the very different phase shift for Si and Ge backscattering which determine two well-separated EXAFS frequencies. The result is interesting: after annealing, a substitutional Si-Ge alloy is formed which is in epitaxy on the Ge(111) substrate. The alloy has bond lengths which are imposed by the substrate, in fact 2.44 Å is the bulk Ge-Ge bond length.[19] Preliminary results on the Ge K-edge SEXAFS for submonolayers and monolayers of Ge onto Si(111) also show Ge-Si intermixing at the monolayer coverage range.[20] The SEXAFS results on system are very important and, once complete will play a key role in the understanding of the strained epitaxial layers, and to understand the X-ray grazing incidence diffraction and standing wave studies on Ge/Si(111) which point to the existence of strained layers which can even be reconstructed in a fashion which is characteristic of the substrate, but not of the adsorbate in its natural crystalline state.[21]

As a short comment on the above reviewed results one can note that: the chemisorption for the near noble metals, those which develop extended interfaces with the group IV semiconductors, develops as a strong interaction with the substrate surface, the deformation of the substrate top planes and the partial penetration into high coordination interstitial sites. The high coordination at bond lengths that are basically identical to those of the stable bulk phases is favoured independent from the fact that the geometrical cage be a precursor of the compound or not, and independent on the second shell coordination (which largely influences instead the electron states observed with spectroscopy). This suggests that the chemical bond structure between the near noble metal and silicon which is established at the chemisorption stage is basically the final one in terms of orbital hybridisation and of charge transfer. The ionicity versus bond length correlation based on Pauling electronegativity confirms this trend. The determination of the chemisorption interstitial position, its geometry, and the bond distances are very important inputs for theoretical calculations of the density of electron states and of the charge distribution at the ultrathin interfaces. Recent developments of PDOS calculations for silicide-like interfaces have modelled the bonding geometry by considering metal occupation of the adamantane interstitial cage of the diamond semiconductors. The sixfold subsurface interstitial site, and the deformation of the Si-Si bonds in the top Si(111) double layer are good explanations of the removal of the sp^3 hybridisation of the Si atoms which has been measured by Cooper minimum photoemission, and by Auger lineshape spectroscopy. The chemisorption stage can be therefore truly considered the precursor stage of silicide formation, since only small bond-angle variations intervene within the first shell and the silicide nucleation stage corresponds to the adding of the external shells of neighbours at the right distance and composition. Kinetic conditions and relative abundances of the atomic species may lead to different stoichiometries of the interface compounds.

SEXAFS , insofar has been a time-expensive probe and systematic studies with the variation of all the growth parameters, have not been attempted or have partially failed for expiration of the attributed beam time. Nevertheless it is not merely prosaic to foresee higher availability of SR beams of higher quality in the near future, and more clever detection arrangements. SEXAFS might not become a routine diagnosis technique, but can certainly develop quickly towards a role of key technique for the study of the local coordination at all growth stages and conditions of a forming interface. It will also continue to be the unique test for the development of the understanding of the XANES data.

LIST OF INTERFACE SYSTEMS STUDIED BY SEXAFS

Ag-Si(111): the submonolayer interface at room temperature was studied by polarisation dependent SEXAFS on the Ag L_3 edge both in the Auger (Ag $L_3M_{4,5}M_{4,5}$) and total yield mode.[22] The adsorption site for 0.3 ML Ag/Si(111) was derived to be the sixfold hollow with the Ag atom basically in the top Si plane and a bond length of

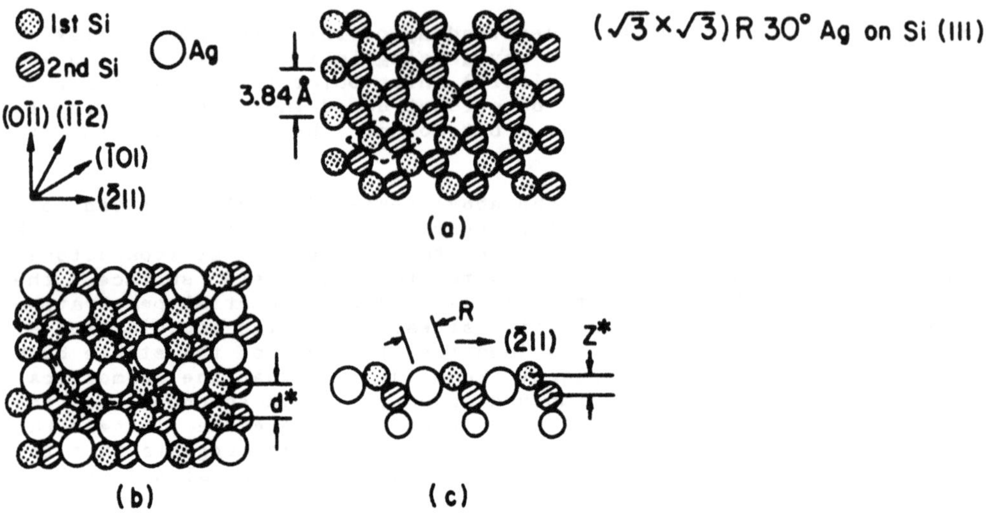

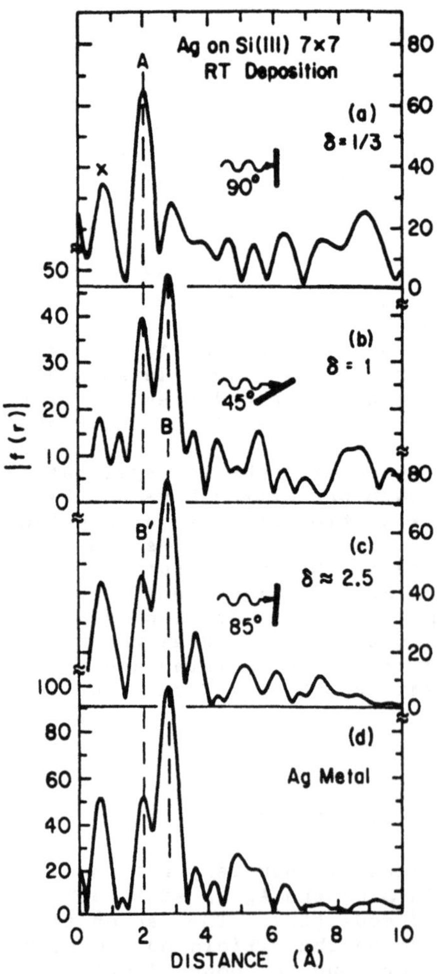

figure 8: Ag L_2 edge SEXAFS for Ag monolayers on Si(111)7x7, and model for the √3x√3 reconstruction which is obtained after annealing. The Ag-Si bond distance found is 2.48± 0.05 Å. The peak B represents the Ag-Ag coordination in unreacted Ag clusters at the surface for coverages exceeding 1 ML. [22,23]

2.48±0.05Å. After annealing of 0.5 ML and higher coverages the $\sqrt{3}x\sqrt{3}$ phase is formed and was found to be due to Ag atoms embedded in the sixfold interstitial site between the top and second Si layer at a distance of 2.48±0.04 Å.(fig. 8) For coverages of 1 ML or higher the $\sqrt{3}x\sqrt{3}$ phase coexists with Ag clusters having basically the bulk Ag bond length. The site assignment is sound in spite of the relatively large error bars on the distances (due to the limited data range and to residual noise in the spectra) since it is independently based on the polarisation-dependent relative SEXAFS amplitudes).

Pd-Si(111): preliminary SEXAFS data in the Auger and TY mode were obtained for submonolayers (unpublished) and 1.5 monolayers of Pd on Si(111) at nominal room temperature.[13] The comparison of the SEXAFS spectrum of the submonolayer and monolayer interface with those for bulk Pd_2Si show strong similarities, with quantitative differences in the coordination numbers of first and second nearest neighbours possibly due to disorder in the intermixed phase.

Ni-Si(111): see previous section and figure 3

Ni-Si(100): see previous section

Ni-SiO_2: preliminary TY SEXAFS data (obtained at LURE, unpublished) on the Ni K-edge for Ni thin films on SiO_2 show the disordered nature of the metal overlayer at low temperature and the formation of Ni-silicide-like regions after annealing, possibly due to diffusion of Ni through pinholes of the SiO_2 layer to access and react with elemental Si.

Pt-Si(111): see previous section and figures 4-7

Co-Si(111): submonolayers and monolayers of Co deposited at RT on Si(111) have been independently investigated by Auger SEXAFS (unpublished)[24] and TY SEXAFS (unpublished, fig. 1).[25] Both experiments show a behaviour similar to the Ni/Si(111) interface formation at low coverages. The higher coverages show an evolution from $CoSi_2$ phase to a Co rich layer. Very detailed XANES spectra show the evolution from $CoSi_2$ to pure Co also in the near edge structures.[17] A systematic work on the Co/Si(111) interface [31] has been done with the SEXAFS-related SEELFS technique.[32]

Cu-Si(111),(100): preliminary TY SEXAFS data on the chemisorption of Cu on Si(111) (Cu K-edge) suggest a similar site and bond length as for Ni/Si(111) (unpublished).[25] The study of the thermal agglomeration of the Cu/Si(111) and Cu/Si(100) interfaces in the 25-50 ML range is underway.[26]

Al-Si(111): inconclusive data were obtained on the Al K edge by Auger and TY SEXAFS for submonolayers and monolayers of Al onto Si(111).[27] The experiment failed due to the instability of the interface versus diffusion and oxidation. New sample preparations were needed every 1.5 hours (from Auger spectroscopy). Due to the long acquisition time required for obtaining a sufficient statistics at the time of the experiment (1982) the data were integrated over several (>10) sample preparations, adding systematic errors of reproducibility on coverage and evaporation rate. This experiment is a case of SEXAFS experiment that suffered just from the scarce beam intensity and acquisition efficiency. It might become feasible in the near future.

Au-Si(111),(100): experiments on the surface reconstructions of Au on Si(111) √3x√3, 5x1, 6x6, and on Si(100) are underway. Preliminary data show the stability of the Au-Si bond length in all the reconstructions.[28]

Au-SiO_2: SEXAFS spectra for a thin layer of Au on polycrystalline SiO_2 were obtained to demonstrate the feasibility of SEXAFS in the fluorescence mode on the Au L_3 edge.[29] The spectrum resembles that of metallic Au, so that surface agglomeration can be guessed.

V-Si(111): SEXAFS data on V/Si(111) have been reported [30] which proposedd the V-Si first-neighbour distance for 0.5 ML V on Si(111)7x7 to be 2.5±0.05 Å, and revealed the formation of VSi_2 for annealings of the interface at 550 C for 5 minutes.

Si-Ge(111): see previous section

Ge-Si(111): see previous section

Metals and semiconductors on III-Vs; studies on these systems are certainly forthcoming. The absence insofar of interface work on compound semiconductors is due to two experimental difficulties: 1) the poor focusing characteristics of the existing X-ray beam lines oblige to use relatively large crystals in order to exploit efficiently the photon beam. Due to the fact that only the cleaved (110) surface of the III-V compounds is well controlled from the geometrical and compositional point of view, large cleaves are needed, but are obtained with difficulty. Then, in case of TY measurements, the III-Vs are less favoured than Si since their high average atomic weight means high photoionization cross section. As a result the contrast between a monolayer of metal adsorbate and the background is much higher in the case of Si, than, for the same adsorbate, in the case of GaAs or InP.

Multilayer systems: SEXAFS work is announced on metal/semiconductor multilayer systems. By using the FY technique one can recover the interface signal with sufficient sensitivity since many interfaces contribute.
A list of other SEXAFS references is found in [3].

REFERENCES

1) P.A. Lee, P.H. Citrin, P. Eisenberger, and B.M. Kincaid; Rev. Mod. Phys. 53, 769 (1981), and references therein.
2) J. Stöhr, in X-ray Absorption: Principles, Applications Techniques of EXAFS, SEXAFS and XANES, edited by R. Prins and D. Koninsgsberger (Wiley, New York)
3) P.H. Citrin, Journal de Physique (Paris) C8, colloque 8, tome 47, (1986) p. 437, proceedings of the Intl. Conf. on EXAFS and Near edge Structure IV, Fontevraud, France, 1986.
4) D. Chandesris, Journal de Physique (Paris), Ecole d'Aussois de Rayonnement Synchrotron, Aussois, France 1986. (in french)
5) C. Calandra, O. Bisi, and G. Ottaviani, Surface Science Reports 4, 271 (1985).
6) G. Rossi, Surface Science Reports (1987).
7) P.H. Citrin, Phys. Rev. B31, 700 (1985).
8) F. Comin, L. Incoccia, P. Lagarde, G. Rossi, and P.H. Citrin, Phys. Rev. Lett. 54, 122 (1985).
9) P. Roubin, D. Chandesris, G. Rossi, and J. Lecante, M.C. Desjounqueres, and G. Treglia, Phys. Rev. Lett. 56, 1272 (1986).
10) K. Baberschke, U. Döbler, L. Wenzel, D. Arvanitis, A. Baratoff, and K.H. Rieder, Phys. Rev. B33, 5910 (1986).

11) C.R. Natoli and M. Benfatto, Journal de Physique (Paris) C8, colloque 8, tome 47, (19860 p. 11; proceedings of the Intl. Conf. on EXAFS and Near edge Structure IV, Fontevraud, France, 1986.
12) T. Ohta, H. Sekiyama, Y. Kitajima, H. Kuroda, T. Takahashi, and S. Kikuta, Jpn. J. Appl. Phys. 24, L475 (1985).
13) G. Martens and P. Rabe, Phys. Status Solidi (a) 58, 415 (1980).
14) D. Chandesris, J. Phys. (Paris) C8, 479 (1986).
15) F. Comin, J. Rowe, and P. Citrin; Phys Rev Lett 51, 2402 (1983).
16) E.J. van Loenen, J.F. van der Veen, and F.K. Goues, Surf. Sci. 157, 1 (1985).
17) G. Rossi in the XANES, XARS article in this school proceedings.
18) G. Rossi, D. Chandesris, P. Roubin, and J. Lecante; Phys Rev B 34, 7455 (1986); J. Phys. (Paris) C8, 521 (1986).
19) J.C. Woicik, R.S. List, B.B. Pate, and P. Pianetta, J. Phys. (Paris) C8, 497 (1986).
20) F. Comin, G. Rossi, and D. Chandesris, unpublished results.
21) M. Sauvage-Simkin, this school proceedings.
22) J. Stöhr, R. Jaeger, G. Rossi, T. Kendelewicz, and I. Lindau, Surf. Sci. 134, 813 (1983).
23) J. Stöhr, and R. Jaeger, J. Vac. Sci. Technol. 21, 619 (1982).
24) F. Comin, and P. Citrin, to be published.
25) G. Rossi, D. Chandesris, and J. Lecante, unpublished results.
26) M. Sancrotti, and G. Rossi, unpublished results.
27) J. Stöhr, R. Jaeger, G. Rossi, T. Kendelewicz, and I. Lindau, unpublished results.
28) M. Handbucken, G. Le Lay, and G. Rossi, unpublished results.
29) S.M. Heald, E. Keller, and E.A. Stern, Phys. Lett. 103A, 155 (1984).
30) S.J. Morgan, A.R. Law, W.G. Hezzender-Harker, R.H. Williams, R. Mc Grath, I.T. Mc Govern, and D. Norman, VUV 8 Intl. Conf. Lund 1986, Abstract vol. II, p.566, ed. by P.O. Nilsson.
31) E. Chainet, M. De Crescenzi, J. Derrien, T.T.A. Nguyen, and R. Cinti, Surf. Sci. 168, 801 (1986).
32) M. De Crescenzi, Surf. Sci, 162, 838 (1985).
33) D. Chandesris, P. Roubin, G. Rossi, and J. Lecante, Surf. Sci. 169, 57 (1986).
34) P. Roubin, thèse d' etat, Orsay 1987.

XANES and XARS for Semiconductor Interface Studies

G. Rossi

Laboratoire pour l'Utilisation du Rayonnement Electromagnetique, CNRS, CEA, MEN; Université de Paris-Sud, F-91405 Orsay, France

DEFINITIONS

XANES: X-ray Absorption Near Edge Structure
or NEXAFS: Near Edge X-ray Absorption Fine Structure

THE PHENOMENON: Modulations of the X-ray absorption cross section in the energy region 15-100 eV above a characteristic edge in condensed matter

THE MEASUREMENT: X-ray absorption as a function of the photon energy hv;
Particle impact ionisation as a function of E_p
High energy resolution is required (2-3 eV)

THE DETECTION: Fluorescence yield,
electron yield
Auger elastic electron yield
stimulated desorption of ions, neutrals

THE INTERPRETATION: information on the geometrical arrangement of the environment of the photo-excited atom.
SCATTERING THEORY: interference between the photoelectron that leaves the excited atom and the back-diffused amplitude due to multiple scattering events with the neighbouring atoms which are strong scatterers for low and intermediate photoelectron energies.
GENERAL: measurment of the density of final electron states accessible via dipole transitions.

ABSTRACT

An empirical, qualitative approach to the analysis of the X-ray near edge spectra of metal-semiconductor and semiconductor-semiconductor systems is proposed in this paper.

INTRODUCTION

The extended X-ray absorption spectra are characterised by a photoelectron energy dependent scattering power function of the localised charge on the atoms in condensed matter. For photoelectron energies higher than 50-100 eV the scattering power of the atoms is generally low enough that only single scattering events dominate: this is the EXAFS region which contains information on the two-body (pair) correlation function.[1] At lower photoelectron energy, i.e. closer to the absorption edge, the atomic scattering power increases and multiple (n) scattering events start to dominate the spectra containing information on the n-body correlation function, i.e. on the geometrical environment of the absorber atom.[2] At very low photoelectron energies, tending to zero kinetic energy, i.e. in the edge region, n tends to infinity, and we can conveniently think of the X-ray absorption spectrum in terms of the partial (due to the transition selection rules) density of final electron states of the medium.[3] The XANES technique attempts to retrieve geometrical information from the spectra between ~15 and ~100 eV above the characteristic absorption edge using the concepts of scattering theory.[2] The XARS technique attempts to retrieve information from the edge peaks on the empty density of states using the concepts of one electron dipole transitions and of band theory.[3,4]

XARS: X-ray Absorption Resonance Spectroscopy
or White Lines Spectroscopy

THE PHENOMENON: resonant edge transitions from deep core levels

THE MEASURE: X-ray absorption as a function of the photon energy hν;(high resolution is required, <3 eV)
Particle impact ionisation as a function of E_p

THE DETECTION: Fluorescence Yield, electron yield,
Auger elastic electron yield,
stimulated desorption of ions, neutrals

THE INTERPRETATION: X-rays: transitions to the empty density of states dominated by one electron dipole matrix elements, modified by many-body effects (core relaxation, post collisional)
K, L1, M1...: 1s,2s,3s...-> np
$L_{2,3}$ edges: 2p -> nd, (n+1)s with $\delta j=0,\pm 1$
i.e. sensitivity to the spin-orbit d PDOS
if $\delta_{spin\ orbit} \approx$ Width of the d band
Electron impact: transitions to empty DOS
matrix elements? multipole transitions?

XANES ANALYSIS AND EXAMPLES

The exact analysis of XANES has been developed within the framework of multiple scattering theory by Natoli and co-workers.[2] The full multiple scattering calculations of the absorption coefficient for hypothetical geometrical arrangements of atoms are compared with the experimental data in order to find the best model of the atomic "cage" surrounding the absorber. Such a first-principles approach to the calculation of the total absorption coefficient is quite expensive in terms of computer time and has not been available so far for routine analysis of experimental data on unknown systems. The total absorption coefficient has been expressed as a series of infinite partial contributions each being due to a particular order of scattering.[2]

$$\alpha_F = \Sigma_{n=0}^{\infty} \alpha_n$$

where α_0 represents the atomic absorption coefficient, $\alpha_1 = 0$,

α_2 represents the partial EXAFS contribution, $\alpha_{n>2}$ represents the nth-order scattering partial contribution. The experimental data are compared to the sum of partial absorption coefficients of sufficient order, typically ≤4.[2] This approach may become affordable for analysis of large sets of data in the near future.[5]

Here we propose a purely empirical analysis of the data based on the ideas that: 1) The XANES is determined by the geometrical arrangements of the atoms around the absorber, i.e. the first neighbour cage, and second neighbour cage mostly. 2) The symmetry of the cage determines the shape of the XANES, whilst the interatomic distances determine the energy spacing between the XANES features. 3) The local configuration of an atom at an interface could resemble, or differ from, the local environment of the same atom in a crystalline standard compound of the same constituent atomic species; the comparison of the XANES spectrum of the unknown system (e.g. the interface) with the "standard" XANES (e.g. a silicide), may give quite strong, although non quantitative, evidence.

The comparison of the spectra can be done graphically by observing the difference spectra between the XANES relative to a standard environment and an unknown environment, or can be done by analysing the oscillations of the absorption coefficient from the edge up to 100-150 eV by means of Fourier transformation in K space. This last approach was followed to analyse the XANES-SEXAFS data of crystalline, amorphous and ion-amorphised Si by Comin et al. [6], and is briefly recalled below. It must be noted that the data are not analysed as SEXAFS, but are just Fourier transformed to bring out the main XANES-SEXAFS frequencies, and then the ion-amorphised spectra are compared empirically to the crystalline and amorphous standards.

<u>The depth analyses of ion-amorphised Si:</u>

The Si K-edge X-ray absorption spectra were extended only up to 90 eV above the threshold: a limit imposed by the photoelectron elastic peak which reaches at this energy the same final state kinetic energy as the detected Si $L_{2,3}VV$ Auger peak. The detection of Auger peaks of low kinetic energy (Si $L_{2,3}VV$ ≈90 eV), high kinetic energy (Si KLL ≈1615 eV), and of the total electron yield (TY) allowed one to be selectively sensitive to the near surface region (5-7 A), subsurface medium depth region (~25 A), and bulk region (>50 A) respectively. The relative differences sought were the medium range order differences between the surface and the subsurface regions of ion-sputtered Si samples. Some of the data are summarised in Figs. 1 and 2. The "standard" crystalline and amorphous spectra are shown in Fig. 1 along with the XANES-SEXAFS oscillations extracted from the background. The differences in the two standards are brought out in the Fourier transform (Fig. 2) where a first peak corresponding to the first neighbour coordination is seen in both cases, but the large peak corresponding to second and third neighbour contributions, is only visible in the crystalline spectrum. The amorphous material does not have order in the interatomic spacings beyond the first neighbours' shell. The other Fourier spectra in figure 2 are those obtained from ion-amorphised Si. The result of the comparison indicates that the subsurface regions of Si are amorphised after high doses of ion-sputtering, showing XANES-SEXAFS features similar to the truly amorphous evaporated Si. The surface region, on the other hand, (poly)crystalline order since the most surface sensitive spectra (labelled LVV) reproduce the bulk crystalline standard spectrum. This fact was suggested to be due to surface reconstruction.[6]

<u>The Co/Si(111) interface:</u>

The interface between Co and Si(111) has received considerable attention due to the epitaxial growth of $CoSi_2$ onto Si(111) after

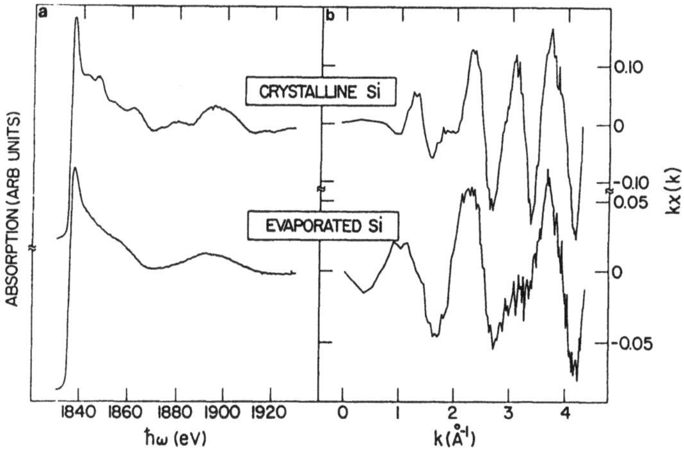

Fig. 1(a). K-edge absorption spectra from single crystal and evaporated (amorphous) Si measured in the Auger mode by monitoring the intensity of the Auger KLL electron peak. (b) Corresponding data after background subtraction and conversion into k space.[6]

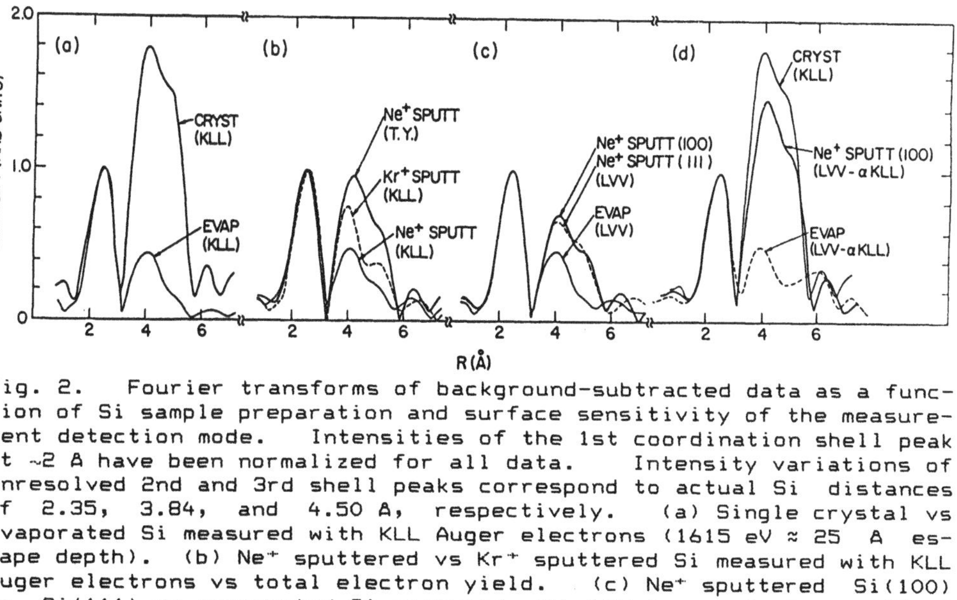

Fig. 2. Fourier transforms of background-subtracted data as a function of Si sample preparation and surface sensitivity of the measurement detection mode. Intensities of the 1st coordination shell peak at ~2 Å have been normalized for all data. Intensity variations of unresolved 2nd and 3rd shell peaks correspond to actual Si distances of 2.35, 3.84, and 4.50 Å, respectively. (a) Single crystal vs evaporated Si measured with KLL Auger electrons (1615 eV ≈ 25 Å escape depth). (b) Ne^+ sputtered vs Kr^+ sputtered Si measured with KLL Auger electrons vs total electron yield. (c) Ne^+ sputtered Si(100) or Si(111) vs evaporated Si measured with LVV Auger electrons (92 eV ≈ 5 Å escape depth). (d) Bold line: Fourier transform of raw difference spectrum for Ne^+ sputtered Si. Dashed line: same procedure applied to evaporated Si. Light line: data from single crystal Si.[6]

mild annealing, and the application of such junctions in permeable metallic base transistors.[7] A SEELFS study has been published,[7] and two SEXAFS studies are announced [8,9]. The chemisorption site was established to be the sixfold interstitial subsurface layer, likewise the Ni/Si(111), Pt/Si(111), and Ag$\sqrt{3}\times\sqrt{3}$/Si(111).[1] Figure 3 displays the TY high resolution (<3 eV) XANES data for 2 ML Co/Si(111), 4 ML Co/Si(111) at room temperature, for an ultrathin annealed (diffused) Co/Si(111) interface, and for the $CoSi_2$ standard.[9] The XANES structures in the $CoSi_2$ are quite prominent and

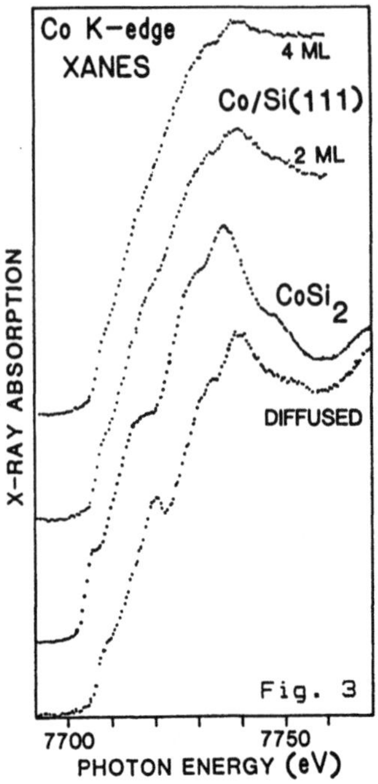

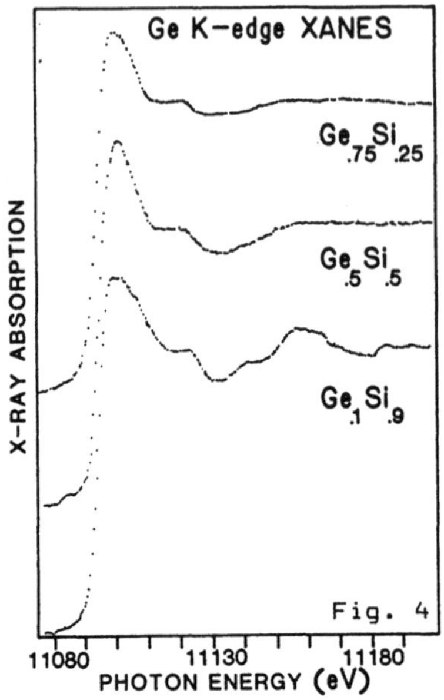

Fig. 3. XANES spectra obtained in the total electron yield mode for in situ grown Co/Si(111)7x7 interfaces and standard silicide.
From the bottom up: Co diluted in Si(111) after annealing; $CoSi_2$; 2 monolayers Co/Si(111) at room temperature; 4 monolayers of Co/Si(111) at room temperature. See the text for discussion. [9]

Fig. 4. XANES spectra obtained in the total electron yield mode for substitutional Ge-Si alloys. From the bottom up: $Ge_{0.1}Si_{0.9}$; $Ge_{0.5}Si_{0.5}$; $Ge_{0.75}Si_{0.25}$. The XANES structures of the diluted Ge in Si (bottom spectrum) resemble (apart from the energy scale) those for crystalline Si (see Fig.1). The XANES for the Ge-rich sample (top spectrum) resembles that for pure Ge (not shown). This simple example shows the sensitivity of the XANES to the geometrical and compositional details of the atomic cage surrounding the absorber atom.[10]

extend from the absorption onset up to the plateau at 80 eV above threshold. From these curves it is evident how important the photon energy resolution is for XANES. The $CoSi_2$ XANES are also observed in the annealed thin interface: this sample did not show any Co trace at the surface and represents the thermal dilution of Co in the substrate. From the XANES it results that the diffused Co is found in the same local environment as in $CoSi_2$. The same structures are also observed in the 2 ML interface, but less pronounced. This is compatible with two hypotheses: either the local Co environment is not quite identical to the $CoSi_2$ so that the structures are smeared due to, for example, disorder in the bond angles, or two types of Co are present, possibly $CoSi_2$-like sites and Co-Co coordination sites. The 4 ML spectrum is intermediate between the interface features and the XANES of pure metallic Co. This result indicates the saturation of the interface reaction at room temperature with unreacted Co piling up at the surface.[9]

The Ge-Si system:

A standard Ge K-edge XANES data base for the Ge-Si system is reproduced in Fig. 4. The data were obtained in the TY mode from substitutional Ge-Si alloys of known concentrations.[10] The Si environment of the most diluted Ge-Si alloy, and the Ge environment of the most concentrated one (in Ge) give quite dramatically different XANES spectra, with a continuous variation at intermediate concentrations. The study of the Ge/Si interface would be strengthened by the possible comparisons with the XANES of the standards.[10]

EXPERIMENTAL REMARKS

The detection of XANES is basically identical to that for SEXAFS described in [1] and references. The more stringent requirements of monochromator resolution are over-compensated by the short data range which allows quicker measurements than SEXAFS and the possibility of avoiding, or limiting to an acceptable level, the phenomenon of Bragg glitches in the spectrum.[1] It is, in fact, relatively easy to choose the incidence angle of the X-rays on the sample such as to displace the Bragg diffraction conditions to energies outside of the ≈100 eV wide XANES range. The case of the Ge-Si crystalline samples shown in Fig. 4 is indicative: the XANES are "clean" from Bragg glitches, whilst long range EXAFS scans on the same samples were very difficult to measure due to the many large Bragg peaks and dips.[10] We believe that the exploitation of XANES data will play a prominent role in the study of the local geometry of atoms at interfaces.

XARS ANALYSIS AND EXAMPLES

Resonances appear in the edge X-ray absorption spectra which are related to the presence of a high density of empty states of appropriate symmetry with respect to one electron dipole transitions from the initial state. The dipole selection rules allow transitions from the atomic-like core initial state to the part of the final state wave-functions that reflects the proper symmetry. Therefore the excitation of a $2p_{1/2}$ or $2p_{3/2}$ core level (L_2 or L_3 thresholds) in a d-metal will couple with the $nd_{3/2}$,(n+1)s or $nd_{3/2,5/2}$,(n+1)s part of the empty density of states, respectively. Likewise s symmetry initial states (K edges, L_1,M_1,N_1,O_1 edges) will be coupled at threshold energy with the p-like portion of the empty density of states near the Fermi level.[4] The amplitude, A, of the resonances can be written as $A \propto |M(E)|^2 (E)$ where the squared matrix element represents the overlap of the final-state wave-function and the initial core-state wavefunction, and (E) is the partial dipole selected empty DOS. This dipole matrix element produces a projection effect of the final state partial density of states on the core hole which gives a "local" character to the DOS information contained in the XARS data.[11] Noble metals with fully occupied d-bands do not show $L_{2,3}$ resonances, but smooth absorption onsets, since the edge transitions couple only to delocalised s-like states. The chemical dependent changes in the amplitude of the $L_{2,3}$ resonances have been systematically investigated for classes of compounds of the near noble metals [12-14] and trends in the filling of the d-bands of Pd and Pt compounds have been suggested based on a quantitative exploitation of the $L_{2,3}$ edge resonance amplitudes.[14,15] In insulators and semiconductors the edge resonances may be due to photoexcitation into bound atomic-like states which do not correspond to the photoionization threshold and can be understood as Rydberg-like states that collapse in a gap of the density of states because of the creation of the core hole: they are sometimes called excitons.[16]

The main issue in the understanding of the edge resonances in metallic systems, or "white lines" lies in the role of the medium during the photoionization, or photoexcitation process, i.e. the role of the delocalised charge which may screen the core hole varying the final state energy of the photoelectron, or spreading it in a relatively wide interval, and the role of external localised shells which may participate in very fast decay processes, like Auger or Super-Coster-Kronig transitions which also change the final state of the primary photoelectron via relaxation of the density of states (occupied and unoccupied) or post-collisional effects.[11,17] A simple picture of the edge resonances which would allow safe interpretation of all X-ray absorption spectra at threshold is not available and the role of the core hole relaxation and other many body effects on the energy, lineshape and amplitude of the final state peaks can be dramatically different from material to material. In many relevant cases, though, the XARS spectra are well accounted for by just considering one electron dipole transition from the initial state, and therefore information on the partial (angular momentum and spin-orbit character) empty DOS close to E_F can be retrieved. This represents a most useful investigation tool for solid materials, and, specifically for metal-semiconductor interfaces.[3] The response of the electron gas to the creation of a deep core hole is understood as the localisation of a screening cloud of s-like symmetry, [18] therefore greatly affecting the absorption spectra of simple metals, but, for the same reason, being relatively less important in the case of d-metals and f-metals where the dominant contributions in the occupied and unoccupied DOS are of d-character and d and f-character respectively. The understanding of the L-edge resonances of near-noble metal silicides has recently been improved by the comparison of the experimental data on the L_3 and L_1 edges of Pd and Pd_2Si with a one-electron calculation of the white lines obtained by deriving the transition matrix elements between the initial state wave-function and the final state wave-functions of the metal and of the silicides that were calculated in the linear muffin tin orbital (LMTO) framework.[18] The obtained X-ray absorption resonance peaks and lineshapes (shown in Fig. 5 from ref. 18) compare favourably with the experimental data [19] confirming that for this class of materials (therefore for the near-noble metal-semiconductor interfaces) the XARS data are a direct probe of the core hole projected empty DOS.

<u>The Pd-Si system:</u>

An increase of the Pd L_2 and L_3 resonance amplitude relative to the continuum absorption amplitude well above the edge (30 eV) was observed in the Pd-silicides (Pd_2Si and PdSi) with respect to Pd metal.[15] The direct consequence of this finding was the confirmation of the hybrid-orbital bonding picture of the near noble metal silicides (and interfaces) since the formation of bonding Pd d-Si p states implies the existence of a partially empty antibonding density of states having the same orbital mixture.[20] Therefore one explains the increased white lines with respect to the pure metal case (only 0.35 unoccupied d-states) by the higher overall density of unoccupied states having d character, spread over a few eV above the Fermi level. The possible role of relaxation effects on these edges was stressed in ref. 19, but the above-mentioned calculations (Fig. 5)[18] strongly support the basic validity of the one-electron DOS reasoning in this case.

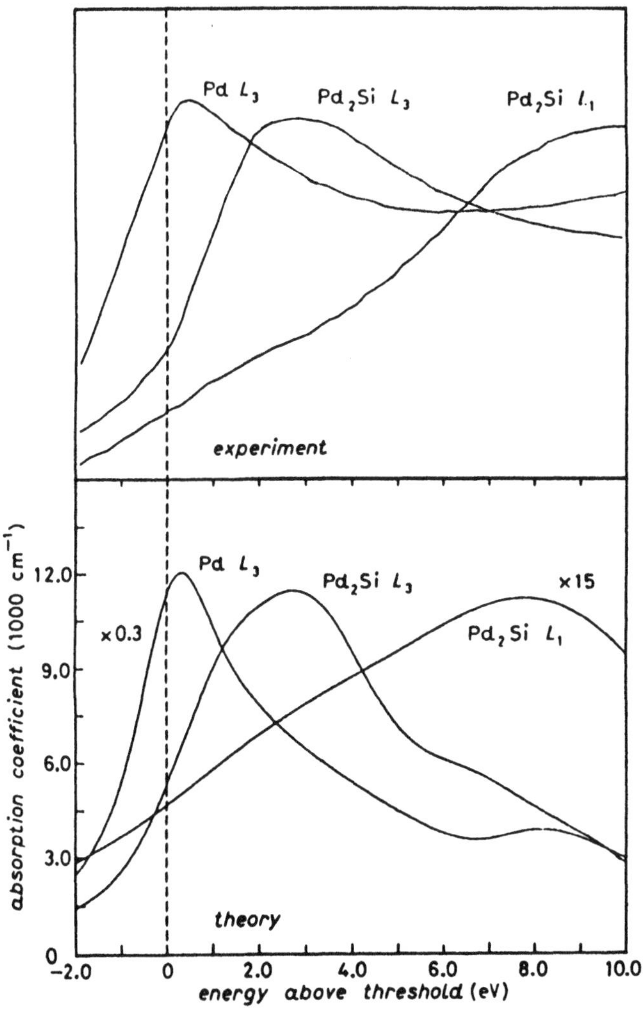

Fig. 5. Comparison between the experimental (upper panel) and theoretical (lower panel) L_1, L_3 X-ray absorption edge of Pd_2Si. XAS L_3 edge of pure Pd is also shown.[18] See text. Experimental data from Ref.19.

The Pt-Si system:

Similar results to the Pd-Si case are obtained on PtSi and Pt/Si interfaces with respect to Pt. Both Pt L_3 and L_2 edges show a higher resonance to continuum absorption ratio in the Pt-Si reacted systems than the pure metal. The non-statistical intensities of the Pt L_2 and L_3 edge resonances, which are well known both in the pure Pt solid and atomic cases and are due to the $5d_{5/2}$ character of the d band empty states (0.3 hole), are strongly reduced in the Pt-Si interface phases and in PtSi (Fig. 6).[3,21] A sharp white line appears also on the Pt L_2 edge, indicating the redistribution of the population of the spin-orbit components of the d-like density of states in the silicide. Figsures 6 and 7 show the Pt $L_{2,3}$ edges for the early stages of growth of the Pt/Si(111) interface, i.e. for a bonding configuration which is precursor of the silicide bonding,

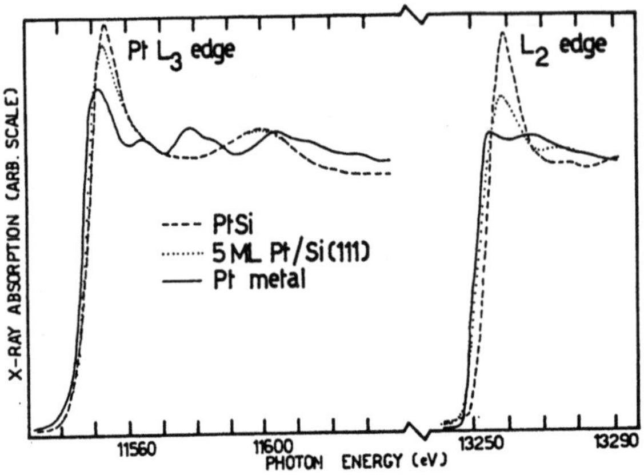

Fig. 6. XARS for L2,3 transitions for metallic Pt (solid); 5 ML/Si(111) interface (dotted) and for bulk PtSi (dashed). The spectra are normalised to each other below the edges and in the continuum adsorption 40 eV above the edges. This procedure shows up the amplitude and width of the white lines (i.e. the resonant absorption to empty 5d states) versus the absorption to non d empty states. The appearance of white lines on the L2 edge for the interface and silicide shows that the total d-valence band width has increased with respect to the spin orbit interaction, and the filling of the d-band is very different from the case of pure Pt.[21,22]

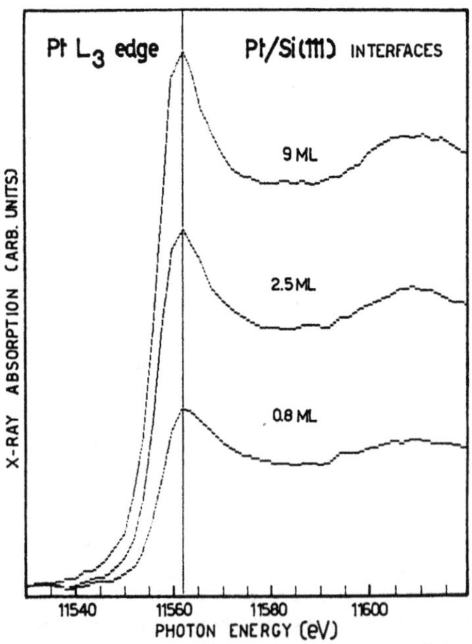

Fig. 7. Pt L_3 XARS for the early stages of growth of the Pt/Si(111)7x7 interface at room temperature. The spectra are in the true relative amplitudes.[22]

as is also obtained from the structural SEXAFS results [1,21,22]. A recent study has exploited the projection effect of the density of states on the initial core hole, by measuring the Pt XARS on L and on M edges.[23] The idea is that by projecting the final DOS onto different initial state wave-functions (2p, 3p ...) which have different spatial extension one could obtain a sort of zoom-lens effect on the local DOS at the Pt site. In spite of the low resolution of the data, the broadening of the bonding part of the d-p hybrid wavefunction of the silicide, whose electron density is large in between the ligands, can be guessed from the $M_{2,3}$ edge data, whilst the narrower distribution of the more localised d-like states are better seen in the $L_{2,3}$ edge resonances.[23]

<u>Ni_2Si :</u>

Ni $L_{2,3}$ edge data have been obtained by means of electron energy loss spectroscopy on Ni_2Si [24]. The results are a peak of empty d DOS in the silicide which is found at 0.7 eV higher energy than in Ni (L_3 edges), and a lowering of the peak intensity which is compatible with the broadening of the empty d state distribution in the antibonding silicide states. The same features have been observed by the authors in X-ray absorption, supporting the dipole one electron transition interpretation of EELS data in this energy range (Ep=1950 eV). Similar results have been obtained on the Co/Si(111) interface and on $CoSi_2$ also by means of EELS.[25]

<u>f metals and rare earth/silicon interfaces:</u>

$L_{2,3}$ XARS of the rare earths and of the rare earth intermetallic compounds has developed in recent years as a key method for the study of the valence configuration of those rare earth materials which present the remarkable phenomenon of valence band instability, or dynamical-mixed valence.[26] The phenomenon arises from the mixing of the shallow localised charge contained in the 4f subshell, and the sd valence charge: the hybridised f-d states allow part of the charge to fluctuate between a localised, and a band configuration, with a typical fluctuation time of 10^{-13}-10^{-14} s. [27]. The time average of such phenomena yields a fractional valence number for such systems, i.e. a statistical value for the number of electrons in the delocalised valence states which is between the two extreme configurations for the rare earth ion, typically 2^+ and 3^+, or 3^+ and 4^+. The X-ray absorption phenomenon takes place on a 10^{-16} s time scale, it is therefore fast with respect to the valence fluctuation time and consequently XARS allows one to take snap-shots of a frozen statistical distribution of rare earth ionic configuration, measuring two well resolved absorption resonances (separated in energy by 7-8 eV) that correspond to the two 4f initial state configurations of the ion : $4f^n$ and $4f^{n-1}$. Therefore XARS is a direct way to observe the valence configuration of mixed valence solids (and interfaces). The issue of the influence of the many-body problem on L edge XARS of valence mixed-valent materials is delicate, possibly even more so than in d metals, since the Coulomb attraction between the open 2p hole and the localised 4f charge might be, in principle, large enough to change dynamically the final state configuration of the photoexcited material.

$L_{2,3}$ edges of the rare earth elements have been fully explained by one electron calculations that considered all electron states not directly involved in the photoionisation process (i.e. the 2p initial states and the 5d final states) as spectators, i.e. disregarding many body effects.[28] The problem that arises when dealing with mixed valence systems is that the lineshape of the $4f^n$ and $4f^{n-1}$ resonances are not known a priori, and are different in general with respect to one another. They partially overlap since the empty 5d DOS energy width is of the same order of the energy separation of the two edge resonances. Furthermore, other symmetry components of the final states, which we may conveniently call XANES oscillations, may start to complicate the absorption spectrum 5-10 eV above threshold and therefore overlap the $4f^{n-1}$ resonance lineshape.[26] Because of these difficulties the accuracy of the mean valence determination from L edge XARS is to be established from case to case based on all elements of analysis: good empty DOS standards, experimental XANES standards, [26] chemical and structural homogeneity of the sample (in fact, expecially at interfaces, one could have coexistence of truly divalent environments and mixed valence environments, so that the L edge XARS spectrum would contain two divalent, largely ovarlapping resonance lineshapes, one from the divalent phase and one from the mixed valent phase, [22,29]). Besides all these difficulties, all belonging to the one-electron approach, there remain the many-body effects: the large core hole screening and post collisional effects, which modify the peak-shapes and produce extra peaks (satellites) in the photoemission spectra,[30,31] are dramatically reduced in XARS because the absorption resonances occur at threshold energy with the final state electron remaining in the neighbourood of the excited atom (zero kinetic energy) therefore contributing to the screening of the core hole, in a quasi-neutral atom configuration.

In many relevant cases an accurate one-electron analysis accounts for the data.[26] It is in this framework that the analysis of the XARS results for the Yb/Si interfaces has been proposed.[3,29]

The Yb/Si(111)7x7, Yb/Si(100)2x1, and Yb/a-Si interfaces:

The Yb L_3 edge resonance for 1 ML Yb/Si(111)7x7 at room temperature (threefold chemisorption sites available) is compared to the bulk fcc Yb resonance in Fig. 8.[29] The difference in the

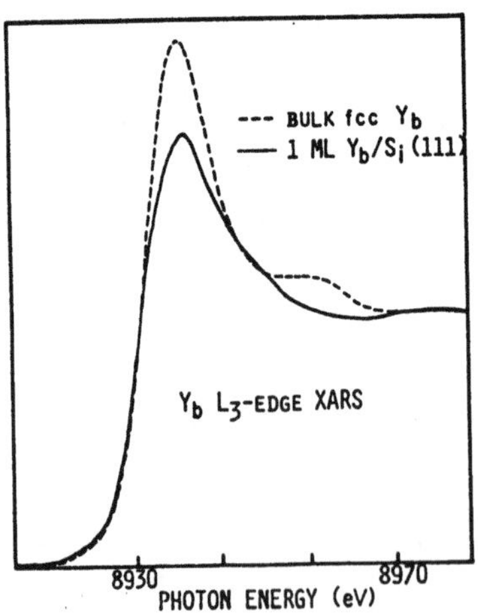

Fig. 8. Yb L3 edge X-ray absorption resonance for fcc Yb (dashed curve) and for 1 ML/Yb(111)7x7 at RT. The lineshapes have been normalised before the edge, and in the continuum absorption 40 eV above threshold. The L3 resonance for 1 ML Yb/Si(111)7x7 only shows a divalent component (likewise in Yb metal) but the relative white-line to continuum amplitude of the spectrum and the width of the peak indicate a redistribution of the empty 5d DOS at the interface.[29,22]

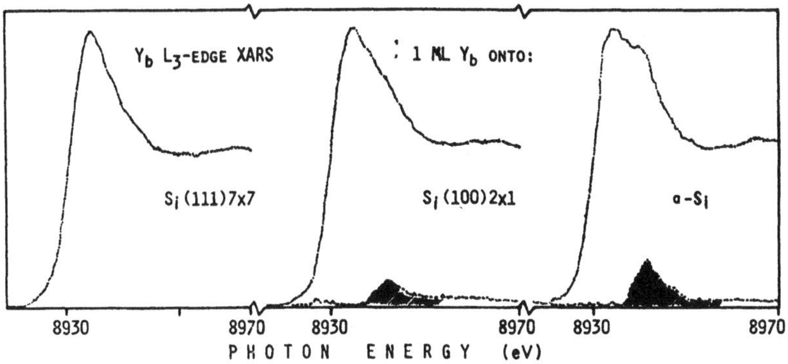

Fig. 9. X-ray absorption Yb L3 resonances for 1 ML Yb on three Si surfaces: 1 ML Yb/Si(111)7x7 (left), 1 ML Yb/Si(100)2x1 (center) and 1 ML Yb/a-Si (right). The shaded curves are obtained by subtracting the divalent Yb white line from the interface spectra, they represent the three-valent component that is present at the Yb/Si(100) and Yb/a-Si interfaces.[29]

lineshape is attributed to the different Yb 5d DOS distribution which in the pure metal corresponds to the almost empty 5d band, while in the Yb-Si bonds it represents the antibonding Yb 5d-Si 3p hybrid states.[22,29] This lineshape is then compared to that obtained for the same amount of Yb chemisorbed onto other Si surfaces: the Si(100)2x1 (fourfold hollow sites available with possible coordination with up to 6 Si nearest neighbours) and a-Si (high Si coordination sites available) in Fig 9. Both the Yb/Si(100) and Yb/a-Si interface show clearly mixed valence behaviour, as is directly visible in the data (second peak at 6.5 eV higher energy) and is shown more clearly by subtracting the divalent lineshape from the measured lineshapes, and obtaining the Yb 3^{+} contributions. One observes that the mean valency seem to correlate with the higher Yb-

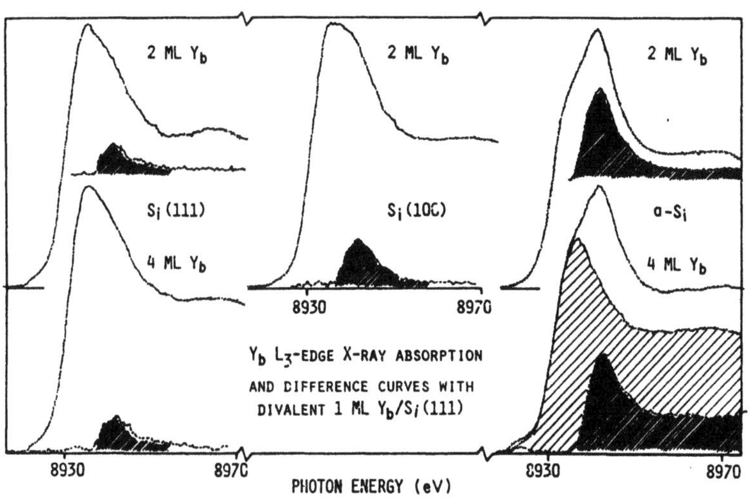

Fig. 10. Yb L3 XARS for 2 and 4 ML/Si interfaces: Yb/Si(111) (left), Yb/Si(100) (center), Yb/a-Si (right). The dashed curves are the trivalent components of the x-ray absorption resonances, shifted by 6.5 eV to higher binding enery than the divalent components.[29]

Si coordination possible on the three different types of chemisorption sites, reflecting the different local potential experienced by the 4f electrons in the relative bonding configurations. In this respect these data show how XARS of rare earth atoms on surfaces can possibly be the most sensitive probe for the local potential at chemisorption sites. The number of bonds formed, and the relative energy released to the system, also appear to drive the growth of the three interfaces for higher Yb coverages (Fig. 10) giving a moderate mixed valency at the quickly saturated Yb/Si(111) interface, a larger degree of reactive intermixing at the Yb/Si(100) interface, and a silicide-like valency (~2.3 as for $YbSi_{\sim 1.8}$) for 4 ML Yb on the amorphous Si surface.[29]

Acknowledgements:

Part of the work reviewed here and in [1] has been done by the SEXAFS collaboration at LURE, i.e. by D. Chandesris, P. Roubin, J. Lecante and the author. The operation of the facilities at LURE is acknowledged. LURE is jointly supported by the CNRS, the CEA and the MEN.

REFERENCES

1) See the article: SEXAFS for SEMICONDUCTOR INTERFACE STUDIES by G. Rossi, in this same volume, and the references quoted therein.
2) C.R. Natoli, in EXAFS and Near Edge Structure, edited by A. Bianconi, L. Incoccia, and S. Stipchich, Springer Series in Chem. Phys. vol 27 (1983); C.R. Natoli and M. Benfatto, Journal de Physique (Paris) C8, 11 (1986), proceedings of the IV Intl. Conf. on EXAFS and Near Edge Structure, Fontevraud,France 1986; M. Benfatto, A. Bianconi, J. Garcia, A. Marcelli, M. Fanfoni, and I. Davoli; Phys. Rev. B (1986).
3) G. Rossi, P. Roubin, D. Chandesris, and J. Lecante, Surf. Sci. 168, 787 (1986).
4) M. Brown, R.E. Peierls, and E. A. Stern, Phys. Rev. B15, 738 (1977).
5) M. Benfatto and C.R. Natoli, private communications.
6) F. Comin, L. Incoccia, P. Lagarde, and P.H. Citrin, Phys. Rev. Lett. 54, 122 (1985).
7) J. Derrien; Surf. Sci. 168, (1986).
8) F. Comin and P. Citrin, private communications
9) G. Rossi, D. Chandesris, P. Roubin, and J. Lecante, unpublished results
10) F. Comin, G. Rossi and D. Chandesris, unpublished results
11) E.A. Stern and J. Rehr; Phys. Rev. B27, 3351 (1983).
12) J.A. Horsley; J. Chem. Phys. 76, 1451 (1982).
13) A.N. Mansour, J.W. Cook, and D.E. Sayers; J. Phys. Chem. 88, 2330 (1984).
14) T.K. Sham; Phys. Rev. B31, 1888, and 1903 (1985).
15) G. Rossi, R. Jaeger, J. Stöhr, T. Kendelewicz, and I. Lindau; Phys. Rev. B27, 5154 (1983).
16) see for example F. Bassani and M. Altarelli, in: Handbook on Synchrotron Radiation, Vol.1. Ed. E.E. Koch (North-Holland, Amsterdam, 1983).
17) G. Van der Laan, B.T. Thole, J. Zaanen, G.A. Sawatzky, J.C. Fuggle, R.C. Karnatak, and J.M. Esteva; J. Physique (Paris) C8, 997 (1986).
18) O. Bisi, O. Jepsen, O.K. Andersen; Europhys. Lett. 1, 149 (1986).

19) M. De Crescenzi, E. Colavita, U. del Pennino, P. Sassaroli, S. Valeri, C. Rinaldi, L. Sorba, and S. Nannarone; Phys. Rev. B32, 612 (1985).
20) C. Calandra, O. Bisi, and G. Ottaviani; Surf. Sci. Rept. 4, 271 (1985); and contributions by C. Calandra to this book.
21) G. Rossi, D. Chandesris, P. Roubin, and J. Lecante, Phys. Rev. B34, 7455 (1986); Journal de Physique (Paris) C8, 521 (1986) proceedings of the IV Intl. Conf. on EXAFS and Near Edge Structure, Fontevraud, France 1986.
22) G. Rossi; Surf. Sci. Rept. (1987).
23) C. Carbone, J. Nogami, I. Lindau, I. Abbati, L. Braicovich, L.I. Johansson, and G. Majni; Thin Solid Films, unpublished.
24) U. del Pennino, C. Mariani, S. Valeri, G. Ottaviani, M.G. Betti, S. Nannarone, and M. De Crescenzi; Phys. Rev. B34, 2875 (1986); Surf. Sci. 168, 204 (1986).
25) J. Derrien, private communication.
26) G. Krill; J. Physique (Paris) C8, 907 (1986), proceedings of the IV Intl. Conf. on EXAFS and Near Edge Structure, Fontevraud, France 1986.
27) See for example: Valence Fluctuations in Solids, Ed. by L. Falicov, W. Henke, and M. Mapple; North-Holland, Amsterdam (1981).
28) G. Materlik, J. Muller, and J. Wilkins; Phys. Rev. Lett. 50, 267 (1983).
29) G. Rossi, D. Chandesris, P. Roubin, and J. Lecante; Phys. Rev. B33, 2926 (1986).
30) A. Kotani, and T, Jo; J. Physique (Paris) C8, 915 (1986), proceedings of the IV Intl. Conf. on EXAFS and Near Edge Structure, Fontevraud, France 1986.
31) O. Gunnarson and K. Schönhammer; ibid. p. 923.

Recent Progress in Electron Spectroscopy: Application to the Local Geometry Determination at Surfaces and Interfaces

J. Derrien

Centre National de la Recherche Scientifique, Laboratoire d'Etudes des Propriétés Electroniques des Solides, associated with Université Scientifique, Technologique et Médicale de Grenoble, B.P. 166, F-38042 Grenoble Cedex, France

1. INTRODUCTION

The understanding of surface and interface phenomena is based on a fair correlation between the surface and interface physicochemical and crystallographic properties and their electronic properties. To look at the solid surfaces and interfaces, on a microscopic scale, the physicists and chemists nowadays are using a great variety of very sensitive techniques, which are actually very sophisticated and rather expensive.

Conventionally, Auger electron spectroscopy and electron diffraction techniques give information on the physicochemical and crystallographic behaviour of the surfaces and interfaces whereas photoemission spectroscopy reflects the densities of occupied states of energy lower than that of the Fermi level ($E \leq E_F$).

One can easily see that at least two kinds of techniques are to be developed :

a - In the case of a surface or interface without a long-range order, where the diffraction techniques are unsuccessful, short-range order techniques are to be improved.

b - The band structure of a solid cannot be fully understood without information on the empty states extending above the Fermi level ($E \geq E_F$). This is the reason why inverse photoemission or BIS technique which probe empty states are now applied in Surface Science /1/.

In this aim, we have proposed recently /2,3/ two simple techniques probing the local geometry of the sample. They are based on fine structures observed in electron energy loss and Auger spectra respectively.

In the first part of this lecture, the physical principles of extended electron energy loss fine structure technique (EELFS), its capabilities and limits will be discussed. Particular attention will be paid to the local order of the initial stages of metal-semiconductor interface formation.

In the second part of this lecture, we show evidence of extended fine structures observed above core-valence-valence Auger transitions in Auger spectra of several materials. We tentatively suggest their underlying physical origins and deduce lattice parameters from these extended fine Auger structures (hereafter called EXFAS), conferring therefore to the conventional Auger technique new capabilities as a surface-sensitive local order probe.

2. EELFS PRINCIPLES

Inner shell electrons of a material can be excited not only by the absorption of X-rays but also by the inelastic scattering of energetic electrons. Therefore if one analyzes the energy distribution N(E) of electrons inelastically reflected by a solid surface, one may observe, with a medium energy resolution analyzer, several regions in the spectrum :

a - The backscattered electrons which are reflected without losing energy. This is the so-called elastic peak.

b - The inelastically scattered electrons which produce valence shell excitations such as surface or bulk plasmons or interband transitions (energy loss ≤ 30 eV).

c - The inelastically scattered electrons which produce inner shell excitations losing energies necessary to produce K, L, M... shell ionisations.

The number of electrons in each of these regions is proportional to the probability of each of these loss events.

Let us recall that the probability of X-ray absorption by a core electron is proportional in the dipole approximation to /4,5/

$$| \langle f | \mathbf{r} . \varepsilon | i \rangle |^2 \rho(E_F) \tag{1}$$

where $|i\rangle$ is the inner shell initial state, $|f\rangle$ the final unoccupied states. **r** is the position operator, ε is the electric field vector and $\rho(E_F)$ is the final state density.

In loss spectra, the inelastically scattered electrons lose a small energy loss ΔE as compared to their incident energy. Thus one may apply the Born approximation in describing classically the diffusion process between incident electrons and sample atoms. The matrix element of the transition rate is

$$| \langle f | e^{i\mathbf{q}\mathbf{r}} | i \rangle |^2 \rho(E_F) \tag{2}$$

where $\hbar\mathbf{q}$ is the momentum transfer between the incident and backscattered electron, $|i\rangle$ and $|f\rangle$ are the initial and the excited final state of the atom. If $\mathbf{qr} \ll 1$, then (2) can be read as

$$q^2 | \langle f | \mathbf{r} . \varepsilon_q | i \rangle |^2 \rho(E_F) \tag{3}$$

where ε_q is the unit vector in the q direction. Equation (3) means that for small **qr** values, the dipole approximation is valid and the matrix element for electron diffusion is similar to the X-ray absorption process one (compare (1) with (3)). This means also that in loss spectra, in an extended energy region far from core ionization edges, oscillating fine structures may be observed. These fine structures arise from the same interference process between the outgoing and backscattered part of the excited electrons as for the photoelectron in extended X-ray absorption fine structure (EXAFS) case /4,5/. The oscillatory part $\chi(E)$ of the loss spectrum of a core ionization region therefore may be described by the same expression used in EXAFS :

$$\chi(k) \approx \sum_j N_j A_j(k) \sin [2 k R_j + \phi_j(k)] \tag{4}$$

where R_j is the radial distance of the N_j backscattering atoms j from the central absorbing atom. $A_j(k)$ is the backscattering amplitude and $\phi_j(k)$ is the total phase shift of the backscattered electron which experiences during its path both the absorbing and the backscattering atom potential. k is the wave vector of the excited electron whose energy is referred to the core edge. Equation (4) is strictly valid for s symmetry core level excitation and for polycrystalline samples /5/ although one can easily demonstrate that it can be extended within certain limits to p symmetry core level excitation. Performing a Fourier transform of the experimental oscillations $\chi(k)$, one can readily obtain R_j of various atomic shells provided that the phase shift $\phi_j(k)$ is known /6,7/. The analogy between EELFS and EXAFS has been stressed here by a classical formulation - Equations (1) to (3) -. Recently, MILA and NOGUERA have achieved to derive an analytic expression of the current detected in loss spectra and they have shown that the signal contains effectively oscillating EXAFS-like factors /8/. So far EELFS technique has been applied to study clean surfaces, oxidation process of a surface, metallic cluster growth /9,10/. Let us illustrate the potentiality of EELFS in the local order determination of Co deposited on the Si(111) surface /2/.

3. LOCAL ORDER OF THE Co-Si INTERFACE AS INVESTIGATED BY EELFS TECHNIQUE

The Co-Si interface has been extensively studied recently by several surface techniques because under appropriate conditions, a nearly perfect epitaxial $CoSi_2$ layer can be grown on top of the Si(111) substrate, conferring to the system numerous advantages, both fundamental and technological, as compared with other metal-silicon systems. Review papers on the Co-Si interface formation are given in Ref. /11,12/. The potentiality of this interface in the elaboration of fast transistors is also described therein and in Ref. /13/.

3.1 Experimental procedure

All experiments were performed in a UHV chamber equipped with LEED, AES and UPS techniques. The Si(111) wafers were prepared by argon ion bombardment and annealed under UHV ambient ($\sim 10^{-10}$ Torr) in order to get a clean and ordered 7x7 Si(111) surface. Co evaporation was done by means of a Co bar heated with electron bombardment. Deposit thickness was monitored by a quartz crystal device. The same cylindrical mirror analyzer (CMA) was used to record AES and EELFS spectra. The incident electron energy was around 2-3 KeV depending on the core level edges to be monitored. The incident beam current was about 1 μA to 10 $\mu A/mm^2$ and did not seem to affect the probed surface. Modulating the analyzer voltage with a $\sim$ 8-10 V peak to peak ac voltage degraded the intrinsic energy resolution of the CMA ($\Delta E/E \sim 0.5$ %) but improved the signal-to-noise ratio. This fact did not influence greatly the detection of the oscillating structures whose period should extend over $\sim$ 50 eV. The second derivative $\chi(E) = d^2N(E)/dE^2$ of the electron distribution N(E) was detected for EELFS spectra in order to enhance the oscillations versus the background. The results discussed here arise mainly from the Co L_{23} deep core edge. A possible interference between oscillations due to the L_2 and L_3 edges separated in our experiment by the spin orbit splitting $\sim$ 15 eV may induce uncertainty in the distance determination. This has been checked with a simulation method and the accuracy in the bond length determination could be estimated to be $\sim$ 0.05 Å.

3.2 The Co-Si room temperature interface

Sequential Co coverages at 2, 4, 6 and $\sim$ 200 Å produce EELFS spectra of the Co L_{23} core edge region, as shown in Figure 1. The low coverage range (2-6 Å) reflects the initial stages of the interface formation and the 200 Å film represents Co metal as checked by other techniques (LEED, ARUPS, XPS, AES, work function, ...) /11,12/. The EELFS features extend to about 300 eV from the L_{23} core edge. Fourier transform, following the same standard procedure as for EXAFS /4,5/ has been applied to analyze the observed fine structures. The magnitudes of the Fourier transforms (F(R)) of these structures are reported in Figure 2.

It is also note worthy that the distances R found in the F(R) have to be corrected for a ΔR due to the phase shift $\phi_j(k)$ in equation (4), in order to get the distances in the real space around each Co absorbing atom. The phase shift $\phi_j(k)$ has been calculated for several elements /6,7/. For the Co-Si system, we found that the Co-Co and Co-Si L_{23} phase shifts amount to 0.22 Å and 0.36 Å respectively. These phase shifts have been deduced experimentally using a 200 Å Co film and a 100 Å $CoSi_2$ film as standard for Co metal and $CoSi_2$ respectively. The local order results (Figure 2) show unambiguously a silicide compound formation at low coverage (2-6 Å) mainly because of a strong decrease of the first nearest neighbour distance as compared with the Co metal distance (200 Å). $R_{Co\ metal} - R_{Interface} \approx 0.26 \pm 0.03$ Å. Moreover, the reversed sign of the oscillations, between 25-100 eV above the core edge, observed for low coverage film as compared with the Co thick film, pleads for the predominance of light backscattering (Si) atoms around the central (Co) atom /6/. We interpret therefore this peak as due to the first Si coordination shell around the central Co atom. Corrected for the phase shift, it gives a Co-Si distance of 2.38 ± 0.05 Å which is very close to the Co-Si distance in $CoSi_2$ (2.32 Å). It might reflect the incorporation of first Co atoms (2 Å) into the Si substrate with the Co environment very close to that in $CoSi_2$. Then with subsequent Co deposits (4 Å and 6 Å) the silicide becomes Co richer (CoSi and Co_2Si) due to a low interdiffusion of Co and Si atoms at room temperature. Finally, with higher coverage, the film evolves towards an almost pure Co film as found by other techniques /11,12/.

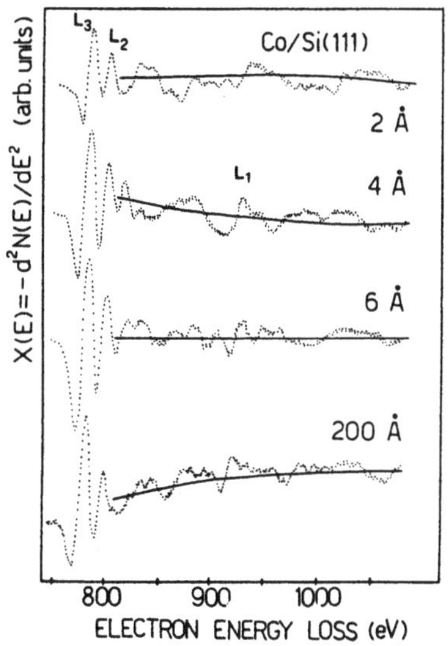

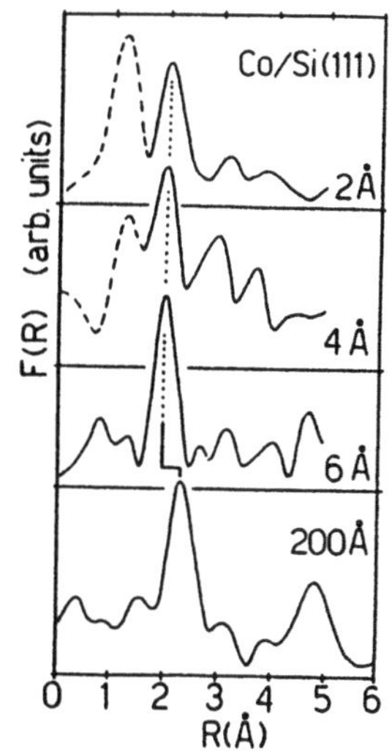

Fig. 1 : Extended fine structures observed in loss spectra above the L_{23} edges of Co deposited on Si(111) at room temperature. Dots are raw data, solid lines the background used for the EELFS analysis. Loss spectra are not normalized between them.

Fig. 2 : Magnitude of Fourier transform of structures in Figure 1. At low coverage (2-6 Å), a silicide compound is found with a mean average nearest-neighbour distance about 2.02 + 0.36 (Co-Si phase shift correction) = 2.38 ± 0.05 Å. Peaks at low distance (≈ 1.2 Å) (dotted line) are mainly due to background procedure and should be ignored. At high coverage, a pure Co film is found displaying a nearest-neighbour distance of 2.28 + 0.22 (Co-Co phase shift correction) = 2.50 ± 0.05 Å.

3.3 The Co-Si annealed interface

It has been demonstrated that when Co-Si(111) wafers are annealed under UHV ambient, interdiffusion of the two species occurs and Co_2Si, CoSi, $CoSi_2$ are formed at progressively higher temperatures. The formation temperatures of these various phases range about 300, 400 and 550°C ± 100°C respectively, depending on the thickness of the initial Co deposit, the annealing time (usually several minutes for thin films). These phases show a well-defined chemical composition as one can check with the Co/Si Auger peak ratio. We heated a 33 Å Co-Si(111) sample at successively 300, 400 and 500°C for a few minutes, checked the Co/Si Auger ratio to make sure of the formation of Co_2Si, CoSi and $CoSi_2$ respectively. LEED observation showed also for the final phase $CoSi_2$ a very sharp (1x1) diagram reflecting the quasi perfect epitaxy of $CoSi_2$ on Si(111). EELFS spectra recorded at these phases are reported in Figure 3. Also are drawn the magnitudes of Fourier transforms of these spectra. Using the final phase

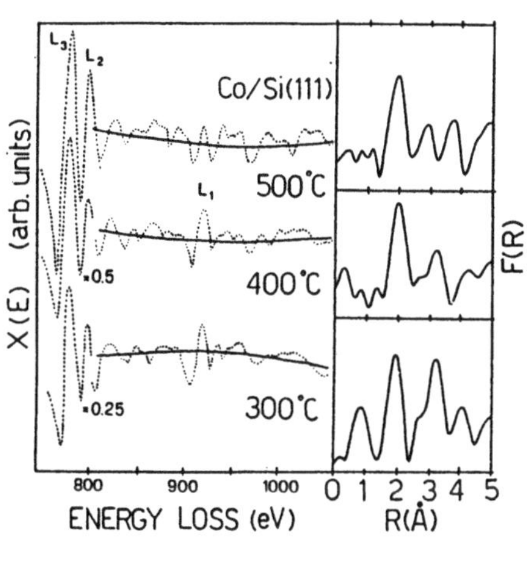

Fig. 3 : Extended fine structures observed on a Co-Si annealed interface (left side) and the corresponding Fourier transforms (right side). Co_2Si, CoSi and $CoSi_2$ are formed at progressively higher temperatures. The mean average nearest-neighbour distances of these phases are observed at 2.33, 2.38 and 2.32 ± 0.05 Å respectively.

$CoSi_2$ as standard (33 Å of Co react with Si to give rise to a ~ 100 Å $CoSi_2$ film) and assuming the Co-Si distance to be 2.32 Å as given by X-ray crystallographic data on bulk $CoSi_2$, we deduce the Co-Si phase shift to amount to 0.36 Å. Assuming the phase transferability to be valid for various silicides, we find from Figure 3 a distance of 2.33 ± 0.05 Å for the Co_2Si phase and 2.38 ± 0.05 Å for the CoSi phase. All these values, measured by EELFS technique agree with crystallographic values within the experimental accuracy (~ 0.05 Å).

4. EXTENDED FINE AUGER STRUCTURE (EXFAS)

Auger electron spectroscopy (AES) is the most currently used technique in order to get information on the chemical composition of the outermost layers of the surface. However the AES technique so far could not provide easily structural information on the investigated samples. The lack of this structural potentiality can now be bridged over thanks to the findings we present in this lecture. We demonstrate here that fine structure related to a core-valence-valence (CVV) Auger transition are detectable in Auger spectra /3/. They extend on several hundreds of eV above the investigated CVV Auger lines. We interpret these extended fine Auger structures (EXFAS) as originated from the same interference process which produces extended X-ray absorption fine structures (EXAFS) usually observed in the X-ray photoabsorption spectra. From the Fourier analysis of our EXFAS spectra, we obtain the radial atomic distribution F(R) of the samples, conferring therefore to the conventional AES technique new capabilities as a surface-sensitive structural probe. This is a new approach to local geometry determination to be compared to other fine structures observed in energy loss spectra (EELFS).

Figure 4 schematizes various fine structures found in the electron yield distribution, near the core edge region (as used in EELFS technique) and above Auger transition (as used in EXFAS technique discussed in this section).

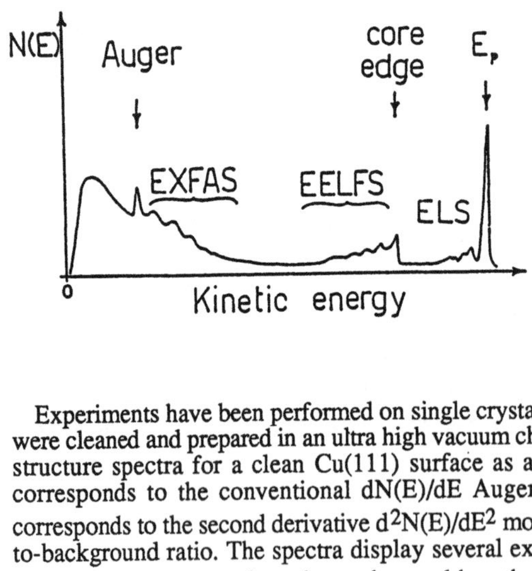

Fig. 4 : Schematic energy distribution N(E) of electrons backscattered from a surface. Primary energy E_p.

Experiments have been performed on single crystals, polycrystals and evaporated films. All samples were cleaned and prepared in an ultra high vacuum chamber. Figure 5(a) shows the extended fine Auger structure spectra for a clean Cu(111) surface as a function of the kinetic energy. The upper part corresponds to the conventional dN(E)/dE Auger mode (first derivative N'(E)). The lower part corresponds to the second derivative $d^2N(E)/dE^2$ mode (N"(E)) in order to improve the fine structure-to-background ratio. The spectra display several extended fine features far from the M_{23} VV Auger lines. These structures have been observed by other authors /14/ who associated these features to a diffraction effect and no quantitative structural analysis has been achieved.

In contrast, we assign these oscillating structures to the same interference process giving rise to EXAFS oscillations. Using therefore the EXAFS analysis procedure, the raw data were background substracted (dotted line in Figure 5(a)) and Fourier integrated choosing the E_0 threshold energy at the inflection point of the M_{23} VV first derivative Auger line. Figures 5(b) and 5(c) show the magnitude of the Fourier transforms F'(R) and F"(R) spectra of the N'(E) EXFAS and N"(E) EXFAS respectively. The first peak in the two F(R)'s is located at 2.20 ± 0.03 Å and once corrected for a ΔR due to the phase shift $\phi_j(k)$, should correspond to the 2.55 Å distance between the first nearest-neighbours atoms in the Cu(111) f.c.c. sample.

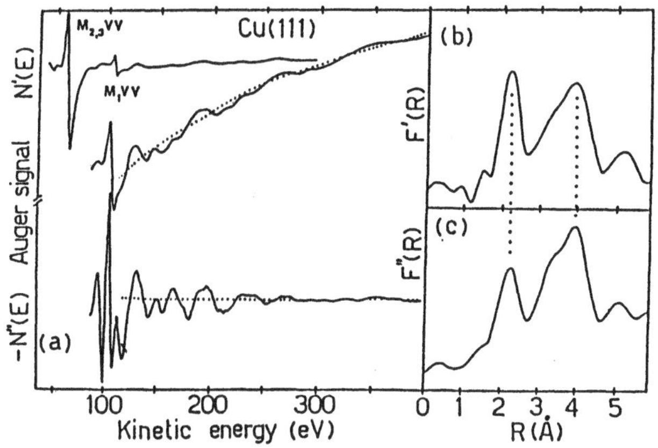

Fig. 5(a) EXFAS of Cu(111) detected in the first N'(E) and second N"(E) derivative mode. (b) F'(R) deduced from N'(E). (c) F"(R) deduced from N"(E).

Similar results for Cr, Fe, Co, Ni are also observed. Figure 6 shows the Fourier transform of EXFAS recorded for a cobalt film above the MVV Co Auger lines in both the first and second derivative mode. In order to compare with EXAFS /15/ and EELFS experiments /16/ above the same M_{23} edge, we also draw in Figures 6(c) and (6d) the corresponding F(R). The agreement is excellent and the first peak in the F(R) reflecting the first neighbours in the h.c.p. cobalt is located at 2.05 ± 0.03 Å for all three spectroscopic techniques (EXFAS, EXAFS and EELFS). Corrected for the phase shift, it gives exactly the 2.50 ± 0.03 Å distance as found from X-ray diffraction on the h.c.p. cobalt.

We summarize in Figure 7 the mechanisms underlying the physical origin of the fine structures observed in Auger spectra. Figure 7(a) recalls the well-known EXAFS mechanisms. Figure 7(b) explains the usual Auger process. In this CVV Auger transition, one (valence) electron fills an initial core hole and a second electron (Auger) is emitted from the solid.The kinetic energy of the escaping Auger electron is given by

$$E_k\,(CVV) \approx E\,(C) - E\,(V_1) - E\,(V_2) \qquad (5)$$

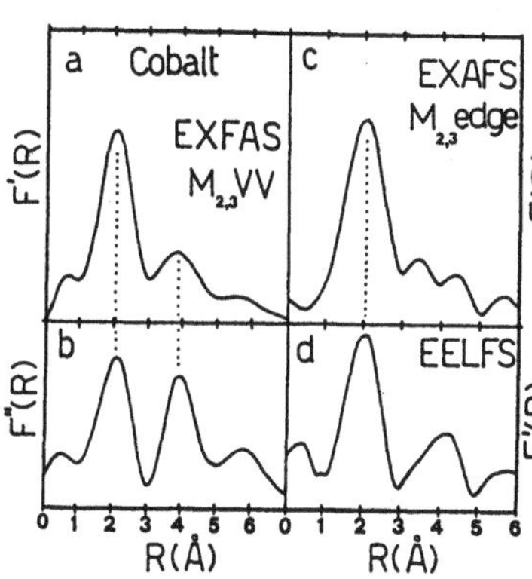

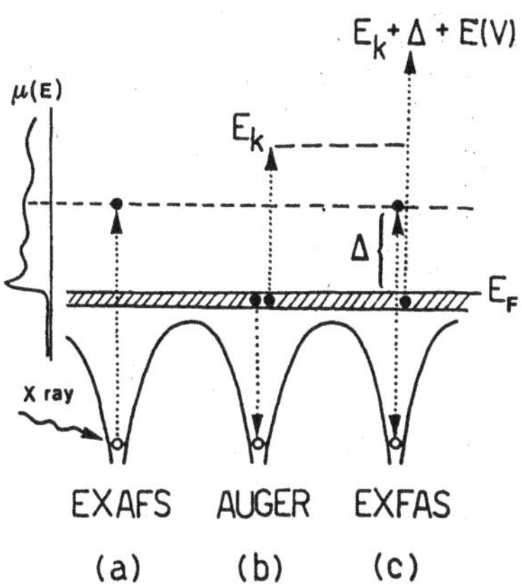

Fig. 6 : (a) and (b) F'(R) and F"(R) of the cobalt EXFAS above the $M_{2,3}$ VV transition deduced from N'(E) and N"(E). (c) F(R) of the synchrotron radiation EXAFS signal as reported in /15/ above the same edge. (d) F'(R) of the EELFS signal detected in the N'(E) mode. All three techniques provide the same nearest neighbours distance.

Fig. 7 : Schematic energy diagrams of the excitation process in EXAFS (a), Auger (b) and extended fine Auger structure (EXFAS) (c) spectroscopies involving the same core level. Note that the Δ states are filled by the core electrons excited by the primary electron beam.

neglecting the final state hole-hole interaction. The initial core hole in the Auger process was created under primary electron irradiation (energy E_p).

The ejected core electron could fill either the empty states Δ above E_F (excited states in the sample) or/and get out into the vacuum. Now if the first electron filling the initial core hole does not come from the valence band as in the CVV Auger process (Figure 7(b)) but from another state at energy Δ above E_F (Figure 7(c)) initially filled by the ejected core electron, then the escaping Auger electron will gain an energy $\Delta + E(V_1)$ according to

$$E_{k'}(C\Delta V) = E(C) + \Delta - E(V_2) = E_k + \Delta + E(V_1)\,. \qquad (6)$$

Moreover, if the excited Δ states of the first electron are EXAFS-like modulated in energy, then the measured Auger kinetic energy of the second (valence) electron will contain entirely the EXAFS structures. So we are probing the EXAFS-like interference effect experienced by the first electron in its excited state Δ through the emission of the second (valence) electron. All excited Δ states above E_F with an energy lower than $\Delta_{max} = E_p - E(C)$ can be filled by this technique. More detailed calculations will be published elsewhere /10/.

5. CONCLUSION

We have proposed here two simple in situ techniques to look at the local geometry of analyzed samples. Both offer good surface sensitivity, applicability to either ordered or disordered surfaces. Moreover they are relatively simple, laboratory-based apparatus and therefore deserve to be improved in the near future.

1. See lectures by F. Himpsel. This School.
2. E. Chaînet, M. De Crescenzi, J. Derrien, T.T.A. Nguyen, R. Cinti, Surface Science 168, 801 (1986).
3. M. De Crescenzi, E. Chaînet, J. Derrien, Solid State Comm. 57, 487 (1986).
4. See lectures by G. Rossi. This School.
5. See review paper by P.A. Lee, P.H. Citrin, P. Eisenberger and B.M. Kincaïd, Review Mod. Phys. 53, 769 (1981).
6. B.K. Teo, P.A. Lee, J. Am. Chem. Soc. 101, 2815 (1979).
7. A.G. McKale, G.S. Knapp, S.K. Chan, Phys. Rev. B33, 841 (1986).
8. F. Mila, C. Noguera, Proc. of the IV EXAFS Conf. (Fontevraud, France 1986), to be published in J. de Physique, Colloques.
9. M. De Crescenzi, Surface Science 162, 838 (1985).
10. J. Derrien, E. Chaînet, M. De Crescenzi, C. Noguera, Proc. of the IX ECOSS Conf. (Luzern, 1987), to be published in Surface Science.
11. J. Derrien, Surface Science 168, 171 (1986).
12. J. Derrien, F. Arnaud d'Avitaya, Proc. of the IVC-10, ICSS-6 Conf. (Baltimore, 1986), to be published in J. of Vac. Sci. Technol. A.
13. See lectures by E. Rosencher. This School.
14. L.McDonnel, B.D. Powell, D.P. Woodruff, Surface Science 40, 669 (1973).
15. M. Fanfoni, S. Modesti, N. Motta, M. De Crescenzi, R. Rosei, Phys. Rev. B32, 7826 (1985).
16. E. Chaînet, M. De Crescenzi, J. Derrien, Phys. Rev. B31, 7469 (1985).

On the Use of Electron Microscopy in the Study of Semiconductor Interfaces

J.-P. Chevalier

C.E.C.M.-C.N.R.S. 15, rue G. Urbain, F-94407 Vitry Cedex, France

Introduction

The electron microscope is, by essence, ideally suited to the study of localised microstructures such as interfaces. Formally, this is because the phase information of the electron waves scattered by the specimen is maintained in the formation of the image (e.g. see /1/).

In modern electron microscopes the spatial resolution attainable is now in the 0.2 nm range, in the high-resolution phase contrast mode of imaging. It is however essential to realise that this resolution is the resolution in the image, and not necessarily the resolution of the microstructural feature of interest in the specimen. Several problems intervene in the relation between the image and the object microstructure. Firstly, it is obvious that the image has only 2-dimensions, whereas the object is in 3-dimensions. Thus in the imaging operation, we inevitably average over the specimen thickness, and not necessarily in a linear manner. Furthermore, the image and object are related by the mode of contrast which is chosen by the operator. Although much work on interfaces has been carried out in the high-resolution mode, other contrast modes can, at a lower resolution, provide information on strain fields and on variations in composition, for example.

The aim of this extended abstact, which corresponds to 2 hours of over-dense lectures, is to point out several domains of application of electron microscopy to semiconductor interfaces and to provide the reader with a reasonable list of references for future reading. It should be realised that this field is actively researched and that furthermore there are very complete text books on electron microscopy in general (e.g. /2/) or more specifically on high resolution /3/ or analytical electron microscopy /4/. Recently some high-resolution results have been reviewed /5,6/ and Hutchison /5/ has proposed some useful classification for crystal -crystal semiconductor interfaces. The reader should also be aware of the excellent series of conference proceedings entitled "Microscopy of Semiconducting Materials", /7,8,9/.

Finally, and although this will not be further discussed, the specimen preparation in the " transverse section geometry ", whereby an electron thin region at the level of the interface, which in this geometry is now parallel to the electron beam, is obtained, is the absolute prerequisite for all subsequent work /10,11/.

Electron Microscopy and Interfaces

When specifically applied to interfaces, the electron microscope can provide information on the following microstructural features:

- epitaxial relations and misfit defects (e.g. dislocations and twins)
- interface structure, intermediate phases, interfacial precipitates
- phase transitions at interfaces
- concentration gradients perpendicular to the interfaces
- structure and morphology of very thin deposits.

These problems cannot necessarily all be solved at the same limit of resolution (0.2 nm) that can be reached in high-resolution electron microscopy. Notably, problems relating to concentration gradients are more difficult, and a resolution limit of several nanometres is more realistic.

Bearing in mind this list of problems, we can consider the different classes of interfaces, as previously proposed /5/. If we first consider crystal-crystal interfaces, the following possibilities can occur, with the two crystals having :
- same or similar structures and similar parameters (little or no misfit)
- same or similar structures and different parameters (misfit problems)
- different structures and similar parameters (epitaxy ?)
- different stuctures and different parameters (epitaxy and misfit).

The first case is typically that of quantum well structures and superlattices in systems such as GaAs/(Ga-Al)As. Here the problem we would like to solve concerns the interface morphology (roughness) and the possible concentration gradients (abruptness). Usual [110] high-resolution images do not yield any such information clearly, since the image contrast changes across the interface are slight /12,13/. Dark-field images (see /2/) using the {200} reflections, whose structure factor is dependent on the difference between the atomic scattering factors for (Ga-Al) and As, are more sensitive (e.g. /14,15/), as are the high-resolution images at the [100] orientation (/14/ and see Fig.1), for the same reasons. Using such images it is possible to evaluate interface roughness, but very difficult to have any accurate notion of abruptness. For the latter a new technique consists in measuring the displacement of thickness fringes (see /2/) across an interface, and this is very sensitive to the change in composition (/6,16/ and see Fig. 2).

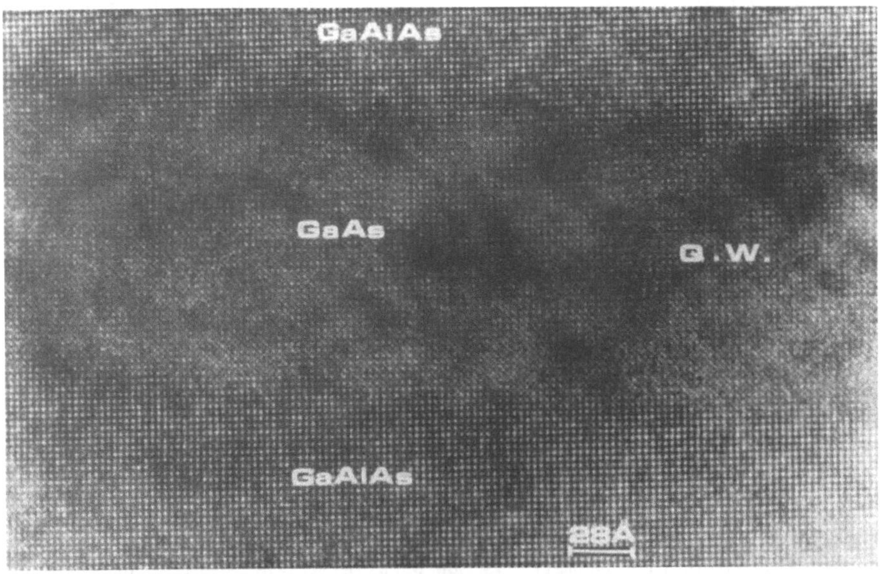

Figure 1 : [100] High-resolution electron micrograph of a GaAs/(Ga-Al)As quantum well, with an Al content of about 30%. Note that the interfaces appear rough, and that the (200) fringes are considerably weaker in the GaAs layer. Unpublished work, courtesy of P.Ruterana.

When we have similar structures, in epitaxial relation, but with different lattice parameters, the misfit that occurs has to be accommodated. An example where this occurs is the system Si/GaAs. Figure 3 , which is a "weak-beam" dark field image, shows such a case, and dislocations can be clearly seen to run through the Si layer /17/. Such contrast modes should not be neglected since they do not only demonstrate the presence of dislocations, but also enable their nature to be determined, their density to be estimated and they further provide some insight on the propagation of defects through the Si layer. Another example of this kind of problem is the system (Cd,Hg)Te/(100)GaAs, where the 15% mismatch is accommodated by a series of regularly spaced dislocations at the interface /5/.

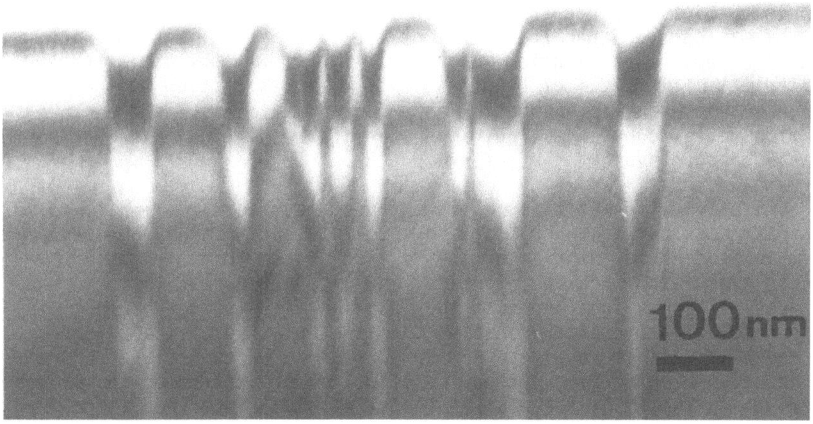

Figure 2 :Bright field image, beam along [100], of GaAs with irregular sequence of GaAlAs layers. The fringe position is very sensitive to composition changes. Micrograph courtesy of C.J.Humphreys (see /6/).

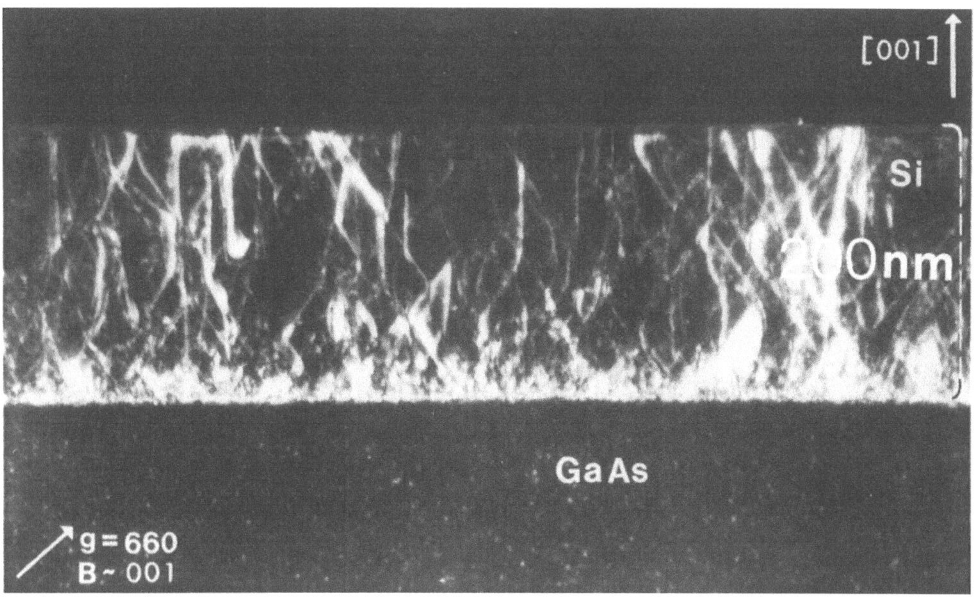

Figure 3 : Weak-beam dark field image, beam along [110] , of 200nm of Si on (100) GaAs. Dislocations start at the interface and are seen to run through the whole layer. Micrograph courtesy of C.W.T.Bulle-Lieuwma (see /17/).

A good example of the third class of interfaces concerns the problem of metal silicides on silicon. Here the structures are different, but the lattice parameters are close. Here, Cherns /18,19/ has been able to determine, through high-resolution electron microscopy together with image calculations, which model for the interfacial structure was more appropriate.

In the cases where both structure and parameters are different, high-resolution microscopy has been used on the systems CaF_2, BaF_2, and SrF_2 on various semiconductor substrates /20/. Here, depending on the misfit, the interface either contains regular dislocations, or appears to be incoherent. Another much studied system is that of Silicon on Sapphire, where for the $(100)Si//(01\bar{1}2)Al_2O_3$ with $[001]$ Si// $2\bar{1}\bar{2}0$ Al_2O_3 epitaxial relation, microscopy has shown that misfit can be relieved by either twinning /21/ or by possible bond reconstruction at the interface /22/. Further work has shown that a perfect Si lattice can be restored by Si implantation followed by annealing, and here high resolution has revealed the presence of an amorphous layer at the interface./23/.

Finally, a completly different class of interfaces also has to be mentioned. These are the interfaces between a crystalline substrate and an amorphous layer. This has been much studied in the case of the Si/SiO_2 system (e.g. /24,25,26/) Figure 4 shows an example of such an interface, and it can be seen that the residual roughness is very slight and that the interface is also very abrupt. In such cases high-resolution microscopy enables the measurement of oxide thickness to be readily made, as well as roughness and abruptness to be appreciated. More thorough work has also been carried out on the effect of projection on the apparent roughness perceived in the image, suggesting that the actual interface is rougher.

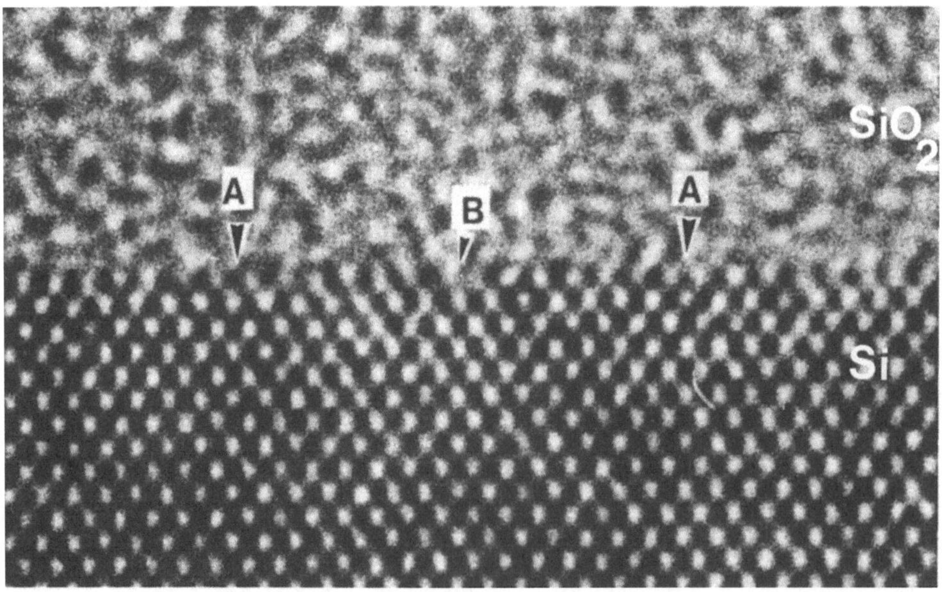

Figure 4 : High-resolution image of a Si/SiO_2 interface. Micrograph courtesy of C.d'Anterroches (see /25/).

Recently results have been obtained on the GaAs/oxide /27,28/, on the InP/SiO_2 /29/ and on the $GaAs/Si_3N_4$ /28,30/ systems. For these systems, the surface preparation appears to determine the nature of the interface. It is also worth mentioning the work of Krivanek and Lilienthal who have shown /31/ that it was possible to apply microanalytical techniques to this kind of problem ($InP/oxide/SiO_2$).

Concluding Remarks

It should, hopefully, be apparent that electron microscopy is a powerful tool for the investigation of the microstructure of interfaces. Most of the artefacts linked to specimen preparation and to electron imaging are now very well known. The point which is crucial to all these studies lies in the mastering of the preparation of cross-sectional specimens.

References

1. Principles of Optics, M.Born and E.Wolf, Pergamon Press (1975).
2. Electron Microscopy of Thin Crystals, P.B.Hirsch,A.Howie,R.B.Nicholson, D.W.Pashley and M.J.Whelan, R.E.Krieger (New York, 1977).
3. Experimental High Resolution Electron Microscopy, J.C.H.Spence, Clarendon Press (Oxford, 1981).
4. Introduction to Analytical Electron Microscopy, J.J.Hren,J.I.Goldstein and D.C.Joy, Plenum Press (New York, 1979).
5. J.L.Hutchison Ultramicroscopy 15 51 (1984).
   Ultramicroscopy 18 349 (1985)
6. C.J.Humphreys p.105 of Electron Microscopy 1986, ed T.Imura,S.Maruse and T.Suzuki, pub. Jap. Soc. Electron Microscopy (Tokyo,1986)
7; Microscopy of Semiconducting Materials 1983, Conference Series N° 67 pub. I.o.P. (London, 1983)
8. Microscopy of Semiconducting Materials 1985, conference Series N° 76 pub. I.o.P. (London, 1985).
9. Microscopy of Semiconducting Materials 1987, to be published.
10. M.S.Abrahams and C.J.Buiocchi Appl.Phys.Lett. 27 235 (1975).
11. M.Dupuy J.Microsc.Spectrosc.Electron. 9 163 (1984)
12. H.Ichinose,T.Furuta,H.Sakaki and Y.Ishida, p.1485 of Electron Microscopy 1986 ed. T.Imura,S.Maruse and T.Suzuki, pub. Jap.Soc. Electron Microscopy (Tokyo,1986)
13. N.Tanaka and K.Mihama, p.1493 of Electron Microscopy 1986 (see ref /12/).
14. H.Ichinose,T.Furuta,H.Sakaki and Y.Ishida, p.1483 of Electron Microscopy 1986 (see ref. /12/).
15. B.C.de Cooman,N.-H.Cho,Z.Elgat and C.B.Carter Ultramicroscopy 18 305 (1985).
16. H.Kakibayashi and F.Nagata Jap.J.Appl.Phys. 25 1644 (1986).
17. C.W.T.Bulle-Lieuwma,P.C.Zalm and M.P.A.Viegers, p.123 of ref./8/.
18. D.Cherns,G.R.Anstis,J.L.Hutchison, and J.C.H. Spence Phil.Mag. A46 849 (1982).
19. D.Cherns,C.J.D.Hetherington and C.J.Humphreys Phil.Mag. A49 165 (1984).
20. J.M.Gibson, Ultramicroscopy 14 1 (1984).
21. J.L.Hutchison,G.R.Booker and M.S.Abrahams, p.139 of Inst.Phys.Conf.Series N°61 I.o.P. (London, 1981).
22. F.A.Ponce Appl.Phys.Lett. 41 371 (1982).
23. D.J.Smith,L.A.Freeman,R.A.McMahon,H.Ahmed,M.G.Pitt and T.B.Peters, p.83 of ref. /7/ (1983).
24. O.L.Krivanek and J.H.Mazur Appl.Phys.Lett. 37 392 (1980).
25. C.d'Anterroches J.Microsc.Spectrosc.Electron. 9 147 (1984).
26. S.M.Goodnick,D.K.Ferry,C.W.Wilmsen,Z.Lilienthal,D.Fathy and O.L.Krivanek Phys.Rev. B 32 8171 (1985).
27. O.L.Krivanek and S.L.Fortner Ultramicroscopy 14 121 (1984)
28. P.Ruterana,J-P.Chevalier,P.Friedel and N.Bonnet to be published in ref./9/.
29. Z.Lilienthal,O.L.Krivanek,J.F.Wagner and S.M.Goodnick Appl.Phys.Lett. 46 889 (1985).
30. P.Ruterana,P.Friedel,J.Schneider and J-P.Chevalier Appl.Phys.Lett. 49 672 (1986).
31. O.L.Krivanek and Z.Lilienthal Ultramicroscopy 18 355 (1985).

Analytical Scanning Electron Microscopy Under Ultra High Vacuum

P. Morin

Département de Physique des Matériaux, Université Lyon I,
43, Boulevard du 11 Novembre 1918, F-69622 Villeurbanne Cedex, France

1. INTRODUCTION

The usual techniques of electron beam surface investigation such as AES, LEED, RHEED, ELS have been improved in recent years by important developments ; these techniques can now be performed in an actual Ultra High Vacuum Scanning Electron Microscope (UHV-SEM) with a focused beam at the submicronic scale. The flexibility of the optical column of the microscope gives suitable beams allowing us to perform these techniques in the same area with appropriate detectors. A UHV chamber makes it possible to observe clean surfaces without contamination. To enhance surface information, low beam energy must be used. But it is well known that for an optical system when the beam size or the beam energy decrease the beam intensity decreases too ; compared with macroscopic investigation higher performance guns and more efficient detectors are therefore needed. Another method to improve the detector is by using low beam intensity to reduce the irradiation damage by the electron beam. These techniques must be consistent with in situ preparation facilities e.g. ionic bombardment, specimen heating, fracture attachment, gas introduction devices, evaporating systems... So a preparation chamber is often coupled to UHV-SEM. The purpose of this paper is to point out the new possibilities given by UHV-SEM at medium and low beam energy ; the reflected modes of conventional STEM /1/ or TEM /2/ are not developed. However, the field of investigation of these techniques is very wide, involves the topographic, chemical, electronic, crystallographic observations ; so we will limit our study to the focused aspects in surface applications. In the present paper the surface information is examined according to three principal modes : secondary, spectroscopic (Auger, EELS) and diffracting (RHEED, LEED, low losses).

2. SECONDARY MODE

Almost all secondary electrons coming from the surface reach the detector, these electrons having a very low energy of some eV, the efficiency of the detection is thus good. Moreover, the secondary intensity is in the same range as the beam intensity. Consequently the secondary mode can be performed with a low beam intensity (10^{-10} - 10^{-13} A). The spatial resolution in this mode is given by the beam diameter. The depth concerned by the secondary emission is about 100Å below the surface. The contribution of the surface can be increased by using very low incident electrons whose penetration depth is smaller than the escape depth of secondary electrons ; at 100eV the mean free path of electrons is minimum and close to 4Å. The secondary signal contains a lot of information (topographic, chemical, crystalline, potential, ...) it is thus well appropriate to exhibit surface inhomogeneities, but the determination of the origin of the contrast is not always easy. This mode is not suitable for quantitative approach.

2.1 The topographic contrast is the most currently used in secondary mode. For surface studies this contrast enables us for example to observe three-dimensional growth as soon as the crystals reach a size in the 100Å range. VENABLES and co-workers /3, 4/ have studied Ag crystal growth on Si. They have shown that the crystal shapes are very different on Si(100) and Si(111). SEM observations as a function of the temperature and deposition rate can be interpreted in terms of nucleation theory /5/. From the shape of microcrystals of Au/graphite, Pb/graphite and Pb/Ge(111) surface energy anisotropy and adhesion energy have been studied /6, 7/. At low beam energy very fine topographic structures can be displayed. ICHINOKAWA /8/ and co-workers observed steps of several monolayers on Si(111) with a beam energy of 250eV and a beam size of 600Å.

2.2 The chemical contrast results from the change of the secondary yield according to the material. By this effect, it is possible to observe superlattice interfaces structure of III-V semiconductors /9/. Very thin layers (< 100Å) are observed after surface etching following a bevel angle prepared by chemical etching or ionic bombardment ; a bevel magnification of one thousand can be reached. BRESSE /9/ observed 100Å layer of InGaAsP in a InP matrix on a chemical bevel and 150Å layers of GaAs/GaInAs superlattice on an ionic bombardment bevel. Chemical contrast based on the variation of the work function has been developed by FUTAMOTO et al /10, 11/ to make surface layers visible. This technique consists of biasing the sample around -500V and using the normal secondary electron collector. The sensitivity is better than 0.1ml for Ag evaporated through a mask of holes on a Si(111) substrat.

2.3 Crystallographic contrast in secondary mode in a usual SEM is due to the electron channeling effect /12/. This contrast is weak, some percentage points of the secondary signal. In UHV-SEM a strong crystalline contrast appears on clean surfaces in two situations : with a beam in glancing incidence and with a beam of very low energy. In the case of beam in glancing incidence HEMBREE and COWLEY /13/ observe an intense contrast between grains of various orientations on a clean pyrolytic graphite with a beam energy of 15keV and a beam diameter of 500Å. These authors explain this contrast by a surface resonance stage mechanism.

The second case of strong crystallographic contrast has been demonstrated by ICHINOKAWA et al /14/ on a surface of Al polycrystalline specimen after cleaning (ionic bombardment and annealing). The highest contrast is obtained with a beam energy of 230eV. This contrast is interpreted in terms of penetration depth of the incident beam caused by the electron channeling effect. These methods are useful to reveal crystallographic disorientations of the surface with a sensitivity better than 1°.

3. SPECTROSCOPIC MODE (Auger, ELS)

3.1 Auger Spatial Resolution

The high spatial resolution is the main advantage of Auger spectroscopy over the other surface techniques such as XPS or UPS, ... In usual scanning conditions the Auger spatial resolution depends essentially of two parameters : the beam diameter and the escape area of backscatterd electrons. Indeed the contribution of backscattered electrons to the Auger signal is far from negligible. At the first approximation, the backscatter escape area can be estimated, by supposing that the diameter of this area is equal to the half of the range R of the incident electrons /15/. From HOLLIDAY and STERNGLASS /16/ $R = (0.05/\rho)\ E^{1.65}$ (μm), where ρ is the volumetric weight (g/cm^3 and E is the beam energy (keV). To reduce this broadening effect lower energy beam must be used, but unfortunately the beam size is wider. The beam diameter given by an optical system can be calculated according to the energy beam, the aberration coefficients, brightness of the gun, the beam intensity and the focusing distance /17/. From the expression of the beam diameter and of backscatter escape area as functions of the beam energy we can deduce an optimal beam energy given the smaller analytical spatial resolution /17/. The best result is given by a probe-forming system equipped with a field emission gun, magnetic lenses and the smaller realistic focusing distance of 1cm. On copper sample, which is a sample of medium density, the spatial resolution is about 500Å with a beam energy of 5keV and a beam intensity of 10nA. If the beam intensity could be reduced down to 1nA ultimate resolution close to 100Å would be reached.

3.2 Auger Images

Auger images are obtained by modulating the brightness of each point according to the amplitude of a selected Auger peak. Images are collected at two energies : A on the peak and B on the background at higher energy ; the resulting image is given by A-B. A more suitable procedure has been proposed by HARLAND and VENABLES /18/ : three images are recorded, A and B as previously then C equally spaced on the background at higher energy. This supplementary image allows us to extrapolate the background under the peak with the algorithm : A - 2B + C. The topographic effects can be reduced by dividing the Auger image by the background image. The corresponding signal processes are respectively (A-B)/(A+B) and (A-2B+C)/(2B-C).

3.3 Auger energy shifts

Though the chemical shifts in AES (1-10eV) are greater than in XPS, this effect has been principally used in XPS thanks to a high-resolution spectrometer. But now, as the inadequate CMA is being increasingly replaced by a better analyser on the Auger systems, it is becoming possible to display submicronic modifications of the chemical binding on the surface.

The measure of the energy of Auger peaks is also sensitive to the position of the Fermi level in the gap of a semiconductor, since the analyser is referenced to the Fermi level of the sample. This effect has been used by PANTEL et al /19/to observe the potential contrast on unbiased p-n junctions of micro-electronic devices, with an accuracy within 20meV. The junction depth is measured with a resolution of 150Å. It is possible to study the surface band bending by this method.

3.4 REELM

Reflected electron energy loss microscopy /20/ is an imaging technique, whose contrast mechanism is based on the variation in the electron energy-loss spectra across the surface. The electron beam energy must be compatible with the electron spectrometer possibilities, thus in the 0.1 - 2.5keV range. The principal characteristics of this method are : i) a lateral resolution limited only by the incident electron beam spot ; ii) a depth resolution that can be varied from one monolayer to 30Å by changing the beam energy. The physical information given by low electron energy loss spectroscopy is of the same nature as that obtained by fast electron loss spectroscopy in Scanning Transmission Microscopy /21/. The spatial resolution is not so good but solid sample can be observed. REELM applications have been demonstrated by BEVOLO /20/ in the field of microanalysis of hybrids (SiH_2, GdG_2n ThH_2, ...) with a spatial resolution of 1000Å.

4. DIFFRACTING MODE

4.1 Micro-RHEED

RHEED observations can be performed in an UHV-SEM by adding a screen in the column axis. The beam is focused on the sample and the illumination beam angle 2α give the angular resolution. In these conditions the beam diameter d on the sample can be expressed as

$$d \simeq \frac{4}{\pi} \frac{1}{\alpha} \sqrt{\frac{I}{\beta}}$$

where I is the beam intensity and β the beam brightness. This expression shows that the possibilities of observing small areas depend essentially on of the beam brightness therefore on the type of gun used. We have d = 0.4μm for a field emission gun and with I = 100nA and α = 10^{-3}. Using a channel plate multiplier the beam intensity can be reduced to lower than 0.1nA /8/ ; the beam diameter is then limited by the optical properties of the column in the some 100Å range. The detector used by COWLEY and co-workers /22/ makes it possible to select one diffracted beam, the corresponding diffracting area can thus be displayed by scanning the beam across the surface. Such observations have been performed by HEMBREE and COWLEY /13/ with a minimum probe size of 300Å.

4.2 Micro-LEED

As for RHEED technique-focused LEED is obtained in UHV-SEM with a focused beam and an appropriate detector. This technique has been recently developed by ICHINOKAWA et al /8/. Thanks to a high spatial resolution at low energy (600Å at 250eV) they have observed atomic steps on Si(111) on images from a diffracted beam. The crystal substructures and orientations of grains can be determined with a spatial resolution of 500Å. The co-existence of different substructures on an Au evaporated surface of Si(111) has been observed with the same resolution.

4.3 Electron Channeling Imaging

ECI /23, 24/ is a method used to display emergent crystalline defects at the surface on solid samples in a SEM. The origin of the contrast is a consequence of the channeling effect. The channeling state of the beam is changed by the bending of the lattice planes near the crystalline defects. To increase the contrast we need to select the high energy backscattered electrons by means of an energy filter. A field emission gun is also needed to obtain good resolution (< 100Å) with a beam intensity of some nA. By this method it is possible to observe individual defects such as dislocations, stacking faults or precipitates.

5. CONCLUSION

Analytical scanning electron microscopy under UHV involves the focused and scanned electron beam surface techniques. The most are developed from macroscopic electron beam surface techniques such as LEED, RHEED, AES, ELS. They make possible investigation of the topographic, chemical, electronic and crystallographic nature of the surface at the submicronic scale. To date there is no commercial apparatus capable of performing all these techniques, though they are compatible. They are developed in improved Auger microprobes or in dedicated laboratory apparatus.

1. J.M. Cowley: SEM 1982/I, p. 51
2. K. Yagi: SEM/1982/IV, p. 1421
3. J.A. Venables, D.R. Batchelor, M. Hanbücken, C.J. Harland and G.W. Jones: Phil. Trans. R. Soc. Lond. A 318, 243-257 (1986)
4. M. Hanbücken, T. Doust, O. Osasona, G. Le Lay and J.A. Venables: Surf. Sci. 168, 133-141 (1986)
5. J.A. Venables, G.D.T. Spiller and M. Hanbücken: Rep. Prog. Phys. 47, 399-459 (1984)
6. J.C. Heraud and J.J. Metois: Surf. Sci. 128, 334-349 (1983)
7. J.J. Metois and G. Le Lay: Surf. Sci. 133, 422-442 (1983)
8. T. Ichinokawa, Y. Ishikawa, M. Kemmochi, N. Ikeda, Y. Hasokawa and J. Kirschner: Surf. Sci. 176, 397-414 (1986)
9. J.F. Bresse: SEM 1985/IV, p. 1465
10. M. Futamoto, M. Hanbücken, C.J. Harland, G.W. Jones, J.A. Venables: Surf. Sci. 150, 430-450 (1985)

11. J.A. Venables: Proc. XIth Int. Cong. on Electron Microscopy, Kyoto, p.75 (1986)
12. L. Reimer: in Scanning Electron Microscopy, Springer Ser. in Optical Sciences, Vol.45 (Springer, Berlin, Heidelberg, 1985) p. 341
13. G.G. Hembree and J.M. Cowley: SEM 1979/I p.145
14. T. Ichinokawa, C. Legressus, A. Mogami, F. Pellerin and D. Massignon: Surf. Sci. 111, L675-L679 (1981)
15. J.I. Goldstein and H. Yakowitz: in Pratical Scanning Electron Microscopy (Plenum Press, New-York, 1975) p.60
16. J.E. Holliday and E.J. Sternglass: J. Appl. Phys., 30, 1428 (1959)
17. P. Morin: J. ofElec. Spect. and Related Phenomena, 37, 171-179 (1985)
18. C.J. Harland and J.A. Venables: Ultramicroscopy 12, 9-18 (1985)
19. R. Pantel, F. Arnaud d'Avitaya, Ph. Ged and D. Bois: SEM 1982/II p.549
20. A.J. Bevolo: SEM 1985/IV p.1449
21. A.G. Nassiopoulos and J. Cazaux: Surf. Sci. 165, 203-220 (1986)
22. J.M. Cowley, G.C. Albain, G.G. Hembree, P.E.H. Neilsen, F.A. Kock, J.D. Landry and H. Schuman: Rev. Scient. Instrument, 46, 826-829 (1975)
23. P. Morin, M. Pitaval, D. Besnard, G. Fontaine: Phil. Mag., 40, 511 (1979)
24. P. Morin, M. Pitaval, M. Tholomier and G. Fontaine: J. Microsc. Spectrosc. Electron., Vol. 6, 257-265 (1981)

Scanning Tunneling Microscopy and Spectroscopy

F. Salvan

Laboratoire de Physique des Etats Condensés, U.A. CNRS 783,
Faculté des Sciences de Luminy, Département de Physique,
Case 901, F-13288 Marseille Cedex 9, France

Scanning Tunneling Microscopy (S.T.M.) is a real-space surface imaging method with atomic or subatomic resolution in all three dimensions which has been developed by G. BINNIG and H. ROHRER. Besides imaging the surface topography, it also provides useful information on its electronic properties as well as spectroscopic images. The principle and operation modes of tunneling microscopy and spectroscopy will be described and illustrated by applications to Si surfaces and interfaces.

1. INTRODUCTION

In the last few years, Gerd BINNING and Heinrich ROHRER have developed Scanning Tunneling Microscopy (S.T.M.) into a powerful high-resolution surface probe /1/. This new physical tool allows real-space imaging of geometric and electronic properties of surfaces where details on the atomic scale are revealed. Despite the large number of potential applications, we will focus here on surface and interface physics which is the topic of this Winter School. Because of the very local character of the analysis, S.T.M. is complementary of most of the surface sensitive techniques which probe either the geometric or the electronic structures of surfaces while averaging the information on a large number of atoms ($\sim 10^9$). In the following chapters, I will describe first the principle of S.T.M. operation on the basis of a one-dimensional model of tunneling,and then discuss a more refined theoretical approach which is the basis of tunneling spectroscopy. The illustrative examples will concern Si(111) surfaces and metal overlayers.

2. S.T.M. OPERATION

The underlying physical basis of S.T.M. operation is vacuum tunneling of electrons between two metallic electrodes brought in close proximity. Tunneling allows electrons to penetrate into classically forbidden regions of space where the particle potential energy exceeds its total energy. This phenomenon is a consequence of the wave-like nature of electrons and is common on the atomic scale. Let us briefly recall the conclusions of the well-known one-dimensional model depicted in Fig. 1.

2.1. One-Dimensional Model of a Tunnel Barrier

The electrodes with characteristic work functions ϕ_1 and ϕ_2 are separated by a distance s sufficiently small that electron wave functions overlap in the classically forbidden vacuum gap. A bias voltage V allows electrons to tunnel from occupied states of the negatively biased electrode to empty states of the counter electrode.

Calculation of the current I in the limit of low voltage V and low temperatures T gives

$$I \propto V \exp - Ks$$

where the inverse decay length K is

$$K = \frac{2m}{\hbar}\left(\frac{\phi_1 + \phi_2}{2}\right)^{1/2} .$$

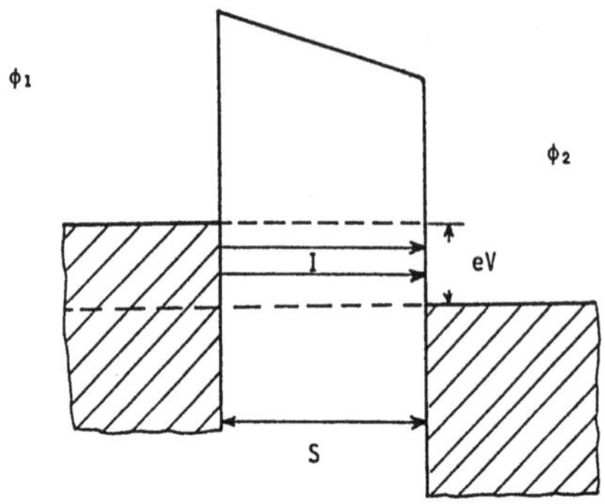

Fig. 1 One-dimensional model of a tunnel barrier

For usual ϕ(∿4 eV), a variation Δs ∿ 1 Å changes I by a factor of 10. The first convincing experimental realization of electron tunneling in these conditions was reported in 1982 by BINNING, ROHRER, GERBER and WEIBEL /2/. The extreme sensitivity of the tunnel current as a function of the distance is the basis of S.T.M. operation where one of the electrodes is a non-isolating sample and the other electrode a fine metallic tip.

2.2. S.T.M. Operation

In S.T.M., an atomically fine metallic tip is brought in close proximity (5-10 Å) to the sample and biased in order to allow electron tunneling across the potential barrier. When the tip is scanned along the surface, the overlaps of the surface and tip wave functions vary from place to place and cause change in the tunnel current I. Monitoring these variations during the scan allows real-space imaging of the geometric and electronic properties of the surface. A schematic view of the experimental set-up is shown in Fig. 2.

The tip is mounted on a system of piezotransducers. Motions parallel to the surface (x,y) and normal to it (z) are acted by voltages V_x, V_y and V_z. This schematic view shows how the constant current imaging mode is achieved via the control of the z motion by a feedback loop C,so that the current I is kept constant during the scans. Three-dimensional images of the topography z(x,y) of the surfaces are displayed by mappings of $V_z(V_x,V_y)$ as shown in the insert.

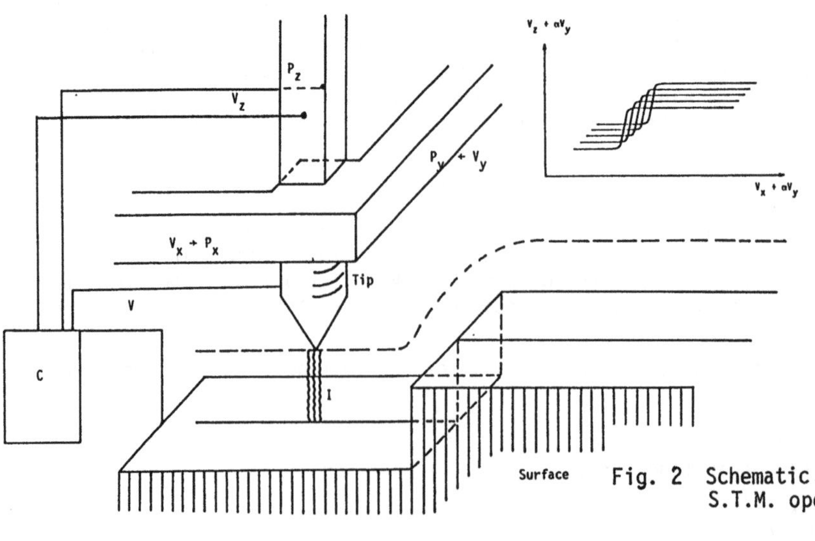

Fig. 2 Schematic diagram of S.T.M. operation

Other imaging modes can be used. They are discussed in a review article by HANSMA and TERSOFF /3/. A series of experimental realizations are described by BINNIG and ROHRER /1/.

2.3. The Si(111) 7x7 Surface

A famous example of surface structure determination by tunneling microscopy is the Si(111) 7x7 reconstruction. BINNIG, ROHRER, GERBER and WEIBEL published in 1983 a now well-known real space image of the surface unit cell /4/. The convergence of many surface-sensitive techniques allows now some quasi-definite conclusions about this surface, the main one being the validity of the dimer adatom stacking fault model proposed by TAGAYANAKI /5/.

Two grey tone pictures of the 7x7 reconstruction obtained on the same sample for different tunneling conditions by BERGHAUS, BRODDE, NEDDERMEYER and TOSCH /6/ are presented in Fig. 3a (sample voltage + 2 V) and Fig. 3b for the reverse polarity. Although the general characteristics of the 7x7 unit cell are visible (twelve adatoms, deep minima at the corners of the cell, secondary minima along the edges...), some details like the inequivalence of the right and left halves of the cell are enhanced in Fig. 3b for electron tunneling from sample to tip. This illustrates an important point to realize at this stage, namely that *STM images depend on tunneling conditions, so that images of a given surface obtained in different tunneling conditions may display different features which cannot be interpreted in the simple 1D model we have mentioned.* A theory taking into account the special geometry of the tunneling conditions is required. We also expect that the transmittivity of the tunnel barrier depends on the overlap of the electron wave functions in the vacuum gap, so that a complete theory will have to take into account the electronic structures of the tip and the surface.

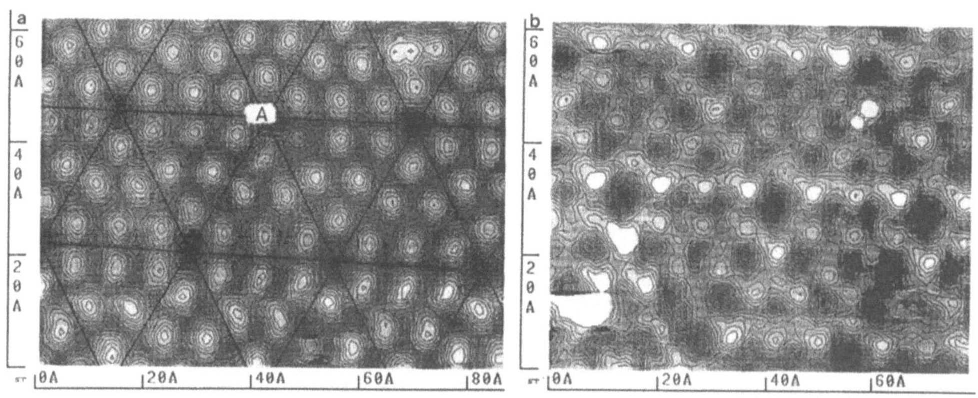

Fig. 3 (a) S.T.M. image of Si(111) 7x7 at sample bias U = + 2 V. The differences of the lines of constant height are 0.2 Å.
(b) S.T.M. image of the same sample obtained at sample bias U = - 2 V.

Courtesy of BERGHAUS, BRODDE, NEDDERMEYER and TOSCH, University of Bochum.

3. THEORY

We will briefly comment on TERSOFF and HAMAN /7/ adaptation of the Bardeen transfer Hamiltonian method /8/ to the S.T.M. geometry. This approach provides a good insight on the physics of S.T.M. imaging.

The tunnel current is calculated in terms of electronic transitions between the unperturbed eigen states of the right and the left electrode coupled by a current operator. TERSOFF and HAMAN considered a particular model for the tip, assuming it spherical and characterized by a radius R and a center of curvature at $\vec{r}_0$. The surface wave functions describe motion parallel to the surface as well as the evanescent wave corresponding to motion perpendicular to the surface.

In the limits of low voltages and temperatures, the tunneling current is proportional to the surface local density of states (LDOS) ρ_S at E_F evaluated at $\vec{r}_0$

$$I \propto V \rho_S(\vec{r}_0, E_F)$$

Generalization for tunneling at higher bias voltages, necessary if one wants to interpret and compare images obtained with different tunneling conditions, suggests that the differential conductivity $\partial I/\partial V$ will reflect the local density of states at the surface at E_F - eV, evaluated at $\vec{r}_0$, as this is schematized in Fig. 4.

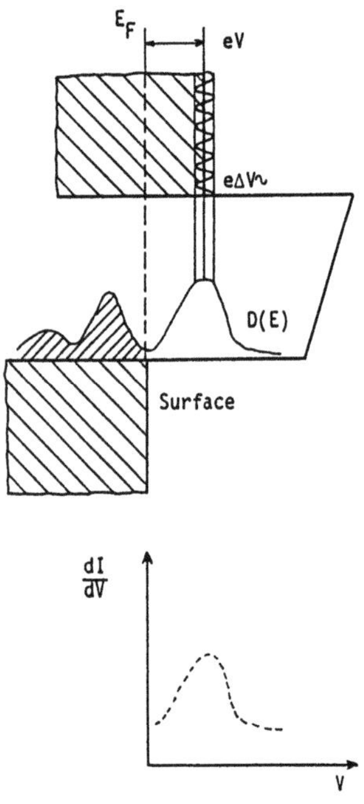

LANG /9/ has shown that it is in fact $\frac{\partial \text{ Log } I}{\partial V}$ which reflects the LDOS, as it has been suggested and evidenced by FEENSTRA /10/. *Thus S.T.M. images obtained in given tunneling conditions (I,V) are in fact contour maps of the local density of electronic states at E_F - eV.* This result constitutes a good theoretical support for the development of tunneling spectroscopy (T.S.) and scanning tunneling spectroscopy (S.T.S.), first impulsed by BINNIG et al. /11/ and followed more recently by HAMERS et al. /12/, through the development of Current Imaging Tunneling Spectroscopy (C.I.T.S.). Spectacular spectroscopic images of surface electronic features have been achieved through these approaches.

Fig. 4 Principle of tunneling spectroscopy

4. TUNNELING SPECTROSCOPY AND SPECTROSCOPIC IMAGES

Conventional tunneling spectroscopy has brought very useful information on the electronic and chemical properties of the tunneling electrode /13/. The advantages of performing T.S. with a tunneling microscope are multiple, since it allows a determination of the local density of both occupied and empty electronic states with atomic resolution. It can be achieved in well-defined tunneling conditions and finally provides spectroscopic images.

4.1. Constant Current Mode Spectroscopy

$\frac{\partial I}{\partial V}$ spectra are obtained by slowly sweeping the voltage V with a superposed small and fast modulation ΔV. The resulting current modulation is detected by a lock-in amplifier as in conventional tunneling spectroscopy. This is schematized in Fig. 4. Since the current is kept constant while the ramp voltage is swept, the gap distance s changes and s(V) characteristics can be simultaneously recorded. Typical spectra have been obtained by BINNIG et al. /11/ on a number of surfaces. They display general features of the density of empty electronic states and the relation with inverse photoemission has been discussed.

For tunneling through surface states with energy $E < \phi$ (work function), empty electronic surface states can be detected. ϕ is the onset of a series of oscillations obtained for $E > \phi$. They are discussed in terms of geometric resonances of the electrons along their path between the surface potential and the image potential. SALVAN et al. /13/ have shown how T.S. can be sensitive to the presence of an adsorbate and its ordering at the surface as illustrated in Fig. 5.

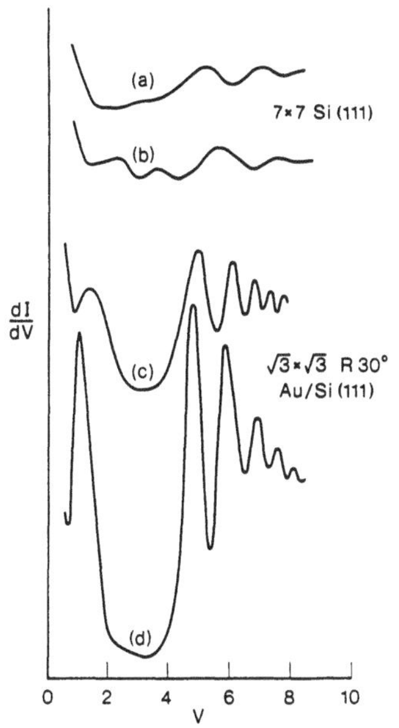

This has been done for Au deposited on a 7x7 Si(111) surface and the relation with inverse photoemission has been studied by NICHOLLS et al. /14/. BECKER et al. /15/ have reported the different electronic structures of the two halves of 7x7 unit cell which reflects the existence of the difference of stacking in the two halves of the reconstructed cell.

Fig. 5 Tunneling spectra of (a) and (b) a Si(111) 7x7 surface ; (c) Au/Si(111) $\sqrt{3}x\sqrt{3}$ R(30°) ; (d) Au/Si(111) 6x6.

4.2. Scanning Tunneling Spectroscopy (S.T.S.)

It is possible to get an image of the spatial distribution of a given electronic structure of the LDOS at a well-defined energy $E_0 = eV_0$ by mapping $\partial I/\partial V$ (x,y) contours corresponding to the bias potential V_0. This has been done by BINNIG et al. who have reported the first image of the localization of the Si dangling bond (D.B.) orbitals assigned to the structure at - 0.8 V of the direct photoemission spectrum /11/. This is shown in Fig. 6 where a comparison of the S.T.M. and the S.T.S. images at different voltages shows how the D.B. are distributed on the rest atoms at the surface.

A major drawback of tunneling spectroscopy performed in the constant current mode is related to the variation in gap distance when the voltage V is changed. Also spectroscopic images acquired at various bias voltages contain a mixture of geometric and electronic structure information since the tip follows different contours for different voltages. In order to circumvent this drawback, other spectroscopic modes have been developed.

4.3. Current Imaging Tunneling Spectroscopy (C.I.T.S.)

These methods tend to acquire electronic images at a given tip to surface distance. The feedback control loop used for constant current mode is thus turned off part of the time by means of a sample and hold circuit. When the feedback loop is active, a constant voltage V_0 is applied to the sample ; when inactive, the position of the tip is held stationary, while the tunneling current I is measured at various different bias voltages. Thus I(v) characteristics at each point of the topographic image corresponding to the appropriate regulating voltage V_0 can thus be obtained. So one acquires simultaneously current images at different gap voltages and at the same tip to surface distance as well as a topographic image.

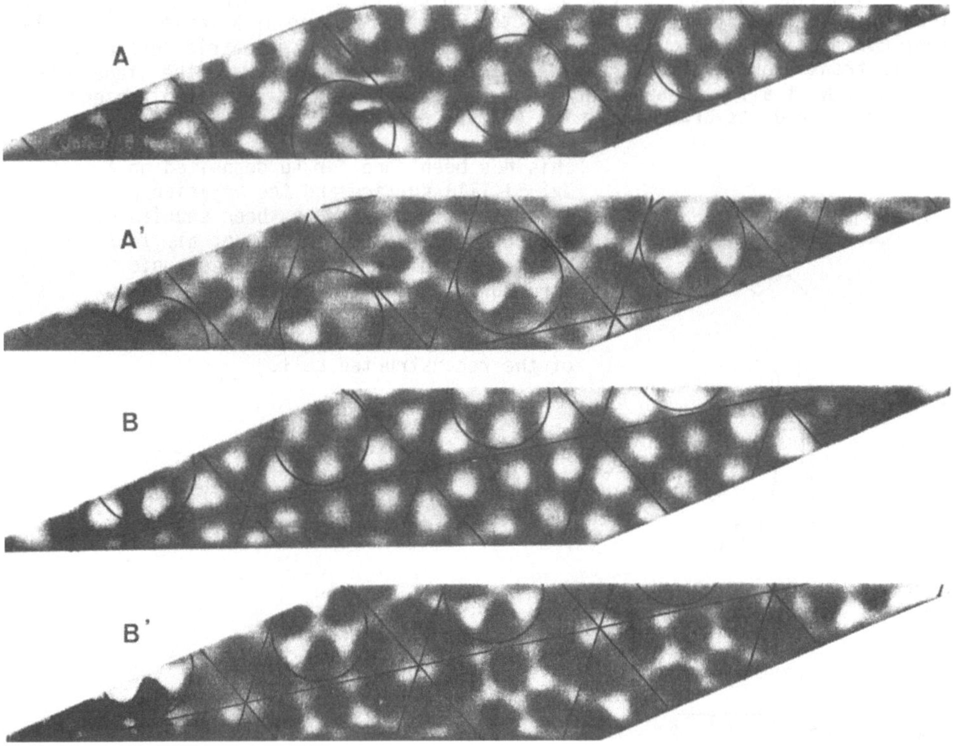

Fig. 6 S.T.M. images of a 7x7 Si(111) surface at sample bias voltages - 1 V (A) and - 0.81 V (B), and the corresponding S.T.S. images A', B' illustrating the localization of the D.B. states in the unit cell.

Courtesy of G. BINNIG and H. ROHRER, IBM Zürich Lab.

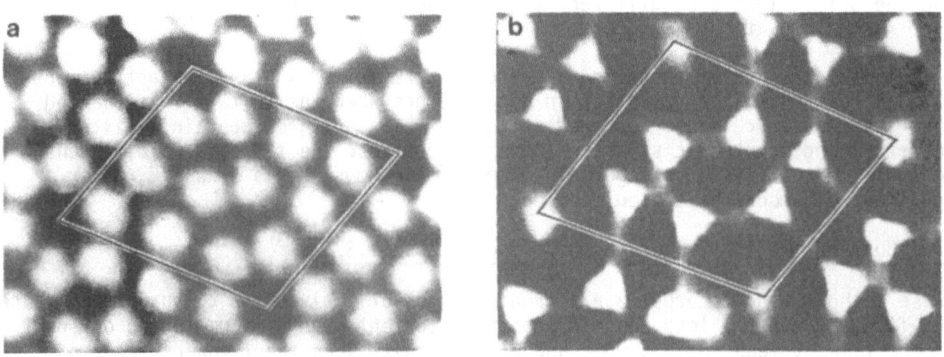

Fig. 7 Si(111) 7x7 surface
a) The 7x7 unit cell topographic image at sample bias voltage + 2V.
b) CITS of occupied dangling bond states at - 0.8 V.

Courtesy of R.J. HAMERS, IBM Yorktown Heights.

A confirmation of BINNIG'S assignment on the localization of the dangling bond states in the topography /1,11/ has been obtained by HAMERS et al /12/. A critical analysis of the different current imaging spectroscopic mode is found in the work of BERGHAUS et al. /6/.

Spectacular results on the determination of electronic and topographic structures of various semiconductor surfaces have been obtained by FEENSTRA and his group who use a slightly different spectroscopic procedure. They thus provide strong support for the PANDEY'S π bonded chain model /16/ in their work on Si(111) 2x1 surfaces /10/. Images of GaAs(110) surfaces where gallium and arsenic atoms are identified /17/ have been obtained by this group as well as the presence of oxygen atoms on a GaAs surface /18/.

This short lecture was concerned with a brief overview of the main results on Scanning Tunneling Microscopy and Spectroscopy. Many problems like the physics of the tip and its relation with the resolution had not been reported. These points are mentioned in the review articles referred in /1/ and /3/. An important conclusion to be drawn from the works of these last two years is that tunneling images are mainly spectroscopic images and reflect the topography only for adequate tunneling parameters.

The author wishes to thank P. DUMAS, M. HANBÜCKEN, H. NEDDERMEYER and S. ROUSSET for useful discussions. Some of the illustrations were kindly provided by the authors which are greatly acknowledged.

References

1. G. Binnig and H. Rohrer : Workshop on Scanning Tunneling Microscopy held in Oberlech (Austria) under the auspices of the I.B.M. Europe Institute, july 1985. The proceedings of this meeting have been published in two issues of the I.B.M. Journal of Research and Development. Most of the topics we are interested in are reported in these proceedings. I.B.M. Journal of Res. and Dev. 30 (1986) n° 4 and n° 5.
2. G. Binnig, H. Rohrer, Ch. Gerber and E. Weibel : Appl. Phys. Lett. 40 (1982) 178.
3. P.K. Hansma and J. Tersoff : J. Appl. Phys. Lett. 61 (1987) R_1.
4. G. Binnig, H. Rohrer, Ch. Gerber and E. Weibel : Phys. Rev. Lett. 50 (1983) 120.
5. K. Tagayanaki : Ultramicroscopy 16 (1985) 101.
6. Th. Berghaus, A. Brodde, H. Neddermeyer and St. Tosch (to be published).
7. J. Tersoff and D.R. Hamann : Phys. Rev. Lett. 50 (1983) 1998.
8. J. Bardeen : Phys. Rev. Lett. 6 (1961) 57.
9. N.D. Lang : Phys. Rev. B 34 (1986) 5947.
10. J.A. Stroscio, R.M. Feenstra and A.P. Fein : Phys. Rev. Lett. 57 (1986) 2579.
11. G. Binnig, H. Fuchs and F. Salvan : Verhandl. der Deutschen Physikalischen Gesellschaft(VI) 20 (1985) 898.
12. R.J. Hamers, R.M. Tromp and J.E. Demuth : Phys. Rev.Lett. 56 (1986) 1972.
13. F. Salvan, H. Fuchs, A. Baratoff and G. Binnig : Surf. Sci. 162 (1985) 634.
14. J.M. Nicholls, F. Salvan and B. Reihl : Phys. Rev. B 34 (1986) 2945.
15. R.S. Becker, J.A. Golovchenko, D.R. Hamann and B.S. Swartzentruber : Phys. Rev. Lett. 55 (1985) 2032.
16. K.C. Pandey : Phys. Rev. Lett. 47 (1981) 1913, and 49 (1982) 223.
17. J.A. Stroscio, R.M. Feenstra and A.P. Fein : Phys. Rev. Lett. 58 (1987) 192.
18. J.A. Stroscio, R.M. Feenstra and A.P. Fein : Phys. Rev. Lett. 58 (1987) 1668.

Field Emission Microscopy for Analysis of Semiconductor Surfaces

V.T. Binh

Département de Physique des Matériaux (UA CNRS),
Université Claude Bernard Lyon 1, F-69622 Villeurbanne, France

1. INTRODUCTION

Field emission microscopy is eminently suited to observations and analysis of semiconductor surfaces. Field emission diagrams give a rapid qualitative control of the surface: cleanliness, adsorption zones, degree of coverage, etc... Since field emission depends primarily on the wave functions of the electrons as they extend into vacuum, these studies can be completed with interest in a following step by Fowler-Nordheim I-V characteristics analysis. Much more informations can be obtained if an energy distribution device is coupled, allowing for example the determination of the surface density of states of semiconductors.

However, these studies will be reliable only if the following major conditions are obeyed: (1) the tip geometry should be known, be simple and reproducible in order to permit calculations of the applied electric field and the current density, and (2) the surface of the tip cathode must be free of any impurity, or covered with a controlled concentration of known adsorbates.

This article will give a very brief review of the salient features of the field emission microscopy and field emission spectroscopy of semiconductors. It will also describe a tip sharpening method to obtain controlled geometry and cleanness of tips of simple semiconductors, and will present some results obtained with Si tips.

2. FIELD EMISSION MICROSCOPY

We give here only a very brief survey, since extensive reviews exist /1-3/. To introduce the field emission microscopy, we will considered first a metal cathode for simplicity. After that, the field emission current from a simple semiconductor will be reviewed.

Field emission consists of electron tunnelling through a surface potential barrier deformed by the application of a high electric field (Fig. 1).

The field emission current density is therefore proportional to the electron density of states at the surface of the solid, and to the transmission coefficient of the potential barrier. It is governed by the Fowler-Nordheim equation /4/. For a metal surface, this equation may be written as

$$\ln (J/F^2) = \ln B - (6.8\times10^7 \Phi^{3/2} / F) \qquad (1)$$

where J is the current density in $A.cm^{-2}$, F is the applied field in $V.cm^{-1}$, Φ is the work function in eV, and B a constant.

Equation (1) indicates that: (1) the field required for an appreciable emission current is around $3\text{-}5\times10^7$ V/cm, (2) for a metal, this current is practically

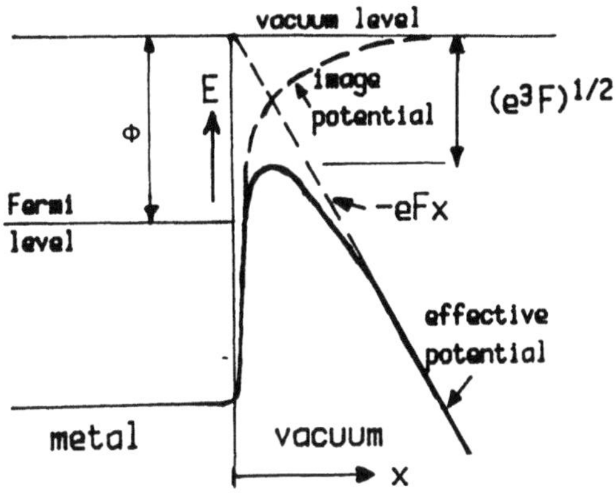

Fig. 1. Schematic diagram of the potential barrier seen by an electron in a field emission experiment (metal cathodes)

independent of the temperature if we do not considered the thermionic contribution, and (3) it is extremely sensitive to the work function.

The requisite applied field is obtained by applying a differential potential V between a cathode, which is a tip, and an anode, which is usually a fluorescent screen. The relation between F and V is given by

$$F = V / \alpha r \tag{2}$$

where r is the tip radius, and α is a geometrical constant usually around 5. Equation (2) indicates that modest voltages, around a few kV, suffice to obtain the necessary fields if the tip apex radii are around 1000 Å (100 - 2000 Å).

The electrons emerging from the tunnelling barrier follow the lines of force which diverge radially from the tip. Consequently, the magnification factor M is given by

$$M = D / \beta r \tag{3}$$

where D is the tip to screen distance which is usually a few centimetres, and β is a compression factor due to the tip geometry, usually taken equal to 1.5. Magnifications of 10^5 to 10^6 are currently obtained.

By the same calculations, the lateral resolution, which is limited by the transverse momentum distribution of the Fermi gas relative to the emission direction, is estimated to be about 20-30 Å.

The field emission micrograph is then obtained as the juxtaposition of dark regions, which correspond to planes with high work function (small emissiom currents), and bright regions, which are planes of low work function (i.e. high emission currents).

Adsorption leads to small changes in Φ, which are proportional to adsorbate coverage,

$$\Delta\Phi = 4\pi P c \tag{4}$$

where P is the dipole moment per ad-complex, and c is the concentration of the adsorbates in particles.cm^{-2}. It is this fact which makes field emission so sensitive to coverage changes, which can be detected visually on the micrographs.

I-V measurements through a "probe hole" opened in the screen (originally suggested by MULLER /5/) then allow the determination of local changes of Φ, that is, variations of the adsorbate concentrations from a region on the tip of about

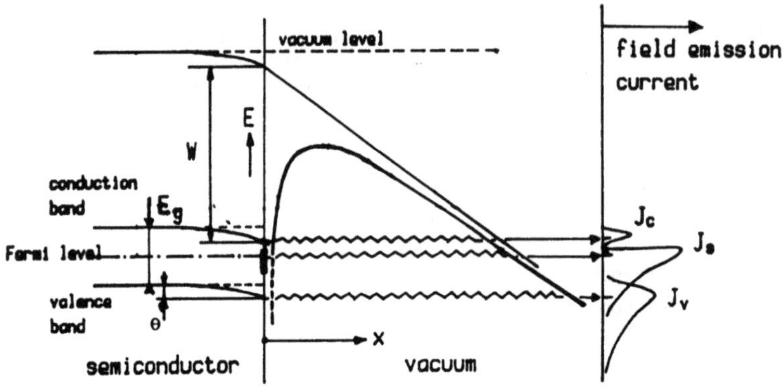

Fig. 2. Schematic description of field emission from a semiconductor

50 A diameter. Variations of a few hundredths of a monolayer of adsorbates are readily measured.

Differences between field emission from a metal surface and a semiconductor surface are: (1) For a semiconductor, tunnelling electrons can come from the conduction band, the valence band and the surface states. The total emission current is then the sum of these three contributions. (2) Depending on the surface charge density, the potential distribution could be modified by field penetration (Fig.2).

Initially, a simple approach to the emission current from a n-type semiconductor, neglecting the field penetration and considering only the contribution from the conduction band /6,7/, gave

$$J = C(T)\ \exp(-6.8\times10^{7}\ v(u)\ W^{3/2}/F) \tag{5}$$

where C(T) is a function of the temperature T, W is the electron affinity in eV, and v(u) is a function which is related to the lowering of the potential barrier due to the contribution of the image potential with

$$u = 3.62\times10^{-4}\ (F\ (\varepsilon-1)/(\varepsilon+1))^{1/2}\ /\ W\ ;$$

ε is the dielectric constant.

In more detailed calculations the field penetration has been considered, and relations giving the emitted current densities from conduction band J_c, valence band J_v /8/, and from surface states J_s /9,10/ have been derived. The approximate emitted current density from a nondegenerate conduction band at the surface (i.e. $F < \varepsilon^{1/2}1.2\times10^{6}$ V/cm) is given by

$$J_c = K_1\ T^2\ \exp(\theta/kT)\ \exp(-6.8\times10^{7}\ v(u)\ W^{3/2}\ /\ F\ ) \tag{6}$$

where K_1 is a constant, and θ the band bending due to field penetration. If the conduction band is degenerate at the surface, the relation giving the current density is more complicated, as was shown in the original paper of STRATTON /8/, and indicates a non-linear variation of $\log(J_c)$ vs 1/F (Fig. 3).

Relation for the valence band contribution to the current density could be approximated by

$$J_v =(K_2\ t(u)\ F^2/\ (W+E_g))\ \exp(-6.8\times10^{7}\ v(u)\ (W+E_g)^{3/2}\ /\ F) \tag{7}$$

where K_2 is a constant, E_g is the band gap, and t(u) is a tabulated function of u. An example of this emission current is shown in Fig. 3.

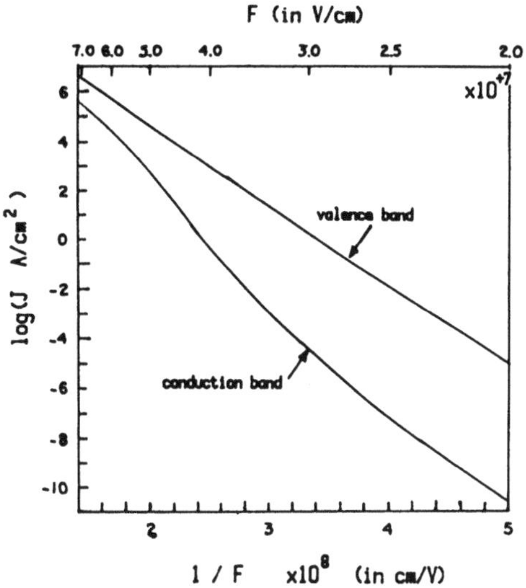

Fig. 3. Field emission currents from the conduction band and from the valence band, at room temperature for n-type Ge (from STRATTON /8/)

The contribution from surface states, J_s, is obtained only by a numerical integration of a rather complicated integral /10/ which is different for each particular case.

Actually, it is rather problematic to predict which of these three currents are the major contributions, because too many unknown parameters are involved. It is worth noting, however, from (6) and (7) that emission current from the conduction band is more temperature dependent than J_v. Experimental I-V measurements will then be of great help in distinguishing the above contributions.

3. FIELD EMISSION SPECTROSCOPY

The energy distribution of the emitted electrons can be measured by positioning an energy analyser just behind the "probe hole" in the screen (Fig. 4). The best performing one, used in field emission is the electrostatic deflection analyser (180° or 135°), which was initially proposed by KUYATT and PLUMMER in 1972 /11/. For a given value of the applied voltage V_a between the two hemispheres, only electrons incident normal to the entrance aperture with an energy E_a

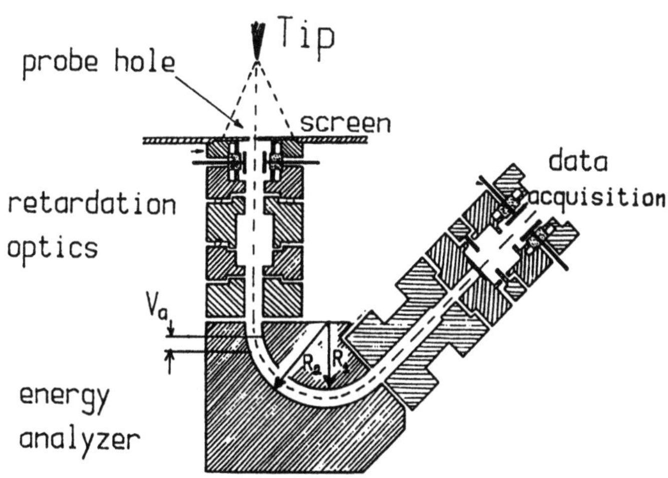

Fig. 4. Schematic diagram of a FEM-FES microscope with a 135° energy analyser (from RIHON /14/)

$$E_A = e\, V_A \,/\, ((R_2/R_1) - (R_1/R_2)), \tag{8}$$

where R_1 and R_2 are the radii of the inner and outer hemispheres, should come out. The energy resolution ΔE_A is about

$$\Delta E_A = E_A\, \omega \,/\, (g\, R_0) \tag{9}$$

where ω is the width of the input and output apertures, R_0 is the mean radius of the hemispheres, and g is a geometrical factor equal 2 for a 180° or 2.4 for a 135° analyser.

Therefore, by retarding the electron energy to between 1-2 eV before entering the analyser, one can measure the energy distribution down to about 3 eV below the Fermi level with an energy resolution of about 20 meV /12/. This interval of energy is well-suited to semiconductor surface analysis, because most of the interesting density of states distribution is within this interval.

Among numerous field emission spectroscopy studies one can cite those of SHEPHERD and PERIA on Ge /13/, RIHON on ZnO /14/ and LEWIS and FISHER on Si /15/. For Ge, the measurements at the centre of the (100) face showed a high-energy peak between 0.6 and 0.7 eV below the Fermi level which was attributed to the presence of a band of surface states, for it was very sensitive to surface conditions, and a second peak at lower energy which corresponded to electrons originating from the valence band. Similar behaviours were reported for ZnO and Si surfaces: electrons from the surface states contribute most of the tunnelling current, though their energy distributions show noticeable differences. However, these results indicate that energy distributions in field emission do not reflect the density of states of the emitter straightforwardly, and also that a better control of the surface cleanness of the tips is needed.

4. PREPARATION OF FIELD EMISSION TIPS OF SILICON

The capabilities of FEM and FES are widely used for metal surface analysis. On the other hand, the amount of such data on semiconductors is still very small and the results have been partially successful /13-19/. The major problems encountered were the production of sharp semiconductor tips, and the production and maintenance of controlled clean surfaces just after either high temperature cleaning procedures or during the following low temperature treatments. Cleaning of a solid surface, in particular for a semiconductor, is a three-dimensional problem. One must control the impurity concentration not only on the 2-D outermost layer of the tip, but also along the third dimension, which is perpendicular to the surface and going into the bulk. Some electronic properties of the semiconductor surfaces are strongly influenced by minute concentrations of impurities just inside the surface.

Preparations of Si tips are usually done by chemical or electrochemical polishings using solutions of HNO_3 and HF, and details are described by numerous authors /15-19/. We have repeated these methods for producing Si tips, and the results lead to the following conclusions, some of which were already stated by other authors. Namely, they are:

(1) Strong tendency of the Si tips to acquire a blade shape during polishing, as consequence of a non-uniform polishing rates.

(2) The resulting tips, just after a first heat treatment in order to clean the surface, generally have an apex diameter greater than one micrometer.

(3) After the first heat treatment in vacuum, the initial conical tip presents a "hole structure", such a structure of the tip is presented in Fig. 5. We interpreted this hole structure as the consequence of evaporation of pouches of silicon compounds, likely silicon oxides, which are created by diffusion of

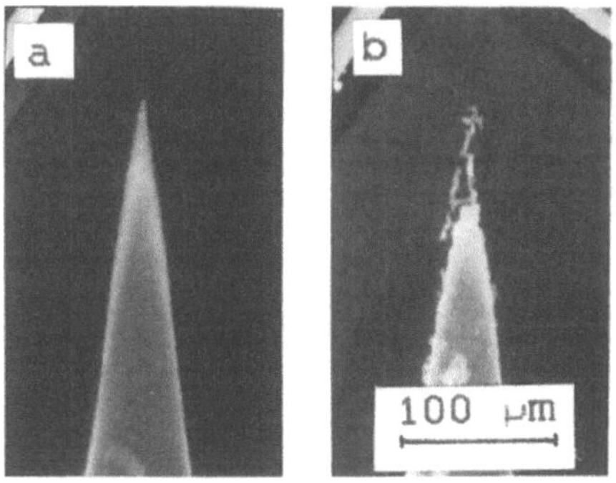

Fig. 5. Scanning electron micrographs of a Si tip, (a) after a chemical etching, (b) followed by a heat treatment in vacuum

impurities (oxygen) along dislocations during the chemical or electrochemical etching. It is therefore likely that such compounds are preferentially evaporated during the first heat treatment, leaving behind the "hole structure" in Si.

(4) When these tips are used in FEM, two consequences are noticed. First, due to the "hole structure" it happens very frequently that more than one tips is emitting electrons, which leads to a superposition of several FE diagrams and difficulties in analysing the I-V characteristics. Second, due to surface segregation of non-controlled bulk impurities towards the surface, it is very difficult to obtain clean diagrams by heat treatments. These bulk impurities can come either from impurities introduced by diffusion during the etching or from the initial doping impurities of the Si wafer.

These conclusions have prompted us to develop an in situ ultra-high vacuum thermal etching based on the principle of a pseudo-stationary tip profile /20,21/. Samples with a section of about 0.5 mm were cut from silicon single crystal wafers, and then heated in ultra-high vacuum by electron bombardment /22/ at temperatures between 0.9 to 0.95 of the melting temperature until the formation of a pseudo-stationary profile (Fig. 6). The requisite heating time is around 50 hours.

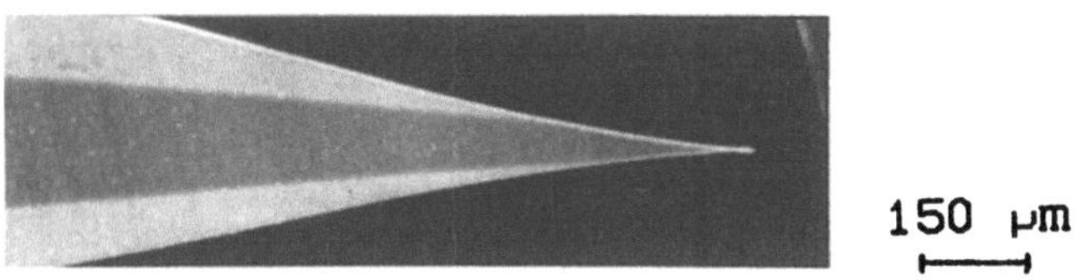

Fig. 6. Scanning electron micrograph of a Si tip obtained by UHV thermal etching

5. FIELD EMISSION OF SILICON

Using the above thermal etching technique, we have obtained systematically clean silicon tips with apex radius of 1000-2000 Å. An example of a FE micrograph of clean Si is presented in Fig. 7. The following characteristics could be observed:

(1) A clean silicon surface presents smooth, but pronounced variations of the emitted currents vs crystallographic orientations (Fig. 7). One can observe such a behaviour around the (011) face of Fig. 7 for example.

(2) In the presence of adsorption, the emitted areas are more localised so the FE micrographs appear more granulous (Fig. 8c).

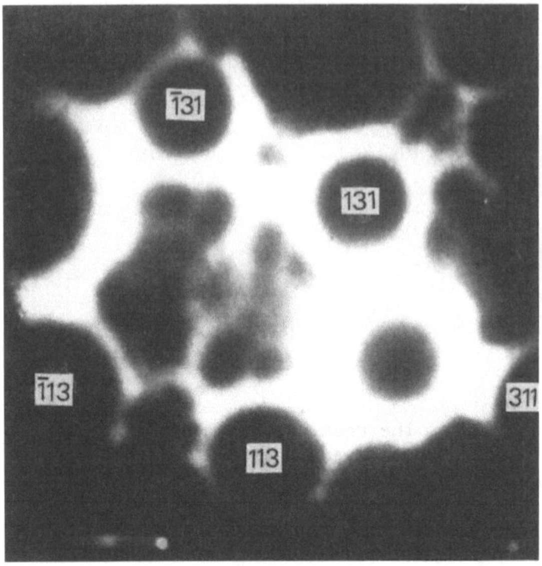

Fig. 7. Field emission micrograph of clean silicon

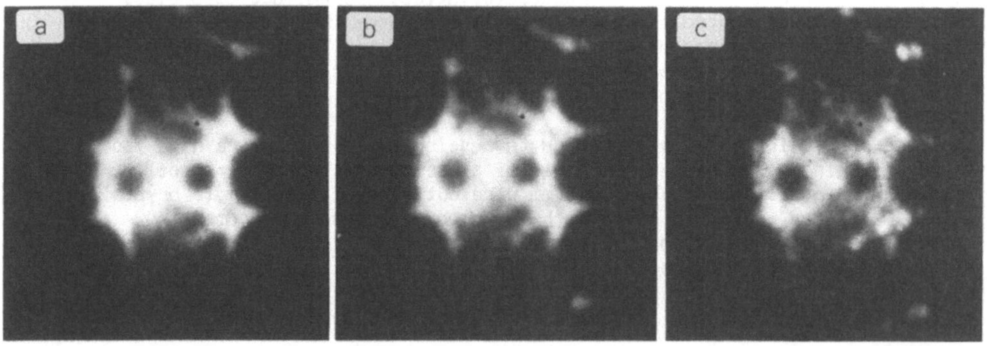

Fig. 8. Field emission micrographs of Si: (a) clean surface; (b) a few minutes at 10^{-7} Torr O_2; (c) 48 h in residual atm. of 10^{-10} Torr

(3) To clean the emitter surface, we must heat it up to temperatures greater than 1500 K. For lower temperatures, the only consequence is a surface migration and clustering of the adsorbates. This could lead to formation of well-defined crystalline clusters at the surface.

(4) Plots of log I vs 1/V (Fig. 9) are linear over five decades. The three Fowler-Nordheim plots are related to the FE micrographs in Fig. 8, and present practically no slope change for observable modifications of the emitted currents as one can see from Figs. 8a-c.

6. CONCLUSIONS

Field emission and field emission spectroscopy from semiconductor surfaces can provide us with information at an atomic scale not only about the surface density of states and the field penetration into the semiconductor, but also about the occupation and mechanism of replenishment of the surface states in the case of non-equilibrium conditions, and thus for controlled clean surfaces or with known adsorbates. One of the major difficulties is now removed: the preparation of

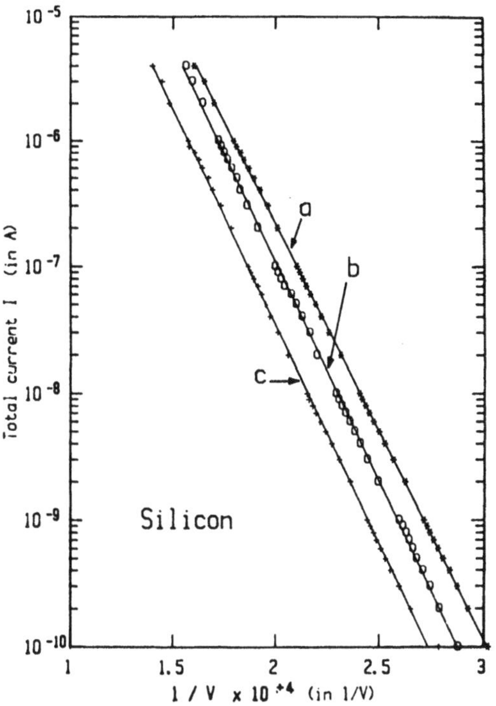

Fig. 9. I-V plots of Si

high-quality semiconductor tips is possible in a reproducible manner by using thermal etching. This leaves us with problems intrinsic to semiconductor surface physics ahead of us.

References:

1. R. Gomer, in "Field Emission and Field Ionization", Harvard University Press, Cambridge, Mass. (1960)
2. R. Gomer, in "Structure and Properties of Metal Surfaces", p.128, Honda Memorial Series on Material Science, Maruzen Lt., Tokyo (1973)
3. A. Modinos, in "Field, Thermionic, and Secondary Electron, Emission Spectroscopy", Plenum Press, New York and London, (1984)
4. R.H. Fowler and L. Nordheim, Proc. Roy. Soc. London A119, 173, 1928
5. E.W. Muller, J. Appl. Phys., 26, 732 (1955)
6. R. Stratton, Proc. Phys. Soc. B68, 746 (1955)
7. G. Busch and T. Fisher, Revue Brown Boveri 45, 532 (1958)
8. R. Stratton, Phys. Rev. 125, 67 (1962)
9. J.W. Gadzuk, J. Vac. Sci. Tech. 9, 591 (1972)
10. A. Modinos, Surface Sci.42, 205 (1974)
11. C.E. Kuyatt and E.W. Plummer, Rev. Sci. Instrum. 43, 108 (1972)
12. J.W. Gadzuk and E.W. Plummer, Rev. Modern Phys. 45, 487 (1973)
13. W.B. Shepherd and W.T. Peria, Surface Sci. 38, 461 (1973)
14. N. Rihon, Surface Sci. 70, 92 (1978); Thesis 1978
15. B.F. Lewis and T.E. Fisher, Surface Sci. 41, 371 (1974)
16. L.A. d'Asaro, J. Appl. Phys. 29, 33 (1958)
17. F.G. Allen, J. Phys. Chem. Solids 8, 119 (1959)
18. E.W. Muller and T.T. Tsong, in Field Ion Microscopy: Principles and Applications (Elsevier, New York) 1969
19. A.J. Melmed and R.J. Stein, Surface Sci. 49, 645 (1975)
20. Vu Thien Binh, A. Piquet, H. Roux, R. Uzan and M. Drechsler, J. Phys. E: Sci. Instrum. 9, 377 (1976)
21. Vu Thien Binh and R. Uzan , Surface Sci. 179, 540 (1987)
22. J.L. Souchière and Vu Thien Binh, Surface Sci. 168, 52 (1986)

Surface and Interface Studies with MeV Ion Beams

C. Cohen

Groupe de Physique des Solides de l'E.N.S., Université Paris VII,
Tour 23, 2 Place Jussieu, F-75251 Paris Cedex 05, France

MeV ion backscattering and nuclear microanalysis are very sensitive tools which, coupled to Auger electron spectroscopy, provide quantitative information on elemental composition of solid surfaces. The channeling phenomenon enables the measurement of a few hundred's of Å displacements of atoms from lattice sites and the analysis of surface crystallographic structures, epitaxial layers and interface defects. The principles of such experiments are presented and examples are given which illustrate the possibilities of nuclear techniques in various domains related to surface physics.

I INTRODUCTION

In most of the techniques devoted to surface studies, particles whose range is of the order of a few monolayers, and corresponds hence to the depth domain which is to be analyzed, are used. This is the case for instance in low-energy ion scattering experiments, in low-energy electron diffraction (L.E.E.D.) measurements, in Auger or photon-induced electron spectroscopy (in the two latter cases low-energy electrons are detected). Similarly, Rutherford backscattering (R.B.S.) and nuclear reaction analysis (N.R.A.) using light ions in the 100 keV - 3 MeV domain provide information about regions of a solid corresponding to the range of the incoming ions i.e. a few microns. The depth resolution associated to the particle energy loss is, usually, in these experiments between 100-1000 Å. These techniques do not appear hence to be very well adapted for the study of the very surface regions. They moreover present two serious misadvantages : i) involving nuclear collisions, they are insensitive to the chemical environment and will hence not provide any information about electronic structure, ii) they require a much heavier equipment (accelerator beams) than the standard surface analysis techniques.

Despite of all these problems, MeV light ions have been more and more used in the past few years in surface physics. The reason for this situation is that, if they cannot be substituted to the standard surface techniques, nuclear techniques provide very useful complementary information, in an unambiguous way. For instance, in elemental analysis it is very important to couple to Auger spectroscopy a method which can determine absolute quantities with high precision. The use of nuclear analysis for this purpose will be illustrated in section II. In surface crystallography, channeling experiments provide information about atomic displacements, either static (defects, relaxation), or dynamic (thermal vibrations). Information about the lattice sites occupied by adsorbed species can also be reached. All these results are complementary to those obtained by L.E.E.D. which, in a first approach, are mainly indicative of the crystal surface symmetries. The principles and some examples of channeling surface studies will be presented in section III. Finally one can take advantage of the large range of energetic light particles in matter to study the structural properties of epitaxial layers. The rapid development of molecular beam epitaxy calls, in particular, for such studies. In this field, the standard surface techniques are not adapted as soon as the epitaxial film corresponds to more than a very few monolayers. On the contrary, channeling experiments may characterize both the bulk structure of the epitaxial film and the interfacial disorder (strains induced by lattice parameter mismatch for instance). This type of study will be presented in section IV.

This paper is of course much too short to intend to be a complete review. Details about backscattering and nuclear reaction analysis can be found for instance in ref /1/, about channeling for material analysis in ref /2/ and about channeling studies of surface structures in ref /3/.

II USE OF RUTHERFORD BACKSCATTERING AND NUCLEAR REACTION ANALYSIS FOR THE STUDY OF ELEMENTAL COMPOSITIONS OF THIN FILMS

When a beam of light ions of incident energy E_0 around 1 MeV penetrates into a solid, the particles are progressively slowed down and, if one neglects the fluctuations in energy loss, an energy $E(x)$ can be associated to a penetration depth x. The particle range is of the order of a few microns and in the major fraction of the path the inelastic processes (target atoms ionisation and electronic excitation) are mainly responsible for the slowing down. The associated stopping power is now well known. At a given energy (i.e. at a given penetration depth) a particle has a well-defined cross section for a given close nuclear encounter event such as a backscattering elastic collision or a nuclear reaction with a target atom.

II.1 Backscattering (R.B.S.)

We are considering elastic events. Consequently, the energy E' after the event of a particle of Mass M_1 and initial energy E which is deflected by a target atom at rest depends only on the deflection angle θ in the center of mass system and on the target atom mass M_2. One has for instance at $\theta = 180°$

$$E' = KE \text{ with } K = \left(\frac{M_2 - M_1}{M_2 + M_1}\right)^2 . \qquad (1)$$

Of course, the requirement is $M_2 > M_1$ and K is an increasing function of M_2. The energy analysis of the backscattered particles will hence provide information about the various components of the target.

In what concerns the cross sections associated to backscattering events, one has to consider that such events correspond to collisions with impact parameter of the order of b, minimum distance of approach of a nucleus in an elastic collision. One has for a Coulomb potential

$$b = \frac{2Z_1Z_2e^2}{\mu\, v^2} \qquad (2)$$

where Z_1 and Z_2 are the atomic numbers of the incident particle and of the target atom respectively, e the electron charge, μ the reduced mass and v the velocity of the incident ion. For MeV light particles one finds $b \simeq 10^{-3}$ Å, i.e. very small as compared to any electronic screening radius. Consequently one can neglect as a first approximation screening corrections and consider that the event is a pure Coulomb interaction between two point charges with the associated Rutherford cross section

$$\frac{d\sigma}{d\Omega} = \frac{b^2}{16} \times \frac{1}{\sin^4 \frac{\theta}{2}} . \qquad (3)$$

This cross section depends on the energy, which, as indicated above, varies along the particle path into the target. A schematic backscattering experiment and the corresponding spectrum are shown in Fig. 1 from ref /2/. The depth resolution depends on the energy resolution of the analysis, on the angles of incidence and detection, on the stopping power (i.e. on the particle nature and energy and on the target composition) and also on the penetration depth as the dispersion in energy loss increases with the latter. The near surface depth resolution may reach 10 Å (with 100 keV protons analyzed electrostatically) but, in standard experiments with surface barrier detectors, resolutions of $\simeq$ 100 Å are typical.

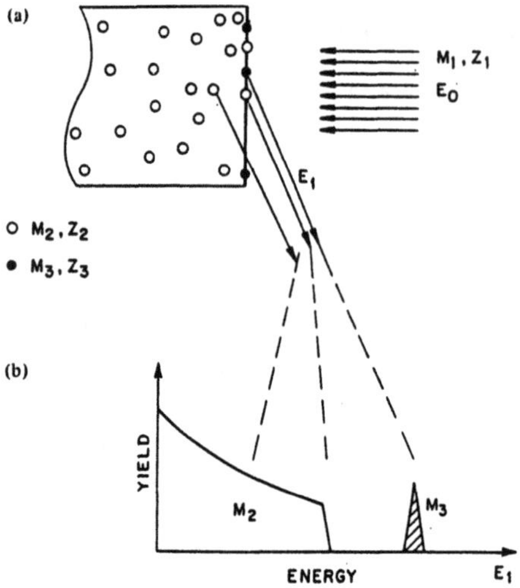

Fig. 1 Schematic backscattering experiment and corresponding spectrum. The target consists of a heavy impurity at the surface of a lighter substrate. From ref. /2/

Assume now that the target contains in its near surface region traces of impurities. They will contribute to a backscattering spectrum by peaks at energies given by relation (1) and integral proportional, via relation (3), to the impurity amount. Such a situation is also illustrated in fig. 1 which corresponds to the favourable case of heavy impurity on a lighter target. In this situation the sensitivity is of the order of 10^{-2} to 10^{-3} monolayer. The use of a standard containing a well-known amount of atoms which do not have to be of same nature than the analyzed impurity allows one to obtain, via relation (3) a quantitative measurement of the impurity amount with an absolute precision of 3 %.

The situation is much less favourable in the case of impurity lighter than the target atoms. The contribution to the spectrum of particles backscattered on such impurity atoms is superimposed on the signal from substrate atoms and the sensitivity is poor (note also that for light atoms the Rutherford cross section is low). In such a situation the study of nuclear reaction products is of great help.

II.2 Nuclear reaction analysis (N.R.A.)

With MeV protons or deuterons, the cross sections for nuclear reactions on light elements (up to sulfur) are high and the analysis of the reaction products whose energy is characteristic of the nuclear reaction induced provides the quantitative measurement of the amounts of light impurities. The calibration requires here the use of a standard containing a well-known amount of atoms of the same nature as those analyzed. There is again, as in the case of backscattering, an energy-to-depth relation via the stopping powers which can be used to extract concentration profiles, but, this time, the depth resolution is poor, of the order of 1000 Å.

In some cases, narrow resonances of cross sections can be taken advantage of for depth profiling with better resolution. One scans the incident beam energy E_0 in an energy domain containing the resonance energy E_R and simply measures the overall nuclear reaction yield at a given detection angle. The principle is schematized in fig. 2 from ref. /4/ for an isolated resonance. For $E_0 < E_R$ there is no nuclear reaction. At $E_0 = E_R$ the reaction takes place with surface atoms. For $E_0 > E_R$ the reaction takes place at a depth such that, due to energy loss, the particle energy $E(x) = E_R$. The near surface depth resolution for these measurements is mainly determined by the resonance energy width and by the stopping power, and can be of the order of 100 Å.

One major advantage of nuclear reaction techniques is that they are specific to a given isotope. Isotopic tracing may then be used to study atomic transport mechanisms in thin film growth. This is illustrated in fig. 3 from ref. /5/.

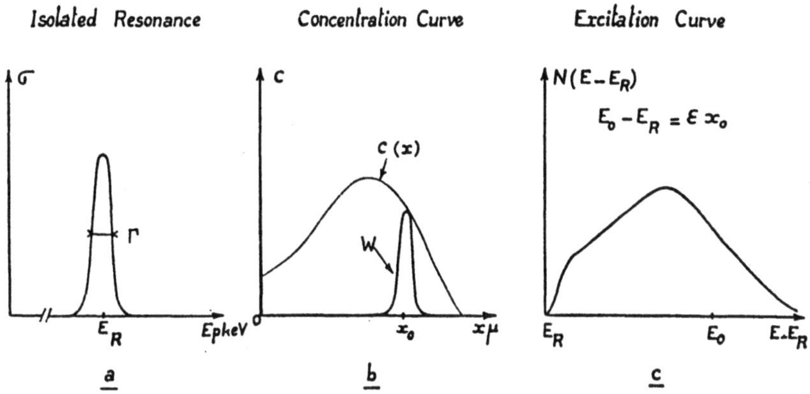

Fig. 2 Principle of concentration profile measurements using narrow resonances of nuclear reaction cross sections. From ref. /4/

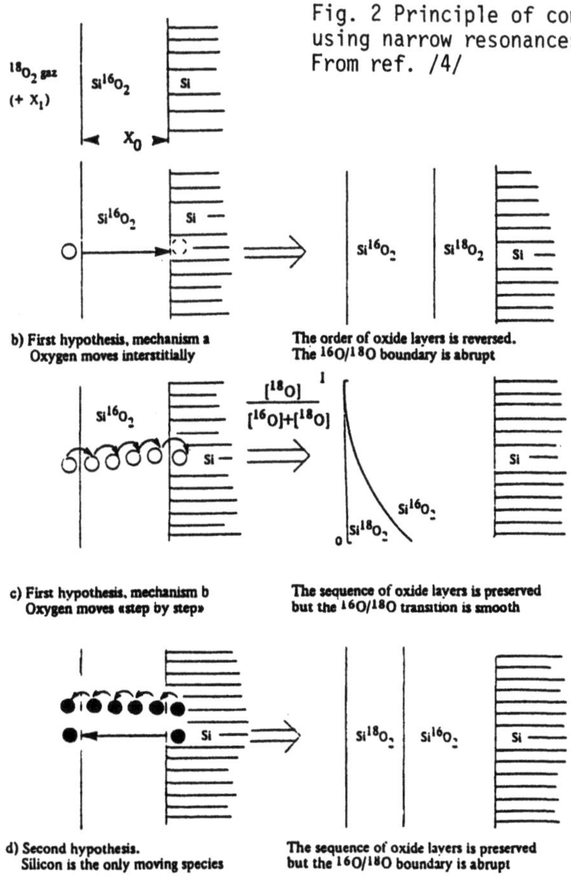

Fig. 3 Illustration of how the labelling technique can be used to identify the mobile species during oxide growth.
a) An oxide layer of thickness X_0 is grown. The subsequent oxidation will take place in an $^{18}O_2$ gas and will contribute an additional oxide layer of thickness X_1.
b) Illustration of the first hypothesis, mechanism a. Oxygen moves interstitially.
c) Illustration of the first hypothesis, mechanism b. Oxygen moves step by step.
d) Illustration of the second hypothesis. Silicon is the only moving species. From ref. /5/

In this figure the mechanism of thermal oxidation of silicon in dry oxygen atmosphere is elucidated. Subsequent oxidation in ^{16}O and ^{18}O atmospheres is performed and the ^{18}O depth profile is analyzed in detail by scanning across a resonance of the $^{18}O(p,\alpha)^{15}N$ nuclear reaction at proton energy of 629 keV. The result is that the main mechanism is interstitial oxygen motion (mechanism b), but that some step by step oxygen motion (mechanism c) also takes place.

II.3 Elemental composition at solid surfaces

The most used surface technique for these studies is Auger spectroscopy which is very sensitive (often less than 10^{-2} monolayer) and, due to the very small escape depth of the Auger electrons (∿ 10 Å), senses exclusively the very surface region. The main limitation is due to the fact that the cross sections for Auger electron production are not well known for all elements and are, in particular, very dependent of the matrix composition. Quantitative information is hence hard to reach and the comparison with R.B.S. and N.R.A. measurements, which in many cases can detect impurity contents down to 10^{-2} equivalent monolayer with a ± 2-3% absolute precision allows one to calibrate Auger results. In some cases, when the impurities searched for remain on the sample after air exposure (for instance metallic deposits), there is no need to couple directly an ultra high vacuum system equipped with Auger spectroscopy to an accelerator. As the nuclear measurement will be insensitive to the sample contamination during its exposure in air, it can be performed ex situ. This may also be the case for the study of the first oxidation steps if ^{18}O is used in the preparation chamber, in so far that isotopic exchange with ^{16}O in air is negligeable.

As the escape probability of Auger electrons varies rapidly with depth, the detected signal intensity depends strongly on the depth location of the emitting atom. This feature is currently used to check whether a deposition or an adsorption process corresponds to a bi-(layer by layer) or tridimensional (formation of aggregates) mechanism. In the first case for instance, the Auger signal varies linearly with coverage during the completion of a monolayer and a decrease of the slope is observed each time a new layer begins to grow. The comparison of such Auger measurements with R.B.S. or N.R.A. analyses gives with high precision the value of the saturation coverage, i.e. the number of adsorbed atoms per/cm^2 in a completed layer.

We will now illustrate the usefulness of nuclear techniques in surface characterisation by an example which has some technological implications : the study of silicon surface cleaning by very low energy ion bombardment. It has been demonstrated /6/ that silicon can be cleaned under ultra high vacuum by using Ar beams with energies as low as 70 eV. When the bombarding energy is decreased, both surface damage and Ar incorporation also decrease significantly. However, even at 70 eV some damage and incorporation remain for room-temperature bombardments. Bombardments at higher temperature were hence attempted in order to check whether some improvement took place, related to disorder annealing and Argon desorbtion. The results are summarized in fig. 4. The Auger analyses are encouraging : the Ar signal decreases continuously when the temperature during bombardment is raised and, above 400° C, no Ar is detected. However, the Ar content measured by R.B.S. on the same samples is independent of the bombardment temperature and corresponds to a constant value of 1.3×10^{14} Ar/cm^2. In the same time, the disorder in the surface region as checked by channeling experiments (see section III) increases markedly with bombardment temperature. The interpretation of all these results is as follows : when the temperature is raised during bombardment, defect-assisted Argon diffusion takes place, and Ar penetrates in the first 50 Å of the Si crystals. In turn, the Ar incorporated atoms stabilize a significant amount of Si defects in this region, even though the Ar concentration is small. The higher is the implantation temperature, the deeper the Ar ions penetrate and the stronger is the crystal perturbation. At the same time of course the very surface Ar concentration decreases as evidenced by the lowering of the Auger signal. Such a study is a good illustration of the complementarity of the information provided by Auger spectroscopy and nuclear techniques.

III STRUCTURAL STUDIES BY CHANNELING

III.1 Channeling phenomena

When a beam of particles is aligned with a major axial or planar orientation in a single crystal, spectacular effects occur and, in particular, the close nuclear encounter probability decreases by one (planar case) to two (axial case) orders of magnitude. We will discuss here the axial situation which is more often taken advan-

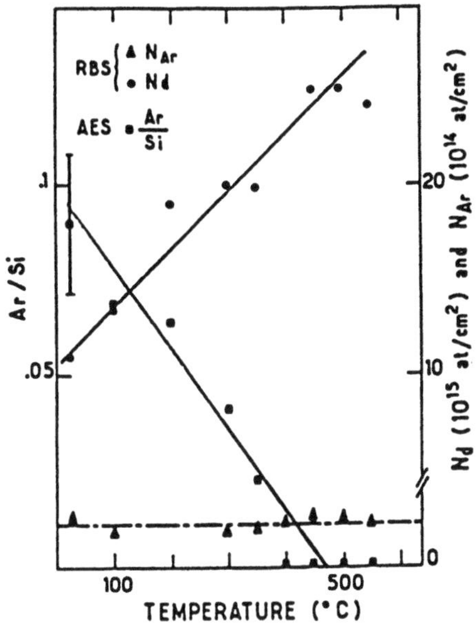

Fig. 4 Temperature dependence for Si samples bombarded with 100 eV Ar^+ and heated during bombardment of : the Ar/Si ratio from Auger spectra (■) the Ar incorporation measured by RBS (▲) ; and the overall bombardment-induced disorder (●). From ref. /6/

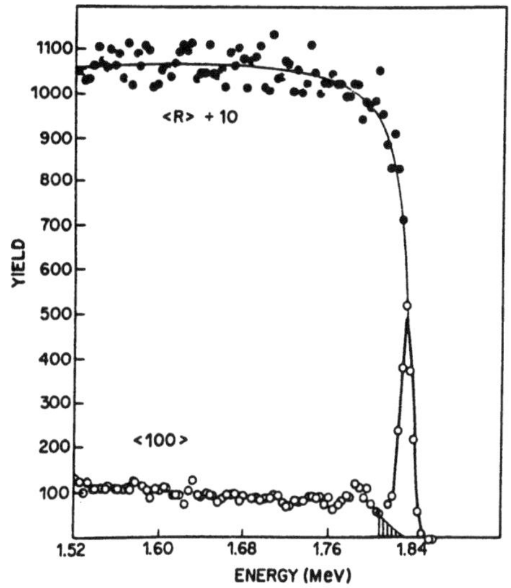

Fig. 5 Backscattering spectra for 2.0-MeV ^{4}He incident on a clean W(001) surface for the beam aligned with the <100> axis (O) and for the beam aligned away from any major crystallographic direction (●). Note the non-channeling spectrum (R), termed random incidence, has been reduced by a factor of 10. From ref. /2/

tage of for structural studies. Typical backscattering spectra on a single crystal for aligned and random incidences are shown in fig. 5 from ref. /2/.

In the axial alignment geometry the consecutive collisions of a particle with the atoms of a string are highly correlated for particles entering the crystal at distances of a row greater than the standard deviation ρ associated to the atom thermal vibration perpendicularly to the string. Thus, as a first approximation, the motion of a particle in a transverse plane perpendicular to the rows of atoms is governed by a continuous axial potential given by

$$U(r) = \frac{1}{d} \int_{-\infty}^{+\infty} V(\sqrt{z^2 + r^2}\,)\, dz \tag{4}$$

where d is the interatomic distance along the row, r the distance from the row in the transverse plane, and V a screened Coulomb interatomic potential. In the small angle approximation, this implies the conservation of the "transverse energy" $E_\perp$ given by

$$E_\perp = U(r) + E\varphi^2 = U(r_i) \tag{5}$$

where φ is the angle with the string at distance r and r_i the entrance distance from the string.

The consequence of such a transverse motion is that the particle flux in the crystal is not uniform. The quantity P(r,z), probability for a particle to be at distance r from a string at a penetration depth z in the crystal is difficult to calculate but, at great z, it tends to become independent of z : a statistical equilibrium is reached and all the allowed configurations in the transverse phase space are equiprobable for a particle of given transverse energy. One has, introducing conditional probabilities ,

$$P(r,z\,|E_\perp) \to P_\infty\,(r|E_\perp) = \begin{cases} \dfrac{2\pi r}{A} & E_\perp > U(r)\,, \\ 0 & E_\perp < U(r)\,. \end{cases} \tag{6}$$

A is the accessible area in the transverse plane for a particle of transverse energy $E_\perp$, i.e.

$$A = \pi r_{E_\perp}^2 \qquad \text{with} \quad U(r_{E_\perp}) = E_\perp\,. \tag{7}$$

And, finally, the flux at statistical equilibrium, $P_\infty(r)$ is given by

$$P_\infty(r) = \int_{E_\perp} g(E_\perp)\, P_\infty\,(r|E_\perp)\, dE_\perp \tag{8}$$

where $g(E_\perp)$ is the transverse energy distribution which is determined by the uniform flux at the entrance of the crystal.

Associating to a string of atoms a unit cell in the transverse plane centered on the string and of radius r_0 given by

$$\pi\, r_0^2 = \frac{1}{Nd} \tag{9}$$

where N is the number of atoms per unit volume, equation 8 leads to

$$P_\infty\,(r) = \frac{2r}{r_0^2}\,\mathrm{Log}\,\frac{r_0^2}{r_0^2 - r^2}\,, \tag{10}$$

i.e. a flux zero on the string and infinite for $r = r_0$. As a matter of fact, when the angular divergence of the beam is taken into account, the shape of the flux profile is smoother than indicated by equ. 10.

III.2 Lattice location at statistical equilibrium

Assume a distribution of impurities $h(\vec{r})$ in the transverse plane. The close nuclear encounter yield (for instance a backscattering event on an impurity atom), normalized to the yield which would correspond to a uniform particle flux in the crystal ("random situation") is given by

$$Y = \frac{1}{P_{unif}(\vec{r})}\int_0^{r_0} 2\pi r h(\vec{r})\, P_\infty\,(\vec{r})\, dr \tag{11}$$

with $P_\infty\,(\vec{r}) = \dfrac{1}{2\pi r}\,P_\infty\,(r)$ and the uniform flux $P_{unif}\,(\vec{r}) = \dfrac{1}{\pi r_0^2}$.

Then using equation (10), (11) becomes

$$Y = \int_0^{r_0} 2\pi r h\,(\vec{r})\,\log\frac{r_0^2}{r_0^2 - r^2}\,dr\,. \tag{12}$$

The measurement of the backscattering yield on the impurity provides hence, via equ. (12), information on its site distribution in the crystal. Practically, one has to perform the measurements as a function of the tilting angle between the incident beam and the crystal axis and, often, to check around various crystallographic directions.

Here we have assumed the statistical equilibrium flux distribution to be valid. This implies that the impurity is distributed over large depths (at least 1000 Å) in the crystal. However, such an assumption is also valid if one wants to locate impurities adsorbed in the back side of a crystal of thickness small as compared to the particle range. The particles reaching the backside are at statistical equilibrium, and relation (12) remains valid for locating the adsorbed impurities. This has been done for instance in ref. /7/ where deuterium adsorbed on nickel crystal surfaces has been located 0.5 Å above the Ni surface and where the thermal vibration amplitude of the adsorbed atoms has been determined.

However, in most surface structure studies, one is dealing with thick crystals which cannot be crossed by the beam. One has hence to study the entrance surface at which statistical equilibrium is not reached and the above equations are no more valid. We shall now discuss this situation in detail

III.3 Study of surface structures by channeling experiments

These studies require the coupling of an ultra high vacuum set-up to an accelerator. In order to keep the surface contamination below a few 10^{-2} monolayer during the few hours required for a set of experiments, and assuming a sticking coefficient of 1 for the molecules of residual gas, one can evaluate that the pressure must be around 1×10^{-10} torr. The goniometric chamber must be equipped with an ion gun for surface cleaning, a L.E.E.D. Auger system in order to get information on the surface crystallographic structure, and cleanliness. A heating system is required in order to anneal the sample after ion bombardment. Cooling may also be necessary if structural changes as a function of temperature are expected, or if surface thermal vibrations are studied. The chamber is also usually equipped with gas introduction lines and evaporation systems, in order to study the behaviour of deposits or adsorbed species. Finally, the detectors for R.B.S. or N.R.A. are often movable in order to align the exit direction studied with a crystallographic axis.

The possibility of surface studies using channeling is related to the fact that when a beam of particle is aligned with a major crystallographic axis, the backscattering yield is much higher for the very first surface layers than in the bulk. Consider the situation represented in fig. 6.

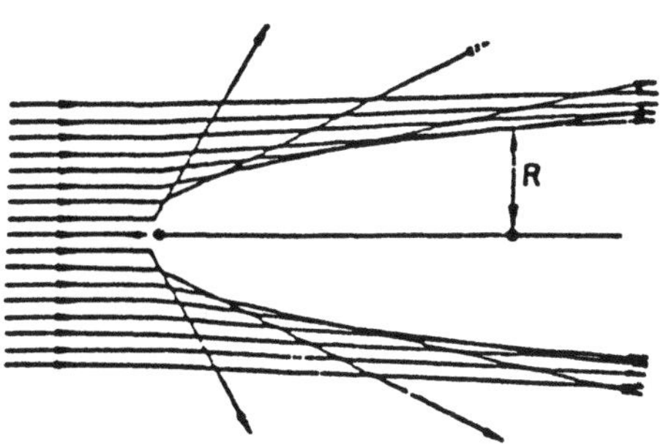

Fig. 6 Formation of a shadow cone at the second of a pair of atoms aligned with an incident ion beam. The shadow cone radius at the second atom is denoted R.

Here a uniform and parallel beam of particles meets an atom; let us ask for the flux distribution on a screen placed at distance d of this atom, d being measured along the incident beam direction. We shall assume for the sake of simplicity that the interaction is described by a Coulomb potential, that the small-angle approximation applies and assimilate the laboratory system to the center of mass system (light particles hitting a heavy atom). With these assumptions a particle with impact parameter P_1 is deflected by an angle

$$\theta = \frac{b}{P_1} . \qquad (13)$$

The impact parameter P_2 with respect to the center of the screen is hence

$$P_2 = P_1 + \theta d = P_1 + \frac{db}{P_1} . \quad (14)$$

Differentiating equ. 14 with respect to P_1 one finds that there is a minimal value R for P_2 given by

$$R = 2\sqrt{bd} = 2(Z_1 Z_2 e^2 d/E)^{1/2} . \quad (15)$$

If a second atom is placed on the screen with mean position at the center, but with a probability distribution characterized by a standard deviation ρ, the backscattering probability on this atom will of course be small if $R > \rho$, and even if this is not the case, will in all cases be smaller than on the first atom. Consider now a whole string of atoms parallel to the incident beam, with interatomic distance d. ρ is now the r.m.s. thermal vibration perpendicular to the string. The third atom of the string will be more shadowed than the second and so on. Calling Y_i the backscattering yield on the ith atom, normalized to the yield on the first (unshadowed) one, let us consider the sum

$$n = \sum_i Y_i . \quad (16)$$

If this sum converges rapidly, the backscattering probability as a function of depth will fall to zero and other backscattering events will only take place at much greater depthes when the particles reach another string. But by this time they have lost some energy and a R.B.S. spectrum will present in its high-energy part a marked peak ("surface peak") whose integral is proportional to n (given by equ. 16). The value n "number of atoms per row" corresponding to the surface peak will of course depend on the relative value of the parameters ρ and R, i.e. on the shadowing efficiency. A remarkable result of Monte Carlo simulations is that, for a "perfect" string n is a universal function of the ratio $\frac{\rho}{R}$ (see fig. 7 from ref. /8/).

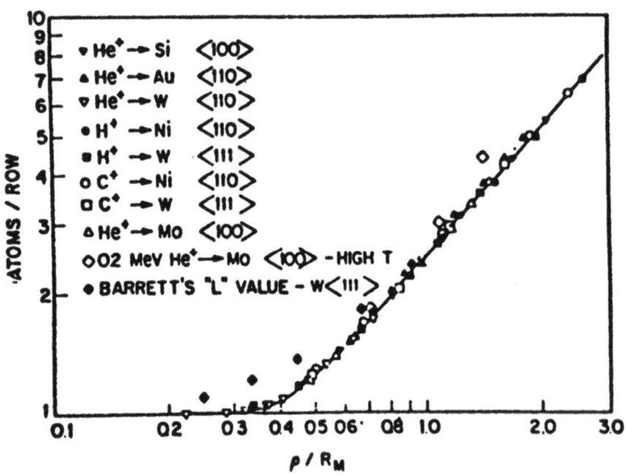

Fig. 7 Calculated normalized backscattering probability expressed as the effective number of atoms/row contributing to the surface peak versus ρ/R. From ref. /8/

In the simulations a description of the crystal (mean atomic position and thermal vibrations) is proposed, and the only physical data required to simulate a channeling experiment is the interatomic potential between an incoming ion and a crystal atom. As this potential is known with good precision, a comparison between simulations and experimental results is a good check of the crystal description assumed. For instance, surface relaxation can be evidenced by studying the number of atoms per row $n(\varphi)$ corresponding to the surfacepeak, as a function of tilting angle with respect to an axis not perpendicular to the crystal surface. If the first surface plane is relaxed, the first atom will be displaced from the string and will shadow the second one in a direction slightly different from the bulk axial direction. The curve $n(\varphi)$ will hence be asymmetrical and shifted, as indicated in the simulations of fig. 8 from ref. /9/ which demonstrate that relaxations of the order of 0.05 Å can easily be measured.

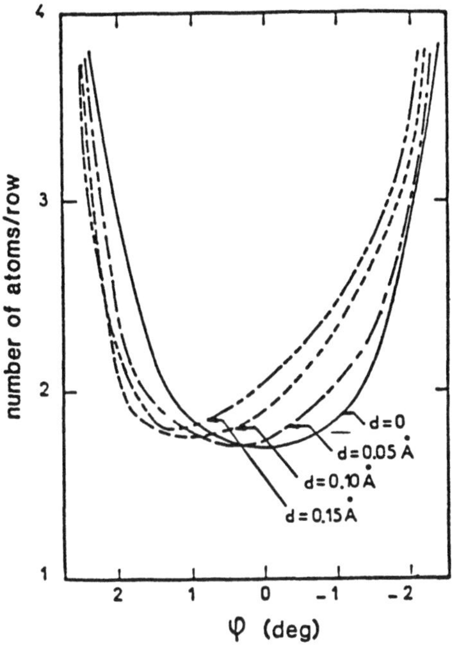

Fig. 8 Monte Carlo simulations of angular scans across the <110> axis of a Cu crystal for 200 keV ^{4}He ions. d is here the amplitude of the relaxation assumed for the first atom, perpendicularly to the string. From ref. /9/

One can also localize the site occupied by an adsorbed atom by studying its shadowing effect on the first crystal layer. It may be convenient in some cases to perform double alignment experiments and to study surface structures by varying the exit direction near a crystallographic axis, the incoming beam being kept parallel to another axis. With this method, the bulk background becomes negligible and angular scans can be performed over wide domains. However, such experiments usually require small detection solid angles and hence high beam doses which may induce some damage. A system developed by the F.O.M. group /10/ which couples an electrostatic analyzer and an energy position detector allows however one to perform such experiments with low beam doses and remarkable depth resolution (of the order of 10 Å).

As seen in Fig. 7 the number of atoms per row n is an increasing function of the ratio $\frac{\rho}{R}$. In order to obtain specific information about the surface plane, n has to be small and this ratio should be minimized. This can be done either by minimizing ρ (i.e. cooling the crystal) or by increasing R (lowering the energy or increasing d, see equ. 15). The two procedures are not equivalent : one gains in sensitivity to small displacements by lowering ρ. However, in double alignment experiments it is convenient to use low-energy beams (200 keV ^{4}He for instance) in order to increase the backscattering cross section and lower both beam damage and experimental time. Still, damage can remain a serious problem in some cases if spontaneous annealing of the surface does not take place.

It must also be pointed out that when a surface is reconstructed in such a way that the atoms occupy a great number of different sites, a channeling experiment will see this plane as randomly filled : we are reaching the limits of the technique. However, valuable information has been obtained by channeling on (100) and (111) Si reconstructed surfaces /3/.

A good illustration of the possibilities of channeling surface studies is given in fig. 9 from ref. /11/ where angular scans n(φ) were registered across the <110> axis normal to a Pt crystal surface. In order to fit the experimental data, the correlation between atom thermal vibrations had to be taken into account : a simulation neglecting correlation would estimate the shadowing effect lower and predict too large values of n(φ).

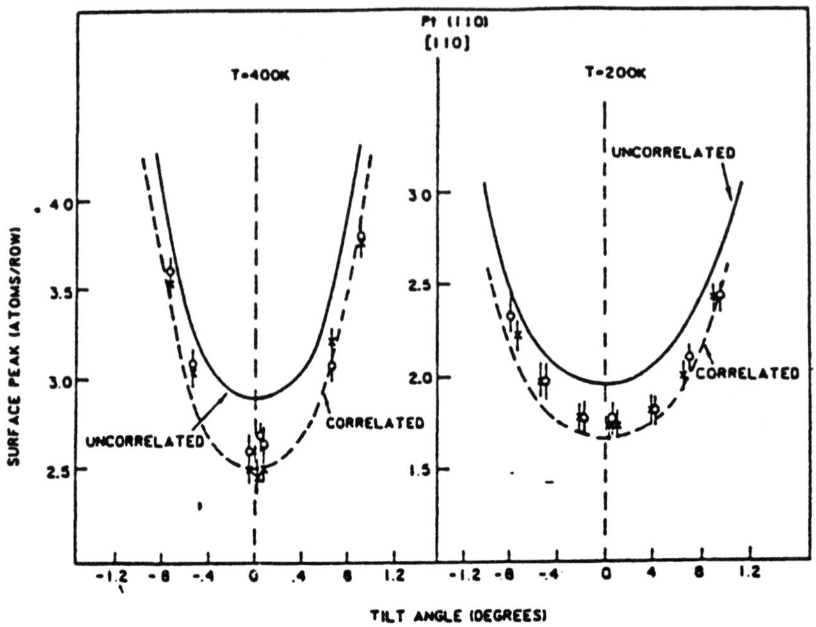

Fig. 9 Experimental and calculated angular dependence of the backscattering yield around the <110> axis of a Pt crystal for 2.00 keV 4He ions. The solid line is calculated without correlations, whereas the dashed line includes the correlations. From /11/

In fig. 10 from ref. /12/, the number of atoms per row measured in <100> aligned spectra on a (100) Pt crystal is reported as a function of crystal temperature. The experiment was performed under a partial H_2 pressure of 10^{-8} torr in the chamber. At low temperature, hydrogen is adsorbed at the surface. The sharp increase of the slope n(T) between 200 and 250 K is due to the fact that at these temperatures hydrogen desorption takes place together with a bidimensional phase transition from the (1x1) to the (5x20) reconstructed structure. In the latter situation, the surface plane is seen as randomly filled in a channeling experiment.

More generally, as the number of atoms per row corresponding to the surface peak in an ideal crystal can be calculated, channeling experiments are able to evidence the appearance of surface disorder. Surface premelting on (110) Pb crystals has been observed in this way at temperature 20K below the bulk melting temperature /13/. Also, the very first steps of oxidation of metal on semiconductor surfaces can be followed. Comparing the surface peak change induced by the oxidation, which gives the number of displaced substrate atoms to the oxygen content as measured for instance using the $^{16}O(d,P)^{17}O^*$ nuclear reaction, one gets information both on the oxide composition and on the interfacial disorder which may be induced by the oxide growth.

Stepped structures are particularly well adapted for the study of surface atomic displacements, using channeling /9/. Vicinal surfaces, i.e. with an orientation close to a low-index plane, provide in a natural way for atomic steps of known direction and density, with terraces of quasi identical width. A great variety of fundamental surface problems are related to the existence of such surfaces : stability ; influence of adsorbed species ; preferential adsorbtion along the step edges and relation with catalysis ; disorder in the terraces induced by interactions between edge of steps ; possible specific relaxation and enhanced thermal vibrations of the loosely bound step atoms etc.

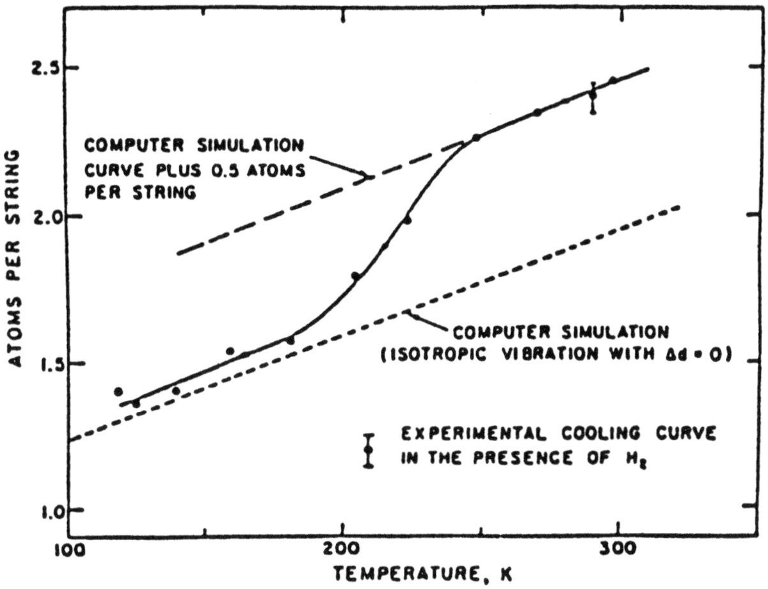

Fig. 10 Temperature dependence of the <100> surface peak on Pt(100) surface, using 2.0 MeV ^{4}He ions. From ref. /12/

The geometry of a channeling experiment performed on these structures is schematized in fig. 11 from ref. /9/. The incident beam is aligned with an axis perpendicular to the terraces, and particles backscattered at 90° along an axis <h k l > in the terraces are detected. Call n(h k l) the number of atoms per row corresponding to the surface peak and n_T (h k l) the number of atoms per row along the same direction corresponding to the terrace width ; the surface peak will represent a number of atomic planes equal to the ratio n/n_T which may be significantly smaller than 1. One obtains in this way exclusive information about the surface plane without any bulk contribution, as demonstrated by the spectra of fig. 12 from ref. /9/ which correspond to experiments performed on (16 1 1) Cu (i.e. 5° off the (100) plane in the <0 $\bar{1}$ 1> zone axis).

The contribution of the edge atoms to the surface peak can reach 50 %. Their position can hence be studied with precision. A relaxation of the edge atoms, perpendicular to the terraces, can be evidenced by recording angular scans across the <h,k,l> axis, the detector being moved vertically. A relaxation of terrace edges in the terrace plane can be studied by keeping the detector fixed at 90° and rotating the crystal in its own plane. In the experiments on (16 1 1) Cu, no step edge rela-

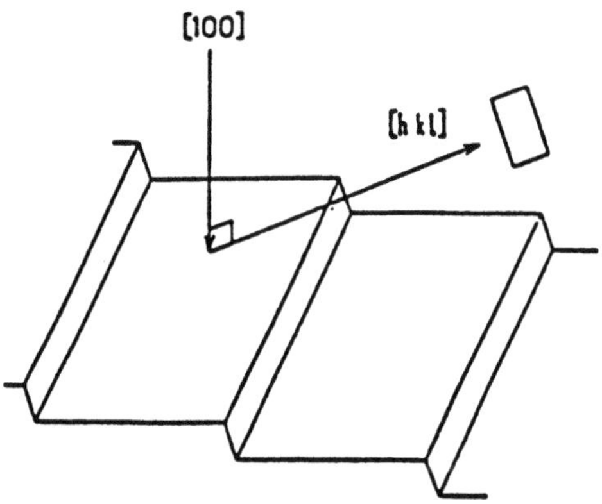

Fig. 11 Principle of a 90° double alignment experiment on a stepped structure. From ref. /9/

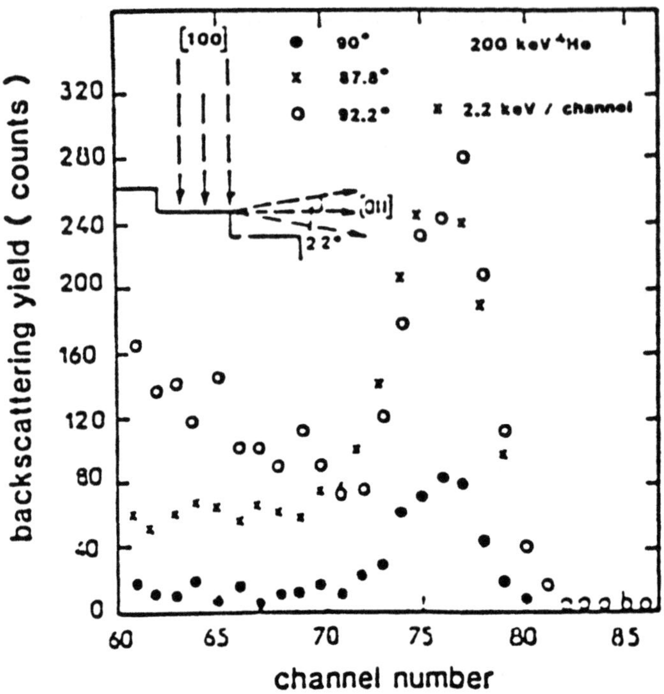

Fig. 12 R.B.S. spectra recorded in or near the 90° double alignment geometry on (16 1 1) Cu stepped structures. The surface peak of the double aligned spectrum corresponds to 0.3 atomic plane. The two surface peaks recorded with the detector at ± 2.2 deg. off the <011> axis are identical in width and integral, evidencing the fact that surface roughness is very low and does not affect the experiments. From ref. /9/

xation was evidenced, but surface disorder all over the terraces was observed. This disorder cannot be evidenced by "standard" channeling experiments which are not enough sensitive ; it depends strongly on the crystal temperature and may be related to a roughning transition. Recent experiments performed by the Authors of ref. /9/ on (4 1 0) Cu evidence this time a significant step edge relaxation : a vertical and horizontal contraction are observed for clean surfaces while vertical and horizontal expansions are seen when oxygen is adsorbed at saturation coverage on the surface. The inversion of the step edge vertical relaxation with oxygen coverage is illustrated by the angular scans of fig. 13.

IV STUDY OF EPITAXIAL LAYERS AND OF INTERFACE STRUCTURES

The very first steps of epitaxy, i.e. the growth of the first 1 - 2 monolayers are to be studied in an ultra high vacuum system by comparing the results of both classical surface techniques and channeling, as described in the preceding section. On the other hand, if one wants to look at thicker epitaxial layers the standard surface techniques become useless, while channeling remains particularly well adapted. Moreover, these thicker layers, although often prepared in U.H.V. conditions can be studied ex-situ and the coupling of U.H.V. systems with an accelerator is not, in this case, an absolute necessity.

Epitaxial metal silicides, which are of great importance in microelectronics (contacts), and which offer the possibility of generating metal - semiconductor interfaces with regular lattice structures enabling calculations of the electronic structure, have been extensively studied by channeling. For these studies the ideal complementary technique is electron microscopy.

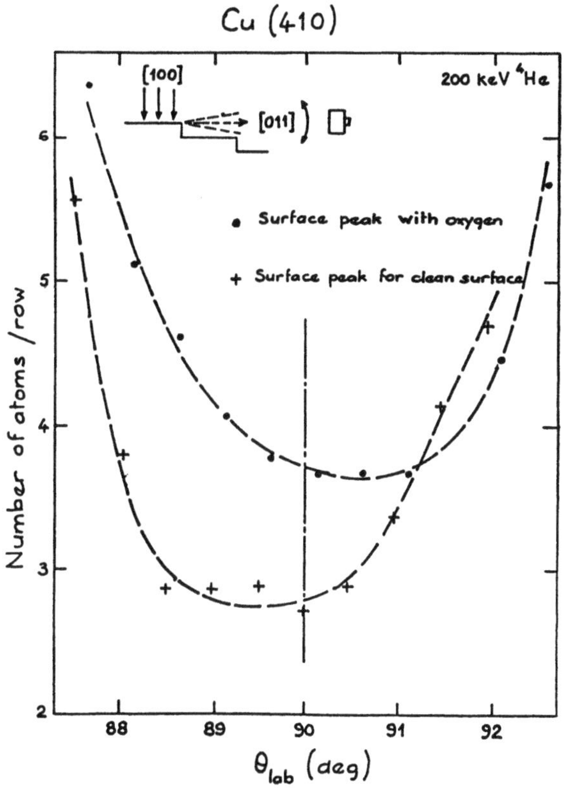

Fig. 13 Evidence for vertical edge of step contraction on clean (4 1 0) Cu surface (+) and for vertical edge of step expansion under oxygen coverage (•). In absence of any relaxation the scans would be symmetrical with respect to the axis drawn on the figure.

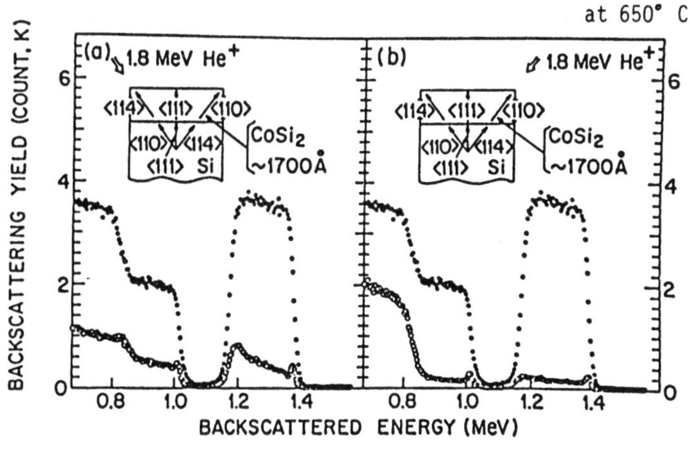

Fig. 14 Channeling and random R.B.S. spectra of a 1700 Å thick $CoSi_2$ on (111) Si formed by molecular beam epitaxy codeposition at 650° C. From ref. /14/

A typical result is presented in fig. 14 from ref. /14/. It corresponds to cobalt silicide $CoSi_2$ grown by molecular beam epitaxy on a <111> Si substrate. Channeling measurements have been performed both along the <110> and the <114> axes of the Si sample. These two axes are both in the (110) plane, they have the same polar angle (35.3°) with respect to the <111> axis normal to the crystal surface, but have a 180° difference in azimuthal coordinate. The spectra of fig. 10a show that when the beam is aligned with the <110> major axis of the substrate the channeling in the epitaxial $CoSi_2$ is poor. On the contrary (fig. 10b) when the <114> higher index axis of the substrate is studied, one sees a very strong channeling effect on $CoSi_2$. This experiment demonstrates that the $CoSi_2$ film consists of a single crystal in epitaxy

<111> / <111> with the Si substrate but with a 180° rotation. The <114> axis of the film is hence the <110> axis of the substrate and vice versa.

Our last illustration is relative to an R.B.S. and channeling study of superlattices. InAs-GaSb superlattice samples consisting of 400 Å alternate layers formed by molecular beam epitaxy have been studied in ref. /15/. The R.B.S. spectra presented in fig. 15 have been obtained by choosing the beam energy and the backscattering angle, in such a way that the energy of particles detected after backscattering on Sb from the ith GaSb layer is equal to the energy of particles backscattered on In from the $(i-1)^{th}$ InAs layer. In other terms, the energy loss in one layer compensates the difference in kinematic coefficients for Sb and In. One obtains in this way an oscillatory spectrum which reflects the alternate composition of the superlattice and enables the study of the indepth periodicity.

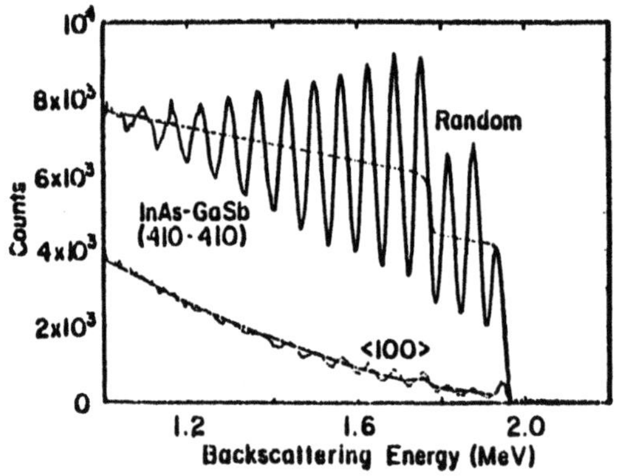

Fig. 15 Backscattering spectra for 2.23 MeV ^{4}He ions incident along the <100> channeling and random directions of an InAs-GaSb superlattice with 20 periods of 410 Å InAs and 410 Å GaSb. For comparison, backscattering measurements from a GaSb single crystal are also given as the dashed curves. From ref. /15/

Channeling experiments performed on these samples demonstrate that the dechanneling is low along the <100> growth direction. On the contrary, strong dechanneling is observed along the <110> direction. This dechanneling is probing beam energy-independent and such a behaviour implies the presence of stacking faults-type defects. The interpretation is that, despite the very good matching of the InAs and GaSb bond lengths, a significant mismatch has to be compensated at each interface where InSb and GaAs bonds exist with a 7% bond length difference. This mismatch is compensated by expansions and contractions along the <100> direction which do not affect the channeling along this axis but are seen as stacking faults along the <110> axis at 45°. Dechanneling occurs hence at each interface crossing. The dechanneling curves may be fitted by Monte-Carlo simulations. One gets in this way detailed information about the nature of the interfacial defects.

Many other experiments have been performed on strained superlattices or epitaxial bilayers on which the tetragonal distortion has been determined by comparing angular scans corresponding to the consecutive layers, these scans being taken across axes not perpendicular to the crystal surface.

It must be pointed out that compound superlattices are particularly sensitive to radiation damage and problems may arise during analysis. However, if one takes care to change often enough the beam impact point on the sample surface, the overall damage per unit area remains sufficiently low. More generally, it can be demonstrated that in mostly all surface and interface channeling studies, radiation damage, if not always completely negligible, does not have a dominant influence on the results.

ACKNOWLEDGMENTS

This work has been partially supported by the French Centre National de la Recherche Scientifique under Greco N° 86.

REFERENCES

1. W.K. Chu, J.W. Mayer and M.A. Nicolet : Backscattering Spectrometry, Acad. Press, New York, San Francisco, London, 1978
2. L.C. Feldman, J.W. Mayer and S.T. Picraux : Material Analysis by Ion Channeling, Acad. Press, New York, London, Paris, San Francisco, Sao Paulo, Sidney, Tokyo, Toronto, 1982
3. J.F. Van der Veen : Surface Sci. Reports, 5, 199 (1985)
4. G. Amsel, J.P. Nadai, E. d'Artemare, D. David, E. Girard and J. Moulin : Nucl. Instrum. and Meth. 92, 481 (1971)
5. S. Rigo, Silicon Films on Silicon, in "Instabilities in Silicon Devices" G. Barbottin and A. Vapaille editors, Elsevier Science Publishers B.V. North Holland, p. 5, 1986
6. P. Rabinzohn, G. Gautherin, B. Agius and C. Cohen : J. Electrochem. Soc., 131, 905 (1984)
7. I. Stensgaard: Nucl. Instrum. and Meth., B15, 300, (1986).
8. I. Stensgaard, L.C. Feldman and P.J. Silverman : Surf. Sci., 77, 513 (1978)
9. J.C. Boulliard, C. Cohen, J.L. Domange, A.V. Drigo, A. L'Hoir, J. Moulin and M. Sotto : Phys. Rev. B, 30, 2470 (1984)
10. R.G. Smeenk, R.M. Tromp, H.H. Kersten, A.J.H. Boerboom and F.W. Saris : Nucl. Instrum. and Meth. 195, 581, (1982)
11. D.P. Jackson, T.E. Jackman, J.A. Davies, W.N. Unertl and P.R. Norton : Surf. Sci., 126, 226 (1983)
12. P.R. Norton, J.A. Davies, D.P. Jackson and N. Matsunami : Surf. Sci., 85, 269 (1979)
13. J.W.M. Frenken and J.F. Van der Veen : Phys. Rev. Lett., 54, 134 (1985)
14. R.T. Tung, J.C. Bean, J.M. Poate, J.M. Gibson and D.C. Jacobson : Appl. Phys. Lett. 40, 684 (1982)
15. W.K. Chu, F.W. Saris, C.A. Chang, R. Luedke and L. Esaki : Phys. Rev. B, 26, 1999 (1982)

Surface Characterization by Low-Energy Ion Scattering

E. Taglauer

Max-Planck-Institut für Plasmaphysik,
D-8046 Garching, Fed. Rep. of Germany

1. Introduction

Since its first introduction by SMITH /1/ low-energy ion scattering has developed into a well-established method for investigating the composition and structure of solid surfaces. The specific features of the method lie in its sensitivity to the atomic masses in the topmost layer of a surface and in the capability to analyse the positions of these atoms relative to each other, i.e. to characterize the local atomic arrangement /2/.

Generally in low-energy ion scattering (LEIS or ISS for ion scattering spectroscopy) ions with kinetic energies between 0.5 and 2.5 keV are used, the ionic species being either noble gas (He^+, Ne^+) or alkali ions (Li^+, Na^+). From the relevant physical parameters involved some basic features of low-energy ion scattering can be considered. The kinetic energies of the order of a kilo electron volt are large compared to thermal energies of surface atoms (about 0.03 eV), i.e. interaction with phonons is negligible. This is further illustrated by considering collision times, which from ion velocities of some 10^7 cm/s and atomic dimensions turn out to be of the order of 5×10^{-16} s, again short compared to thermal vibration periods of about 10^{-13} s. Consequently, the probing ion "sees" a rigid arrangement of surface atoms in which thermal motions are only detectable as quasi-static disordering. The de Broglie wavelength of the projectiles in the considered mass and energy range is very small (10^{-2} Å) and also the distance of closest approach to the target atom is much smaller (about 0.2 Å) than a typical lattice parameter of, say, 3 Å. Consequently diffraction phenomena are not to be expected in LEIS and the impinging ion can be considered to interact mainly with individual atoms on the surface. Information about surface structure must arise from peculiarities of the collision processes as is discussed in the following sections. Finally, the ionisation energies of noble gas atoms (He 24.6 eV; Ne 21.6 eV) are much larger than typical work functions, which are about 4 to 5 eV for many metals and semiconductors, whereas the ionisation potential of alkali atoms is of the same order (Li 5.39 eV; Na 5.14 eV; K 4.34 eV). Therefore neutralisation of the impinging ions in the collision process can be expected to be much more important for noble gas than for alkali ions.

The physical parameter range being sketched, in the following sections the principles of the method, experimental techniques, analyses of surface composition and structure are outlined using illustrative examples.

2. Principles and Techniques

Low-energy ion scattering is a conceptually and experimentally fairly simple surface analysis technique. It does, however, require the usual conditions for surface analysis, such as ultra-high vacuum and adequate sample preparation, because of its extreme sensitivity to the topmost atomic layer. A schematic of a typical ISS set-up is shown in Fig. 1. Its main components are the ion source, preferably with mass separator, the specimen holder and manipulator and the energy analyzer and detector system. Important geometrical parameters are the laboratory scattering angle θ , the angle of incidence ψ and for mono-crystalline samples the azimuthal angle φ which determines the position of the scattering plane relative to the crystal surface.

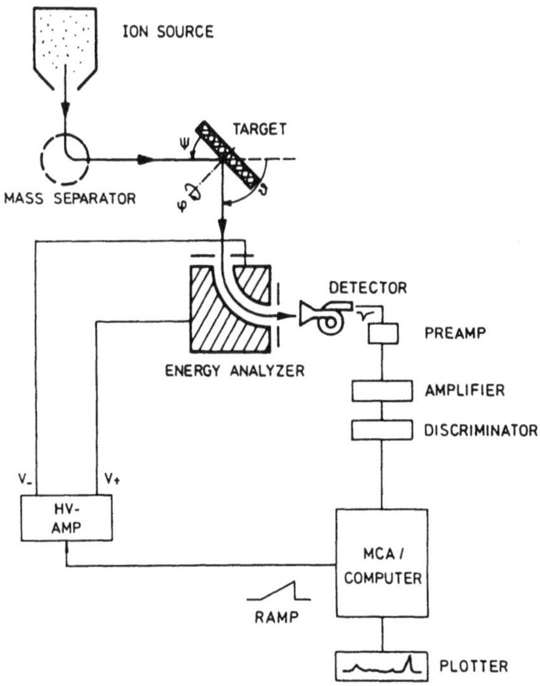

Fig. 1 Schematic of a typical experimental set-up for ISS using an electrostatic energy analyzer

Other experimental arrangements use cylindrical mirror analyzers (CMA) /3,4/, which have a larger acceptance solid angle, but less angular variability. Also time-of-flight techniques are applied /5,6/, they offer the possibility of also detecting neutral backscattered particles, but they are usually operated above 2 keV primary energy.

A typical energy spectrum of the backscattered ions is shown in Fig. 2 for Ne^+ scattering from a Zn doped GaAs sample. The intensity maxima correspond to Ne^+ scattering from Ga, As and Zr respectively. Their position in energy can be calculated from conservation of energy and momentum in a binary collision to be

$$E/E_0 = K^2(\theta, A) \quad (1)$$

where the kinematic factor K depends only on the scattering angle and the mass ratio $A = M_2/M_1$ of the projectile (M_1) and target (M_2) atoms,

$$K = \left[\cos\theta + (A^2 - \sin^2\theta)^{1/2}\right]/(1+A) \quad (2)$$

for $A > 1$. The validity of this binary collision concept is fairly well established for the ISS situation /7/.

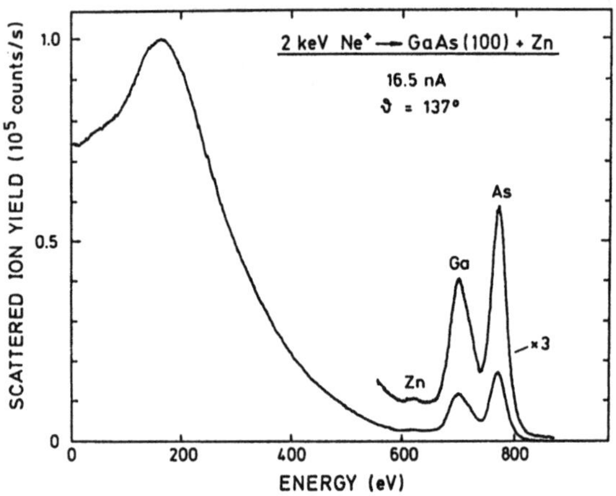

Fig. 2 Energy spectrum of 2 keV Ne^+ ions backscattered from a GaAs(100) surface at a scattering angle of 137^0

The intensity in the scattering peaks is determined by the scattering cross section σ and the probability P that the ion escapes neutralisation during the collision process. Thus, for scattering from an atomic species with surface density N into the solid angle $\Delta\Omega$ we obtain

$$I = k \cdot I_0 \cdot N \cdot (d\sigma/d\Omega) \cdot \Delta\Omega \cdot P \quad (3)$$

where k is an apparatus factor containing transmission and detection probabilities and I_0 is the incoming ion flux. According to (3) the surface density N can be determined if the scattering cross section and the neutralization probability are known. The differential scattering cross section depends on the energy, the scattering angle and the nuclear mass and charge numbers of the collision partners. $d\sigma/d\Omega(E,\Theta,Z_1,Z_2)$ can be sufficiently well calculated with appropriate interaction potentials such as screened Coulomb potentials

$$V(r) = Z_1 \cdot Z_2 \cdot e^2 \cdot r^{-1} \cdot \phi(r) \quad (4)$$

using the Molière approximation to the Thomas-Fermi screening function $\phi(r)$ /7/. Calculated scattering cross sections for helium with three different energies as a function of the target atom nuclear charge number are plotted in Fig. 3. The strongest variation with Z_2 is with light elements.

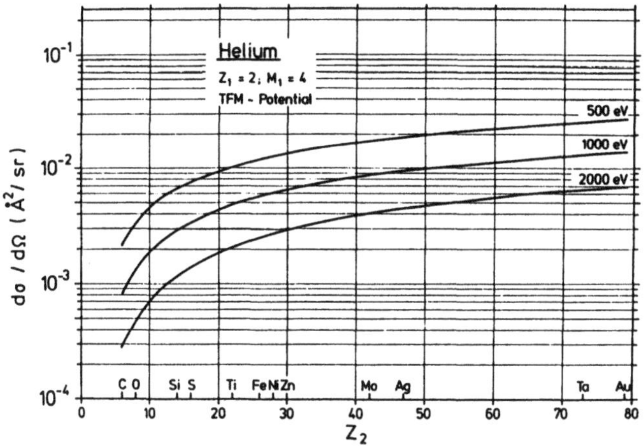

Fig. 3 Calculated dependence of the scattering cross section on the target atomic number for three different primary energies. The scattering angle is 137°

However, the method is not absolutely quantitative because the factor P is not equally well known. This is demonstrated in Fig. 4 in which the experimentally determined scattered ion yields corresponding to the situation of Fig. 3 are plotted. Considerable variations due to different neutralization probabilities are observed. Thus for quantitative surface composition analysis proper calibration is necessary.

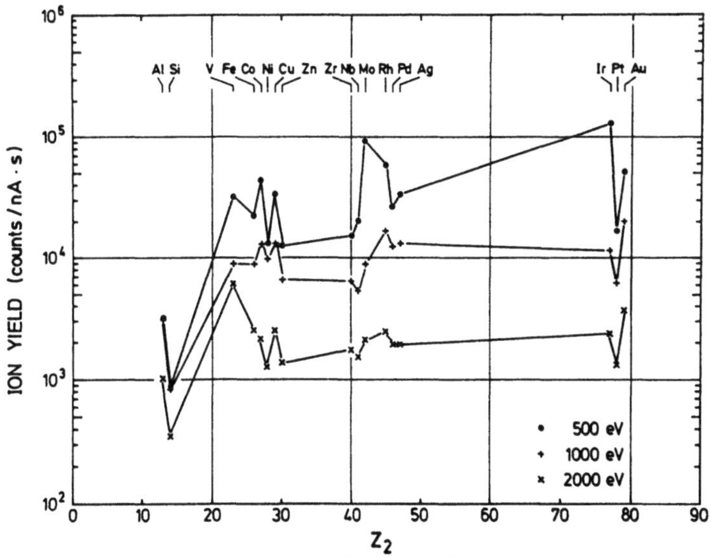

Fig. 4 Yields (counts/s) of He^+ ions scattered from various elemental surfaces, given in relation to the target current. Data taken with a cylindrical mirror analyzer (Θ = 137°) at three different primary energies

Ion survival probabilities have been found to be as low as 10^{-2} or less for He^+ /8/ and Ne^+ /9/ scattering from metals and semiconductors /10/, their value can also depend on the particle trajectory. Some of these problems can be overcome by using alkali ions /8,9,11/ which have survival probabilities of the order of one. Their spectra exhibit a considerable amount of multiple scattering. This can be partly reduced by applying large scattering angles /12/ which are of particular interest in structure analysis as treated in Sect. 4.

3. Surface Composition

Surface composition analysis by ISS has been successfully applied to a variety of problems such as adsorption layers /13/, alloy segregation /14/, supported catalysts /15/, semiconductor surfaces /16/, etc. In general various atomic species contribute to the scattering signal in proportion to their density in the surface atomic layer according to (3). As an example the surface composition changes with temperature of a P-doped GaAs sample are shown in Fig. 5. This kind of material is used for polarized photo-electron sources /17/. Oxygen deteriorates its efficiency and can be removed by heating, but at the same time the P dopant is lost at the surface. At higher temperatures (above 500°C) segregation of Ga occurs. This behavior is demonstrated in Fig. 5 which also shows that the increase of the Ga signal due to segregation to the surface simultaneously leads to the decrease of the underlying As signal. That is, the signal from the second atomic layer, or a substrate in general, is shadowed by the segregated or adsorbed overlayer. Therefore the intensity of the substrate S can be written as decreasing with the adsorbate density N_A:

$$I_S = k \cdot I_0 \cdot (N_S - \alpha N_A) \cdot (d\sigma/d\Omega) \cdot P_S \qquad (5)$$

where α is a shadowing factor which depends on the size of the adsorbate species and the scattering geometry.

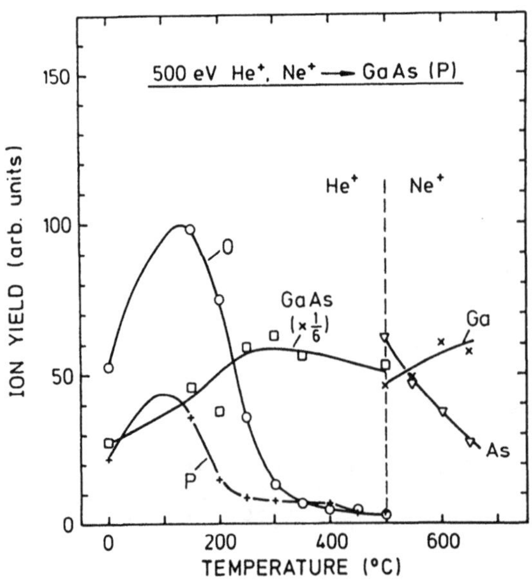

Fig. 5 Variation of the surface composition of a GaAs(P) sample with temperature. Heating up to 300°C leads to removal of O and also P. Above 500°C Ga segregates to the surface. In the lower temperature range He^+ scattering is used in order to detect the lighter elements O and P; in the higher temperature range Ne^+ scattering is used to obtain better mass resolution and thus separate Ga and As peaks

The shadowing effect is best illustrated with the concept of a shadow cone which is the cone-shaped volume behind a scattering atom into which the primary projectile flux cannot penetrate due to the deflection by the scattering atom. This is illustrated in Fig. 6. The left hand side shows a number of He trajectories whose envelope forms the shadow cone, the right hand side shows the flux distribution at various distances behind the scattering Ni atom. The calculation includes thermal vibrations.

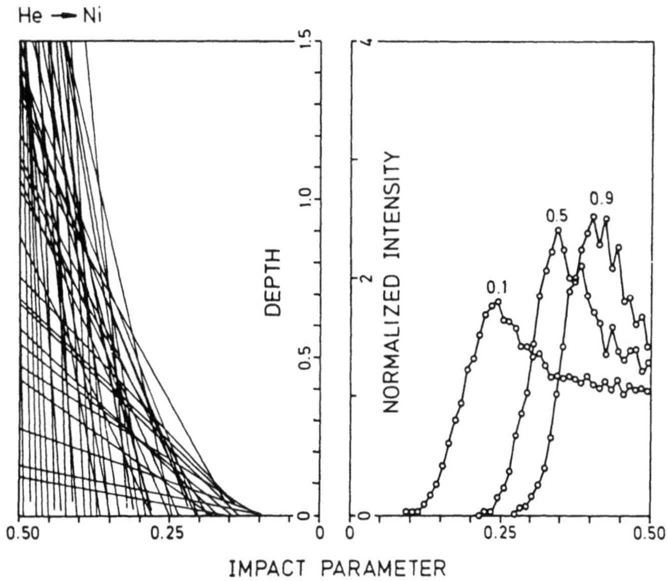

Fig. 6 Calculated particle trajectories (left) and flux distributions (right) showing the shadow cone formed by a Ni atom exposed to the flux of He particles impinging from the bottom upwards. The Ni atom is at the origin in both diagrams. Intensity distributions are given for three distances behind the Ni atom, the unit length on both axes is 3.52 Å. Target temperature corresponds to 300 K

The shadow cone concept not only illustrates how substrate atoms can be effectively shadowed by a - not necessarily complete - layer of adsorbate atoms. It can be even more quantitatively exploited for surface structure determination as shown in the following section.

4. Surface Structure

A particularly successful method for structure determination with ISS is the use of large scattering angles (ideally close to 180°, in practice values above 145° are sufficient), i.e. very small impact parameters. In this case the shadow cone concept for individual atoms can be applied. The technique was first introduced by M. AONO et al. /18/ and named impact collision ion scattering spectroscopy, ICISS. The scattering geometry is shown in Fig. 7. When the angle of incidence is raised, a surface atom

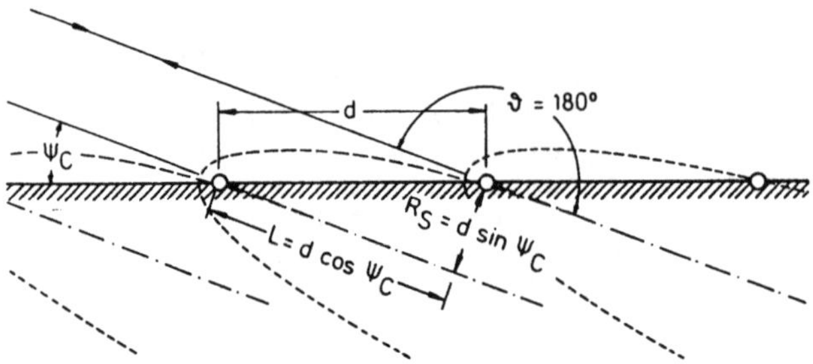

Fig. 7 Shadow cones and geometry for ICISS

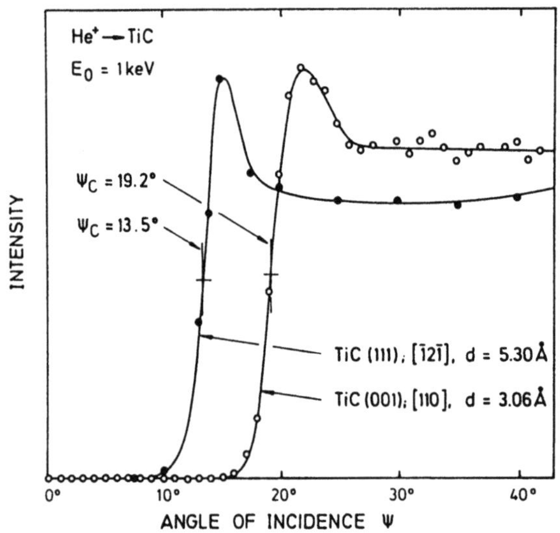

Fig. 8 Intensity distribution as a function of ψ in the ICISS technique ($\Theta = 163^{0}$) for scattering along two different crystallographic directions on TiC crystal surfaces /18/

can move out of the shadow cone of the preceding atom and thus passes through the primary ion flux distribution as given on the right hand side of Fig. 6. The corresponding scattering intensity dependence on ψ for a TiC single crystal surface is shown in Fig. 8. The steep intensity increase is connected to the critical angle ψ_c at which a surface atom is located at the edge of the shadow cone of the preceding atom. For a known surface geometry, i.e. atomic distance d, the shadow cone radius R_s can be calibrated and then applied to other structures for measuring the unknown distances of neighboring atoms. By this means a number of reconstructed surfaces /19/, adsorbate positions /20/ and also thermal amplitudes /21,22/ have been determined.

The variation of ψ_c with atomic distance d is plotted in Fig. 9 for 1 keV He^+ and Ne^+ scattering from Ni. From the slope of these curves the accuracy of the determination of atomic distances can be obtained. Around a typical value of d = 2.5 Å for Ne a slope of $\Delta \psi_c / \Delta d = 10^0/1$ Å is found. That is, if ψ_c can be determined with an accuracy of 1^0, the accuracy for d is 0.1 Å.

For further illustration the determination of the reconstructed (2x1) structure of Si(001) is shown in Figs. 10 and 11 /23/. The azimuthal intensity variation at an angle of

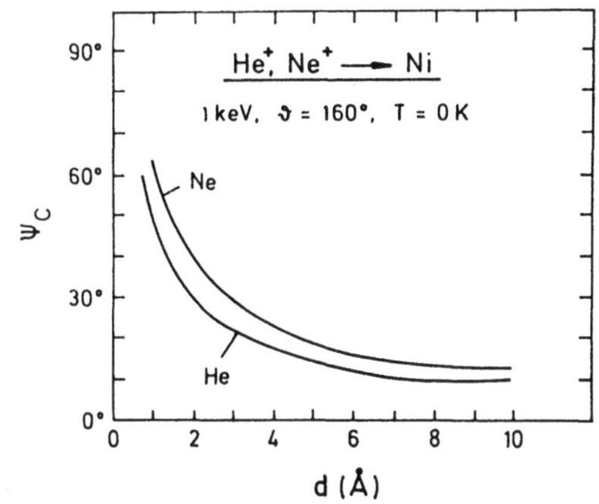

Fig. 10 Azimuthal intensity variation for 1 keV He^+ scattering from Si(001); $\psi = 4^0$, $\Theta = 164^0$ /23/

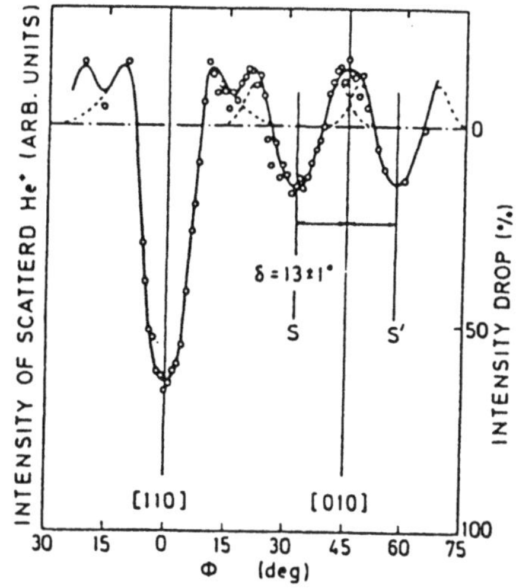

Fig. 9 Calculated dependence of the critical angle ψ_c on the atomic distance d

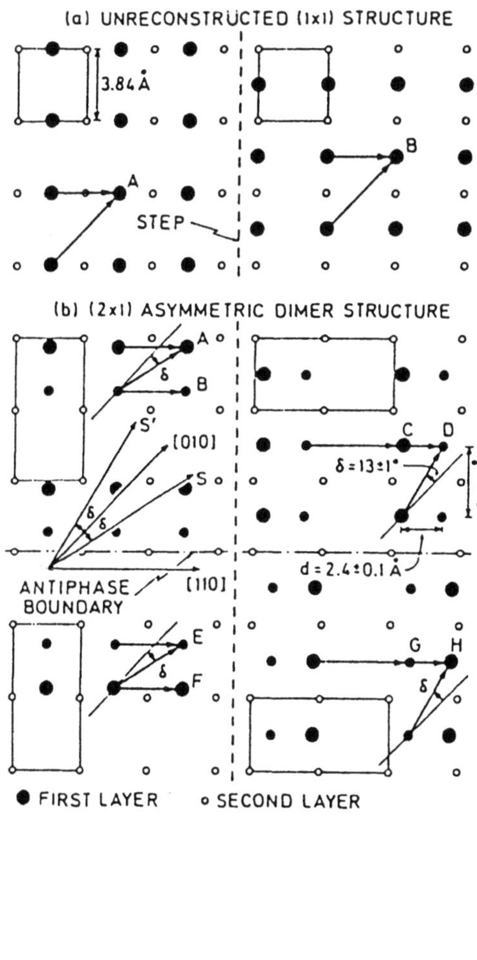

Fig. 11 Model of the (2x1) asymmetric dimer structure of Si(001) /23/

incidence $\psi = 4^{o}$ is shown in Fig. 10. The intensity drops due to shadowing at $\varphi = 0^{o}$, 32° and 58° can be explained by the structure shown in Fig. 11. An asymmetric dimer structure with an intradimer distance of 2.4 Å parallel to the surface is in agreement with the azimuthal dependence shown in Fig. 10. The shadowing effects in [110] direction and at azimuths S and S' at angles of $\pm 13^{o}$ off [010] can be ascribed to the atomic positions in Fig. 11. The size of each intensity drop is determined by the number of atoms contributing to the shadowing effect, labelled A to H, this also gives quantitative agreement between the model and the experimental result.

5. Concluding Remarks

Low-energy ion scattering is a very useful method for determining atomic surface compositions and particularly also surface structures in terms of relative atomic positions. Its strength lies in the application to fairly well defined surface systems and not so much in routinely analysing "technical" surfaces of unknown condition. Since it is not an absolutely quantitative method, relative measurements during variation of one parameter of interest are preferable. The technique of backscattering at large scattering angles can be very successfully applied to the analysis of surface structures with an accuracy of 0.1 Å. The use of alkali ions or neutral particle detection in time-of-flight devices considerably reduces the ion fluence (and thus surface damage) necessary to record one spectrum. But these techniques lose to some extent the exclusive sensitivity to the outermost atomic layer which is provided by the effective neutralization of noble gas ions which penetrate below the surface atomic layer and undergo multiple scattering processes.

References

1. D.P. Smith: J. Appl. Phys. 18, 340 (1967)
2. E. Taglauer: Appl. Phys. A38, 161 (1985)
3. H.H. Brongersma, T.M. Buck: Surf. Sci. 53, 649 (1975)
4. E. Taglauer and W. Heiland: In Applied Surface Analysis, ed.by T.L. Barr, L.E. Davis (American Soc. for Testing and Materials, Philadelphia 1980) p. 111
5. T.M. Buck, G.H. Wheatley, G.L. Miller, D.A.H. Robinson, Y.-S. Chen: Nucl. Instr. Meth. 149, 591 (1978)
6. S.B. Luitjens, A.J. Algra, E.P.Th.M. Suurmeijer, A.L. Boers: J. Phys. E: Sci. Instrum. 13, 665 (1980)
7. W. Heiland, E. Taglauer: In Methods of Experimental Physics, Vol. 22, ed. by R.L. Park, M.G. Lagally (Academic Press, Orlando 1985) p. 294
8. E. Taglauer, W. Englert, W. Heiland, D.P. Jackson: Phys. Rev. Lett. 45, 579 (1981)
9. G. Engelmann, E. Taglauer, D.P. Jackson: Nucl. Instr. Meth. Phys. Res. B13, 240 (1986)
10. A. Richard, H. Eschenbacher: Nucl. Instr. Meth. Phys. Res. B2, 444 (1984)
11. S.H. Overbury: Nucl. Instr. Meth. Phys. Res. B2, 448 (1984)
12. H. Niehus, G. Comsa: Nucl. Instr. Meth. Phys. Res. B15, 122 (1986)

13. E. Taglauer, W. Heiland, J. Onsgaard: Nucl. Instr. Meth. 168, 571 (1980)
14. T.M. Buck: In Chemistry and Physics of Solid Surfaces IV, ed. by R. Vanselow and R. Howe, Springer Ser. Chem. Phys. 20 (Springer, Berlin, Heidelberg 1982) p. 435
15. H. Jeziorowski, H. Knözinger, E. Taglauer, C. Vogdt: J. Catal. 80, 286 (1983)
16. H.H. Brongersma, P.M. Mul: Surf. Sci. 35, 355 (1973)
J. Onsgaard, W. Heiland, E. Taglauer: Surf. Sci. 99, 112 (1980)
17. J. Kirschner, H.P. Oepen, H. Ibach: Appl. Phys. A30, 177 (1983)
18. M. Aono: Nucl. Instr. Meth. Phys. Res. B2, 374 (1984)
19. H. Niehus, G. Comsa: Surf. Sci. 140, 18 (1984)
20. Th. Fauster, H. Dürr, D. Hartwig: Surf. Sci. 178, 657 (1986)
21. G. Engelmann, E. Taglauer, D.P. Jackson: Surf. Sci. 162, 921 (1985)
22. M. Aono, R. Souda: Jap. J. Appl. Phys. 24, 1249 (1985)
23. M. Aono, Y. Hou, C. Oshima, Y. Ishizawa: Phys. Rev. Lett. 49, 567 (1982)

Part IV

Electronic Properties of Interfaces

Band Structure Theory of Semiconductor Surfaces and Interfaces

C. Calandra and F. Manghi

Dipartimento di Fisica, Università di Modena, I-41100 Modena, Italy

1. INTRODUCTION

The purpose of a theory of the electronic properties of surfaces and interfaces is twofold. On one hand it has to account for the surface stability by explaining the occurrence of reconstructed surfaces, the formation of interface compounds, the abruptness of some interfaces and the presence of intermixing in others. To this end one has to determine the total electronic energy in the ground state of the system under consideration with the highest accuracy. On the other hand a good theory has to provide a satisfactory description of the spectrum of the elementary excitations (electron-hole energies, plasmons, excitons...) of the surface or interface investigated. This aim can only be attained by including many-body effects in order to calculate the self-energy and lifetime of a given excitation and to account for spectral lineshapes. In spite of the considerable efforts that have been made in the last few years, this program is far from being completed even for the simplest systems. The difficulties are more computational than conceptual in character: even adopting the simplest approximations, the evaluation of the quasi-particle energy spectrum or the determination of the stable geometry of reconstructed surfaces with a high number of atoms in the unit cell are too time consuming.

In this paper we will review some of the most important progress that has been recently made in the theoretical determination of the electronic structure of surfaces and interfaces. Due to the great variety of systems that have been studied it is not possible to discuss in a single paper all the work that has been done in the field. For this reason we will limit our considerations to some specific systems. For details on other surfaces or interfaces not considered in the present work we refer the reader to review articles which have appeared recently in the literature /1-6/.

2. GENERAL FEATURES OF THE ELECTRON-SOLID INTERACTION

In order to describe the motion of the electrons in a semi-infinite solid, we have to specify the electron-solid interaction. By this term we indicate some sort of potential representing the interaction of a single electron with the rest of the crystal. Due to the many-body character of the crystal hamiltonian such a potential does not have a simple form and is generally non-local and energy dependent. It can be derived from the full many-body hamiltonian under certain approximations which we will detail in this section.

It is interesting to discuss some aspects of this potential separately in order to illustrate the physical consequences of its behaviour. In a very rough way it is possible to distinguish three different contributions to the electron-solid potential: i) the long-range part, giving the behaviour of the potential far from the solid; ii) the short-range part which describes the interaction of the electrons with the screened ions near the surface and inside the solid; iii) the dissipative part, arising from the interaction of the electron with the elementary excitations of the solid.

2.1 The long-range part

In the proximity of a surface the Coulomb field Φ of an electron moving on the vacuum side is modified by the boundary conditions imposed by the Maxwell equations at the surface. As a consequence the self-interaction of the electron in the same Coulomb field is changed. By using classical electrostatics one can easily show that the electron self-interaction near the interface between the vacuum and a medium of dielectric constant ε is given by /7/

$$\phi(z) = -\frac{e^2}{4z}\frac{\varepsilon-1}{\varepsilon+1} = -\gamma(\varepsilon)\frac{e^2}{4z}, \tag{1}$$

where z is the distance between the electron and the solid surface. The dielectric is taken to occupy the half-space $z \leqslant 0$, while the electron is confined to $z > 0$. $\phi(z)$ is often referred to as the potential felt by the electron due to the positive image charge inside the solid and is called "image potential". From the microscopic point of view $\phi(z)$ arises from the interaction between the electron and the charge density fluctuations induced in the solid; it describes correctly this interaction far from the surface, where the external electron can be considered distinguishable from the electrons of the medium. At short distances this approximation breaks down and a quantum mechanical treatment is needed.

An important consequence of this long-range behaviour is the occurrence of "image-like" surface states in a solid. They are eigenstates of the equation

$$\left[-\frac{\hbar^2}{2m}\nabla^2 - \gamma(\varepsilon)\frac{e^2}{4z}\right]\psi(\underline{r}) = E\,\psi(\underline{r}) \tag{2}$$

in the vacuum side. The nature of the boundary conditions at the surface z=0 depends on the type of material. The most general boundary conditions require a continuity of the wavefunction across the surface. However, the main features of the spectrum can be derived by imposing the simplest boundary condition, namely that ψ vanishes at the surface just as it would for a negative electron affinity material /8-10/. Under such a condition the eigenfunctions of (2) take the separable form

$$\Psi_{k_{\shortparallel}n}(\underline{r}_{\shortparallel},z) = \frac{1}{2\pi}\, e^{i\underline{k}_{\shortparallel}\cdot\underline{r}_{\shortparallel}}\,\Theta_n(z), \tag{3}$$

where $k_{\shortparallel}$ represents the wavevector for free motion parallel to the surface and $\Theta(z)$ are solutions of the one-dimensional equation

$$\frac{d^2\Theta_n(z)}{dz^2} + \frac{2m}{\hbar^2}\left[\xi_n + \gamma(\varepsilon)\frac{e^2}{4z}\right]\Theta_n = 0. \tag{4}$$

The eigenvalues ξ_n are hydrogenic in character and are

$$\xi_n = -\frac{e^4 m}{32\,\hbar^2 n^2}\gamma^2 \quad n=1,2,\dots. \tag{5}$$

Image state bands are therefore given by

$$E_{k_{\shortparallel}n} = \xi_n + \frac{\hbar^2 k_{\shortparallel}^2}{2m}. \tag{6}$$

The general expression for $\Theta(z)$ is

$$\Theta_n(z) = (C_n / n!)\, e^{-\tau_n/2} \left(\frac{d}{d\tau_n} - 1 \right)^{n-1} (\tau_n)^n , \quad (7)$$

where $\tau = (e^2 m\gamma / 2n)z$ and C_n is a normalization constant. For n=1 and n=2 we have respectively

$$\Theta_1(z) = (2/b^{3/2})z\, e^{-z/b} , \quad (8)$$

$$\Theta_2(z) = (1/2b)^{3/2}\, z(2-z/b)\, e^{-z/2b} , \quad (8')$$

where b is the characteristic length of the problem and is given by

$$b = \frac{4\hbar^2}{e^2 m\gamma} . \quad (9)$$

It is useful to make an estimate of the binding energy of these image surface states. For silicon $\varepsilon \simeq 12$, $\gamma \simeq 0.846$ and for $k_\parallel = 0$ we have

$$\xi_n = 0.85\, \gamma(\varepsilon)/n^2 \simeq 0.719/n^2 \text{ eV} . \quad (10)$$

Therefore the lowest image surface state lies less than 1 eV below the vacuum level. Normally these states are empty since in most semiconductors the Fermi energy lies around 4 eV below the vacuum level.

Experimental observation of image surface states has been possible up to now on metal surfaces using inverse photoemission /11,12/, tunneling spectroscopy /13,14/ and two-photon photoemission /15/. They have been identified at the Si (111) (2X1) surface by scanning tunneling microscopy /16/. In general the hydrogenic model was found adequate, the experimentally observed deviations from (10) being small.

2.2 The short-range part

We turn now to the discussion of the short-range part of the electron-solid force. This is essentially the same interaction felt by an electron inside the solid, although at the surface it is somewhat modified by the redistribution of the valence charge. Strictly speaking we cannot write down a single-particle potential and single-particle states in a solid, due to the many-body character of the solid hamiltonian. However, for a description of the electronic properties of the ground state of the system we can resort to the density functional theory and derive a set of single-particle equations through which it is possible to determine the total energy and the electron density of the system.

This theory is based on a theorem by Hohenberg and Kohn /17/ stating that all the ground state properties can be expressed as a functional of the electron density $n(\underline{r})$. The total energy is such a functional and is written as

$$E[n] = \int v(\underline{r})\, n(\underline{r})\, d\underline{r} + \frac{e^2}{2} \iint \frac{n(\underline{r})n(\underline{r}')}{|\underline{r}-\underline{r}'|}\, d\underline{r}d\underline{r}' + T_s[n] + E_{xc}[n] , \quad (11)$$

where $E[n]$ is the total energy of a system in the external potential $v(\underline{r})$, $T[n]$ is the kinetic energy of a system of non-interacting electrons with density $n(\underline{r})$. The exchange-correlation energy $E_{xc}[n]$ defined as the difference between $E[n]$ and the first three terms in the previous equation, is a universal functional of $n(\underline{r})$ and contains all the many-body effects beyond the Hartree theory. $E[n]$ achieves its minimum value for the correct ground state energy. To find this minimum energy one can use the theory due to Kohn and Sham /18/ that formally reduces the N-electron problem to the solution of N simultaneous single-particle equations

$$\left\{-\frac{\hbar^2}{2m}\nabla^2 + v(\underline{r}) + e^2\int\frac{n(\underline{r}')}{|\underline{r}-\underline{r}'|}d\underline{r}' + v_{xc}(\underline{r})\right\}\psi_i(\underline{r}) = \varepsilon_i\,\psi_i(\underline{r}), \quad \text{where} \tag{12}$$

$$n(\underline{r}) = \sum_i |\psi_i(\underline{r})|^2 \tag{12'}$$

and v_{xc}, called the exchange and correlation potential, is given by the functional derivative

$$v_{xc} = \frac{\delta E_{xc}}{\delta n}. \tag{12''}$$

Equation (12) is a single-particle equation describing the motion of an electron in the field of the ions and in the presence of the other electrons of the solid. It is seen that the total potential is made up of three contributions: i) the external attractive potential $v(\underline{r})$ of the ions; ii) the Hartree potential giving the electrostatic repulsion caused by the electron distribution in the ground state; iii) the exchange-correlation potential including all the complicated many-body effects arising from the electron-electron interaction.

A few comments on these equations are in order. First we notice that both the Hartree and the exchange-correlation potential depend functionally upon the electron charge density $n(\underline{r})$. This implies that the N single-particle equations are to be solved in a self-consistent way. The requirement of self-consistency is rather important in surface calculations due to the necessity of accounting for the redistribution of the valence charge taking place in the outermost planes.

A second important comment concerns v_{xc}. To derive an expression for this potential we need to know the dependence of E_{xc} on $n(\underline{r})$. An exact expression connects E_{xc} to the pair correlation factor $G(\underline{r},\underline{r}')$ of the electron system /19/

$$E_{xc} = \frac{e^2}{2}\int d\underline{r}\int d\underline{r}'\; n(\underline{r})\, n(\underline{r}')\, G(\underline{r},\underline{r}') \,/\, |\underline{r}-\underline{r}'|. \tag{13}$$

However $G(\underline{r},\underline{r}')$ is not known and one has to use some approximation. Notice that one can alternatively write down E_{xc} in terms of the so-called exchange-correlation hole $n_{xc}(\underline{r},\underline{r}')$ /19/

$$E_{xc} = \frac{e^2}{2}\int d\underline{r}\int d\underline{r}'\; n(\underline{r})\, n_{xc}(\underline{r},\underline{r})/|\underline{r}-\underline{r}'|, \tag{14}$$

which shows that E_{xc} is basically given by the interaction of the electron charge density $n(\underline{r})$ and the charge density of the exchange-correlation hole. Notice that

$$n_{xc}(\underline{r},\underline{r}') = n(\underline{r}')\, G(\underline{r},\underline{r}') \tag{15}$$

and the following sum rule holds

$$\int d\underline{r}'\; n_{xc}(\underline{r},\underline{r}') = 1, \tag{16}$$

stating that the exchange-correlation hole contains one electron.

As to the physical meaning of the one-particle eigenvalues ε_i, we notice that in the density functional theory (DFT) they have no formal justification as quasi-particle energies. Rather they must be seen as ingredients to calculate the total energy of the system. In spite of that they are commonly regarded as single-particle binding energies and used to interpret the experimental data. Indeed some important features of the experiments, for example the shape of the valence bands in XPS, have been succesfully interpreted using the Kohn-Sham eigenvalues. However, significant discrepancies occur in the description of the excited states. For example the band gap of most semiconductors turns out to be nearly 50% lower than in the experiments /20-22/. Of course for the purpose of calculating the total energy of the system in its ground state the DFT is perfectly adequate, once one has obtained a reasonable approximation for the exchange-correlation energy.

2.3 The dissipative part

Unlike Kohn-Sham eigenvalues ε_i, the quasi-particle energies E_i are generally complex and can be obtained by solving the single-particle equation

$$\left(-\frac{\hbar^2}{2m}\nabla^2 + v + e^2\int\frac{n(\underline{r}')}{|\underline{r}'\underline{r}'|}d\underline{r}'\right)\psi_i(\underline{r}) + \int\Sigma(\underline{r},\underline{r}',E_i)\,\psi_i(\underline{r}')\,d\underline{r}' = E_i\psi_i(\underline{r}) . \quad (17)$$

Here ψ_i is the quasi-particle wavefunction and $\Sigma(\underline{r},\underline{r}';E)$ is the non-local energy dependent self-energy operator which is non-Hermitian in character /23/. The imaginary part of E_i corresponds to the damping of the excited states due to the electron-electron interaction. We can write (17) in the following form:

$$\left(-\frac{\hbar^2}{2m}\nabla^2 + v + e^2\frac{n(\underline{r}')}{|\underline{r}'-\underline{r}|}d\underline{r}' + v_{xc}\right)\psi_i + \int[\Sigma(\underline{r},\underline{r}',E_i)-v_{xc}\delta(\underline{r}-\underline{r}')]\times$$

$$\psi_i(\underline{r}')d\underline{r}' = E_i\,\psi(\underline{r}) . \quad (18)$$

The term in square brackets represents the difference between the equations satisfied by quasi-particle states and the Kohn-Sham equations. This part has to be taken into account if one wants to describe the excitations of the system and it is what we call the dissipative part. If it were negligible then the Kohn-Sham equations would give the quasi-particle energy spectrum. However, in the last few years it has become clear that some fundamental properties of a solid, like the band gap of a semiconductor or the shape of the Fermi surface in a metal, require some evaluation of the self-energy /24,25/ and cannot be achieved within the density functional approach. Nearly all the errors occurring in the description of excited states by Kohn-Sham eigenvalues are inherent in the ground state band structure and cannot be simply eliminated by choosing more accurate v_{xc} /25/. This is equivalent to saying that the interaction between the electron and the excitations of the solid is essential in order to reproduce the experimental information.

3. THEORETICAL METHODS

Having outlined the general features of the electron-solid interaction we turn now to discuss the approximations that have been adopted in surface band structure calculations. Most of the work performed up to now on semiconductors surfaces has been based on the density functional theory in the local density approximation (LDA). This is achieved by replacing (15) by the approximate expression

$$n_{xc}^{LDA} = n(\underline{r})\ \ G^h\ (\underline{r}-\underline{r}';\ n(\underline{r})\ ), \tag{19}$$

where G^h is the correlation factor of an homogeneous electron gas of density $n(\underline{r})$. Equation (19) is formally correct in the homogeneous limit when $n(\underline{r})$ is constant. This suggests that the LDA can be a good approximation when the actual $n(\underline{r})$ is slowly varying, i.e. almost constant, with slow variations in space. These conditions are approximately met in many solids, particularly metals, and for this reason the LDA has been widely and successfully used in bulk band structure calculations. However, in the presence of a surface these conditions are not satisfied, due to the rapid variations of the electron charge density near the surface. Inserting (19) into (14) we get

$$E_{xc}^{LDA}\,[\,n\,] = \int n(\underline{r})\ \varepsilon_{xc}\ [n(\underline{r})]\ d\underline{r}\ , \tag{20}$$

where ε_{xc} is the exchange-correlation energy per electron of a homogeneous electron gas of density $n(\underline{r})$. From (20) the exchange-correlation potential is easily derived:

$$v_{xc}^{LDA}\ (\underline{r}) = \varepsilon_{xc} + n(\underline{r})\ \frac{d\ \varepsilon_{xc}}{dn}\ . \tag{21}$$

ε_{xc} has been calculated very accurately for the homogeneous system at different densities. Explicit forms can be found in the literature /26,27/. It has been pointed out by a number of authors that LDA, when applied to surface problems, does not reproduce the long-range behaviour of the image potential /19,28/. Such a long-range behaviour is a consequence of the fact that when the electron moves out of the solid the exchange-correlation hole lags behind giving rise to the Coulomb potential $-e^2/4z$ at large distances. In order to treat both the long-range and the short-range part of the potential on the same footing we need to adopt an approximation for the exchange-correlation hole more realistic than (19). Clearly a calculation of the surface band structure using the LDA does not give the image state bands. However, if one is interested in the filled states or in the lowest empty bands the inclusion of image effects is not essential.

Attempts to go beyond the LDA using a form for v_{xc} which includes the long-range part of the interaction have been based on a non-local approach to n_{xc} proposed by Gunnarsson and Jones /29/. These authors suggest writing

$$n_{xc}^{NLDA}\ (\underline{r},\underline{r}') = n(\underline{r}')\ G(\underline{r},\underline{r}';\ \overline{n}(\underline{r})\ )\ , \tag{22}$$

where $G(\underline{r},\underline{r}',\overline{n}(\underline{r}))$ is an approximate correlation factor of the form

$$G(\underline{r},\underline{r}';\overline{n}(\underline{r})\ ) = c(\overline{n}(\underline{r}))\ \left\{ 1-\exp(\lambda(\overline{n}(\underline{r}))/|\underline{r}-\underline{r}'|^5)\right\}. \tag{23}$$

This analytical form gives the right behaviour far from the surface. This approach has been used by Ossicini and coworkers /30,31/ to study the behaviour of the valence charge density at metal surfaces using a jellium model and in atomic structure calculations. Gies /32/ has applied the results for jellium to evaluate the deviations in the binding energy of image states from the hydrogenic model, using a quantum defect theory. Calculations of the semiconductor bulk band structure using this approach do not lead to significant improvement compared to the LDA results /22/.

Up to now only one calculation of semiconductor surface electronic structure has been performed with a non-local form for the exchange-correlation energy /33/.

Most of the results we will present in the next section have been obtained using the LDA. No theoretical results are available which include self-energy effects.

A common feature of all the calculations performed up to now on semiconductor surfaces is the use of pseudopotential theory (we do not consider here semi-empirical tight-binding calculations since our interest is focused on first-principle approaches). In many cases the effects of non-locality in the pseudopotentials have been neglected and simple local forms have been adopted that reproduce the ion energy level sequence with good accuracy /34,35/. In view of the difficulties of reproducing the band gap of the semiconductors in LDA many authors have adopted the Slater approximation to the exchange-correlation potential /36/. This approximation leads to gaps in better agreement with the experiments, although is probably inaccurate for total energy calculations. Ground state energy minimizations to determine the stable surface or interface geometry have been performed using the LDA approximation. The use of pseudopotentials and plane wave expansion is not appropriate when dealing with transition metal-silicon interfaces; these systems have been thoroughly investigated in the last few years with the purpose of understanding the basic properties of the metallic contact, usually formed by a silicide phase. With only few exceptions most of the work in this field has been performed using semi-empirical tight-binding approaches /37-39/. First-principle calculations of the energetics of the Ni-silicide interface formation have been done with the LAPW method /40/. For details on this work on silicides the reader is referred to reference /3/.

4. TECHNICAL DETAILS

Before presenting the results of theoretical calculations performed on clean surfaces and interfaces it is convenient to devote a few words to the technical aspects of pseudopotential surface band structure calculations. As in any pseudopotential approach one has to solve a single-particle equation for the valence electrons only, the interaction with the ion core being represented by a model potential, often referred to as bare pseudopotential. This is in general non-local in character, being dependent on the electron angular momentum. For many purposes one can avoid the complications due to the non-locality; one can use local pseudopotentials of simple analytical form containing a few parameters adjusted to fit the electronic energy levels of the free ion /34,35/. These potentials have the advantage of converging rapidly enough in reciprocal space to allow an expansion of the wavefunctions in terms of a relatively small number of plane waves. This is the reason why most of the calculations performed up to now rely on local potentials.

The geometry of the surface and of the topmost layers is usually one input of the calculation. In many cases this has been taken from the outcomes of structural LEED analysis or determined by comparing the calculated energy band structure with the experimental spectra obtained by photoemission, inverse photoemission etc. This spectroscopical selection of the structural model does not allow one to determine the stable geometry unambiguously. It is therefore important to adopt a total energy minimization procedure that allows discrimination between the various models giving similar electronic spectra. In the case of the clean cleavage surface of silicon such total energy analysis has been of crucial importance in establishing the origin of the observed reconstruction.

In order to determine the surface band structure one needs a method to deal with the symmetry breaking due to the surface. A very common procedure consists in building up a periodic structure made of slabs separated by regions of vacuum /4/. Each slab is made by regularly stacked atomic planes of the system under consideration and has a surface on both sides. This artificial introduction of periodicity normal to the surface makes it possible to use bulk band structure techniques to derive the surface electronic properties. In particular three-dimensional (3D) plane waves can be chosen as basis functions and the useful

3D reciprocal space representation can be used, the only disadvantage being the large size of the basis set, due to the small dimension of the reciprocal unit cell. Since a good description of the bulk band structure is achieved by including about 80 plane waves with momentum $|K+G|^2 \leq 7.0$ Ry. /35/ it is reasonable to choose the same cutoff value in the surface calculation. This same cutoff corresponds to a basis set of about 1500-2000 plane waves for an elongated unit cell of vertical size of approximately 30 Å (18 Å occupied by atomic planes + 12 Å of vacuum). The standard techniques which allow one to deal with such a large basis set consist first of all in the exploitation of symmetry and in the use of symmetry-adapted combinations of plane waves; in this way it is possible to diagonalize separately different matrices of smaller size. The size of the matrices can be further reduced by using a perturbation method: plane waves of momentum $|K+G|^2 \leq 2.7$ Ry. are treated directly, while the others are introduced as a perturbation, following the partition scheme due to Lowdin /41/.

The slab method can also be used to study either the interface between two solids or the chemisorption problem; in both cases the only difference with respect to the crystal-vacuum interface consists in substituting the vacuum region with one or more atomic planes of the species of interest /42/.

5. RECONSTRUCTION AND STRUCTURAL STABILITY

To see how the theoretical scheme presented in the previous section can be profitably used to predict the stable structure of surfaces we will discuss the results of total energy determinations of the surface relaxation and reconstruction of the cleavage surface of semiconductors. These results have been obtained by using density functional theory in the LDA and the pseudopotential approach to represent the electron-ion interaction. We will confine our discussion to the results for the cleavage surfaces of tetrahedrally bonded semiconductors.

It is well known that these surfaces present structural modifications compared to the ideal termination of the bulk crystal. The most dramatic effects occur in Si (111) which immediately after cleavage shows a 2X1 reconstruction while displaying a 7X7 LEED pattern after thermal treatment. The (110) surfaces of III-V materials like GaAs or InSb retain the periodicity of a parallel plane of the bulk but show a significant amount of relaxation compared to the ideal surface /2/.

The instability of the ideal geometry has been attributed by several authors to the dehybridization of the valence electron orbitals that takes place at the surface /43/. The argument relies upon a description of the electronic structure of the bulk semiconductor in terms of sp^3 bonds. According to this model four equivalent sp^3 hybrids are formed in the bulk on each atom which point toward the four neighbouring atoms: bonding combinations of these hybrids give rise to the valence band states and explain the binding properties of these materials. Cleaving the semiconductor corresponds to cutting one of these bonds leaving the corresponding hybrid dangling into the vacuum. Due to the missing neighbour the sp^3 configuration is no longer energetically convenient for a surface atom and some sort of dehybridization takes place that drives the surface toward a more favourable geometry. As a typical example we may consider the case of GaAs (110) surface where, as a consequence of the two atoms in the surface unit cell being different, we have two kinds of dangling bonds, a filled one located on the group-V atoms and an empty one on group-III atoms, whose energy is approximately that of the atomic sp^3 hybrid. In the ideal geometry these states lie in the band gap close to the bulk band edges. The dehybridization determined by threefold coordination will cause the group-III atoms to change from sp^3 toward sp^2 structure with three equivalent bonds, each having one electron, and an empty orbital perpendicular to the surface, like in $GaCl_3$. For the same reason the group-V atoms will go toward a s^2p^3 configuration similar to what is found in AsH_3, the three bonding orbitals having a more pronounced p character and the dangling bond a more pronounced s character. These changes in the hybridization cause the observed relaxation in which the III atom is pulled toward the bulk and the V atom is pushed out of the surface. Such displacements take place with a significant charge transfer from the group III to group V atoms with respect to the bulk. In particular the ions present in the bulk are converted toward a neutral configuration. The relaxed surface shows a significant corrugation usually referred to as surface buckling. We will not enter here into the details of the possible surface structures of III-V (110) surface for which the reader is referred to the review article by Kahn /2/. Enough to say that experimental work on LEED has allowed to determine these structures very accurately.

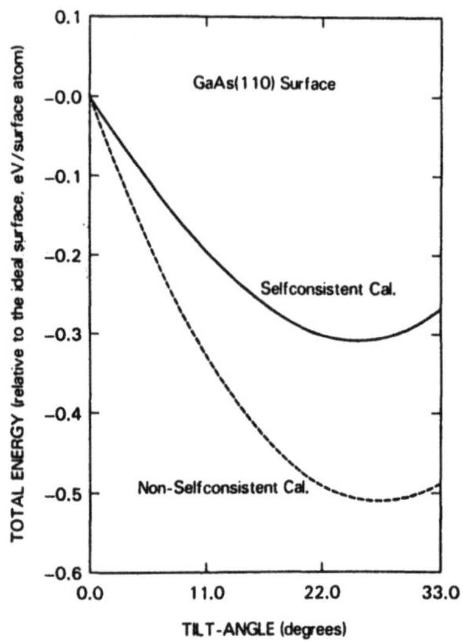

Fig. 1. Total energy of the GaAs (110) surface as a function of buckling relaxation

A direct confirmation of these experimental findings and of the validity of the buckling model has been given by total energy calculations carried out self-consistently using density functional theory and a pseudopotential approach to describe the electron states /44/. The calculated difference in total energy between the buckled and the ideal surface as a function of the rotation angle of the surface bonds is given in Fig. 1 for GaAs and for a particular relaxation model. Similar curves can be drawn for other relaxation models consistent with LEED data and for other III-V compounds confirming that the dehybridization buckling is a rather general feature in semiconductor surfaces. This conclusion does not imply that the hybrid model provides a good description of the electronic structure of the surface. Indeed even for the purpose of describing bulk electronic properties the sp^3 model turns out to be very inaccurate, especially for the conduction band states. This inaccuracy is a consequence of the reduced basis set and of the lacking of self-consistency in the solution of the single particle equations . Similar arguments have been proposed during the 1970s to explain the reconstruction observed in group IV semiconductors, particularly in Si (111). In the unreconstructed model this surface shows one dangling bond per atom with one electron. Since there is one atom per unit cell the surface would be metallic in character, in conflict with what is found experimentally /4/. It is natural to think that the system will lower its energy by some geometrical distortion introducing an energy gap between occupied and unoccupied surface states, as is observed. Such distortions are responsible for the observed reconstruction, in particular the 2X1. In the first model proposed to explain the reconstruction it was supposed that, as a consequence of the dehybridization, half of the surface atoms move toward the bulk and the remainder are pushed out of the surface /45/. As in the case of GaAs this displacements leads to a substantial charge transfer from the atoms pulled down to those displaced outwards. However, unlike what happens in GaAs, the two atoms in the surface unit cell are second-neighbours; this

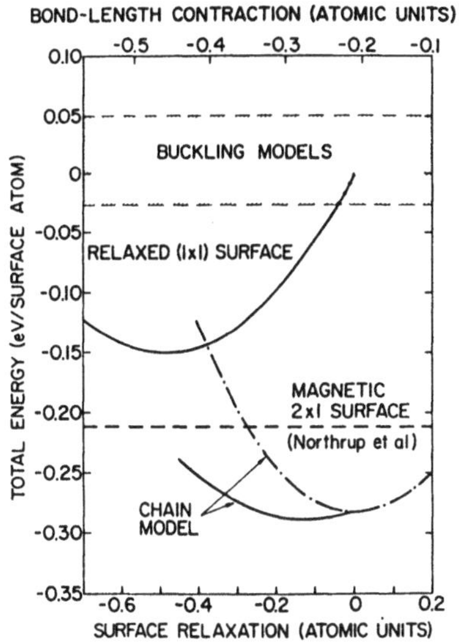

Fig. 2. Total energy of the Si (111)-(2X1) surface as a function of relaxation for different reconstruction models

implies that the dangling bond orbitals do not interact appreciably. This buckling model was used to interpret data but it did not provide a good description of the experimental findings.

A crucial step toward the understanding of the structural properties of the Si (111) surface was performed by Pandey /44,46/ who was able to show that the buckling alone does not stabilize the surface. Indeed he found from accurate self-consistent psedopotential calculations (see Fig. 2) that pure buckling rather than stabilizing the surface raises its energy. To a large extent this is a consequence of the charging of the surface atoms. The transfer of electrons from the Si atoms that go down to those moving outwards creates a Coulomb potential that tends to resist the distortion and to stabilize the ideal model compared with the buckled one. A substantial relaxation is needed to lower the buckled surface energy below the energy of the ideal model. As an alternative and more favourable model, Pandey proposed the so-called π-bonded-chain model /46/. Unlike the previous models based on the dehybridization concept, the Pandey model relies on changes in the surface topology which lead to a configuration where the dangling bonds reside on nearest neighbour atoms and are close enough to interact appreciably and to cause a substantial reduction of the total energy. It must be emphasized that the total energy minimization procedure has been of crucial importance for the theoretical determination of the stable structure. One major objection to this model, when it was first introduced, was that several bonds had to be broken in order to produce the chain geometry and this would be energetically unfavourable. However, by performing accurate total energy calculations Northrup and Cohen /47/ showed that a path exists which allows one to pass from the ideal to the distorted surface by a continuous deforming and reforming of bonds, with a barrier height of about 0.03 eV per surface atom. The energy to surmount this barrier can be easily supplied by the cleaving process. A number of experiments, including optical absorption, LEED and ion scattering, have been performed to check the validity of the π-chain model. Up to now all these experiments appear to be consistent with the model /48/.

6. SURFACE ELECTRONIC STRUCTURE OF III-V SEMICONDUCTORS

The atomic geometry of the non-polar (110) surface of III-V semiconductors has been extensively studied and very well characterized: at the surface the anions move outwards from the ideal position and the cations inwards, retaining a 1X1

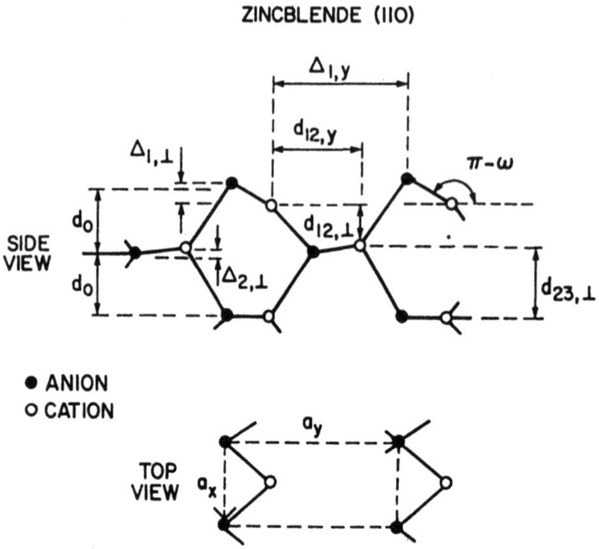

Fig. 3. Schematic drawing of the surface geometry for the (110) surface of zincblende structure semiconductors with an indication of the surface unit cell

configuration but giving rise to a buckled surface. The relative displacement of the cation and anion within the outermost layer defines a rotation angle ω (see Fig. 3) which was reported to vary between 27° and 34.8° involving no bond length change. This model dates back to 1976 and refinements of it have been reported over the years, including a relaxation of the top layer toward the substrate /2,43/, a counter-relaxation within the second layer and the possibility that the relaxations parallel to the surface in the uppermost layer would be reduced relative to those characteristic of a bond-lenghth-conserving rotation /49/. In the following we will consider a recent model based on LEED intensity analysis and a tight-binding total energy minimization method /50/.

The relevant structural parameters are defined in Fig.3 and their values are reported in table 1. This surface distortion corresponds to a tilt angle of about 25° accompanied by a small bond length variation of about 2% caused by a slight reduction of the atomic displacements parallel to the surface with respect to a pure rotation model; notice that the information on the lateral distortion is the most controversial one since different techniques give different results /51/.

Table 1. Structure parameters for the relaxed geometry of (110) surfaces /50/

	GaP	GaAs	GaSb	InP	InAs	InSb
a_0	5.4505	5.6537	6.1180	5.8687	6.0360	6.4782
$\Delta_{1\perp}$	0.63	0.69	0.77	0.73	0.78	0.88
Δ_{12y}	4.242	4.518	4.793	4.598	4.985	5.320
$d_{12\perp}$	1.386	1.442	1.615	1.549	1.497	1.541
d_{12y}	3.196	3.339	3.629	3.382	3.597	3.894

6.1 GaAs(110) and GaP(110).

The first output of a self-consistent surface calculation is the total charge density. Fig. 4 shows the charge density contour plots for the GaP (110) surface

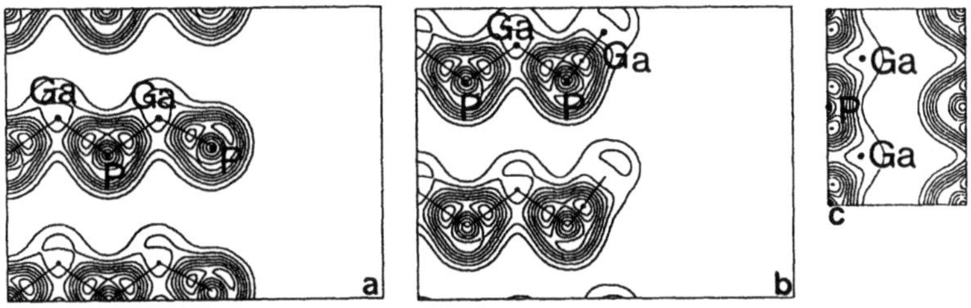

Fig. 4. Total valence charge density of GaP(110) plotted along planes perpendicular to the surface and passing through (a) P , (b) Ga surface atoms and (c) along the ideal surface plane

along three different planes: along the planes perpendicular to the surface and passing through anion (a), cation (b) surface atoms and (c) parallel to the surface at the ideal surface position. The perturbation induced by the surface extends up to two layers and manifests itself only in a change of shape: the pileup of charge along the bonds between anions and cations is essentially the same in the surface as in the bulk; since a bulk-like configuration for the bonding charge is optimal , this fact is a positive test of the greater stability of the relaxed surface compared to the ideal one, where the bonding charge at the surface differs significantly from the bulk value /34/.

The next output of the calculation is the surface energy spectrum. Figures 5 and 6 show the two-dimensional band structure of GaAs and GaP (110). Surface states are identified through the localization of the wavefunction near the surface: an eigenstate is classified as a surface state if its amplitude integrated over a region around the last two atomic planes is larger than twice its localization in the inner planes. Surface bands that become degenerate with the bulk continuum quickly lose their surface character, since they couple with bulk band states, and cannot be classified as surface states according to our criterion. Although these surface resonances may contribute significantly to the surface local density of states, we cannot assign a precise band dispersion, so that we have not drawn them in our band structure. Surface states are labelled A_i and C_i depending on whether they are localized on the anions or cations respectively.

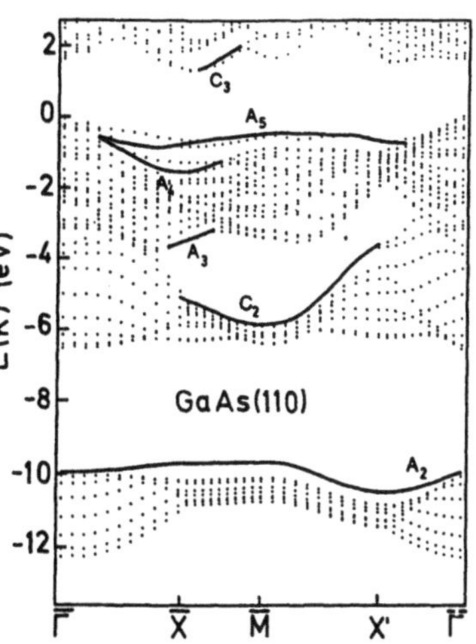

Fig. 5. Surface band structure of GaAs (110). Solid lines indicate surface states. Energy is referred to the top of the valence band

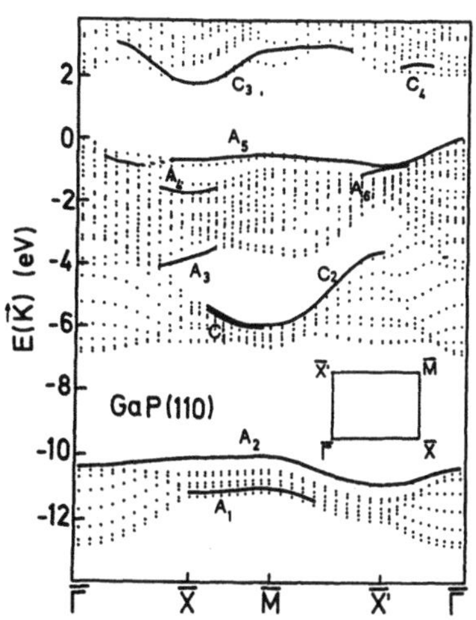

Fig. 6. Surface band structure of GaP (110)

The surface band structures of GaAs and GaP are qualitatively very similar: surface states appear mainly in the gaps and lenses of the projected bulk band structure, as one would expect. The only significant difference between the two systems consists in the energy position of the first unoccupied surface states which in the case of GaP overlaps the optical gap around X point, while for GaAs it is completely outside of it , degenerate with bulk states. This fact is a well-known characteristic of GaP which among III-V semiconductors is the only system where an empty surface state is found inside the optical gap.

A deeper understanding of the surface electronic structure of these surfaces can be achieved by considering the orbital character of the surface states. It can be identified by plotting the square modulus of the wavefunction along appropriate planes. The analysis of these plots along the high symmetry directions of the two-dimensional Brillouin zone is necessary in order to determine the energy band dispersion and to assign a definite orbital character to the bands, especially when bands cross. This "graphic" analysis is also important in order to understand the nature of the chemical bonds between surface atoms . In the following we will present such an analysis for the case of GaP.

Let us start by considering the surface charge density obtained by summing over $k_\parallel$ the square modulus of all the wavefunctions localized at the surface, up to the Fermi level. The contour plots of this surface charge density are shown in Fig. 7. By comparing Figs. 4 and 7 it appears that the bonding between the surface atoms and the underlying layer is fully accounted for by surface states; this is not true for the bonding in the surface: the differences between 4 (c) and 7 (c) suggest that the charge associated with occupied surface states does not correspond to the total charge localized at the surface and that a non-negligible contribution from bulk states extending up to the surface must be present to explain the bonding between the anions and cations in the surface layer.

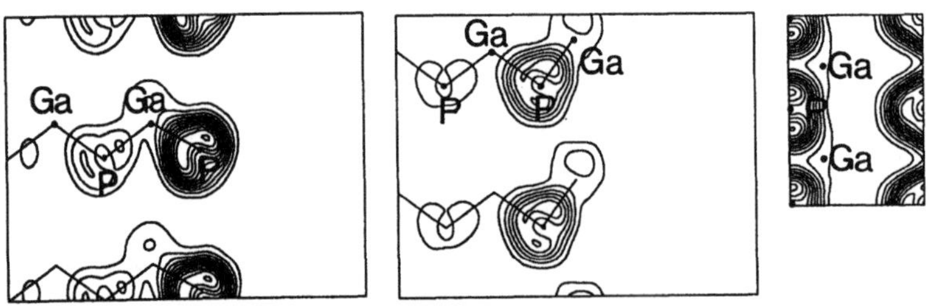

Fig. 7. Surface state contribution to the total valence charge density.

It is possible to identify the contribution of the various occupied surface states to the surface charge density by considering the charge density map of each of them. Figure 8 displays the plots of the valence charge density for each surface band of GaP (110) along the planes perpendicular to the surface. Notice that again we will consider the sum over $k_\parallel$ instead of a single state at a particular k-point. From Figs. 4 and 7 we expect to find a small number of occupied surface states to be cation-derived; in fact there exist only two of them C1 and C2 at the lower edge of the internal gap; moreover C1 is present only in a small region of k-space, around the $\bar{M}$ point. As is shown in Fig. 8, they are predominantly s states localized either at the surface (C2) or at the first sublayer (C1). All the other surface states are anion-derived : A1 and A2 are s states localized, like C2 and C1, at the surface and at the first sublayer; they occupy the lower part of the valence band, A2 overlapping the ionic gap. A3, appearing near the upper edge of the internal gap around $\bar{X}$ point, is a well defined py state localized at the first sublayer. States A4 and A5 which are present near the top of the valence band have a py-pz character and are generally identified as an anion back-bond and dangling bond respectively. A4 is well defined only around $\bar{X}$, while A5 runs all along the lower edge of the optical gap. State A5 together with A2 accounts for a large part of the charge localized at the surface and spilling out of it. The last occupied surface state is A6 existing along the $\bar{\Gamma}$-$\bar{X}$ direction only, with a py-pz character both on surface anions and on anions of the first sublayer. Notice that, for the particular surface geometry we have been considering, state A6 does not have a px character as previously reported both for GaP and GaAs /34,35/; since along this direction px states cannot mix with py and pz, state A6 is in the present calculation totally different from the state previously found. Notice that among these states those with a strong localization in the first sublayer (A1, C2, A3, A4) are not present in the case of an ideal termination of the crystal, appearing as a consequence of the atomic distortion at the surface.

Let us consider now the empty part of the spectrum. As previously mentioned, GaP is known to have an empty surface state inside the optical gap: in fact state C3 overlaps the gap around the X point and along the $\bar{\Gamma}$-$\bar{X}$-$\bar{M}$ direction; since the charge density associated with this state points toward the missing atom, it can be

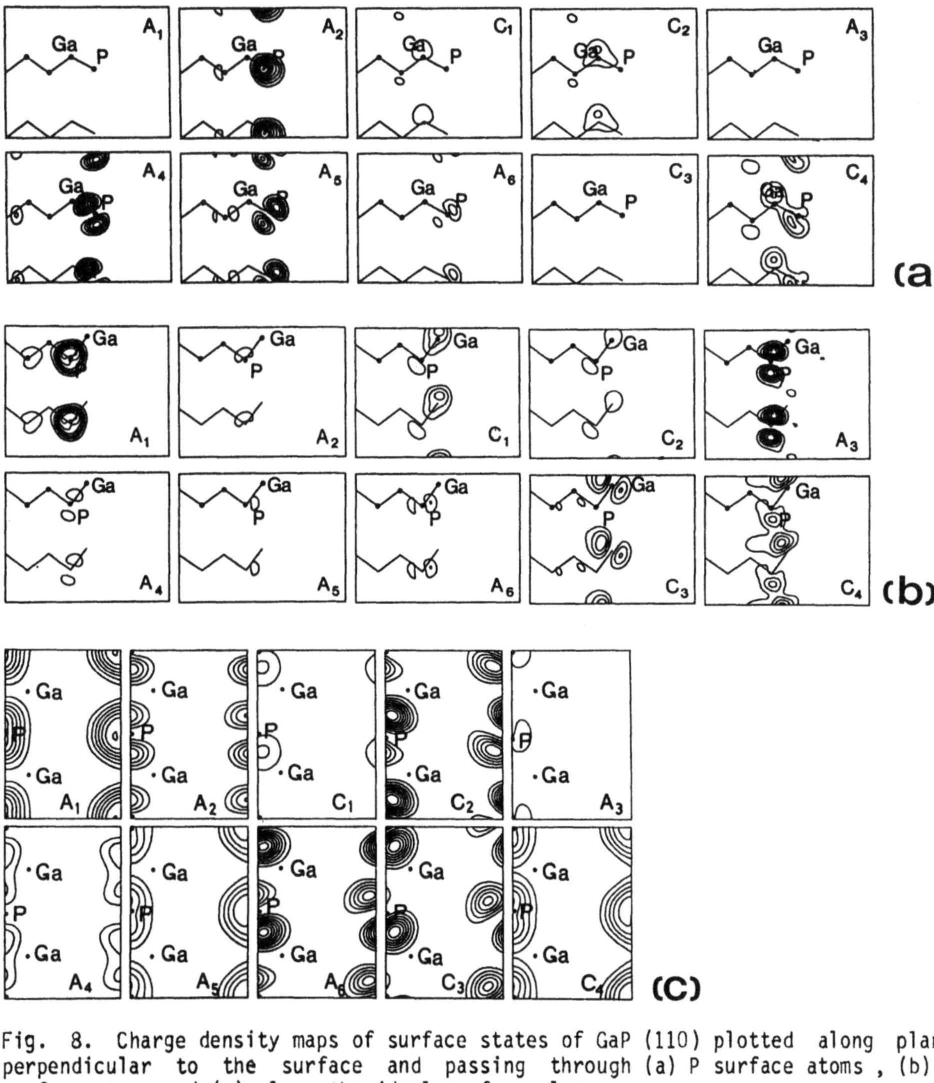

Fig. 8. Charge density maps of surface states of GaP (110) plotted along planes perpendicular to the surface and passing through (a) P surface atoms, (b) Ga surface atoms and (c) along the ideal surface plane

identified as the cation-derived dangling bond. Two other surface states exist in this energy region, labelled C4 and C5; they are weak resonances actually localized both on cation and anion sites with an antibonding character.

Let us conclude this description of the surface energy spectrum of GaP by comparing theory and experiments. Occupied surface states have been detected experimentally by angle-resolved photoemission and resonant photoemission. Surface states are detected between -2 and -0.5 eV from the valence band maximum (VBM) /52,53/ ; in particular from angle resolved photoemission data /53/ it is possible to follow the dispersion of the last occupied surface state along $\bar{\Gamma}$-$\bar{X}$ and Γ-X': it presents a downward dispersion from $\bar{\Gamma}$ to $\bar{X}$ and from $\bar{\Gamma}$ to $\bar{X}'$ of about 0.5 eV in relatively good agreement with our theoretical results on states A5 and A6. Also the state at -2 eV detected by resonant photoemission /54/ is present in the calculated spectra as A3.

Evidence of the occurrence of empty surface states in GaP band gap has been provided by inverse photoemission experiments. Himpsel and coworkers /55/ found a surface state near $\bar{\Gamma}$ at 1.9 eV above the valence band maximum and by comparing their results with the measurements of the Fermi level pinning /56/ they evaluate a band dispersion of about 0.2 eV for the Ga-derived dangling bond band. Although our calculations show a resonance near $\bar{\Gamma}$, not indicated in Fig. 6, there seems to be a discrepancy of about 0.5 eV between theory and experiment. More recent inverse photoemission experiments by Perfetti and coworkers /57/ locate the surface state at higher energies in better agreement wth the theory.
Optical transitions involving surface states have also been measured. These spectra compare favourably with the surface dielectric function calculated on the basis of our theoretical band structure /58,59/.

6.2 Non-local density functional calculations

As previously mentioned LDA has intrinsic limitations particularly evident in surface calculations where it fails to reproduce the correct behaviour of the effective potential far from the surface. The surface states we have been describing in the previous paragraph are however localized so close to the surface (0.5-1 Å from it) to be probably insensitive to this asymptotic behaviour; the success of LDA in accounting for these states suggests that this must be the case and that up to these distances the LDA effective potential and the "true" one must coincide. As mentioned in chapter 3 the long range part of the potential can be reproduced only by adopting approximations beyond the local density. This fact can be understood very easily in terms of the exchange-correlation hole: in LDA the exchange-correlation hole $n_{xc}(\underline{r},\underline{r}')$, describing the depletion of charge at a certain position $\underline{r}'$ due to the presence of an electron at $\underline{r}$, is always centered on the electron (see (19)), even if the electron is out of the crystal surface. In the non-local density approximation (NLDA) of (22) on the contrary the exchange-correlation hole is centered on the electron inside the solid but it remains located at the surface when the electron is in the vacuum region. This behaviour is illustrated in Fig.9 where the exchange-correlation hole obtained from a self-consistent NLDA calculation for GaAs (110) is shown. The exchange-correlation potential originated by this exchange-correlation hole goes to zero in the vacuum region correctly as the image potential; Fig. 10 shows the profile of the exchange-correlation potential obtained for GaAs (110) both by LDA and NLDA: the two potentials are almost identical up to about 1-1.5 Å from the last atomic plane. This is an a posteriori justification for using LDA to study the lower part of the surface energy spectrum; on the other hand it is necessary to perform a NLDA calculation if one is interested in the upper part of the spectrum where surface states are weakly bound and their amplitude is concentrated relatively far away from the surface. In this energy region image states can appear where an electron is bound to the surface through the Coulomb interaction with its own image charge. Up to now only simple models of the kind developed in section 2 have been proposed to described such states /60/; these models cannot of course account for the interaction with the surface lattice, which may be important

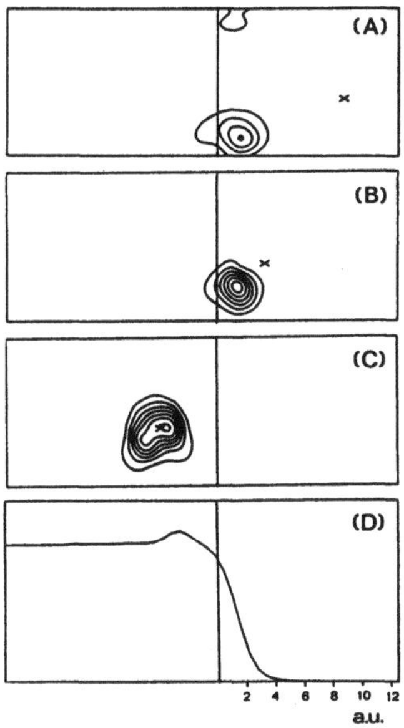

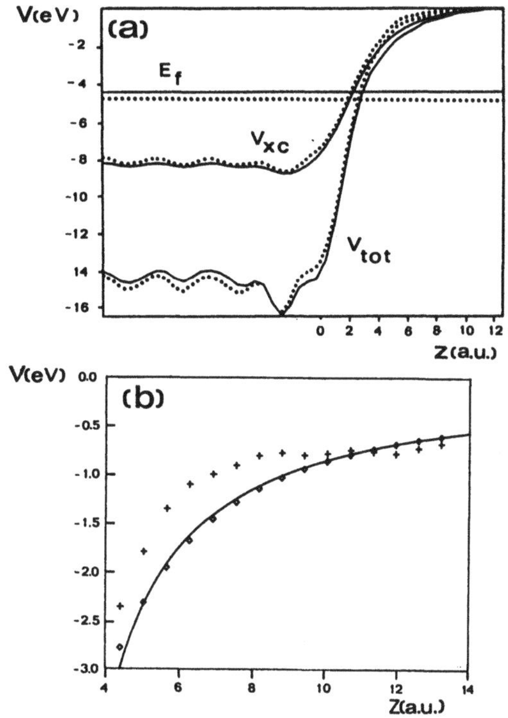

Fig. 9. Contour maps of the exchange-correlation hole in the NLDA for an electron at three different positions, indicated by a cross, outside ((A), (B)) and inside (C) GaAs. The spacing of the contours of (B) and (C) is 0.002 electrons per a.u. and for (A) is 0.00005 electrons per a.u. The profile of the electronic charge is shown in (D). The position of the last atomic plane from which distances are measured is indicated

Fig. 10. (a) Laterally averaged potential from the NLDA (solid line) and LDA (dotted line) calculations for the GaAs (110) surface. The total and the exchange-correlation potentials are plotted as a function of the distance from the last atomic plane. (b) Asymptotic profile of the exchange-correlation potential obtained in NLDA (open diamonds) and LDA (crosses) compared with the image potential (solid line). Distances from the last atomic plane are indicated

in describing the correct dispersion of these states /61/. The NLDA calculation of the surface electronic structure reported here for GaAs identifies an image state along the $\bar{\Gamma}$-$\bar{X}'$ and $\bar{\Gamma}$-$\bar{X}$ directions with a binding energy of about 1.0 eV. Experimental evidence for such a state has been recently provided by inverse photoemission experiments in GaP (110) /62/.

7. ELECTRONIC STRUCTURE OF DEPOSITED MONOLAYERS

One of the main purposes of surface science is to provide an explanation of interface behaviour by studying the early stages of interface formation. This leads to the analysis of the systems obtained by depositing atoms of a given element onto a semiconductor substrate. In many cases the phases obtained at very low coverages do not show any degree of two-dimensional ordering or present interdiffusion phenomena, which cannot be treated theoretically without resorting to rather crude approximations.

To illustrate how the previous theory can be applied to treat the electronic properties of these systems, we present the results of theoretical calculations for a monolayer of Sb deposited on GaP (110). The reason for this choice is twofold. First we notice that this is the only example of an abrupt ordered two-dimensional structure that can be formed onto the cleavage surface of III-Vs. Second it provides a good example of the peculiarities of the chemical bond that links an overlayer to a semiconductor substrate. The choice of GaP as the substrate is not a limitation, since it has been established that such abrupt interfaces can be found on the (110) face of other III-Vs /63-65/. Moreover, it allows one to compare the outcome of the surface band structure calculations for the covered surface with the theoretical results for the clean surface, illustrated in the previous section.

The chemisorption geometry of the Sb overlayer has been established by dynamical ELEED analysis /63/ and semi-empirical calculations of the total energy /65/. Sb atoms are adsorbed on both Ga and P sites giving rise to a 1X1 structure. They form zig-zag chains with bond angle of nearly 90° . The chains are linked to the substrate by strongly covalent directional bonds with a bond angle of nearly 109° . This last value is typical of the sp^3 hybridization, while the angle in the chains is typical of a planar bond with only p orbitals involved. Such features indicate a peculiarity of the chemisorption bond in this system: it cannot be represented either by a sp^3 model, as if the overlayer were continuing the substrate in regular stacking, or by a pure p bond, like the one found in group III hydrides. The situation is somewhat intermediate between the two model descriptions.

Figure 11 shows the calculated surface band structure along the high symmetry directions of the two-dimensional Brillouin zone. Continuous lines indicate the surface localized states, while broken lines refer to weaker resonances. For a detailed description of this energy spectrum we refer the reader to a forthcoming paper /66/. Here we will discuss the main features of the theoretical band structure.

We can distinguish four main groups of bands in Fig. 11. The first is made by bands S1 and S2 . They arise from bonding and antibonding combinations of s orbitals belonging to the Sb atoms. Their dispersion is similar to the one predicted for a single chain, although the different chemical environment of the two Sb atoms in the unit cell opens a gap along the $\bar{X}$-$\bar{M}$ direction. These bands do not contribute significantly to the chemisorption energy, since both bonding and antibonding states are occupied.

A second group of states labelled S3' ,S3 and S4 , is found between -3 and -4.5 eV. From the analysis of their orbital composition they can be shown to derive both from combinations of planar p Sb orbitals and from the bonding of pz Sb

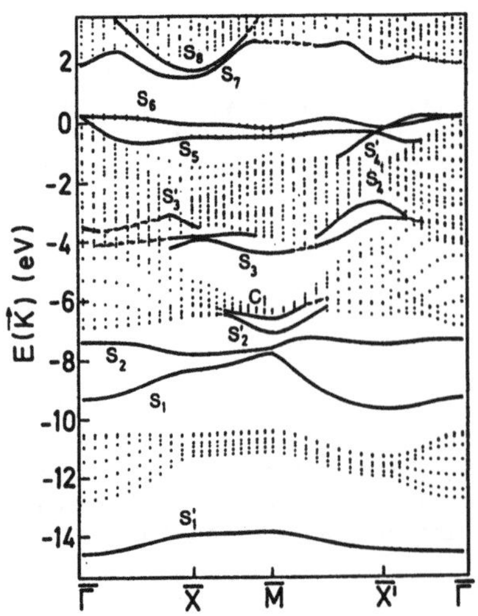

Fig. 11. Surface band structure of a Sb monolayer on GaP (110)

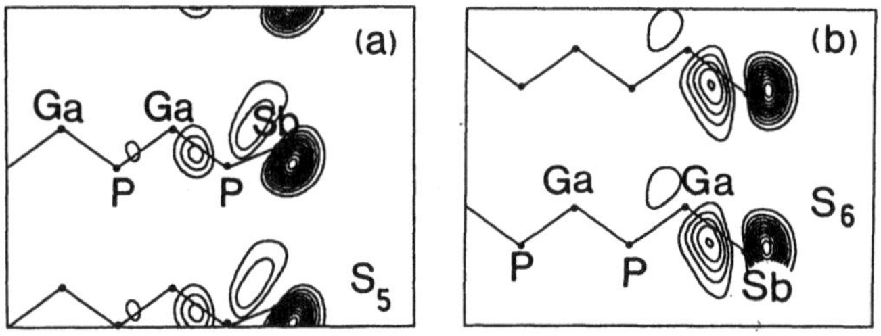

Fig. 12. Contour plots for states S5 and S6 near the $\bar{X}$ point along the $\bar{X}$-$\bar{M}$ direction of the two-dimensional Brillouin zone

orbitals with the P dangling bond states of GaP (110). The dominance of the Sb-P bonding compared to the bonding of Sb atoms with Ga is a consequence of the smaller Sb-P distance and of the strong covalent character of this bond.

The rather flat bands S5 and S6 show a behaviour similar to what has been found for the filled dangling bond of the free surface. This is clearly shown in Fig. 12 giving the valence charge density plots in the direction normal to the surface. In the tight binding study of Mailhiot et al. /65/, they have been identified as bonding states between pz orbitals of the Sb chains. It is not clear at present whether this different interpretation is the consequence of using different theoretical approaches to the calculation. It is more likely that the discrepancy derives from the use of different models to interpret the outcomes of the theoretical calculations.

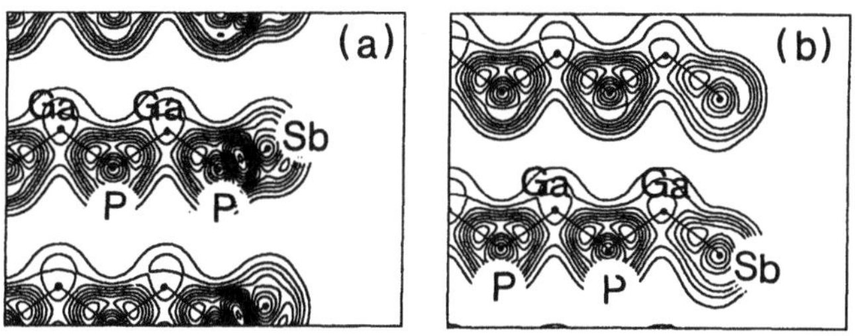

Fig. 13. Contour plots for the total valence charge of the Sb monolayer deposited on GaP (110) drawn along planes perpendicular to the surface and passing through (a) P and (b) Ga surface atoms

The total valence charge, calculated for this interface, provides a clear description of the main features of the chemisorption bond in the system. Figure 13 reports the charge density map for the (110) planes perpendicular to the surface and passing through P and Ga surface sites. It is clear from this figure that the bond between the chain and the substrate is stronger on the P site than on the Ga site. Indeed the Ga-Sb bond looks more ionic in character and similar to what can be found in GaSb. This suggests that Ga dangling bond states contribute more significantly to the empty bands than to the filled ones.

The last group of bands S7 and S8 running along the upper part of the gap and the lower edge of the conduction band are empty states with mixed Sb and Ga character. They are expected to be important in determining the Fermi level pinning at the interface.

A detailed comparison between theory and experiments is not yet available for this interface. However, similar studies performed on Sb/GaAs(110) by angle resolved photoemission have shown that the theoretical description is essentially correct although discrepancies of nearly 0.5 eV may occur in the location and dispersion of the lower bands /67,68/.

8. CONCLUSIONS

We think that the previous discussion of the theoretical methods and the detailed description of the results for some case studies provide an adequate illustration of the power and limitations of the theory presently used to calculate the electronic properties of surfaces and interfaces. Indeed, to give a full account of the theoretical advances in this field, covering all the most significant applications, would be a rather difficult task, in view of the great number of systems that have been investigated and the many interesting results that have been obtained. In spite of that we think that much theoretical work is still needed to achieve a really sound description of the spectroscopical data. As we pointed out previously, the description of the excited states provided by the density functional theory is not adequate and the evaluation of self-energy effects is particularly important for an understanding of the experimental data. From this viewpoint it is not hazardous to say that the theory of surfaces and interfaces is still in its infancy and much progress will be made in the next years.

References

1. L. Brillson : Surf. Sci. Rep. 2, 123 (1982).
2. A. Kahn : Surf. Sci. Rep. 3, 4/5 (1983).
3. C. Calandra, O. Bisi and G. Ottaviani : Surf. Sci. Rep. 4, 271 (1984).
4. M. L. Cohen, in Highlights in the Theory of Condensed Matter, Eds. F. Bassani, F. Fumi and M. P. Tosi (North-Holland, Amsterdam 1985) p. 16.
5. P. K. Larsen, B. A. Joyce and P. J. Dobson, in Dynamical Phenomena at Surfaces, Interfaces and Superlattices, Eds. F. Nizzoli, K. H. Rieder and R. F. Willis (Springer-Verlag, Berlin 1985) p. 196.
6. F. Flores and C. Tejedor : J. Phys. C: Solid St. Phys. 20, 145 (1987)
7. L. Landau and L. Lifschitz, Electrodynamique des milieux continus (Editions MIR, Moscow 1969).
8. M. W. Cole and M. H. Cohen : Phys. Rev. Lett. 23, 1238 (1969).
9. M. W. Cole : Phys. Rev. B2, 4239 (1970).
10. M. Babiker and D. R. Tilley : Proc. Roy. Soc. A378, 369 (1981).
11. V. Dose, W. Altmann, A. Goldmann, U. Kolac and J. Rogosik : Phys. Rev. Lett. 52, 1919 (1984).
12. D. Straub and F. Himpsel : Phys. Rev. Lett. 52, 1922 (1984).
13. R. S. Becker, J. A. Golovchenko and B. S. Swartzentruber : Phys. Rev. Lett. 55, 987 (1985).
14. G. Binnig, K. H. Frank, H. Fuchs, N. Garcia, B. Reihl, H. Rohrer, F. Salvan and A. R. Williams : Phys. Rev. Lett. 55, 991 (1985).
15. K. Giesen, F. Hage, F. J. Himpsel, H. J. Riess and W. Steinmann : Phys. Rev. B33, 5241 (1986).
16. J. A. Stroscio, R. M. Feenstra and A. P. Fein : Phys. Rev. Lett. 57, 2579 (1986).
17. P. Hohenberg and W. Kohn : Phys. Rev. 136B, 864 (1964).
18. W. Kohn and L. J. Sham : Phys. Rev. 140A, 1133 (1965).
19. O. Gunnarsson, M. Jonsson and B. I. Lundqvist : Phys. Rev. B20, 3136 (1979).
20. Z. H. Levine and S. G. Louie : Phys. Rev. B25, 6310 (1985).
21. M. Lanoo, M. Schluter and L. J. Sham : Phys. Rev. B32, 3890 (1985).
22. F. Manghi, G. Riegler, C. M. Bertoni and G. B. Bachelet : Phys. Rev. B31, 3680 (1985).
23. L. Hedin and S. Lundqvist : Solid State Physics 23, 1 (1969).
24. M. S. Hybersten and S. G. Louie : Phys. Rev. Lett. 55, 1418 (1985); Phys. Rev. B34, 5390 (1986).
25. R. W. Godby, M. Schluter and L. J. Sham : Phys. Rev. Lett. 56, 2415 (1986).
26. L. Hedin and S. Lundqvist : J. Phys. C : Solid St. Phys. 4, 2064 (1971).
27. D. M. Ceperley and B. J. Alder : Phys. Rev. Lett. 45, 566 (1980).

28. N. D. Lang : Solid State Physics 28, 225 (1973) and references therein.
29. O. Gunnarsson and R. O. Jones : Physica Scripta 21, 394 (1980).
30. S. Ossicini and C. M. Bertoni : Phys. Rev. A31, 3550 (1985).
31. S. Ossicini, C. M. Bertoni and P. Gies : Surface Sci. 178, 244 (1986).
32. P. Gies : J. Phys. C : Solid St. Phys. 19, L209 (1986).
33. F. Manghi : Phys. Rev. B33, 2554 (1986).
34. J. R. Chelikowsky and M. L. Cohen : Phys. Rev. B20, 4150 (1979) and references therein.
35. F. Manghi, C. M. Bertoni, C. Calandra and E. Molinari : Phys. Rev. B24, 6029 (1981).
36. See the previous reference for a detailed discussion of this point.
37. O. Bisi, L. W. Chao and K. N. Tu : Phys. Rev. B30, 4664 (1984).
38. J. Robertson : J. Phys. C : Solid St. Phys. 18, 947 (1985).
39. O. Bisi and K. N. Tu : Phys. Rev. Lett. 52, 1633 (1985).
40. D. R. Hamann and L. F. Mattheiss : Phys. Rev. Lett. 54, 2517 (1985).
41. P. O. Lowdin : J. Chem. Phys. 19, 1936 (1951).
42. C. M. Bertoni, C. Calandra, F. Manghi and E. Molinari : Phys. Rev. B27, 1251 (1983).
43. C. B. Duke : Applications of Surf. Sci. 11/12, 1 (1982).
44. K. C. Pandey : Phys. Rev. Lett. 49, 223 (1982).
45. D. Hanemann : Phys. Rev. 121, 1093 (1961).
46. K. C. Pandey : Phys. Rev. Lett. 26, 1913 (1981).
47. J. E. Northrup and M. L. Cohen : Phys. Rev. Lett. 49, 1349 (1982).
48. See Lectures by G. Le Lay at this school.
49. M. J. Gossmann and W. M. Gibson : Surf. Sci. 139, 239 (1984).
50. C. Mailhiot, C. B. Duke and D. J. Chadi : Surf. Sci. 149, 366 (1985).
51. C. B. Duke and A. Paton : J.Vac. Sci. Tecnol. B2, 327 (1984); see also L. Smit, T. E. Derry and J. F. van der Veen : Surf. Sci. 150, 245 (1985).
52. F. Cerrina, A. Bommannavar, R. A. Benbow and Z. Hurych : Phys. Rev. B31, 8314 (1985).
53. F. Solal, G. Jezequel, F. Houzay, A. Barski and R. Pinchaux : Solid St. Commun. 52, 37 (1984).
54. F. Sette, P. Perfetti, F. Patella, C. Quaresima, C. Capasso, M. Capozi and A. Savoia : Phys. Rev. B28, 4882 (1983).
55. D. Straub, M. Skibovski, F. J. Himpsel : J. Vac. Sci. Technol. A3, 1484 (1985).
56. A. Huijser, J. van Laar and T. L. van Rooy : Surf. Sci. 62, 472 (1977).
57. P. Perfetti : private communication.
58. P. Chiaradia, G. Chiarotti, F. Ciccacci, R. Memeo, S. Nannarone, P. Sassaroli and S. Selci : Surf. Sci. 90, 70 (1980).
59. F. Manghi, E. Molinari, A. Selloni and R. Del Sole : to be published.
60. M. Weinert, S. L. Hubert and P. D. Johnson : Phys. Rev. Lett. 55, 2055 (1985).
61. N. Garcia, B. Reihl, K. H. Frank and A. R. Williams, Phys. Rev. Lett. 54, 591 (1985).
62. P. Perfetti : private communication.
63. C. B. Duke, A. Paton, W. K. Ford, A. Kahn and J. Cannelli : Phys. Rev. B26, 803 (1982).
64. C. Maani, A. McKinley and R. H. Williams : J. Phys. C : Solid St. Phys. C18, 4975 (1985).
65. C. Mailhiot, C. B. Duke and D. J. Chadi : Phys. Rev. B31, 2213 (1985).
66. F. Manghi, C. Calandra and E. Molinari : Surf. Sci. (in press).
67. P. Martesson, G. V. Hansson, M. Lahdeniemi, K. O. Magnusson, S. Wiklund and J. M. Nicholls : Phys. Rev. B33, 7399 (1986).
68. A. Tulke and H. Luth : Surf. Sci. 178, 131 (1986).

Electronic Properties of Semiconductors: Fermi Level Pinning in Schottky Barriers and Band Line-up in Semiconductors

F. Flores

Departamento de Física de la Materia Condensada,
Universidad Autónoma de Madrid, E-28049 Cantoblanco-Madrid, Spain

Electronic properties of different semiconductor interfaces are discussed in relation to the formation of Schottky barriers and heterojunction band offsets. Different models are presented with emphasis on the induced density of interface states (IDIS) model and the defect model: they are compared with each other and with the experimental evidence.

1. Introduction

Semiconductor interfaces continue to be a challenge for experimental and theoretical physicists, many of their properties being still under debate/1-6/ although they have been profusely analysed for a long time /7-11/. Their importance is connected not only with basic but with applied physics/12/: semiconductor interfaces are at the heart of microelectronic devices.

In this paper, I will address the problem of the electronic properties of semiconductors from the point of view of their barrier formation. Semiconductor device characteristics are mainly dependent on their barrier heights/13/: basic physics tries to explain how these barriers are formed in order to control their values for industrial necessities.

The first semiconductor interface was prepared by F. BRAUN/7/ in 1874. Since then , a lot of work has been done; let us only mention that in the late thirties, the first theories about semiconductor interfaces were proposed/8-9/, and that the pioneering work of BARDEEN/10/ and HEINE/11/ prepared the field for our current understanding of these interfaces.

The perspective on semiconductor interfaces changed in the late seventies with the new experimental tools available to analyse interfaces/13-4/. Many new properties were found and several models were proposed to explain semiconductor interface formation/4-6/.

The purpose of this paper is to present the actual status of the theory about semiconductor interfaces, discussing different models for metal-semiconductor and semiconductor-semiconductor junctions. No attempt is made to present the experimental evidence/1-5/: we only discuss in Sect. 4 the main facts without referring to methods or techniques. This is,therefore, a theoretical presentation where the different models are discussed with respect to the support they obtain from the experimental evidence. Readers interested in experimental information are referred to some review papers/1-5/.

The structure of this paper is as follows: in Sect. 2 we discuss very general concepts about surface and interface dipoles: in fact, semiconductor barrier formation is intimately related to the creation of those interface dipoles. In Sect. 3 we discuss different models for metal-semiconductor interfaces, and in Sect. 4 we consider the case of heterojunctions. In Sect. 5 we summarize the main experimental evidence and compare it with the two models that have received the most widespread acceptance; finally, in Sect. 6, we present recent theoretical results obtained with more elaborate calculations that confirm some of the results predicted by one of the simplest models discussed in Sects. 3 and 4.

2. Surface and Interface Dipoles

Before discussing in detail the formation of semiconductor interfaces, it is convenient to discuss how different effects contribute to the work function and the absolute electronic energy levels of crystal wave functions.

2.1 Metals

Consider the simple Sommerfeld model for a metal surface, with a uniform background of positive charge, n_+, for $z>0$ /14-15/. The surface electronic charge distribution is calculated/14/ using a local density formalism, whereby an electron sees the local potential:

$$V_L(z) = V_{xc}(z) + \phi_e(z); \tag{1}$$

$V_{xc}(z)$ is a short range potential, depending locally on the electronic charge/14/. This charge changes across the surface from a constant value (inside the metal) to zero (in vacuum)/14/ (see Fig. 1b). Notice that $V_{xc}(z)$ is the well-known exchange and correlation potential associated with the local hole appearing around an electron/16/.

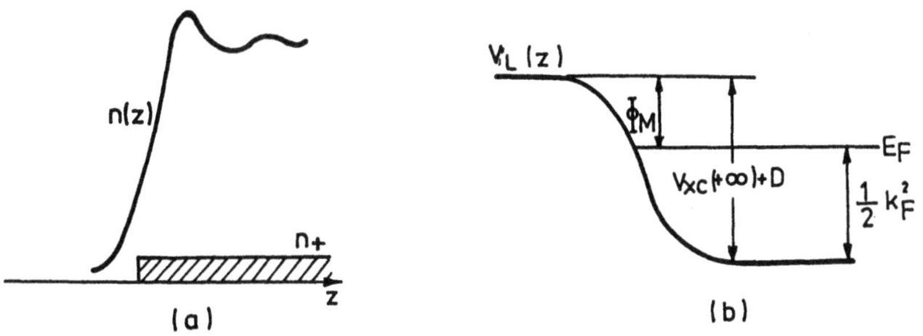

Fig.1(a): Electronic density n(z) near a metal surface; n_+ refers to the uniform positive charge.(b) Local potential for electrons near a metal surface. ϕ_M is the metal work function

In (1), $\phi_e(z)$ is the long range electrostatic potential created at the surface by the lack of local balance between the electronic and the positive charge/15/. Then, an electrostatic dipole appears at the surface, this dipole creating a voltage drop between the vacuum and the metal.

In accordance with this discussion, we can write the metal work function as follows/15/:

$$\phi_M = -(V_{xc}(+\infty) +\tfrac{1}{2}k_F{}^2) + D = -\mu + D \tag{2}$$

where $V_{xc}(+\infty)$ is the exchange and correlation potential inside the metal, μ being the inner contribution to the metal work function (the internal chemical potential). Notice that μ is the bulk contribution to ϕ_M, while D is the typical surface contribution. The important point to notice is the following: μ , being a bulk crystal contribution to ϕ_M, cannot be modified by external methods (unless we use a different crystal); D is, however, a surface contribution which can be controlled by changing surface properties, i.e., by depositing a monolayer of a different material on the metal/14/.

2.2 Metal-Metal Interface

The simplest example of interface formation is provided by the metal-metal case. When two metals having different work functions are in contact, a transfer of charge flows from one metal to the other, this process tending to equalize the two metal work functions/17/. This physical mechanism provides the charge creating an interface dipole, D_I, such that

$$D_I = \phi_{M_1} - \phi_{M_2}. \tag{3}$$

Obviously,this condition imposes that the two metals Fermi levels must be aligned.

2.3 Semiconductor Surfaces

This is a more complicated case due to the energy gap appearing between filled and empty bands. Figures 2(a) and 2(b) show two different cases corresponding to covalent and III-V ionic semiconductors. In (111)-covalent faces, surface states appear /15,18/ in the energy gap, those states pinning the Fermi energy as shown in Fig.2(a). Typically, in III-V ionic (110)-faces no surface states appear in the energy gap/18/ and the bulk bands are flat as in Fig.2(b). For clean (111)-covalent surfaces band-bending near the surface is created by the surface states.

For the ionic case, the Fermi level is a function of the semiconductor doping, and the band structure with respect to the vacuum is defined by the electron affinity, χ (see Fig.2)/19/. In this case, as in metals, there are two contributions to Φ_{SC}, an inner one associated with bulk properties and a surface one related to the interface electrostratic dipole created across the last one or two layers of the semiconductor.

As regards the case of Fig.2(a), a new effect appears: there is a band bending boundary layer near the surface/20/. This effect extends, typically, up to several hundred angstroms into the metal (the typical distance is the Debyelength, L_D, given by $(k_BT / 4 e^2(n_b+p_b))^{\frac{1}{2}}$ where T is the temperature and n_b and p_b the number of electrons and holes in the bulk/20/). What is important to notice is that the whole charge per unit area in the boundary layer is a small quantity compared with the density of states appearing in the surface band/18/. This is related to the fact that the electrostatic voltage drop in the boundary layer (a fraction of an eV) is the same order of magnitude as the dipole appearing across the microscopic interface, while the boundary layer length is several orders of magnitude greater than the interface width. Then, boundary layer effects can be neglected as regards the microscopic interface physics/18/. This means that from now on we shall treat all semiconductor interfaces as having no band bending (similar to the case shown in Fig.2(a)).

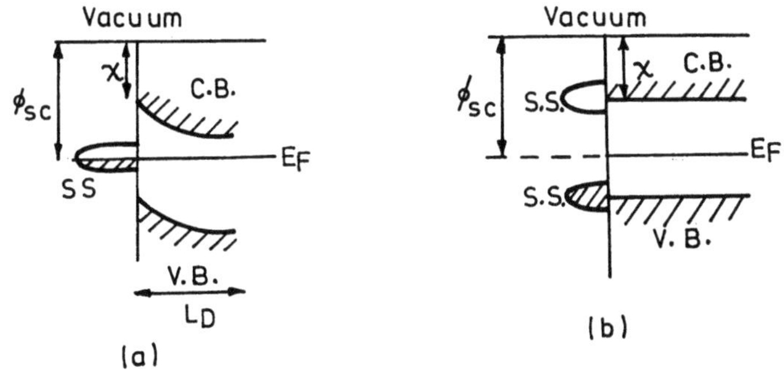

Fig.2(a) Shows a (111)-covalent surface. χ is the electron affinity, ϕ_{SC} the semiconductor work function and C.B., V.B. and S.S. stand for conduction band, valence band and surface states, respectively. (b) Shows a (110)-ionic surface with two dangling bond surface states

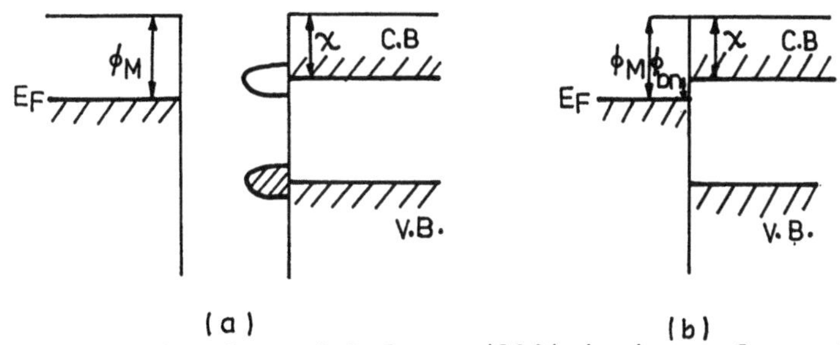

Fig.3: Schottky model for a (110)-ionic surface. (a) Before contact. (b) Intimate contact. ϕ_{bn} is the barrier height for electrons

Still, there is an important difference between (111)-covalent and (110)-ionic surfaces: the high density of surface states appearing in the first case in the energy gap. Then, for typical covalent faces, the work function is well defined and its value is independent of the semiconductor doping (we can say that the doping effects are screened across the boundary layer/18,20/).

3. Metal-Semiconductor Interfaces

Figures 3 and 4 show two typical metal-semiconductor interfaces. In one case, a metal is in contact with a (110)-ionic surface, while in the second case we consider a (111)-covalent semiconductor.

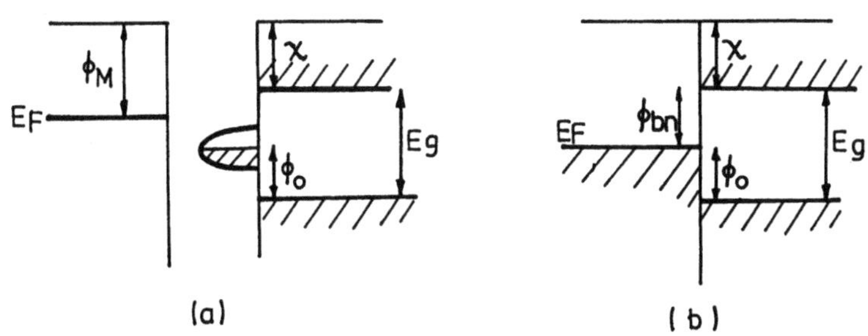

Fig.4: Bardeen model for a (111)-covalent surface. (a) Before contact (b) Intimate contact. ϕ_{bn} is the barrier height

3.1 Schottky Model

This model/9,13,19/ appears in trying to explain the behaviour of the interface shown in Fig.3. In this case, one can argue that electrons at the metal Fermi level cannot be transferred to the semiconductor Fermi level since, in this last case, neither bulk nor surface states exist. This suggests that no interface dipole is induced when the contact is built up and consequently we can obtain the barrier height of the junction for electrons, ϕ_{bn}, as follows (Fig.3):

$$\phi_{bn} = \phi_M - \chi \qquad (4)$$

This equation defines the barrier height of the junction shown in Fig.3, ϕ_{bn} being the parameter defining the interface transport properties/13/.

3.2. Bardeen Model

This is the model/10/ that appears naturally when one considers the case of Fig.4. In this interface, there is a high density of states at the semiconductor surface, and electrons can be transferred from the metal to the semiconductor Fermi level. In other words, the high density of states associated with the semiconductor surface bands behaves locally as a metal; then, when the junction is formed, both Fermi levels end up aligned as discussed in Sect. 2.2.

In this model, the interface barrier height, ϕ_{bn}, is independent of the metal work function, its value being determined by the surface band position. According to Fig.4, ϕ_{bn} is given in the Bardeen model by

$$\phi_{bn} = \chi + E_g - \phi_0 \, , \qquad (5)$$

ϕ_0 defining the semiconductor surface Fermi level with respect to the valence band top.

According with the results of Sects.3.1. and 3.2., metal-semiconductor interfaces would present very different behaviours depending on the semiconductor surface bands. Unluckily, experimental facts (see below) tend to contradict that conclusion, and people were forced to consider other models.

3.3. Defect Model

One of the experimental facts (see below) for metal-semiconductor interfaces that any acceptable model has to explain is that the junction barrier height does not depend very much on the metal deposited on the semiconductor. The defect model/21/ tries to explain this by assuming that the interface Fermi level is pinned by the defects created in the semiconductor by the deposition of the metal. Figure 5 shows schematically a defect level inside the semiconductor energy gap; that level pins the interface Fermi energy, the mechanism being the transfer of charge flowing from the metal to the semiconductor (or vice versa). The induced dipole aligns the metal Fermi energy with the semiconductor defect level. In this case the barrier height is defined by the position of the defect level with respect to the semiconductor conduction band. Thus

$$\phi_{bn} = E_g - \phi_1 \, , \qquad (6)$$

where ϕ_1 is defined in Fig.5.

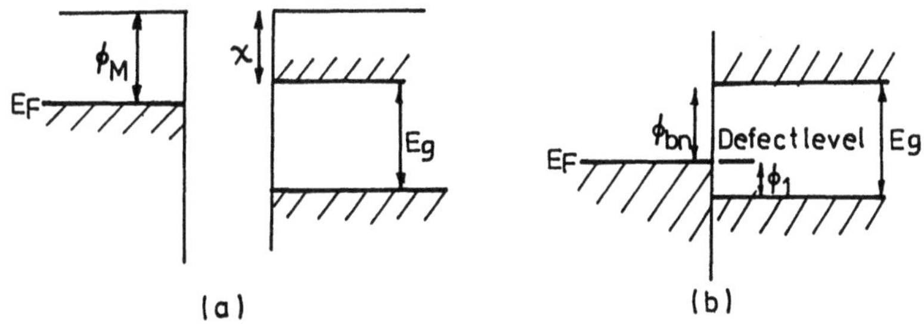

Fig.5:Defect model for a metal-semiconductor interface. (a) Before contact.(b) Intimate contact. The metal deposition on the semiconductor creates the defect.

The general approach to the defect model has consisted in analysing the electronic levels associated with the different defects at the interface/22-23/ and to deduce which one could explain the evidence found for the Fermi level pinning. A main assumption of this approach is that the deposition of the metal on the semiconductor surface can release the energy necessary to create the defects/24/.

A discussion of this model with respect to other models and the experimental evidence will be presented below (Sect.5).

3.4. Induced Density of Interface States Model

Previous models have assumed that the density of states at the interface is given by the superposition of the density of surface states for each independent crystal. This assumption obviously neglects effects induced by the coupling between the metal and the semiconductor. The induced density of interface states (IDIS) model tries to take into account those effects.

HEINE/11/ was the first to note that the metal wave functions near the Fermi level can tunnel into the semiconductor energy gap when both crystals are in intimate contact. On the other hand, the new density of interface states, associated with the tails of the metallic wave functions tunnelling into the semiconductor energy gap, is compensated by a decrease in the density of states in the conduction and valence bands of the semiconductor/25/.Figure 6 illustrates

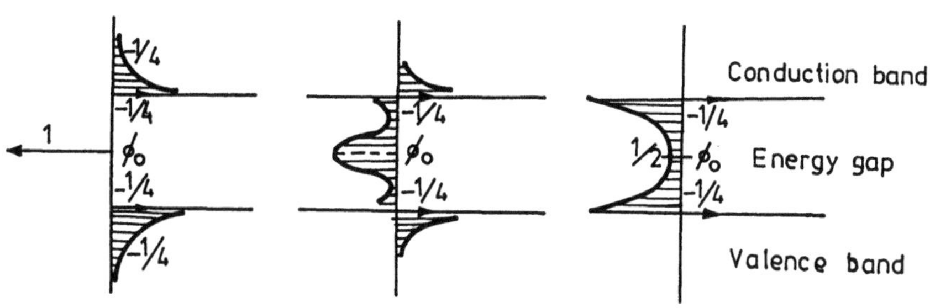

Fig.6: Induced density of states around the main gap of a one-dimensional metal-semiconductor model. (a) For a decoupled interface ($d \to \infty$) a surface state appears at the middle of the gap. (b) As the metal approaches the semiconductor, the surface state broadens. (c) For intimate contact the resonance extends to the whole gap. Notice that in all the cases there appears a 1/4-defect of state at the band edges, and that the total density of interface states amounts to zero.

this for a one-dimensional metal-semiconductor interface; in that figure, the density of interface states is shown for different metal-semiconductor distances. For a decoupled interface ($d\rightarrow\infty$), there is a surface state at the middle of the semiconductor energy gap having a half-occupancy (this simulates a covalent surface). As the metal and the semiconductor approach each other, the surface state broadens and we find a resonance at the mid-gap; eventually, for an intimate contact, the resonance extends to the whole energy gap. The important point about this result is that in all cases, the total interface density of states up to the middle of the gap is zero, with the valence band compensating for the excess of states appearing in the lower half of the energy gap/25/.

These results show that a charge neutrality level for the semiconductor can be defined in such a way that the interface densities of states below that level, for the energy gap and the valence band, compensate each other locally.

In our one-dimensional model, the mid-gap defines the charge neutrality level of the semiconductor/26/, and its role is at the interface equivalent to the Fermi level of a metal. For example, if the metal Fermi level is above it, the electronic wave functions tunnelling into the semiconductor create a dipole tending to restore the equality of both the Fermi level and the semiconductor charge neutrality level. Then, in a zeroth-order approximation the barrier height of the junction is independent of the metal.

Similar arguments can be applied to 3-dimensional crystals by taking an average over the different momenta of the two-dimensional Brillouin zone/26/.

In general, a charge neutrality level can be introduced for a semiconductor, its position being defined by its energy, ϕ_0, with respect to the valence band top (as in Fig.4, later on, we shall see why we are introducing the same symbol in both cases for defining the surface Fermi level). Moreover, at a metal-semiconductor interface there is an induced interface dipole, D, which is given by/26/

$$D = (E_F - \phi_0) / S \qquad (7)$$

E_F being here the Fermi level measured with respect to the valence band top. This equation states that the induced dipole is proportional to the difference between the Fermi level at the interface and the charge neutrality level: D is created by the electronic metal wave functions tunnelling into the semiconductor energy gap.

S in (7) is found/26/ to be about 0.1 for covalent and III-V ionic semiconductors, and about 0.3 for II-VI semiconductors. These values show that the shift of E_F around ϕ_0 must be, at must, about 0.2eV, and in most cases less than 0.1eV for covalent and III-V semiconductors. Notice that for a difference of 0.1eV between E_F and ϕ_0, the induced interface dipole is about 1eV.

We should also mention that the charge neutrality level, ϕ_0, is related to the surface band states of clean surfaces (see Figs.4 and 6). As Fig.6 illustrates, the charge neutrality level coincides with the surface state energy of a covalent semiconductor, a case for which the surface band is locate around the middle of the optical gap (see Fig.2(a)): the reason is that the charge neutrality level is the centre of the resonance resulting from the coupling of the surface state band to the adsorbed metal. This comment can be applied strictly to covalent (111)-surfaces, a case for which there is a half-occupied band at the centre of the optical gap/15/ associated with

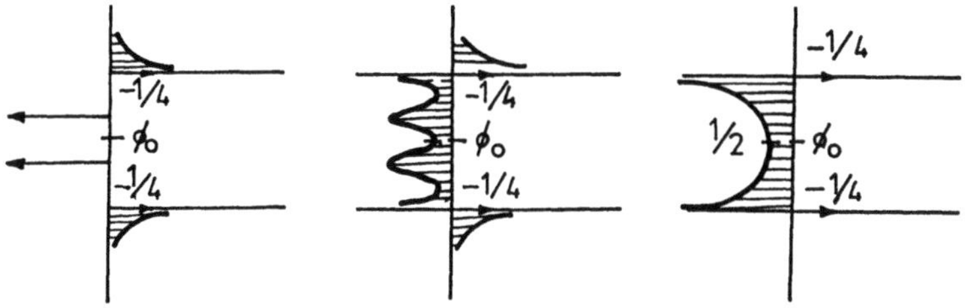

Fig.7: As Fig.6 for a case simulating a (110)-ionic surface

a simple dangling bond. For a (110)-surface, there are two dangling-bonds and things evolve in a metal-semiconductor interface as shown in Fig.7. At a clean surface, there are two dangling-bond surface states symmetrically located with respect to the middle of the optical gap/15/; as the interface is formed the density of states evolves as shown in Fig.7, and the charge neutrality level is defined by the mid-point between the dangling-bond surface states.

TERSOFF/27/ has recently calculated the charge neutrality level of different semiconductors using properties associated with their electronic band structure. In Table I, we collect his results/27/. From these values and (7) we can obtain different barrier heights. In fact, $(E_F - \phi_0)$ can be calculated by noticing that (see Fig.4)

$$D \simeq \chi + E_g - \phi_0 - \phi_M . \qquad (8)$$

Table I

Charge neutrality levels for different semiconductors

	Si	Ge	GaP	InP	AlAs	GaAs	InAs	GaSb
ϕ_0	0.36	0.18	0.81	0.76	1.05	0.70	0.50	0.07

This equation states that the difference between the metal Fermi level and the semiconductor charge neutrality level is almost completely screened by the transfer of charge from one crystal to the other.

Using now (7), we find that

$$E_F - \phi_0 \simeq S\ (\chi + E_g - \phi_0 - \phi_M), \qquad (9)$$

this equation defining the final interface Fermi level.

To be specific, consider now a Ag-Si interface; for this case (taking $S \simeq 0.10$ and using well-known values for the parameters of (9)) we find

$$E_F - \phi_0 \simeq 0.07\text{eV} \quad (\text{Ag-Si})$$

and $E_F = 0.43$. This yields a barrier height of 0.67eV, to be compared with the experimental value of 0.79eV/13/.

4. Heterojunctions

Many of the models applied to metal-semiconductor interfaces have also been proposed for heterojunctions.

4.1. Anderson Model

This is the equivalent/28/ of the Schottky model for metal-semiconductor interfaces. Here, no induced interface dipole is assumed to be created between crystals. This assumption is based on the fact that, for a good match between the lattices of semiconductors forming the heterojunction, no surface states are found in the energy gap. Figure 8 shows the case of two (110)-semiconductor surfaces.

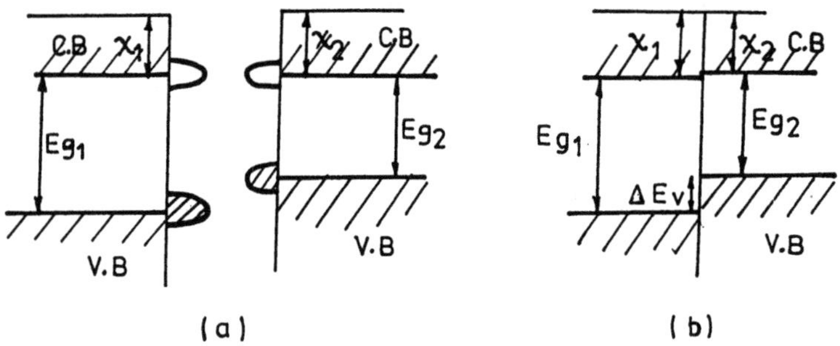

Fig.8: Anderson model for a heterojunction. (a) Before contact. (b) Intimate contact

In the Anderson model the band offset between the valence bands, ΔE_V, at the interface is given by (see Fig.8)

$$\Delta E_V = \chi_1 + E_{g1} - \chi_2 - E_{g2}. \qquad (10)$$

4.2. Defect model

This appears naturally when considering two semiconductors with lattices having a large mismatch/29/. If this is the case, a large number of defects can appear at the interface pinning the heterojunction band offset at a given value.

Actual evidence seems to disprove, however, both the Anderson and the defect models for most heterojunctions (see below).The IDIS model/30-32/ receives its strongest support from the evidence accumulated for heterojunctions. We discuss this model in the next paragraph.

4.3. Induced Density of Interface States Model

The crucial concepts for heterojunctions are the following/30-32/: (i) an interface dipole,D, is induced between the two semiconductors when their respective charge neutrality levels do not coincide. This dipole appears due to the transfer of charge, from one semiconductor to the other, associated mainly with the tails of the wave functions of one semiconductor tunnelling into the gap of the other (see Fig.9).

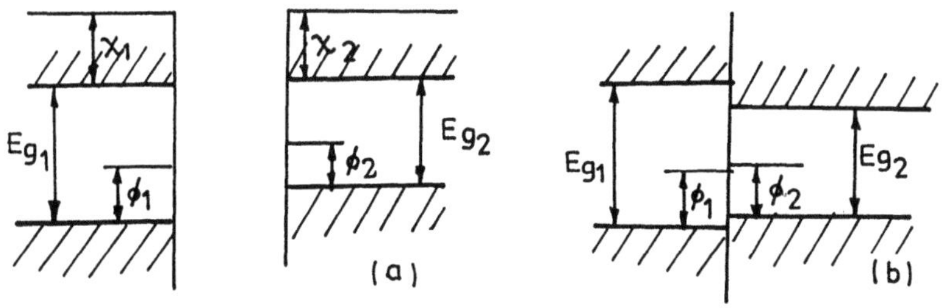

Fig.9: IDIS model (a) Semiconductor levels before contact. (b) Intimate contact. Notice that in this case the charge neutrality levels do not exactly coincide.

(ii) This interface dipole tends to restore a situation for which both charge neutrality levels are equal. The induced dipole is proportional to the charge neutrality level difference, and is approximately given by (see fig.9(a))

$$D = (\Delta E_V + \chi_1 - \chi_2) / S, \tag{11}$$

where S is a screening factor, similar to the one found for metal-semiconductor interfaces. S is related to the interface dielectric constant and, approximately, can be taken as 0.15 for covalent and III-V ionic semiconductors (notice that S must be somewhat larger for heterojunctions/30,33/ since the interface dielectric constant is smaller for this case).

From the values given in Table I and (11) we can find ΔE_V. This can be achieved by noticing that the induced dipole, D, must be practically equal to the difference of energy between the initial positions of the charge neutrality levels:

$$D \simeq \chi_1 + E_{g1} + \phi_1 - \chi_2 - E_{g2} + \phi_2 . \tag{12}$$

From (11) and (12) we find

$$\Delta E_V = \phi_2 - \phi_1 + S(\chi_1 + E_{g_1} - \phi_1 - \chi_2 - E_{g_2} + \phi_2). \tag{13}$$

The best evidence supporting this analysis has been afforded by MARGARITONDO/34/, who has shown that the predictions of this model for the heterojunctions of Si and Ge with other semiconductors are in very good agreement with the experimental evidence (notice that Margaritondo took S=0 in (13)). KATNANI and MARGARITONDO/35/ also found that for many semiconductors there is a kind of transitivity rule (see below). This can be easily explained by means of Fig.9, where all the charge neutrality levels of different semiconductors have been aligned. Moreover, a similar argument can be applied to metals by aligning the semiconductors charge neutrality levels with the metal Fermi energies (Fig.10).

5. Experimental Evidence

I shall not try to discuss experimental methods as applied to metal-semiconductor and semiconductor-semiconductor interfaces. Readers are referred to the following review papers: KENDELEWICZ and LINDAU/4/, BRILLSON/3/, CALANDRA et al /2/, LE LAY/1/, FLORES and TEJEDOR/6/. Let us just summarise the main experimental facts that any model has to explain.

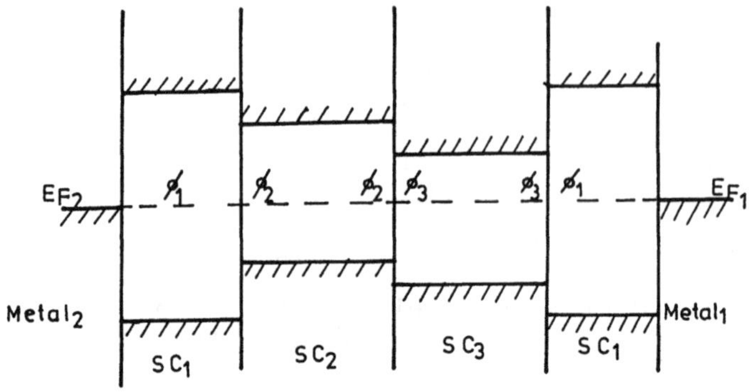

Fig.10: Shows the alignment of the charge neutrality levels and Fermi levels of different metals and semiconductors

As regards metal-semiconductor interfaces, the main facts are the following: (i) the interface Fermi level (and the barrier height) does not depend very much on the metal deposited on the semiconductor; (ii) Schottky barriers seem to be formed at a monolayer coverage with a Fermi level pinning practically coinciding with the case of a thick metal coverage; (iii) chemical reactions do not affect the barrier height of the junction very much.

As regards heterojunctions, the main facts seem to be: (i) the band offsets for different heterojunctions show a kind of transitivity and commutativity rule. Transitivity means that ΔE_V obeys the following rule:

$$\Delta E_V(A-B)+\Delta E_V(B-C)+\Delta E_V(C-A)=0. \tag{14}$$

Commutativity means that

$$\Delta E_V(A-B) = \Delta E_V(B-A); \tag{15}$$

(ii) the band offsets seem to be very little dependent on the face orientation of the heterojunction; (iii) finally, the band offsets for heterojunctions having good lattice matching are not found to be controlled by interface defects/36/.

5.1. A Comparison Between the IDIS Model and the Defect Model

A detailed comparison between IDIS model and the defect model has been published by FLORES and TEJEDOR/6/. Here, we only present a summary of the main points.

As regards the metal-semiconductor interfaces both models seem to offer an appropriate explanation of the main facts. Thus, (i) the independence of the Fermi level pinning with the deposited metal is explained in the following way: (a) the IDISM ascribes that pinning to the semiconductor charge neutrality level; (b) the DM assumes that defects in the semiconductor, induced by the adsorbed metal, are pinning the Fermi level. (ii) The formation of the Schottky barrier by the deposition of a small fraction of a metallic monolayer is explained in both models by assuming the formation of two-dimensional islands. Islands operate as thick metals, pinning the Fermi level at its final value. (iii) The barrier height independence of the chemical reactivity is explained in the IDISM showing that interlayers do not change the Fermi level by more than 0.2 or 0.3eV (see Sect. 6). In the DM, it is assumed that defects are not modified by the reactive interface.

The substantial difference between the IDIS model and the defect model appears in semiconductor-semiconductor interfaces. As explained above, experiments can rule out any effect associated with defects at semiconductor-semiconductor interfaces having a good matching. Moreover, transitivity and correlation with metal-semiconductor barrier heights find a simple and convincing explanation in the IDIS model.

In conclusion, although our present knowledge of different semiconductor interfaces strongly support the IDIS model, more work on more semiconductor interfaces is necessary before a definitive conclusion can be drawn. In particular, for metal-semiconductor interfaces it is necessary to have a better knowledge of the geometry associated with the metallic atoms deposited on the semiconductor, the interdiffusion between the metal and the semiconductor, etc. It would not be surprising if for the metal-semiconductor interface, depending on the different processes of interdiffusion, the two mechanisms assumed to operate in the IDISM and DM were each to play a partial role in the determination of the interface barrier height/6/.

6. Extensions of the IDIS Model

The strong support found for the IDIS model has prompted people to analyse its main properties by means of more accurate theoretical calculations/37-40/.

In its most elaborate forms/37/, the IDIS model has been analysed by means of a self-consistent local density formalism. At an intermediate level of sophistication a self-consistent tight-binding method/38-40/ has been proposed to calculate the induced interface dipoles and the heterojunction band offsets. The advantage of this latter approach is that it keeps a great simplicity and introduces into the calculation the main effects associated with the transfer of charge between crystals.

In Table II, we present recent theoretical results calculated for different (110)-interfaces using the methods mentioned above, within the IDIS model. Notice the good agreement found between different theoretical calculations and the experimental evidence.

The same self-consistent tight-binding method has been applied to analyse: (i) metal-semiconductor interfaces /38/, (ii) the effect

Table II

(a) Tight-binding model; (b) Local density calculations; (c) results obtained using (13) and Table I

Heterojunction	(a)	(b)	(c)	Experiments/5/
GaP-Si	0.64	0.61	0.60	0.8
GaAs-Ge	0.61	0.63	0.55	0.35,0.55
AlAs-Ge	0.94	1.05	0.96	0.95
AlAs-GaAs	0.32	0.37	0.41	0.38.0.55
ZnSe-Ge	1.70	2.17	1.59	1.40,1.52
ZnSe-GaAs	1.01	1.59	1.04	1.10

of interlayers between the crystals forming the interface (either metal-semiconductor interfaces/41/ or heterojunctions/42/), and (iii) the effect of having several layers of one crystal deposited on a semiconductor/43/.

The interested reader is referred to the original papers. Here the main results of the calculations are summarised:

(i)Si-Ag, Si-Al and GaAs-Ag/38,44/ interfaces have been analysed. It is found that the deposition of a metal on the semiconductor introduces a shift in the charge neutrality levels less than 0.2eV. This shift is small but depends on the position where the metal atoms are placed on the surface/45/. This result shows that, as regards metal-semiconductor interfaces, the charge neutrality level is only well-defined within an accuracy of 0.2eV; accordingly, each specific case has to be analysed in detail if a high accuracy for the barrier height is needed (for heterojunctions, the geometry of the interface is unique and we do not find that problem).

(ii) Intralayers have been analysed in the following systems: Si-H-Ag, Si-Cl-Ag; Si-Cs-Ag/41/ and Ge-Al-ZnSe/42/. Changes in the barrier heights have been found, but not larger than 0.2-0.3eV. These changes have been attributed to shifts in the charge neutrality levels of the semiconductors forming the interfaces/42/. In other words, the charge neutrality levels appear to be also dependent on the intralayer deposited at the junction.

(iii) Finally, barrier heights as a function of the number of layers deposited on the semiconductor have been analysed for a GaAs-Ag/43/ interface and a GaAs-Ge/46/ heterojunction. In both cases, it was found that the interface barriers were practically formed with the deposition of only a monolayer (in one case Ag, in the other Ge, were deposited on GaAs).

Acknowledgements

I acknowledge many helpful discussions with C. Tejedor, A. Muñoz and J. C. Durán.

References

1. G. Le Lay, J. Vac. Sci. Technol. B2, 354 (1983).
2. C. Calandra, O. Bisi and G. Ottaviani, Surf. Sci. Rep. 4, 271 (1984).
3. L. Brillson, Handbook of Synchrotron Radiation, Vol.II, ed. G.V. Marr (Amsterdam: North-Holland, 1985).
4. T. Kendelewicz and I. Lindau, Crit. Rev. 13, 27 (1986).
5. G. Margaritondo, Solid State Electron. 29, 123 (1983).
6. F.Flores and C. Tejedor, J. Phys. C (1987)
7. F. Braun, Papp. Ann 153, 556 (1874)
8. N. F. Mott, Proc. Camb. Phil. Soc. 34, 568 (1938).
9. W. Schottky, Z. Phys. 113, 367 (1939).
10. J.Bardeen,Phys. Rev. 71, 717 (1947).
11. V. Heine, Phys. Rev. 138, 1689 (1965).
12. F. Capasso, Surf. Sci. 142, 513 (1984).
13. E. H. Rhoderick, Metal-Semiconductor Contacts (Oxford: Oxford University Press 1978).
14. N. D. Lang, Solid State Physics, Vol. 28, (New York and London: Academic Press, 1973).
15. F. García-Moliner and F. Flores, Introduction to the Theory of Solid Surfaces (Cambridge, Cambridge University Press, 1979).
16. D. Pines, Elementary Excitations in Solids (Benjamin 1964).

17. L. D. Landau and E. M. Lifshitz, Electrodynamics of Continous Media (Oxford, Pergamon Press 1963).
18. F. Flores in Crystalline Semiconducting Materials and Devices, p.397, ed. Butcher, March and Tosi (New York, Plenum Press, 1986).
19. A. G. Milnes and D. L. Fencht, Heterojunctions and Metal-Semiconductor Surfaces (Amsterdam, North-Holland 1971).
21. W. E. Spicer, I. Lindau, P. Skeath and C. Y. Yu, J. Vac. Sci. Technol. 17, 1019 (1980).
22. R. E. Allen and J. D. Dow, J. Vac. Sci. Technol. 19, 383 (1979).
23. M. S. Daw and D. L. Smith, Phys. Rev. B 20, 5150 (1979).
24. W. Mönch, Surface Sci. 132, 92 (1983).
25. C. Tejedor, E. Louis and F. Flores, Solid Stat. Commun. 15, 587 (1974).
26. C. Tejedor, F. Flores and E. Louis, J. Phys. C 11, L19 (1978).
27. J. Tersoff, Phys. Rev. Lett. 32, 465 (1984).
28. R. L. Anderson, Ph.D. Thesis Syracuse University , New York (1960).
29. J. L. Freeouf and J. M. Woodall, Surf. Sci. 168, 518 (1986).
30. C. Tejedor and F. Flores, J. Phys. C 11, L19 (1978).
31. F. Flores and C. Tejedor, J. Phys. C 12, 731 (1979).
32. J. Tersoff, Phys. Rev. B 30, 4874 (1984).
33. J. Tersoff, Phys. Rev. B 32, 6968 (1985).
34. G. Margaritondo, Phys. Rev. B 31 2526 (1985).
35. A. D. Katnani and G. Margaritondo, Phys. Rev. B 28, 1944 (1983).
36. P. Chiaradia, A. D. Katnani, H. W. Sang and R. S. Bayer, Phys. Rev. Lett. 52, 1246 (1984).
37. C. G. Van De Walle and R. M.Martin, J. Vac. Sci. Technol. B 4, 1055 (1986).
38. F. Guinea, J. Sánchez-Dehesa and F. Flores, J. Phys. C 17, (1983).
39. G. Platero, J. Sánchez-Dehesa, C. Tejedor and F. Flores, Surf. Sci. 168, 553 (1986).
40. C. Priester, G. Allan and M. Lanoo, Phys. Rev. B 33, 7386 (1986).
41. J. Sánchez-Dehesa, F. Flores and F. Guinea, J. Phys. C 17, 2039 (1984).
42. J. C. Durán, A. Muñoz and F. Flores (to be published).
43. J. Ortega, J. Sánchez-Dehesa and F. Flores (to be published).
44. G. Platero, J. A. Vergés and F. Flores, Surface Sci. 168, 100 (1986).
45. H. I. Zhang and M. Schlüter, Phys. Rev. B 18, 1928 (1978).
46. A. Muñoz, J. Sánchez-Dehesa and F. Flores, Europhys. Lett. 2, 335 (1986).

Photoemission and Inverse Photoemission from Semiconductor Interfaces

F.J. Himpsel

IBM T.J. Watson Research Center, Yorktown Heights, NY 10598, USA

Two recent developments in the spectroscopy of semiconductor interfaces are discussed: The use of various core level spectroscopies to selectively look at the interface atoms, and the application of inverse photoemission to observe unoccupied electronic states.

The spectroscopy of buried interfaces is still in its infancy. Experimental techniques are discussed which have monolayer sensitivity and are capable of discriminating against the overlayer and the substrate. The use of core level spectroscopy is demonstrated in detail by using the CaF_2/Si(111) interface as a model system. CaF_2 forms a sharp epitaxial interface with Si(111), which makes it not only a good testing ground for interface spectroscopies but also a prime candidate for silicon-on-insulator technology and three-dimensional devices. A microscopic model of the bonding at the interface is obtained by combining results from a variety of photoemission techniques.

Inverse photoemission is a newly-developed technique that measures energy and momentum of unoccupied electronic states by utilizing the time-reversed photoemission process, i.e., the production of ultraviolet bremsstrahlung by low-energy electrons impinging onto a surface. This technique is applied to III-V semiconductors and their interfaces with metals. For thin metal overlayers it is possible to detect unoccupied states in the band gap that are correlated with the pinning of the Fermi level and the formation of a Schottky barrier.

1. Probing the Electronic Structure of Interfaces

The electronic structure of interfaces can be viewed as generalization of the electronic structure of surfaces. All the basic concepts[1] remain the same except that vacuum is replaced by a different material. For example, two dimensional electronic states at an interface can be characterized by their quantum numbers E, $\vec{k}_\parallel$, σ and point group symmetry (E = energy, $\vec{k}_\parallel$ = reduced momentum parallel to the interface, σ = spin). In order to separate them from three-dimensional states of substrate and overlayer one may utilize the fact that the energy of two-dimensional states is independent of the third momentum component $k_\perp$. In cases where the $E(\vec{k}_\parallel,k_\perp)$ band dispersions of the substrate and overlayer are known one can simply look for interface states at $E,\vec{k}_\parallel$ values located in a band gap of bulk states.

In practice it is a lot harder to find well-defined interface states than to detect surface states. One has to cope with preparative difficulties as well as with measurement problems.

Most "ideal" (i.e., perfectly ordered and atomically abrupt) interfaces that have been prepared to date are between similar materials, e.g. AlAs and GaAs. In this case, the changes in chemical bonding are minimal. Combining two slabs of bulk material is a good approximation of the interface, and the physical changes originate mainly from altered boundary conditions for the bulk wave functions at the interface. When producing interfaces between dissimilar materials one has to accommodate differences in bonding, which often result in incompatible crystal structures or in a chemical reaction and interdiffusion between the materials. Even with compatible structures there is a stringent lattice-match requirement. If the lattice constants of substrate and overlayer differ by more than about 4% it is generally not possible to grow more than a few monolayers of epitaxial material before misfit dislocations are formed at the interface. One of the few heterogeneous, but nearly perfect interfaces that have been studied up to now is CaF_2 on Si(111). It will be used as a model for demonstrating the state of the art in measuring interface states.

After overcoming the preparative hurdles one is still faced with measurement problems. Most surface techniques rely on a short mean free path of the probing particles, typically from several atomic layers for electrons to less than a monolayer for atoms and for the tip of a scanning tunneling microscope. With such a short penetration depth it is impossible to look through the overlayer for realistic, buried interfaces such as semiconductor-metal contacts. Often, one can get an idea of the interface electronic structure by depositing very thin overlayers ranging from submonolayer coverage to a few layers. A spectroscopic feature arising from the interface layer increases its intensity up to a monolayer coverage and becomes exponentially attenuated when the first layer is covered by subsequent layers. However, there is evidence in a few cases that the electronic structure of a monolayer changes drastically after further layers are deposited, e.g. for the F-derived states at the CaF_2/Si(111) interface. Also, all technologically important interfaces are deeply buried. Therefore, is important to ask if it is possible to devise experimental techniques which allow us to look through a thick overlayer and still be sensitive to a monolayer of atoms at the interface. Several techniques can be conceived for such a purpose. With various core level spectroscopies (Fig.1) one can utilize the chemical shift of core levels to distinguish interface atoms from substrate and overlayer atoms. The capabilities will be demonstrated below in the context of the CaF_2/Si(111) interface. The simplest variant is to measure the energy shift and the intensity of interface core levels by photoelectron spectroscopy (Fig. 1a). From the shift one can infer the charge transfer and the bonding using the concepts developed for bulk solids and molecules[2]. From the intensity it is possible to determine the number of interface atoms and to ascertain that the interface is atomically abrupt and not smeared out over several layers. For looking at valence states it is useful to measure the optical transition rate between an interface core level and these states. Unoccupied valence states are seen in absorption (Fig. 1b) and occupied valence states in fluorescence (Fig. 1c). For obtaining the most detailed information from core levels it helps to focus on the sharpest core level for each element, as done in Fig. 2 (after Ref. 3). The intrinsic lifetime broadening is mainly determined by Auger decay. For shallow core levels this width is increased by band dispersion. It turns out that the best choice is to use the $1s_{1/2}$, $2p_{3/2}$, $3d_{5/2}$, $4d_{5/2}$, $4f_{7/2}$, $5d_{5/2}$, $5f_{7/2}$ levels after they move down from the valence shell to become core levels.

Another characteristic that distinguishes the interface from the substrate and the overlayer is lower symmetry. The lack of inversion symmetry at an interface allows second harmonic generation which is forbidden for inversion-symmetric bulk solids. This effect has

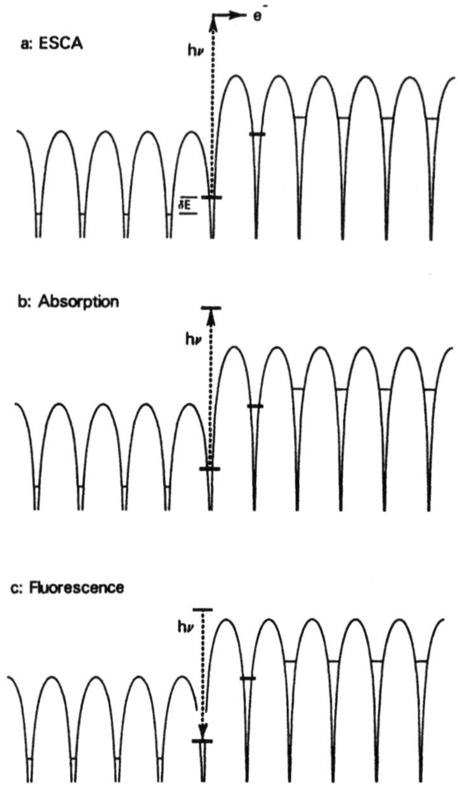

Fig. 1: Application of various core level spectroscopies to probe an interface. Core level shifts (δE) make it possible to tag interface atoms.

The Sharpest Core Levels and their Binding Energies (eV)

H 1s 14																	He 1s 25
Li 1s 55	Be 1s 112											B 1s 189	C 1s 284	N 1s 410	O 1s 543	F 1s 698	Ne 1s 870 2p 22
Na 2p 31	Mg 2p 49											Al 2p 73	Si 2p 100	P 2p 135	S 2p 163	Cl 2p 200	Ar 2p 248 3p 16
K 2p 295 3p 18	Ca 2p 346 3p 25	Sc 2p 399	Ti 2p 454	V 2p 512	Cr 2p 574	Mn 2p 639	Fe 2p 707	Co 2p 778	Ni 2p 853	Cu 2p 933	Zn 3d 10	Ga 3d 19	Ge 3d 29	As 3d 42	Se 3d 55	Br 3d 69	Kr 3d 94 4p 14
Rb 3d 112 4p 15	Sr 3d 134 4p 20	Y 3d 156	Zr 3d 179	Nb 3d 202	Mo 3d 228	Tc 3d 253	Ru 3d 280	Rh 3d 307	Pd 3d 335	Ag 3d 368	Cd 4d 11	In 4d 16	Sn 4d 24	Sb 4d 32	Te 4d 40	I 4d 50	Xe 4d 68 5p 12
Cs 4d 78 5p 12	Ba 4d 90 5p 15	La 4d 103	Hf 4f 14	Ta 4f 22	W 4f 31	Re 4f 41	Os 4f 51	Ir 4f 61	Pt 4f 71	Au 4f 84	Hg 5d 8	Tl 5d 13	Pb 5d 18	Bi 5d 24	Po 5d 31	At 5d 40	Rn 5d 48 6p 11
Fr 5d 58 6p 15	Ra 5d 68 6p 19	Ac 5d 80															

Ce 4d 109 4f 1	Pr 4f 3	Nd 4f 5	Pm 4f 5	Sm 4f 5	Eu 4f 2	Gd 4f 8	Tb 4f 2	Dy 4f 4	Ho 4f 5	Er 4f 5	Tm 4f 5	Yb 4f 1	Lu 4f 7
Th 5d 85	Pa 5d 94	U 5f 1	Np 5f	Pu 5f 2	Am 5f	Cm 5f	Bk 5f	Cf 5f	Es 5f	Fm 5f	Md 5f	No 5f	Lw 5f

Fig. 2

been observed at surfaces [4] and interfaces [5]. It can be converted into a spectroscopy [6] by tuning the incident photon energy $h\nu$ to resonances which occur when either $h\nu$ or $2h\nu$ corresponds to a transition energy between two interface states. Single photon transitions are affected by the crystal symmetry, too. For cubic bulk solids the optical transition matrix element is independent of the polarization of the light (inside the medium) whereas at the interface the perpendicular and parallel components behave differently. Such a polarization modulation can be seen for CaF_2/Si(111) (see below, Fig. 6).

2. The CaF_2/Si(111) Interface: A Model Case

Calcium fluoride has a cubic structure (see Fig. 3) that is matched with the silicon lattice to a high degree (0.6% at room temperature, about 2% at the growth temperature). In addition, CaF_2 is a very stable solid which does not exothermically react with Si. Therefore, it forms an epitaxial, atomically abrupt interface. [7,8]. For the (111) face, which is the stable surface of CaF_2, the surface energy is low enough to provide layer by layer growth on Si(111). (For CaF_2 on Si(100) clustering is observed. Likewise, the reverse interface, i.e. Si on CaF(111), does not exhibit wetting on thermodynamic grounds). In their electronic structure the two materials are quite dissimilar with homopolar versus ionic bonding and band gaps of 1.1eV and 12.1eV for Si and CaF_2, respectively.

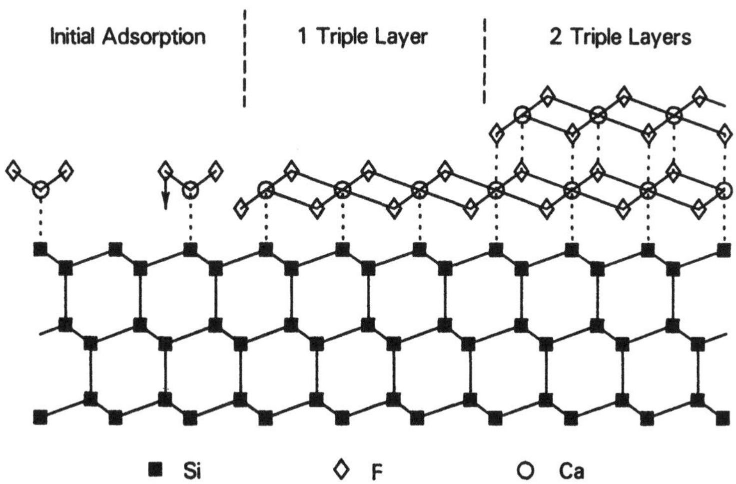

Fig. 3: Sideview of possible CaF_2 /Si (111) interface structures[8] at different stages of growth. The CaF_2 lattice consists of F/Ca/F triple layers.

The results of core level spectroscopy [8-11] at the CaF_2/Si(111) interface are shown in Fig. 4 for the sharpest core levels of each of the three elements (compare Fig. 2). The 2p-levels of Si and Ca are split by spin-orbit interaction. They can be decomposed into the $j = 3/2$ and $j = 1/2$ components using a simple procedure where a fraction of the $j = 3/2$ component is shifted down in energy by the spin-orbit splitting and subtracted from the $j = 1/2$ component. The Si core level exhibits two shifted components in addition to the bulk line, one at higher and the other at lower energy. Ca has a single shifted line, and F no shifted line at all. Core level shifts contain a ground state contribution due mostly to charge transfer [2,12] and a relaxation contribution due to screening of the core hole in the excited state. These two components can be sorted out by measuring the energy shifts of the Auger

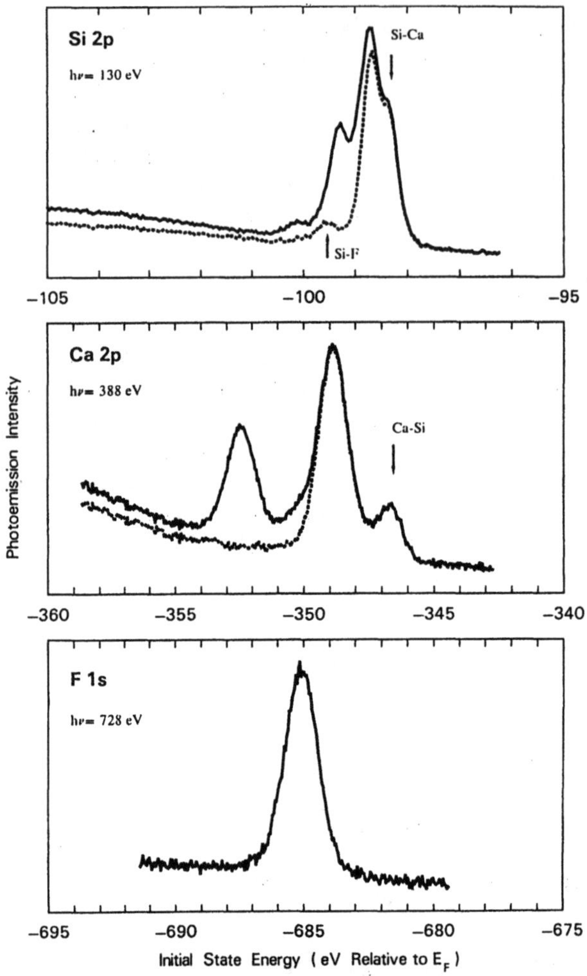

Fig. 4: Core level (ESCA) spectra of 4 triple layers of CaF_2 on Si (111). Shifted interface levels are marked. The dashed lines are for the $2p_{3/2}$ component after numerically removing the $2p_{1/2}$ spin orbit partner. The intense Si - Ca level indicates that the Si surface atoms bond mainly to Ca.

electrons which are created by the core hole decay (Fig. 5). The double charge left after the Auger decay gives rise to four times the relaxation shift compared with that of a single core hole. After subtracting the single hole relaxation, which occurred already before the Auger decay, one obtains a relaxation shift for the Auger electron energy that is three times the relaxation shift of the photoejected core electrons [12,13]. The ground state shift of the Auger electrons, on the other hand, is the same as that of the core electrons. The difference between Auger and core level shifts shows up clearly for the F1s level where no core level shift is detectable but a sizeable Auger shift of 2eV exists. By combining Auger and core level shifts one obtains ground state and relaxation shifts for F and Ca. It comes out that the relaxation shifts are about 1eV upwards for both elements whereas the ground state shifts are opposite (1eV down for F and 1eV up for Ca). For Si there is no sharp Auger transition associated with the 2p level and one would have to resort to the ls level for a

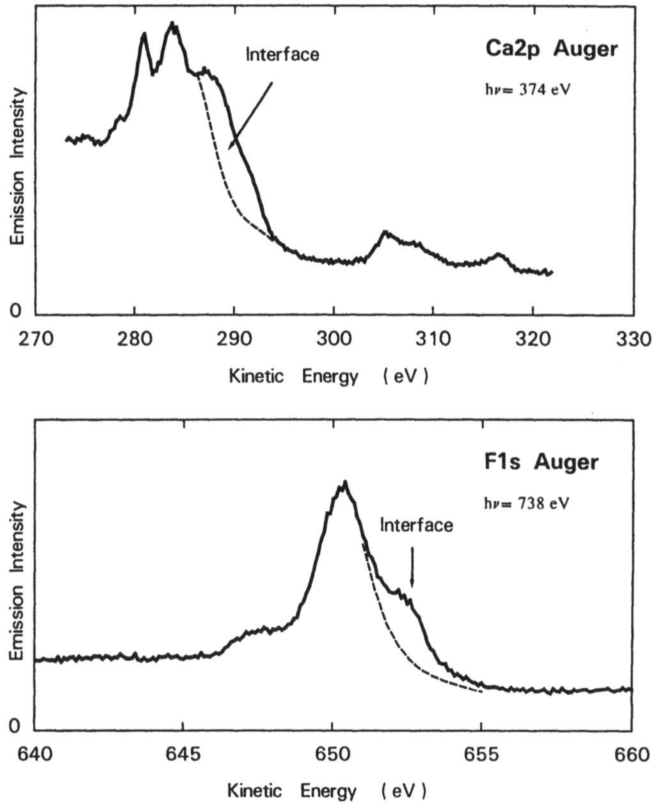

Fig. 5: Auger electron spectra of 4 triple layers of CaF_2 on Si (111). Shifted Auger lines are observed for the interface which are absent in a thick CaF_2 film (dashed lines).

similar decomposition. However, the relaxation contribution will be small, since Si is screened well due to its high dielectric constant, both at the interface and in the bulk. Ca and F on the other hand, are poorly-screened in bulk CaF_2 and well-screened at the interface.

From the ground state shifts one can identify the direction of the charge transfer and draw conclusions about the bonding at the interface. Since a Si(111) surface has one broken bond per surface atom it is natural to ask if Ca or F saturates this bond. This question has not a straightforward answer. Two shifted Si2p core levels coexist at the interface indicating that Ca as well as F bonds to silicon at the interface. The F atom causes a downwards shift of the Si 2p levels because it induces a positive charge on the adjacent Si whereas the Ca atom shifts the Si 2p levels up due to its opposite electronegativity. Judging from the intensities of the shifted Si2p core levels it is apparent that the majority of Si surface atoms bonds to Ca. In fact, the number of Si atoms bonding to F can be further reduced by growing the CaF_2 film at high Si substrate temperatures, close to the desorption point. The Ca core levels exhibit an upwards chemical shift at the interface indicating extra negative charge compared with bulk CaF_2 which has Ca in the 2+ oxidation state. The oxidation state of interface Ca can be determined by measuring the optical absorption from the Ca2p core level as shown in the following paragraph.

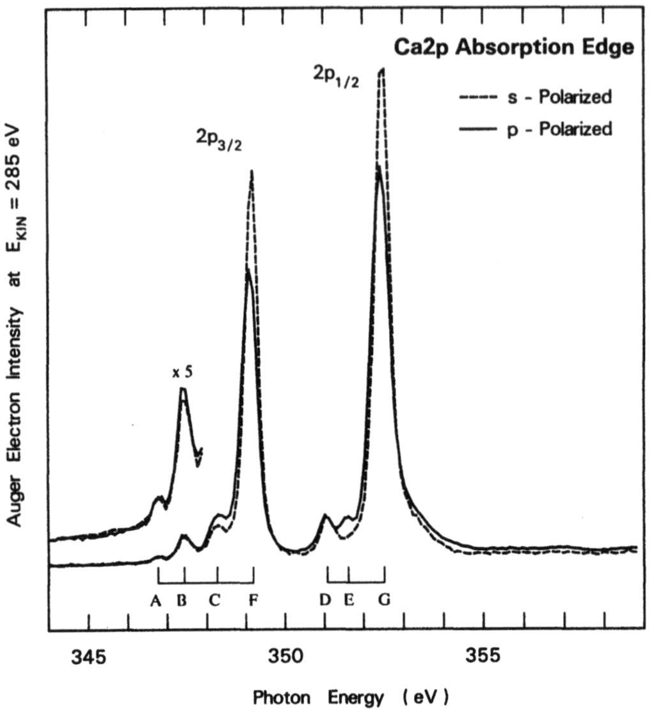

Fig. 6: Optical transitions from the Ca2p core levels to unoccupied valence states for 7 triple layers CaF_2 on Si (111), detected via the Auger electron yield. The multiplet structure of the interface peaks A-E shows that Ca is in the 1+ oxidation state at the interface (after Ref. 9).

Optical transitions from a core level into unoccupied interface states (see Fig. 1b) can be detected via the yield of Auger electrons. The Auger electron yield is proportional to the number of core holes created, i.e., to the number of core level transitions. The sensitivity to the interface is enhanced when energy-shifted Auger electrons coming from interface atoms (see Fig. 5) are selected. At the Ca 2p edge (Fig. 6), one observes prominent transitions into unoccupied states of the Ca^{2+} ion which result in a sharp spin-orbit-split doublet F,G in bulk CaF_2. A number of extra lines (A to E) are observed which are assigned to interface transitions because they are attenuated with increasing film thickness. There is a pronounced polarization dependence [14] indicating the lower symmetry at the interface as opposed to the isotropic bulk. In particular, structure E appears to be absent in s polarization, i.e., it is excited by the component of the electric field perpendicular to the interface. For a close examination of the Ca 2p absorption edge it is useful to start from an atomic picture. [15] Dipole selection rules allow transitions from Ca 2p into unoccupied 3d and 4s states (plus higher-lying states). The 2p-to-4s transitions are expected to be weak compared with the 2p-to-3d transitions since the 4s wave function is much more extended than the 3d wave function in Ca and has little overlap with the 2p core wave function. The multiplet structure of the 2p-to-3d transition gives us a fingerprint of the oxidation state of Ca at the interface. Neutral Ca gives rise to a triplet [15] with two strong lines and a 50-times weaker line. For Ca^{2+} in CaF_2 we expect a similar multiplet structure since it differs from neutral Ca only by the filled 4s shell. Indeed, two prominent lines are observed for thick CaF_2 films. For Ca^{1+}, on the other hand, there exists an unpaired 4s electron which creates a more complex multiplet structure via Coulomb and exchange interaction with the 2p hole and the 3d electron. An estimate for this multiplet can be obtained from calculations for the isoelectronic K atom [15] which give four lines for the $2p_{3/2}$ and three lines for the $2p_{1/2}$ core holes. We observe the same number of interface features assuming that two of them are hidden under the bulk peaks. Indeed, peaks F and G do not disappear in the submonolayer regime as one would expect from a pure bulk feature. The polarization dependence probably originates from an orientation of the spin of the 4s electron perpendicular to the interface caused by bonding with the Si surface atoms.

A picture of the electronic structure at the interface (Fig. 7) arises from the core-level data. The Si atoms are nearly neutral judging from the relatively small Si 2p core-level shift of 0.4 eV. Therefore, one has a half-filled dangling-bond orbital which can interact with the 4s electron of the interface Ca^{1+} by forming a pair of bonding-antibonding states. The occupied bonding state has been seen by angle-resolved photoemission at 1.2eV below the valence-band edge. [9-11] A similar type of bonding has been predicted [16] for Na on Si(111) which is isoelectronic to Ca^{1+}. The electronic structure at the interface is drastically different from bulk CaF_2 with the band gap shrinking from 12.1eV to an estimated 2-3 eV for the bonding-antibonding interface splitting. In fact, the interface is semimetallic because the Fermi level is pinned close to the valence band maximum of Si. The observation of such strong changes in the band structure of an interface layer is encouraging for future fabrication of superlattices or other structures with nearly atomic dimensions. By combining dissimilar materials one may be able to tailor electronic properties near the interface and effectively create new solids.

3. Probing Unoccupied States at Interfaces

Photoelectron spectroscopy is arguably the most versatile and widely-used technique to probe electronic states at surfaces and interfaces. However, it cannot access unoccupied states just above the Fermi level, such as acceptor-like levels. They can be seen only indi-

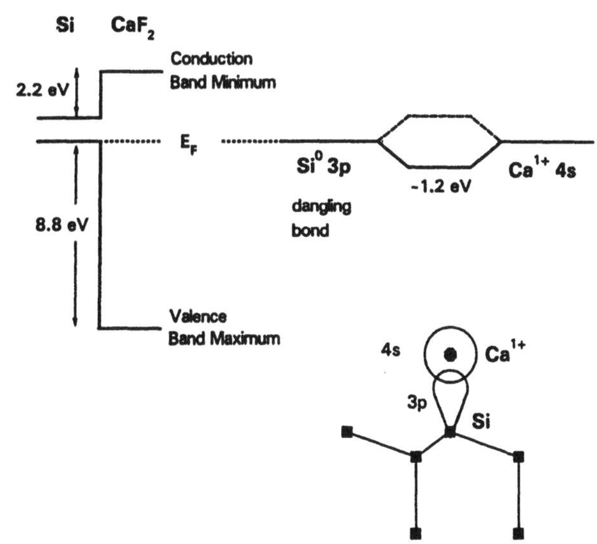

Fig. 7: Energy diagram of the CaF_2/Si (111) interface and model for the bonding obtained from core-level and valence-band photoemission.

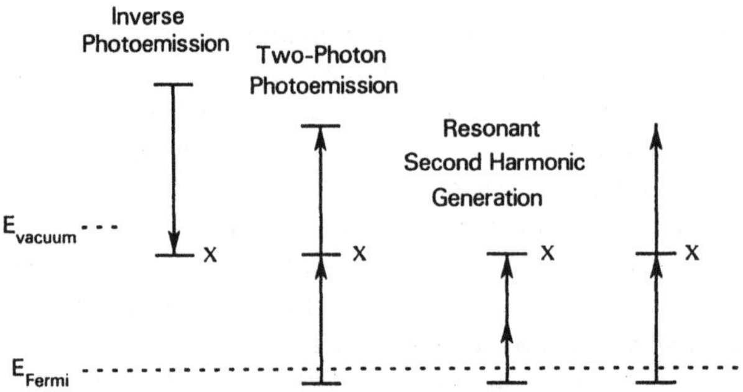

Fig. 8: Various techniques for measuring an unoccupied state (X) at an interface.

rectly by measuring transitions from a core level as described in Fig. 1b. Recently, the efforts to develop new techniques for probing unoccupied states have taken a leap forward and several new methods have been tried out. They are listed in Fig. 8. In two-photon photoemission[17-19] one uses a second photon to promote the electron above the vacuum level where it can be detected by a photoelectron spectrometer. Two-photon photoemission and, even more so, purely optical multiphoton techniques such as second harmonic generation [4-6] can achieve high energy resolution, but are restricted in their use to intermediate states with long lifetimes. Otherwise these processes are so inefficient that they can be observed only at a laser power close to the destruction limit of the sample. There exist a number of single-photon optical techniques that can be applied to surfaces and interfaces such as photovoltage spectroscopy, multiple reflection spectroscopy, and cathodoluminescence spectroscopy. [20] Optical methods give relatively indirect information because they involve a convolution of a number of states, both occupied and unoccupied. The same holds for electron energy loss spectroscopy. Traditional electrical methods (e.g. capacitance/voltage spectroscopy) are also susceptible to ambiguities due to the complex nature of the electrical conduction process, but they achieve a sensitivity for dilute defect states that is unmatched by other spectroscopic techniques. In the following we will concentrate on a multipurpose technique, i.e., inverse photoemission [21].

Inverse photoemission can be viewed as the time-reversed photoemission process (see Fig. 8). Strictly speaking, however, one has an N+1 electron system in inverse photoemission but only N electrons in photoemission, where N is the number of electrons in the sample. Thus, one measures the state of an extra electron injected into the solid. Actually, this situation matches most practical purposes, where the transport of extra charge carriers is considered. Inverse photoemission weighs all states uniformly, independent of their lifetime. By using the conservation laws for energy and momentum one can determine all quantum numbers of an electronic state like in angle-resolved photoemission. [1,21] For example, the energy of an unoccupied state is given by the energy of the incident electron minus the energy of the detected photon (Fig. 8). As reference energy one can use the Fermi level cutoff from a metal sample.

The experimental requirements in inverse photoemission are dictated by the extremely low quantum yield of about 10^{-8} photons per electron, integrated over angle and energy. The setup is relatively simple: an electron source and a photon detector. The electron source is a standard item. The only critical design feature is a high current density (typically 1-10μA/mm^2 at 10 eV) which requires a short path length for the electrons in order to in-

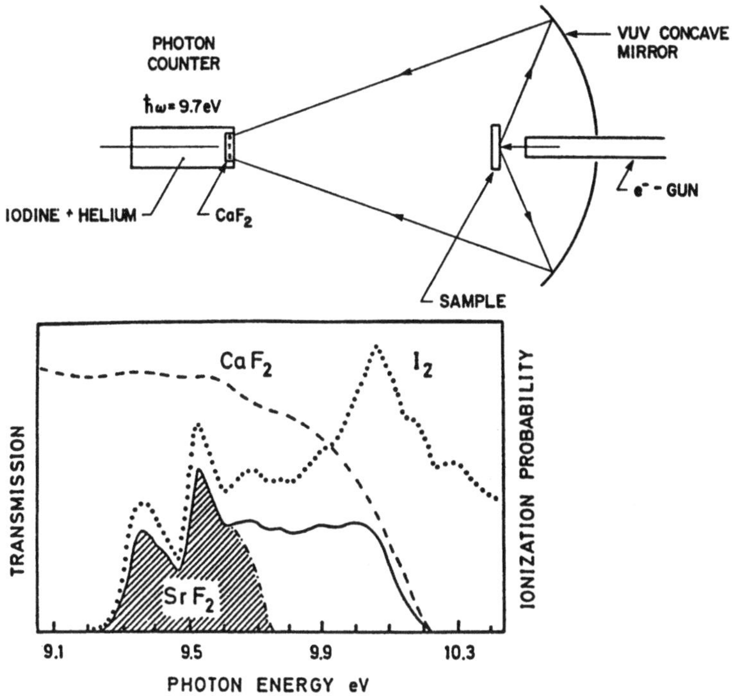

Fig. 9a: Inverse photoemission setup using a Geiger counter (after Ref. 22).

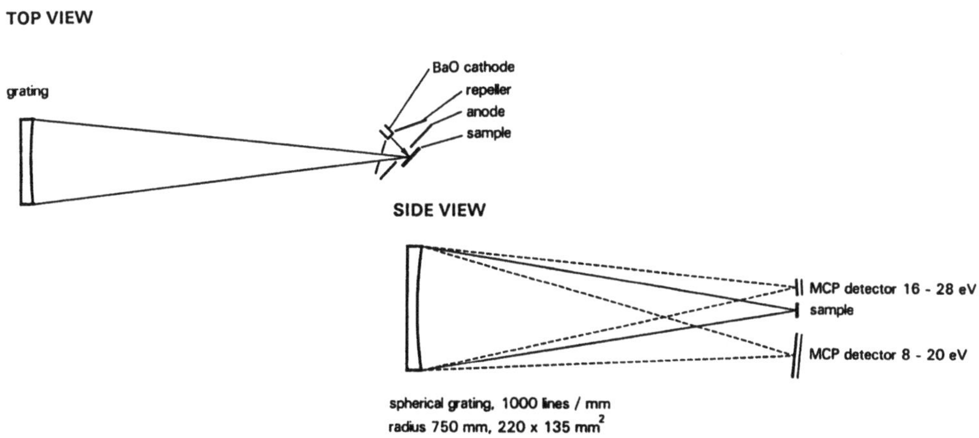

Fig. 9b: Inverse photoemission setup using a spectrograph with position-sensitive detectors (after Ref. 23).

crease the space-charge limit of the current density. Two types of photon detectors have dominated the field although several other imaginative designs exist. A fixed photon energy detector [22] matches the photoelectric threshold of iodine vapor with the cutoff of a window (Fig. 9a). It needs to operate in the so-called isochromat mode where the kinetic energy of the incident electron is swept in order to generate a spectrum. With a photon spectrograph [23] (Fig. 9b), one can vary the photon energy as well as the electron energy, but it is useful to take a photon spectrum with the electron energy fixed. Thereby, it is possible to take advantage of modern multidetection techniques where typically 100 photon energies are detected simultaneously with a channel plate amplifier and a position-sensitive resistive anode detector. The equipment shown in Fig. 9 is representative for the state of the art. Both detection systems have their merits. The fixed photon energy detector can be made very efficient because of a large acceptance angle,but has limited energy resolution and a fixed spectral window around $h\nu = 9.7eV$. The monochromator system achieves better energy resolution (0.3eV) and can be tuned but it is expensive and not as efficient. Generally speaking, the monochromator system is advantageous for band structure measurements where good energy resolution and a tunable photon energy (in order to vary all three k components) are required. The fixed photon energy detector is useful for radiation-sensitive adsorbates.

4. Inverse Photoemission Results from Semiconductor Interfaces

After having gathered experience with bulk semiconductors and surfaces [24] spectroscopists are starting to dabble in more complex systems such as semiconductor interfaces [24,25]. For transition metals on III-V semiconductors it has been possible to detect interface states in the gap [25] and to correlate them with d-like metal-derived levels.

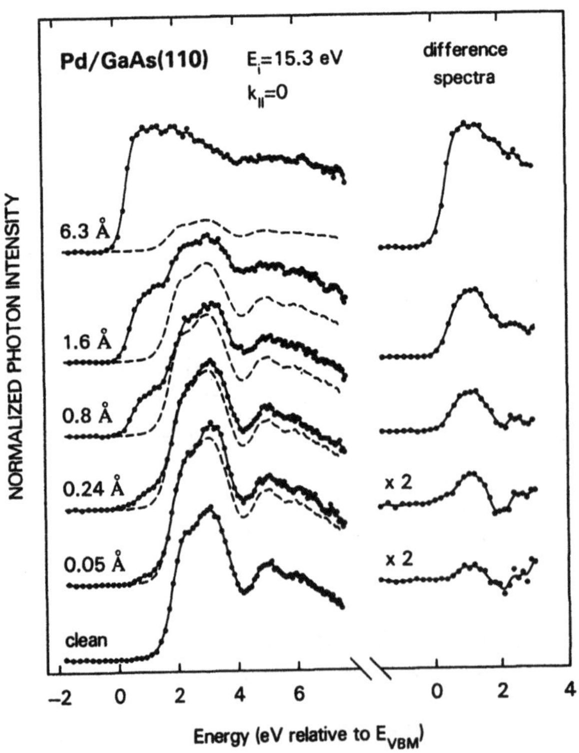

Fig. 10: Coverage-dependent inverse photoemission spectra for Pd on GaAs(110). The raw data are shown together with difference curves which bring out the metal-induced interface states at about 1 e V above the valence band maximum (after Ref. 25).

A sequence of inverse photoemission spectra at various metal (Pd) coverages on GaAs(110) is shown in Fig. 10. Already at a very low coverage of 0.05Å one can detect extra states in band gap (near 1eV). Their intensity increases with coverage without significant line shape changes in the difference spectra until about a monolayer coverage is reached. Beyond that coverage the metal-induced density of states broadens due to the formation of a three-dimensional overlayer. We note that the spectrum from this overlayer is not yet characteristic of the bulk metal since most of the systems discussed here react at room temperature to form a metal-semiconductor compound for a thickness of up to 20Å. Even beyond that thickness the segregation of semiconductor constituents to the metal surface occurs and shows up in the spectra. In the following we adopt a common practice where the metal induced states are determined for a range of overlayer thicknesses (from submonolayer to several layers) and the low-coverage limit is taken as representative for the interface. Such an extrapolation is likely to succeed for reactive interfaces (such as transition metals on III-V semiconductors) but can fail for weakly interacting interfaces (e.g. CaF_2 on Si(111) where the interface structure changes [9,10] when the first overlayer is covered up by a second layer).

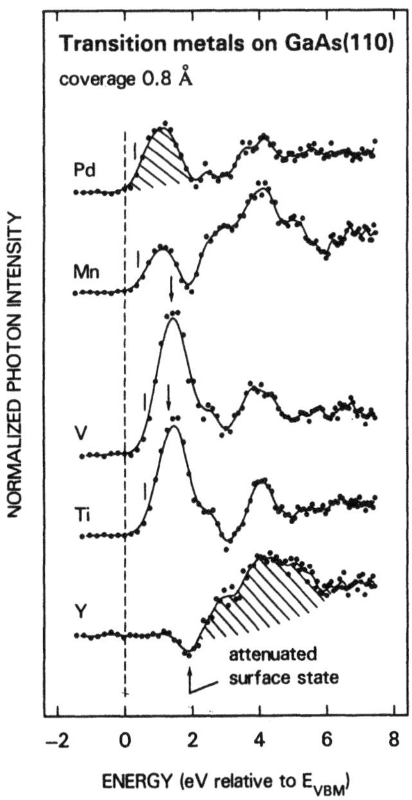

Fig. 11: Difference spectra (metal-covered minus clean) showing unoccupied interface states for a variety of metal/semiconductor combinations. The Fermi level and the known acceptor levels for transition metals substituting the cation are given by tickmarks and arrows, resp., in order to demonstrate that the observed interface states are correlated with the Fermi level pinning position, and the position of nonbonding d-like transition metal states (after Ref. 25).

The difference spectra (metal-covered minus clean) in the low coverage limit are given for a variety of metals on GaAs(110) in Fig. 11 (after Ref. 25). Trends in the position of the interface states relative to the band edges and the Fermi level position can be used to make a connection with various models for the Fermi level pinning. Instead of discussing a large number of proposed mechanisms we concentrate on a model proposed recently [25] for these strongly reacting transition metal/semiconductor interfaces. In this model the gap states seen with inverse photoemission (and their occupied counterparts seen in photoemission) are related to the Fermi level pinning. Their origin is traced to nonbonding (bonding) d-like transition metal states. The d-states move down in energy when going from the left of the transition metal row to the right (compare hatched areas in Fig. 11). For Y the d shell is almost empty, for Pd it is almost filled. In order to test the model for Fermi level pinning we can look if the interface states track the Fermi level pinning position in the gap (see marks in Fig. 11). Indeed, the lower edge of the unoccupied gap states follows the Fermi level E_F and even the maximum in their spectral density is correlated with E_F. For example, E_F is about 0.3eV lower in the gap for Pd on GaAs than for V on GaAs and the center of gravity (close to the peak) of the density of interface states is about 0.3eV lower as well. The second statement of the model can be tested by comparing metal impurity states for dilute systems (e.g. substitutional metal atoms in III-V semiconductors) with the observed interface states. Such levels (where known) are marked in Fig. 11 by arrows. Some conspicuous coincidences lead to a tentative identification of the interface states with such impurity levels. For a conclusive analysis it will be needed to establish the actual chemical environment of the transition metal atoms at the interface (e.g. cation versus anion substitutional sites, complex formation) and the energy levels of these sites. Independent of a detailed analysis there is direct spectroscopic evidence for interface states which can be used to test Fermi level pinning mechanisms.

5. References

1. F.J. Himpsel, Advances in Physics **32**, 1 (1983).
2. K. Siegbahn, Rev. Mod. Phys. **54**, 709 (1982).
3. F.J. Himpsel, Chapter 16 in "Chemistry and Physics of Solid Surfaces VI", ed. by R. Vanselow and R. Howe, Springer, New York (1986), p.435.
4. Y.R. Shen, (Wiley, New York, 1984), Ch. 25; T.F. Heinz, M.M.T. Loy and W.A. Thompson, Pyys. Rev. Lett. **54**, 63 (1985).
5. J.F. McGilp and Y. Yeh, Solid State Commen. **59**, 91 (1986).
6. J.D. Jiang, E. Burnstein, and H. Kobayashi, Phys. Rev. Lett. **57**, 1793 (1986); K. Giesen, F. Hage, H.J. Riess, W. Steinmann, R. Haight, R. Beigang, R. Dreyfus, Ph. Avouris, and F.J. Himpsel, Physica Scripta, submitted.
7. L.J. Schowalter and R.W. Fathauer, J. Vac. Sci. Technol. A **4**, 1026 (1986).
8. F.J. Himpsel, F.U. Hillebrecht, G. Hughes, J.L. Jordan, U.O. Karlsson, F.R. McFeely, J.F. Morar and D. Rieger, Appl. Phys. Lett. **48**, 596 (1986).
9. F.J. Himpsel, U.O. Karlsson, J.F. Morar, D. Rieger, and J. Yarmoff, Phys. Rev. Lett. **56**, 1497 (1986).
10. D. Rieger, F.J. Himpsel, U.O. Karlsson, F.R. Mc Feely, J.F. Morar, and J.A. Yarmoff, Phys. Rev. B**34**, 7295 (1986).

11. Marjorie A. Olmstead, R.I.G. Uhrberg, R.D. Bringans, and R.Z. Bachrach, J. Vac. Sci. Technol. B**4**, 1123 (1986). and Phys. Rev. B, in press.

12. The Madeleung energy also contributes to the ground state core level shift. It does not reverse the sign of the chemical shift but needs to be taken into account for a quantitative analysis.

13. G. Hohlneicher, H. Plum, and H.J. Freund, J. Electr. Spectrosc. Relat. Phenom. **37**, 209 (1985); C.D. Wagner, Farad. Discuss. Chem. Soc. **60**, 291 (1975); G. Hollinger, thesis, Lyon 1979.

14. The residual polarization dependence of the bulk peaks is due to different collection geometries for s and p polarization, which cause different interface-to-bulk intensity ratios.

15. M.W.D. Mansfield, Proc. Roy. Soc. London, Ser. A **348**, 143 (1976), and **346** 555 (1975); J. Barth, F. Gerken, and C. Kunz, Phys. Rev. B **28**, 3608 (1983); J. Zaanen, G.A. Sawatzky, J. Fink, W. Speier, and J.C. Fuggle, Phys. Rev. B **32**, 4905 (1985).

16. J.E. Northrup, J. Vac. Sci. Technol. A **4**, 1404 (1986).

17. R. Haight, J. Bokor, J. Stark, R.H. Storz, R.R. Freeman, and P.H. Bucksburn , Phys. Rev. Lett., **54**, 1302 (1985), and Phys. Rev. B **32** 3669 (1985); R. Haight and J. Bokor, Phys. Rev. Lett. **56** 2846 (1986).

18. J.M. Moison and M. Bensonnssan, Phys. Rev. **B35**, 914 (1987).

19. K. Giesen, F. Hage, F.J. Himpsel, H.J. Riess, and W. Steinmann, Phys. Rev. Lett., **55**, 300 (1985).

20. B.G. Yacobi and D.B. Holt, J. Appl. Phys. **59**, R1 (1986); R.E. Viturro, M.L. Slade, and L.J. Brillson, Phys. Rev. Lett **57**, 487 (1986).

21. F.J. Himpsel, Comments Cond. Mat. Phys. **12**, 199 (1986); V. Dose, Prog. Surf. Sci. **13**, 225 (1983); N.V. Smith, Vacuum **33**, 803 (1983).

22. V. Dose, Appl. Phys. **14**, 117 (1977); A. Goldmann, M. Donath, W. Altmann and V. Dose, Phys. Rev. **B32**, 837 (1985).

23. Th. Fauster, F.J. Himpsel, J.J. Donelon, and A. Marx, Rev. Sci. Instrum. **54**, 68 (1983); Th. Fauster, D. Straub, J.J. Donelon, D. Grimm, A. Marx, and F.J. Himpsel, Rev. Sci. Instrum. **56**, 1212 (1985).

24. F.J. Himpsel and D. Straub, Surf. Sci. **168**, 764 (1986).

25. R. Ludeke, D. Straub, F.J. Himpsel, and G. Landgren, J. Vac. Sci. Technol. **A4**, 874 (1986); W. Drube, F.J. Himpsel, and R. Ludeke, J. Vac. Sci. Technol., submitted.

Photoelectron Spectroscopies: Probes of Chemical Bonding and Electronic Properties at Semiconductor Interfaces

G. Hollinger

Institut de Physique Nucléaire (et IN2P3), Université Université Claude Bernard Lyon-1, 43, Boulevard du 11 November 1918, F-69622 Villeurbanne Cedex, France
and
Laboratoire d'Electronique (LEAME) (UA-CNRS 848), Ecole Centrale de Lyon, F-69131 Ecully Cedex, France

1. INTRODUCTION

Over the last ten years the basic understanding of the microscopic properties of semiconductor interfaces has deepened from various experimental and theoretical aspects. Morphology, atomic geometry, chemical bonding, electronic band states and defects have been examined using a great variety of surface and interface sensitive microscopies, spectroscopies and diffraction techniques. Photoelectron spectroscopies based on both UV and X-ray conventional photon sources and synchrotron radiation have produced substantial progress due to their extensive applicability. In this paper, after an overview of the basic principles of photoemission, I discuss the use of integrated core level and valence band photoelectron spectroscopies to study the formation and properties of semiconductor interfaces.

2. BASIC PRINCIPLES

In photoelectron spectroscopy, photons of energy hν are absorbed in the sample by an optical excitation process. The resultant emission of photoelectrons is measured with an electron analyzer. From the number, energy, momentum and spin of photoelectrons one can gain information on the atomic composition, chemical bonding and electronic states in the solid. The most important parameters in a photoemission experiment are :

i) the photon energy hν and the photon polarization vector,
ii) the photoelectron kinetic energy and its direction of emission.

In angle-integrated photoemission one measures the distribution in kinetic energy of the photoelectrons averaged over all the directions (ideally a 2π solid angle). In practice, only a few spectrometers [1] allow real angle integrated measurements to be performed. Most experiments use Cylindrical-Mirror Analyzers (CMA) or hemispherical analyzers with retarding lenses. CMA's average only the azimuthal orientation (the polar emission angle θ is kept constant) whereas hemispherical analyzers detect only the electrons in a finite solid angle dΩ along a fixed direction. In such experiments angle-integrating conditions are really achieved only when measuring disordered systems (polycrystalline and amorphous layers). For ordered systems (single crystals) some angular effects can distort more or less strongly the ideal averaged electron distribution curve (EDC), depending on the electron energy and the value of the angular acceptance [2,3]. For photon energies higher than 30-40 eV (the X-ray regime) the photoemission process can be described by three independent steps :

i) absorption of a photon and excitation of an electron,
ii) transport of the excited electron to the surface,
iii) escape of the electron into vacuum.

The transport process is governed by inelastic events (fig. 1b) involving single electron or plasmon excitations. This leads to electron mean free paths of ~ 4-40 Å depending on the electron kinetic energy and the nature of the solid (composition and density) [5]. The analyzed depth is directly related to the effective electron mean free path $\lambda \sin\theta$ through the exponential decay low $I = I_0 \exp(-x/\lambda \sin\theta)$ (fig. 1a) which gives the no-loss

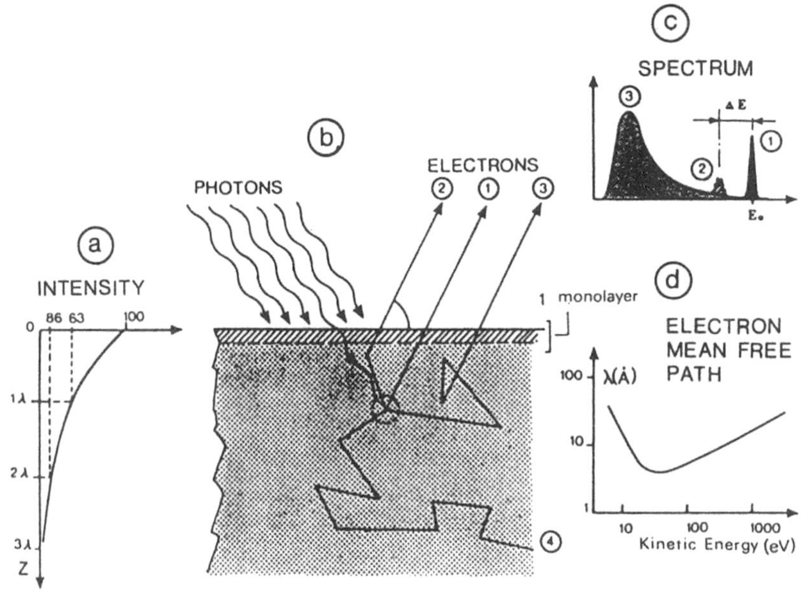

Fig. 1. Schematic diagram showing the three steps of the photoemission process.
a) no-loss intensity decay law
b) inelastic events (refraction at the surface is not shown)
c) schematic energy distribution curve.(1) : primary electrons, (2) and (3) : secondary electrons
d) inelastic electron mean free path as a function of electron kinetic energy.

remaining intensity I of a monoenergetic flux I_0 generated at a distance x from the surface. Consequently, 95% of photoelectrons escaping into vacuum without inelastic scattering (which carry the useful informations) emerge from a depth length smaller than $3\ \lambda \sin \theta$. Therefore, the surface sensitivity can be adjusted by varying the electron kinetic energy or the emitting angle θ.

The relation between the electron distribution curve and the one-electron density of electronic states in a solid is illustrated in fig. 2. The relevant energy conservation equation is

$$h\nu + E_i(N) = E'_k(\ell) + E_f(N-1) \tag{1}$$

where $E_i(N)$ is the total energy of the initial ground state, $E_f(N-1)$ is the total energy of the final state after ejection of one electron from the level ℓ and, $E'_k(\ell)$ is the kinetic energy of the photoelectron ℓ.

The binding energy E_b^v of an electron ℓ relative to the sample vacuum level is defined as

$$E_b^v(\ell) = E_f(N-1) - E_i(N). \tag{2}$$

Substituting (2) into (1) gives

$$h\nu = E'_k + E_b^v(\ell). \tag{3}$$

Because both sample and analyzer are in thermodynamic equilibrium, their Fermi levels are equalized (fig. 3). It is thus clear that binding energies relative to Fermi levels of samples are easy to measure if the electron kinetic energy in the analyzer, E_k and the

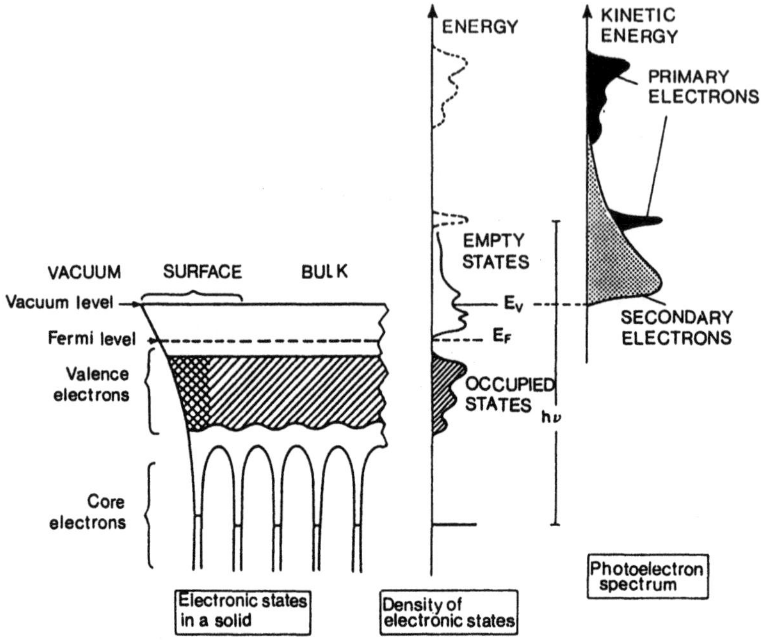

Fig. 2. Energy diagrams of photoelectron spectroscopy

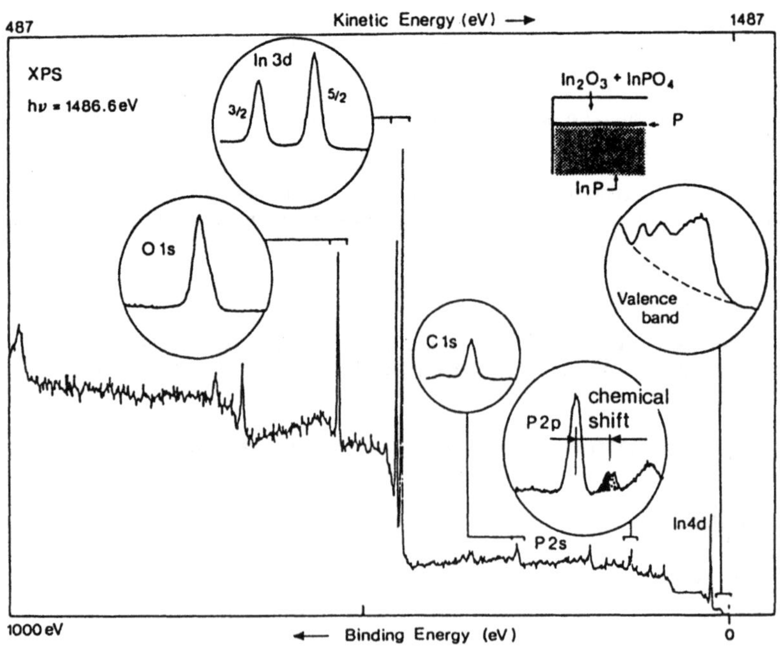

Fig. 3. XPS overview spectrum of a thermally oxidized InP sample.

analyzer work function, ϕ_A are known. Consequently the Fermi level is chosen as reference level for binding energies. Relation (3) thus becomes

$$h\nu = E_k + E_b^F + \phi_A . \qquad (4)$$

Referring binding energies to the sample vacuum level or to some inner potential, "the natural zero level" [6] would necessitate knowing the sample work function (which varies with surface properties) or the inner potential position (not a measurable parameter). Generally, the region probed by photoemission (10-100 A) is much smaller than the thickness of the depletion layer over which band bending occurs in a semiconductor. Thus, any change of the band bending due to surface or interface modifications can be measured by a change in the kinetic energy of the semiconductor substrate. In first approximation, the structures appearing in a photoemission spectrum reflect directly the one-electron density of core and valence states in the solid, as seen for example in fig. 3. Core level electrons from atoms in different environments can have different binding energies. These so-called chemical shifts are essential to extract information on chemical bonding. In addition to primary photoemission features, Auger electrons and characteristic losses can appear in the photoemission spectrum.

In the sudden approximation (Koopman's Theorem), the binding energy is equal to the one-electron orbital energy ε, taken to be positive. These single particle energies are the quantities obtained in electronic structure calculations. In reality the relaxation (screening) of the N-1 orbitals around the ℓ hole tends to lower the final state total energy and a relaxation energy, E_R [7] can be defined as

$$E_b = \varepsilon - E_R . \qquad (5)$$

For closed shell systems (most of the semiconductors and insulators) this one-electron picture is a good approximation. More complex situations with multiplet splitting [7] and multielectronic shake-up satellites (which correspond to transitions between the initial ground state and some excited final states) can be encountered, specially for open shell systems like compounds of transition elements [7].

The behavior of photoionization cross sections can be understood within the one electron, central potential model and the dipole approximation [8]. The total $n\ell$ subshell cross section can be expressed as the summation done over the 2 dipole-allowed values of the photoelectron angular momentum $\ell+1$ and $\ell-1$:

$$\sigma_{n\ell}(E_k) \propto [\ \ell+1\ R^2_{\ell+1}(E_k) + \ell\ R^2_{\ell-1}(E_k)\ ] . \qquad (6)$$

The radial integral $R_{\ell\pm1}(E_k)$ can be evaluated using the radial part of the $\phi_{n\ell}$ initial state orbital and the radial part of the final state-$h\nu$ dependent-$\phi'_{n\ell}(E_k)$ continuum orbital. This explains why photoionization cross sections are characteristic of the orbital and of the photon energy, as seen for the example in fig. 9.

Differential cross sections in a given solid angle depend on the incident photon beam direction for unpolarized light and on the polarization vector direction for polarized light through an asymmetry parameter [8].

3. INSTRUMENTATION : LIGHT SOURCES

3.1 Conventional Laboratory Photon Sources

In the ultraviolet range the HeI and HeII gas discharge line sources are commonly used and have led to the so-called UPS spectroscopy. In addition to the valence bands some shallow core levels can be probed with HeII radiations, for example the In 4d and Ga 3d levels.

Table 1. Characteristics of some conventional photon sources

Photon Source	Main Line Energy (eV)	Satellite Energy (eV)	Relative Intensities	Line Widths (eV)
He Iα	21.21		100	< 0.01
β		23.09	2	
γ		23.74	0.2	
He IIα	40.81			< 0.01
β		48.37		
γ		51.01		
Y Mζ	132.3			0.47
Zr M ζ	151.4			0.77
Mg $K\alpha_{12}$	1253.6		100	0.7
Kα'		1258.1	1	
$K\alpha_3$		1262.0	9.2	
$K\alpha_4$		1263.6	5.1	
Al $K\alpha_{12}$	1486.6		100	0.85
Kα'		1492.2	1	
$K\alpha_3$		1496.2	7.8	
$K\alpha_4$		1498.1	3.3	
Al Kα	1486.6	monochromatized		0.25-0.4
Ag Lα	2984.2	unmonochromatized		2.6
	2984.2	monochromatized		~1.3

In X-Ray photoelectron spectroscopy (XPS) the $K\alpha_{12}$ lines of aluminium or magnesium are used. They allow the measurement of the sharpest core levels for most elements as well as their valence states. One disadvantage with He, AlKα and MgKα photon sources is the existence of higher energy photon satellite lines (see Table 1) which can hinder valence band measurements. They can be eliminated by monochromatization of the main line. The monochromatization of the $AlK\alpha_{12}$ line and of the AgLα line by Bragg reflection (first and second order respectively) from a quartz crystal can reduce the photon width contribution to ~ 0.3 eV and 1 eV respectively. The inherent loss in intensity due to monochromatization can be overcome by multichannel detection.

To bridge the gap between XPS and UPS, Mζ transitions ($4p_{3/2} \rightarrow 3d_{5/2}$) of yttrium and zirconium can be used with a reasonable energy resolution. They allow to enhance the surface sensitivity in measurements of outer core levels [9]. Note however, that these sources have not been widely used because of the sensitivity of their linewidths to the surface oxidation of the anode (fig. 4).

3.2 Synchrotron radiation

Synchrotron radiation emitted by electron storage rings offers a continuum of high intensity ranging from infrared to hard X-rays. In contrast to conventional light sources synchrotron radiation is linearly polarized and has a time structure which permits time of flight or coincidence experiments. The principal advantage of synchrotron radiation is the possibility of tuning the photon energy by means of suitable monochromators. This allows surface sensitivity and photoionization cross sections to be varied. In addition, changing the sample orientation with respect to the polarization vector orientation gives information on the parity and the character of electronic states. Finally, in addition to photoemission, complementary synchrotron radiation techniques like SEXAFS and XANES [10] bring additional information on bonding geometries and unoccupied states.

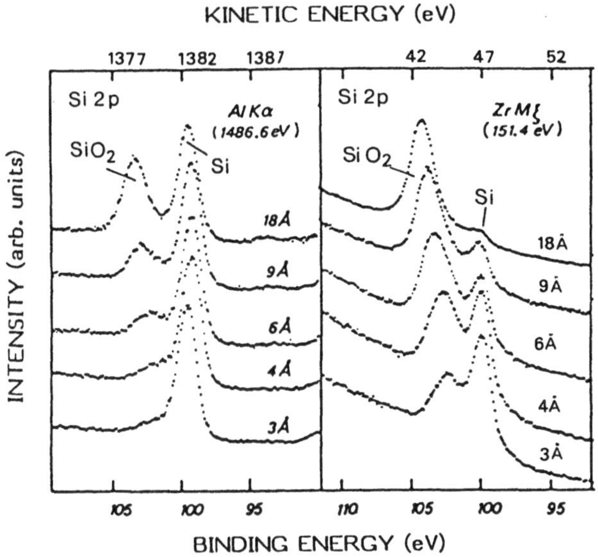

Fig. 4. Variation of surface sensitivity using different photon energies. The surface sensitivity of oxidized Si is enhanced using the ZrMζ source (E_k = 50 eV,

4. CORE LEVEL PHOTOEMISSION

4.1 Analytical Applications

Standard equations for photoelectron spectroscopy quantitative analysis are now well established [11]. They are based on the following relation which expresses the detected photoelectron current from the A_{th} elements in the matrix m:

$$I_A = J\,\sigma_A^{\theta}\ T \int_0^{\infty} N_A(x)\ \exp\,(-x/\lambda \sin\theta)\ dx \qquad (7)$$

where J is the flux of photons, σ_A^{θ} the photoionization cross section of the relevant inner shell in the detection solid angle, T is the transmission factor of the spectrometer, N_A is the atomic composition of atom A in matrix m at depth x and, $\lambda \sin\theta$ is the effective electron mean free path.

By combining such relations it is possible to model intensities, from simple homogeneous standard compounds to complex heterogeneous interfaces.

4.2 Core Level Shifts

The exact value of the binding energy measured for a given element depends on the chemical environment of that element. For example, the Si2p level shifts by ~ 4 eV from Si to SiO_2. The origin of such chemical shifts is generally explained as changes in the valence charge distribution. It can be interpreted qualitatively on the basis of electronegativity and charge transfer. Oxidation corresponds to a decrease of valence electrons thus, the repulsive interaction of a core electron with the remaining electrons decreases, yielding an increase of the binding energy. Such an approach was widely used by Siegbahn et al. [12] to interpret shifts in various organic and inorganic compounds and made the success of ESCA in chemistry and surface chemistry. Quantitatively, in this initial state picture the core level shift for an atom i between its compound and the element i can be expressed using a simple model (which can be also found by simplifying quantum mechanics expressions [13]). The atomic valence electron orbitals define a spherical valence shell (with radius r) of charge q with the core electron residing inside. If a charge transfer Δq occurs the electric potential inside the valence shell is changed and so is the binding energy. In fact the Coulomb interaction of the core electron with the surrounding ions j has also to be taken into account through a crystal (or Madelung) potential. The charge transfer chemical shift can then be expressed as

$$\Delta\varepsilon_{CT} = k\,\Delta q_i + \sum_j \frac{\Delta_{qj}}{r_{ij}} = \Delta q_i \left(k - \frac{\alpha}{R_{ij}}\right). \quad (8)$$

The first term is the atom potential, the second term is the crystal potential. k is a parameter characteristic of the element i which can be approximated by 1/r ; Δq_j and r_{ij} are the charge transfer on the atom j and the distance between atoms i and j, respectively. α is the contribution to the Madelung constant from the particular atom i and R_{ij} is the lattice parameter involved in classical Madelung energy calculations.

Relation [8] indicates that the chemical shift induced by a charge transfer is the result of a partial cancellation of two terms, the atom potential and the crystal potential. Therefore, any direct correlation between shifts and charges can occur only when the Madelung energy variation is negligible. It is the case when different chemical environments exist in the same compound. As an example, on fluorinated silicon surfaces Si atoms bonded to one, two and three fluorine atoms have been detected with corresponding shifts equal to 1.08, 2.09, 3.17 eV respectively (fig. 5) [14]. The chemical shift is directly proportional to the number of Si-F bonds. Such linear behavior was observed in some other silicon systems and the Si 2p shift per Si-F, Si-O, Si-N, Si-C and Si-H bonds was found equal to about 1.1, 0.9, 0.6, 0.5 and 0.3 eV respectively, as expected on the basis of the corresponding electronegativities F (3.95), O (3.5), N (3.1), C (2.5), H (2.1) and Si (1.8).

This simple correlation is no longer true when the chemical shifts of oxides from the principal elements are displayed as a function of the atomic number Z [15] (fig. 6).

There is no direct correlation with electronegativity due to some additional initial state, final state and Fermi level shifts. Rigorously, the core level shift relative to the Fermi level has to be expressed by the following relation :

$$\Delta E_B^F = \Delta\varepsilon_{CT} + \Delta\varepsilon_H + \Delta E_R + \Delta E_F. \quad (9)$$

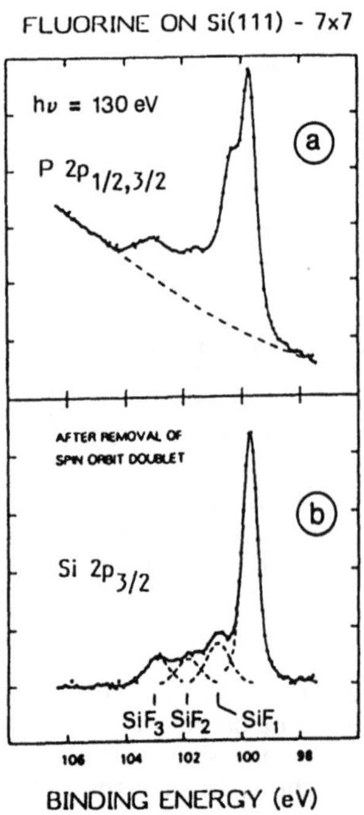

Fig. 5. Si 2p core level spectra for SiF_1, SiF_2, SiF_3 fluorosilyl species. A 1 eV shift is observed for each bound F atom.
(a) $2p_{1/2,3/2}$ raw spectrum.
(b) after numerical removal of the $2p_{1/2}$ component from the background substracted spectrum (after Ref. [14])

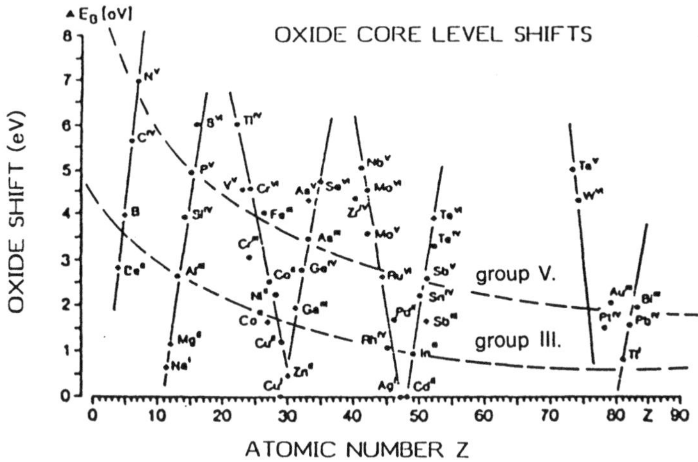

Fig. 6. Oxide-element core level shifts plotted as a function of the atomic number Z (after ref. [15]).

$\Delta\varepsilon_{CT}$, the charge transfer shift, is generally the main factor and has been already discussed above. $\Delta\varepsilon_{H}$ is also an initial ground state factor which accounts for possible band modifications without or with small charge transfers. Such situations can occur in alloys or intermetallic compounds where a rehybridization can modify the position of the centroid of the main valence band and therefore of the core levels relative to the Fermi level. As an example Martin [16] observed shifts to higher binding energies in PdY and Pd_2Y_5 for both Pd and Y core levels with respect to the element values which is unexpected from a charge transfer point of view. Similarly, Grunthaner [17] found for Pt 4f levels (fig. 7) in PtSi and Pt_2Si, shifts in the direction opposite to what would be expected from electronegativity arguments. Such band modifications were also evoked to explain the origin and the sign of surface chemical shifts in transition metals like W and Ta [18].

ΔE_R is the variation of the extra atomic relaxation energy in the final state. The screening of the suddenly appearing positive charge can vary strongly from metals to

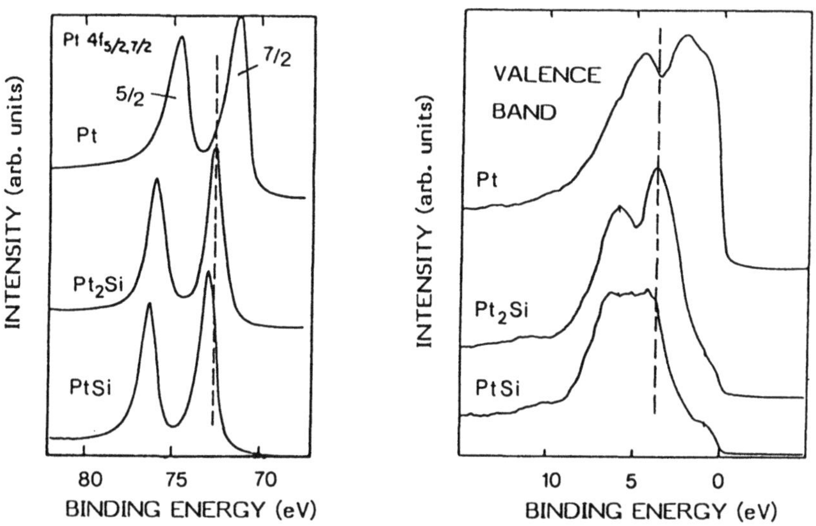

Fig. 7. Pt 4f and valence band spectra for Pt, Pt_2Si and PtSi. The Pt 4f core level shifts follow the variation of the Pt 5d valence band positions relative to the Fermi level (after ref. [17]).

semiconductors and insulators and can give different ΔE_R values. The relaxation contribution can be estimated by comparing photoemission and Auger core level shifts. Due to the two-hole final state the relaxation contribution to the shift is higher in the Auger process than in photoemission [19]. In a very simple model the Auger shift can be expressed as

$$- \Delta E^F_{Auger} = \Delta \varepsilon + 3 \Delta E_R + \Delta E_F . \tag{10}$$

As an example when a 4 eV shift, between Si and SiO_2, is measured in photoemission Si 2p spectra, the corresponding Auger KLL is equal to ~ 8 eV (see fig. 17 in section 5), which gives ΔE_R = 2 eV [20]. Relaxation effects were also used to explain the existence of photoemission and Auger core level shifts measured between thin and thick rare gas overlayers on metals [21].

ΔE_F is the variation of the Fermi level position with respect to the bands. Variable doping or surface Fermi level pinning modifications can induce core level shifts. Fig. 8 shows that Si $2p_{3/2}$ binding energies relative to the Fermi level of highly doped n-type and p-type Si surfaces can span almost the Si band gap [22]. To summarize this point, contributions of different physical origins can induce core level shifts. In general this makes chemical identification in complex systems almost impossible only on the basis of core level shifts. Practically, core level shifts are used to distinguish between different chemical environments and to follow band-bending modifications.

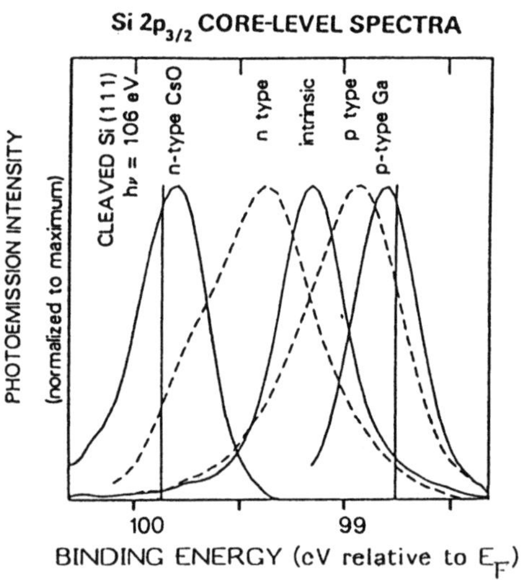

Fig. 8. Si $2p_{3/2}$ core level spectra for intrinsic Si(111)-2x1 surfaces and highly doped n-type and p-type Si(111) surfaces. Flat band conditions are achieved by evaporation of Cs + O and Ga overlayers (after ref. [22]).

5. VALENCE BAND SPECTROSCOPY

Depending on the photon energy one can consider two regimes : the band structure regime (5-20 eV) and the X-ray regime (~ 40 eV and higher).

In the band structure regime EDC's show rich structures due to momentum conservation and final state modulation. Angular resolved photoemission can be used to map two- and three-dimensional energy bands [2].

In the X-ray regime the final states can be described by free electron states and EDC's simply reflect the one-electron density of occupied states modulated by matrix element effects. Since the total density of states can be resolved into partial components of each atom and each type of angular symmetry, one can express the integrated photoelectron spectra as

$$N(E_k) \propto \sum_i \sigma_i(h\nu)\, \rho_i(E)$$

where ρ_i is the partial electronic density of states for orbital i and σ_i the corresponding photoionization cross section.

For photon energies ranging from 40 to 150 eV both bulk and surface states can be probed due to the strong surface sensitivity of photoelectron spectroscopy. This allows one to study surface, adsorbate and interface induced electronic states. In XPS (hν ~ 1000-1500 eV) only bulk features are detected.

Photoionization cross sections show some variations as a function of hν as shown in Fig. 9 [23] . Generally the highest values are obtained near threshold and then the cross sections decrease smoothly with increasing hν. One can take advantage of these variations to identify the bonding character of some particular structures in the electronic density of states. This is illustrated in Fig. 10 for SiO_2. Theoretical total and

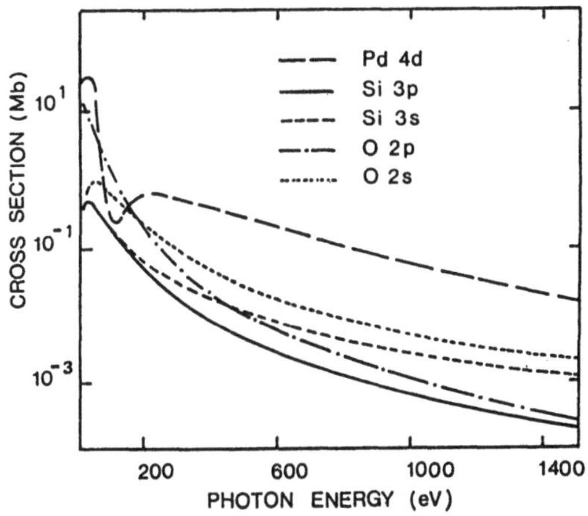

Fig. 9. Evolution of Si 3s, Si 3p, O 2s, O 2p and Pd 4d photoionization cross sections as a function of photon energy (after ref. [23]).

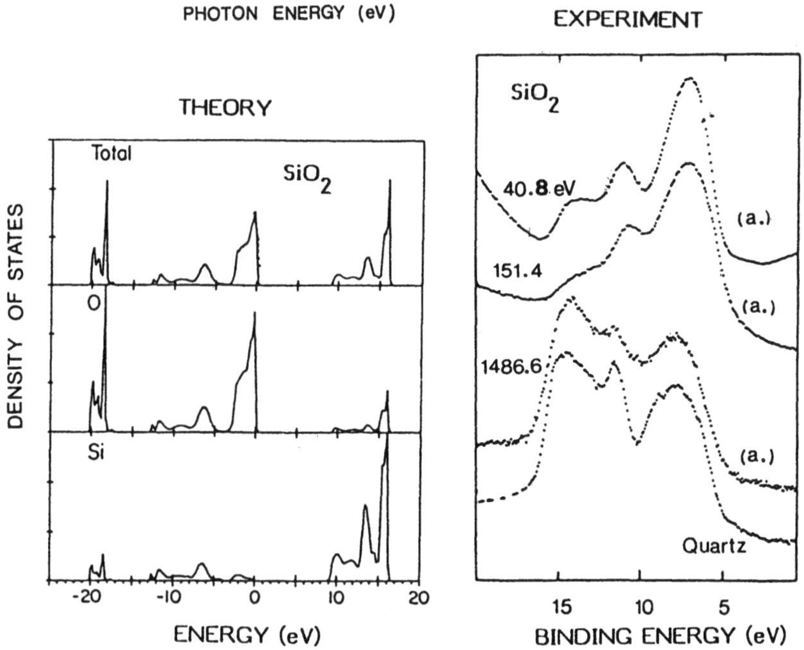

Fig. 10. Valence band spectra for SiO_2 measured at 40.8 eV, 151.4 eV and 1486.6 eV compared to total and partial theoretical densities of states calculated by Robertson [24]. (a) means amorphous.

partial O and Si densities of states are compared to photoemission spectra measured at 40.8 eV and 1486.6 eV. For a photon energy of ~ 40 eV, the σ(O 2p)/σ(Si 3s+3p) photoionization cross section ratio is equal to 7 whereas it is only equal to 0.2 for hν ~ 1500 eV. This means that UPS probes mostly O 2p densities of states near the top of the valence band whereas XPS strongly emphasizes deeper Si s states.

The presence of nodes in the radial part of some atomic orbitals (4d or 5d for example) induces so-called Cooper minima in the photoionization cross section. Using the tunability of synchrotron radiation in the photon energy range hν = 70-200 eV, this was exploited to study the electronic structure of bulk and interfacial compounds (mostly silicides) involving 4d and 5d valence states. An example is given in Fig. 11 for the Pd_2Si interface phase formed by reacting Pd with Si [25]. Photoemission EDC's measured at low photon energy (40-80 eV) or high photon energy (XPS) are dominated by the emission of Pd d states. By tuning the synchrotron light to hν = 130 eV it is possible to detect with equal sensitivity the Pd d states and Si s + p states. This allows to separate the contributions of Si and Pd orbitals and to follow the band modifications in silicide formation.

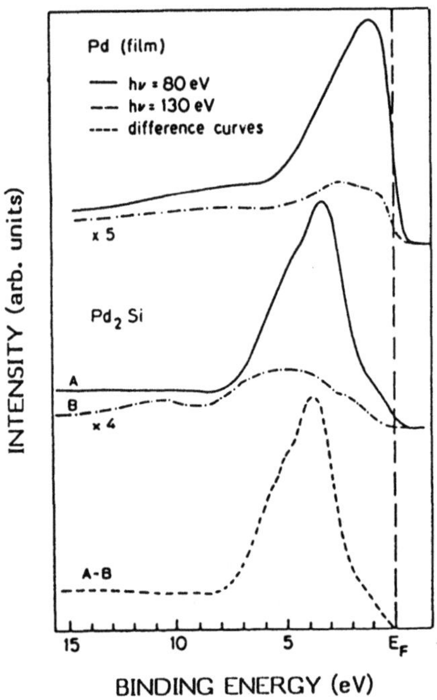

Fig. 11. Valence band spectra for Pd and Pd/Si (Pd_2Si) interface measured for high Pd 4d cross sections (hν = 80 eV) and at the cooper minimum (hν = 130 eV) (after ref. [25]).

The main structures in the density of electronic states in a solid are characteristic of its local order : nature and number (coordination of first neighbors, bond lengths, bond angles, etc ...). Long-range order induces only fine structures. Consequently any structural modification can be associated with a variation in the shape of the density of states and therefore of the photoemission valence band spectra. Theoretical density of state calculations can be used to simulate and identify local order configurations in theory-experiment comparisons. Examples are given by the investigation of local order in amorphous semiconductors [26], the test of different Ni/Si interface models [27] and the understanding of anomalous bonding in interfacial SiO_2 [28]. Note that such an approach is also widely applied in angular photoemission studies to test surface and adsorbate bonding geometry models.

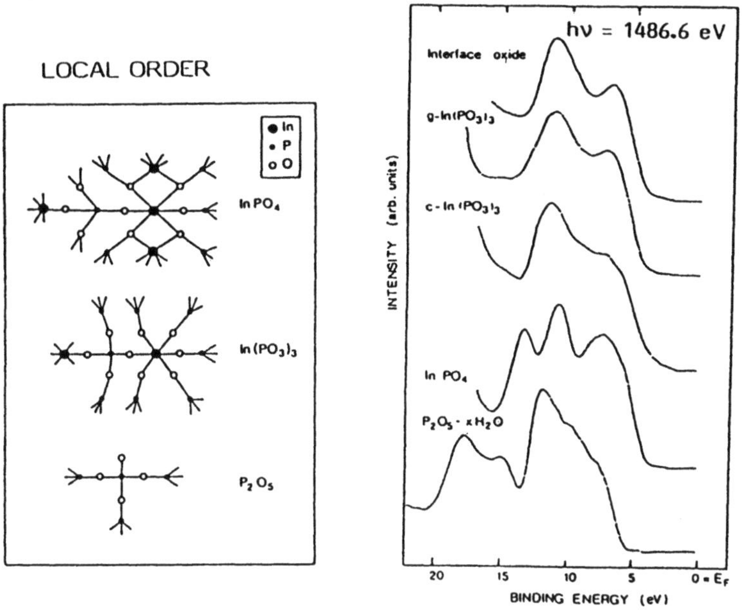

Fig. 12. Local order and valence band spectra for P_2O_5, $InPO_4$ and $In(PO_3)_3$. The top spectrum corresponds to an interfacial anodic oxide identified as an $In(PO_3)_3$-like phosphate (after ref. [29]).

Due to the specificity of the density of states to a given local order, valence band spectra can be used for chemical identification using "finger printing" techniques in comparisons of unknown compound data with standard compound data. Fig. 12 shows that the phosphorous compounds of the In-P-O system, P_2O_5, $InPO_4$ and $In(PO_3)_3$, have very different valence band widths and shapes. This allowed to identify unambiguously $In(PO_3)_3$-like condensed phosphates at the anodic oxide-InP interface [29].

6. SELECTED EXAMPLES

Photoemission spectroscopies are surface-sensitive techniques. Their applicability to interface studies is limited by the effective electron mean free path since signals from the substrate, the interface and the overlayer have to be detected simultaneously. There are two kinds of interface studies. In most cases the interface is grown in situ starting from the clean semiconductor surface. Then, the formation of the interface is followed by analyzing the same surface with overlayers of increasing thicknesses. In the case of interfaces with thick overlayers, some thinning procedures (chemical etching or ion sputtering) can be used to reduce the thickness of the overlayer.

Four kinds of information can be extracted from a photoemission experiment :
i) analytical information (from core level intensities) : atomic and chemical composition, growth mode, diffusion phenomena, ...
ii) chemical bonding (from core level shifts and valence bond spectra) : bonds at the interface or nature of the reacted phases,
iii) electronic bands (from valence band spectra) : electronic structure of the interfacial phases, band discontinuities at the interface,
iiii) electronic defects : pinning position of the Fermi level (band bending), generally correlated to electrical interface properties.

Note that photoelectron spectroscopies do not give access to atomic geometries although some information can be obtained indirectly by comparing valence band spectra and theoretical simulations or more directly using photoelectron diffraction [3]. We can divide semiconductor interfaces in two main categories, nearly perfect abrupt interfaces and thick heterogeneous interfaces. In the first case (for example $NiSi_2$/Si, $CoSi_2$/Si, Ge/GaAs, AlAs/GaAs, CaF_2/Si, SiO_2/Si) we want to know how the overlayer is geometrically and chemically bonded to the substrate and what are the interface electronic states. Such systems can be investigated theoretically, which helps one to understand some basic phenomena like Schottky barrier heights or band discontinuities. In the second case (most metal-semiconductor or oxide-compound semiconductor systems) we want first to characterize the chemical structure and the morphology of the multiphase interfacial layer.

It is not the purpose of this paper to review the great mass of results obtained on semiconductor interfaces with photoemission spectroscopies. The examples discussed below have been selected either because they are illustrating some particular type of information (sections 6.1, 6.2, 6.3) or because I am familiar with them (section 6.4, 6.5). In this latter case I try to emphasize the interest of combining different experimental approaches (XPS - UPS - Synchrotron).

6.1 Schottky Barrier Formation

The Schottky barrier height at metal-semiconductor interfaces is the energy separation between the Fermi level E_F and the semiconductor conduction band minimum at the interface. The microscopic mechanisms which determine barrier heights are still not fully understood. Photoemission was used to correlate interface chemistry and Schottky barrier formation [30-32].

Figure 13 shows how metal-silicon barrier heights can be measured using Si 2p core level spectra. One first measures the Si 2p position for the clean surface, then determines the Si 2p shift induced by the changes in band bending following metal deposition. The change in Schottky barrier $\Delta\phi_b$ is replicated in the shift of the Si 2p core level. Knowing the Fermi level pinning position for the clean surface [22] allows one to determine the absolute barrier height. Such measurements have been carried out with synchrotron radiation (150 meV instrument resolution) for V, Pt, Pd and Ni on Si(111) surfaces [31,32]. Results are illustrated in fig. 13b and 13c. They have demonstrated that the final electrical interface barrier is largely formed with about 1 to 5 Å (depending on the metal) of metal coverage. It was also shown that the evolution of the barrier was independent of the metallic character (not fully developed at such coverage) of the overlayer. Changes in barrier heights as small as 10-20 meV can be detected with high energy resolution synchrotron radiation photoemission. Similar measurements could have been obtained with XPS but probably with less accuracy.

6.2 Modelling Reactive Metal-semiconductor Interfaces

The interfacial zone between a metal and a semiconductor can extend over a thick layer (several tens of angströms) for interfaces where the two materials diffuse or react chemically giving solid solutions or defined compounds. The formation of such heterogeneous interfaces can be followed using core level photoemission taking advantage of the existence of specific shifts for each phase. Well-known examples are metal-silicon interfaces (formation of silicides) Al or Pd-InP interfaces (formation of AlP and Pd_3P respectively with surface segregation of metallic In). J. Weaver and his coworkers have developed a two-phase growth model which quantitatively describes the evolution of the reacted interfaces [33]. In the model the amount of each phase obeys bulk thermodynamic phase rules, a multilayer system is assumed and classical quantitative analysis relations are used. The case of the V/Ge(111) system is presented in fig. 14. Decomposition of Ge 3d core level spectra shows the presence of three distinct phases assumed to be the Ge substrate, V_2Ge_3 and a solid solution of Ge dissolved in vanadium. Curves of growth versus attenuation show that the reacted V_2Ge_3 phase and the solid solution

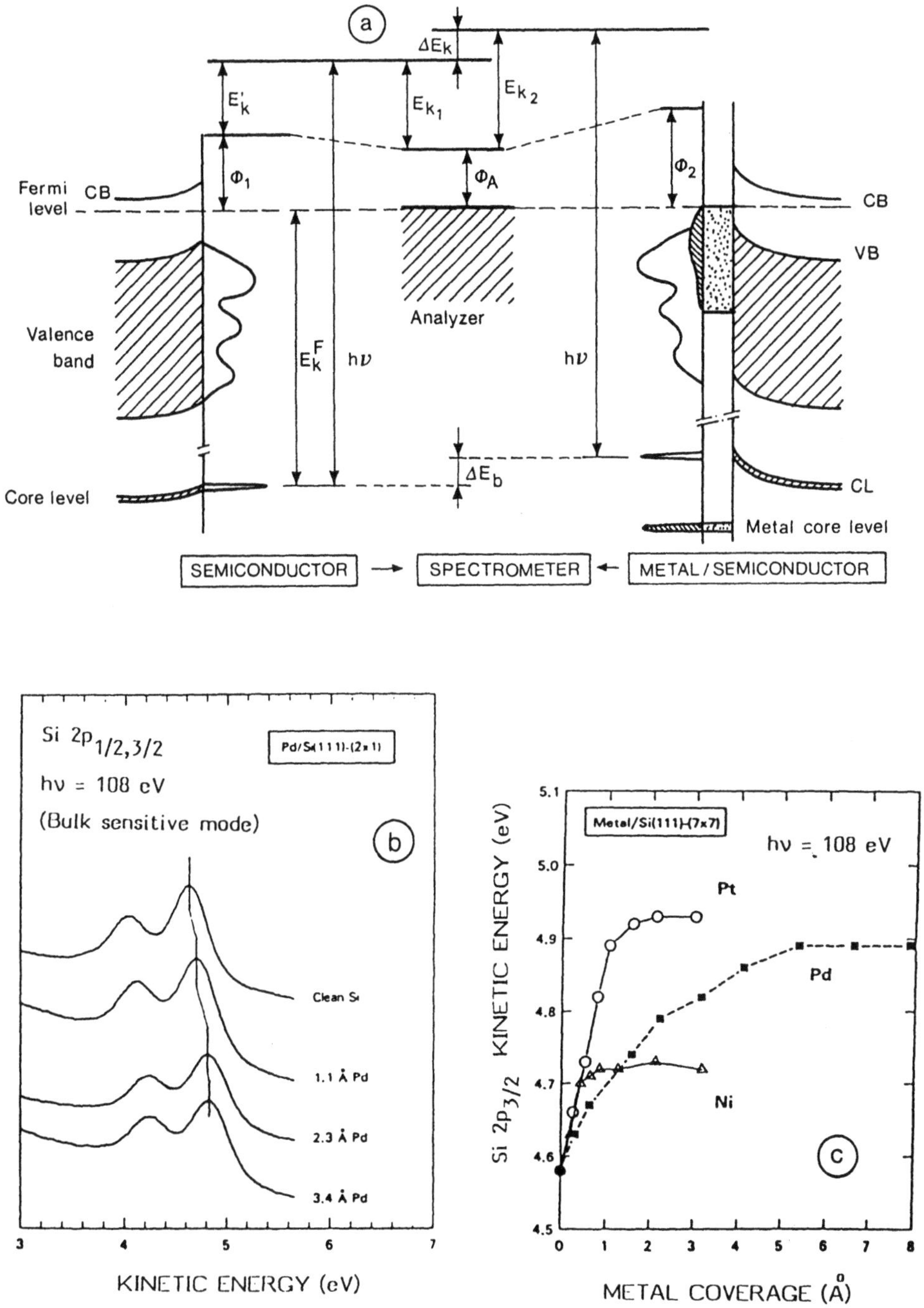

Fig. 13. (a) Schematic energy diagram for a clean semiconductor surface (Si) with the Fermi level pinned by surface states - left - and after deposition of a thin metal overlayer - right -. The samples are in electrical contact with the spectrometer. (b) The Si 2p core level shift measured after deposition of thin Pd overlayers reflects directly the Schottky barrier height variation $\Delta\phi_b$ (after ref. [32]).
(c) Si_o 2p energy dependence on Pt, Pd and Ni coverages. The saturation is reached at a few A metal coverage (after ref. [31]).

appear near 2 Å and 20 Å metal coverage, respectively. The unreacted vanadium film starts to form at about 80 Å coverage. A schematic representation of V/Ge(111) interface evolution is presented on fig. 14c.

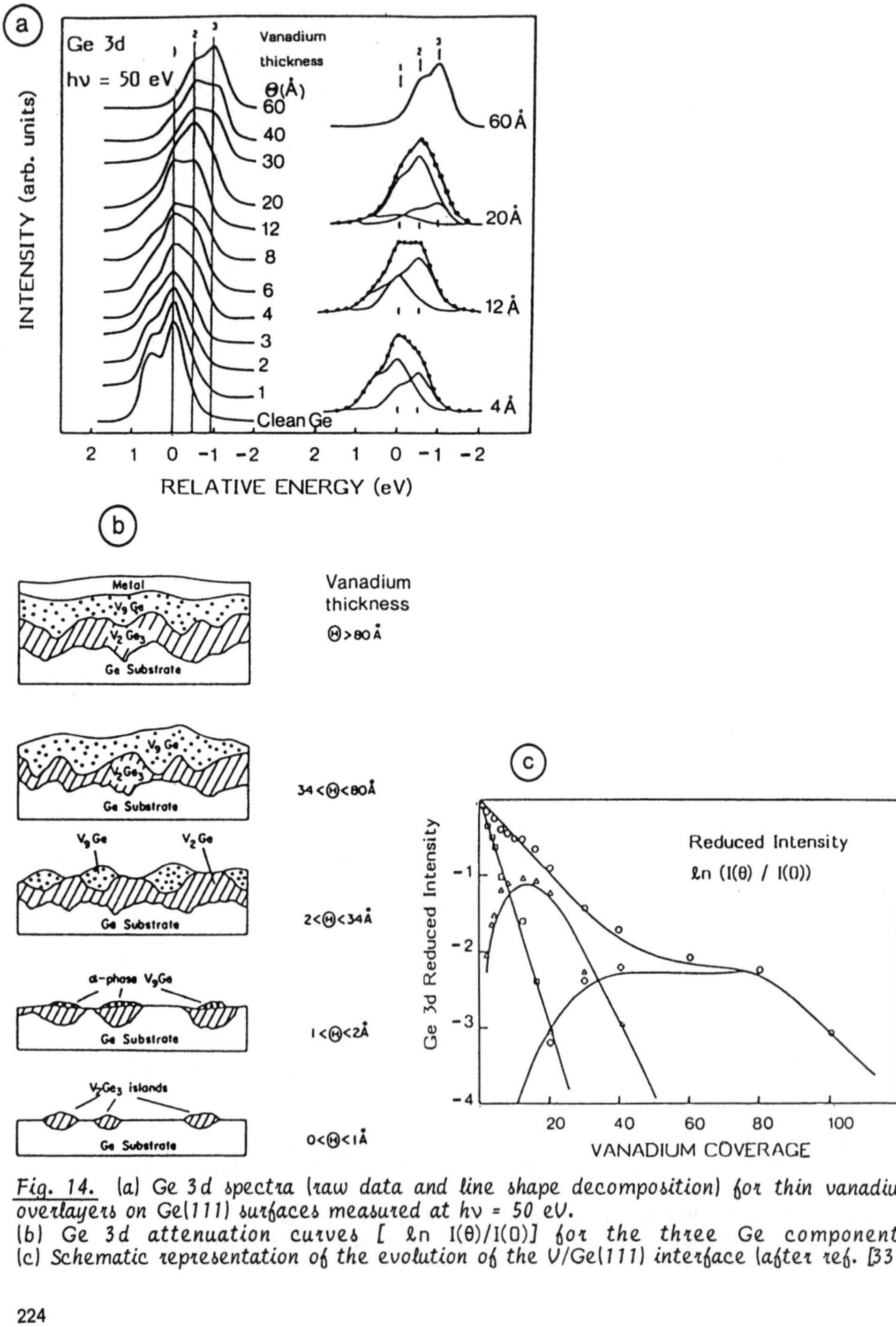

Fig. 14. (a) Ge 3d spectra (raw data and line shape decomposition) for thin vanadium overlayers on Ge(111) surfaces measured at hν = 50 eV.
(b) Ge 3d attenuation curves [ln I(θ)/I(0)] for the three Ge components.
(c) Schematic representation of the evolution of the V/Ge(111) interface (after ref. [33]).

6.3 Band Discontinuities at Heterojunctions

One of the main problem in heterojunction physics is understanding the band line-up, i.e., the distribution of the band gap difference between the valence band ΔE_v and the conduction band ΔE_c (fig. 15). Predicting band offsets has not been solved either theoretically or experimentally within the meV accuracy needed for practical applications. Simple measurements of valence band offsets can be performed with XPS or synchrotron radiation. The principle of measuring valence band offsets with photoemission is illustrated in fig. 15. For large band offsets, valence band spectra from thin overlayers grown on a semiconductor substrate give a direct measurement of the band discontinuity at the interface (fig. 15b). Signals from both overlayer and substrate are distinguishable and the band bending is negligible over the region probed by photoemis-

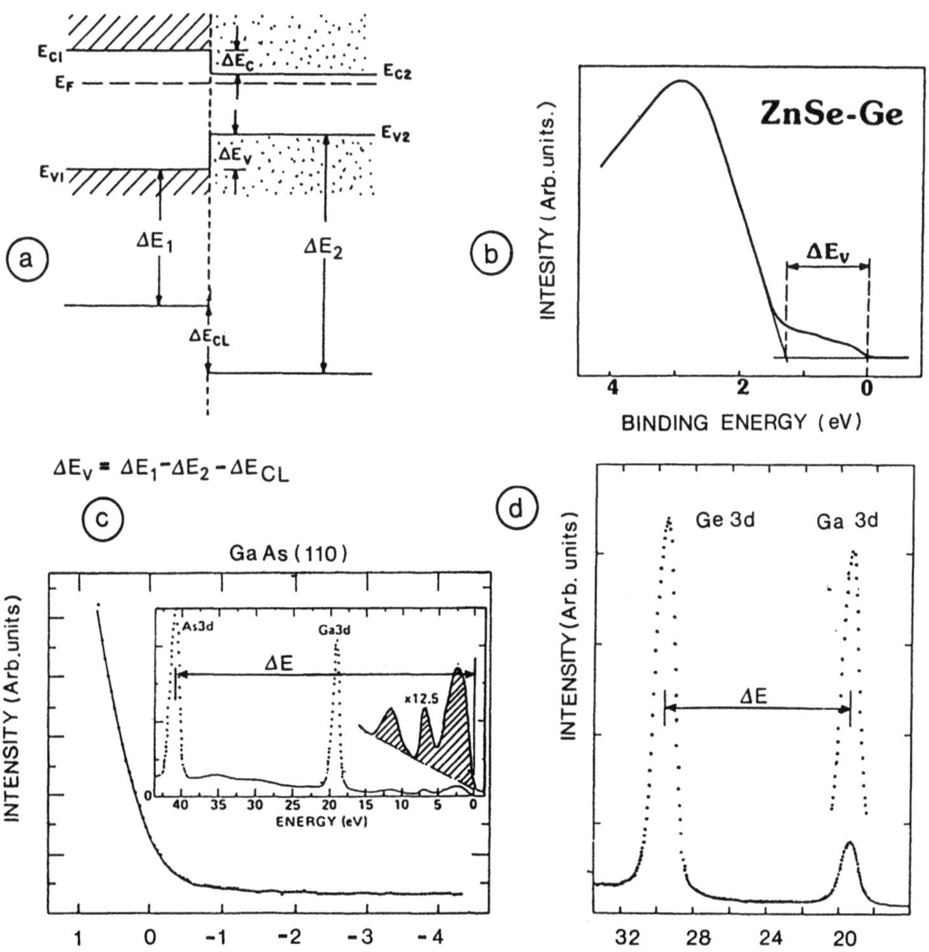

Fig. 15. Determination of heterojunction valence band discontinuities ΔE_v by photoelectron spectroscopy.
(a) Heterojunction band diagram at the vicinity of the interface with negligible band bending.
(b) Direct visualization of the band offset using valence band spectra (after ref. [34]).
(c) & (d) Measurement of substrate-overlayer (Ge/GaAs) core level splittings, ΔE_{CL} knowing the core level to valence band maximum binding energy differences δE precisely (after ref. [36]).

sion. For smaller band discontinuities, the substrate edge and the overlayer edge overlap and have to be measured separately. The distance between the two edges, corrected from some possible band bending (detected by measuring a substrate core level), gives ΔE_v. The weakness of this technique is its low accuracy estimated to be ± 0.1 eV for one interface. A large number of heterojunctions was investigated, most of them by KATNANI and MARGARITONDO [35]. On the basis of their experimental data these authors developed an empirical method to estimate ΔE_v's. Table 2 lists empirical E_v's referred to the valence band edge of Ge, the estimated accuracy if ~ 0.15 eV. The accuracy of the photoemission method was improved by Kraut et al. [36] who were able to locate more precisely the valence band edges in XPS spectra with respect to some core levels using instrumentally broadened theoretical densities of states. By following the formation of the Ge/GaAs(110) interface with Ge 3d, As 3d and Ga 3d core level spectra, they determined the valence band discontinuity to be 0.53 ± 0.03 eV.

Table 2. Position of the band edges are given relative to the valence band edge of germanium. The difference between two terms gives a first-order estimate of the discontinuity ΔE_v (after ref. [35]).

Semiconductor	E_v (eV)	Semiconductor	E_v (eV)
Ge	0.00	CdS	-1.74
Si	-0.16	CdSe	-1.33
α-Sn	0.22	CdTe	-0.88
AlAs	-0.78	ZnSe	-1.40
AlSb	-0.61	ZnTe	-1.00
GaAs	-0.35	PbTe	-0.35
GaP	-0.89	HgTe	-0.75
GaSb	-0.21	CuBr	-0.87
InAs	-0.28	GaSe	-0.95
InP	-0.69	$CuInSe_2$	-0.33
InSb	-0.09	$CuGaSe_2$	-0.62
		$ZnSnP_2$	-0.48

The Ge/GaAs heterojunction has been the most widely studied system because it is lattice matched and is simple to grow (see the review by Ludeke et al. [37]). To understand more deeply the factors which control heterojunction properties, the effects of temperature growth, deposition rate and crystal orientation were studied. For Ge/GaAs(110) interfaces grown at low temperature (t < 300°C) SXPS data indicate abrupt interfaces with no outdiffusion of Ga or As. Above this temperature, out diffusion of As with formation of a superficial $GeAs_x$ phase occurs with traces of Ga into the Ge films. The measured valence band discontinuities vary from 0.25 eV to 0.65 eV depending on the growth conditions whereas a nearly constant pinning position at 0.7-0.8 eV above the valence band edge was measured for n-type GaAs. In contrast, KATNANI et al. [38], who studied the evolution of Ge/GaAs(100) MBE grown heterojunctions, found a 0.47 ± 0.05 eV band offset independent of the surface reconstruction, of the As/Ga ratio and of the Ge overlayer thickness. The photoemission technique was also used to test the validity of band offset commutativity ($\Delta E_v^{AB} = \Delta E_v^{BA}$) and transitivity ($\Delta E_v^{AB} + \Delta E_v^{BC} + \Delta E_v^{CA} = 0$) relationships. As an example TRAN MINH DUC et al. [39] verified the applicability of such rules for heterojunctions based on CdTe, HgTe and ZnTe II-VI semiconductors.

6.4 SiO_2-Si Interface Formation

The SiO_2-Si interface is one of the best interfaces which can be prepared. The glassy SiO_2 film bonds almost perfectly to the Si crystal with a structurally abrupt interface. Interfaces with less than one electrically active defect per 10^5 interfacial bonds have been fabricated in Si MOSFET technology.

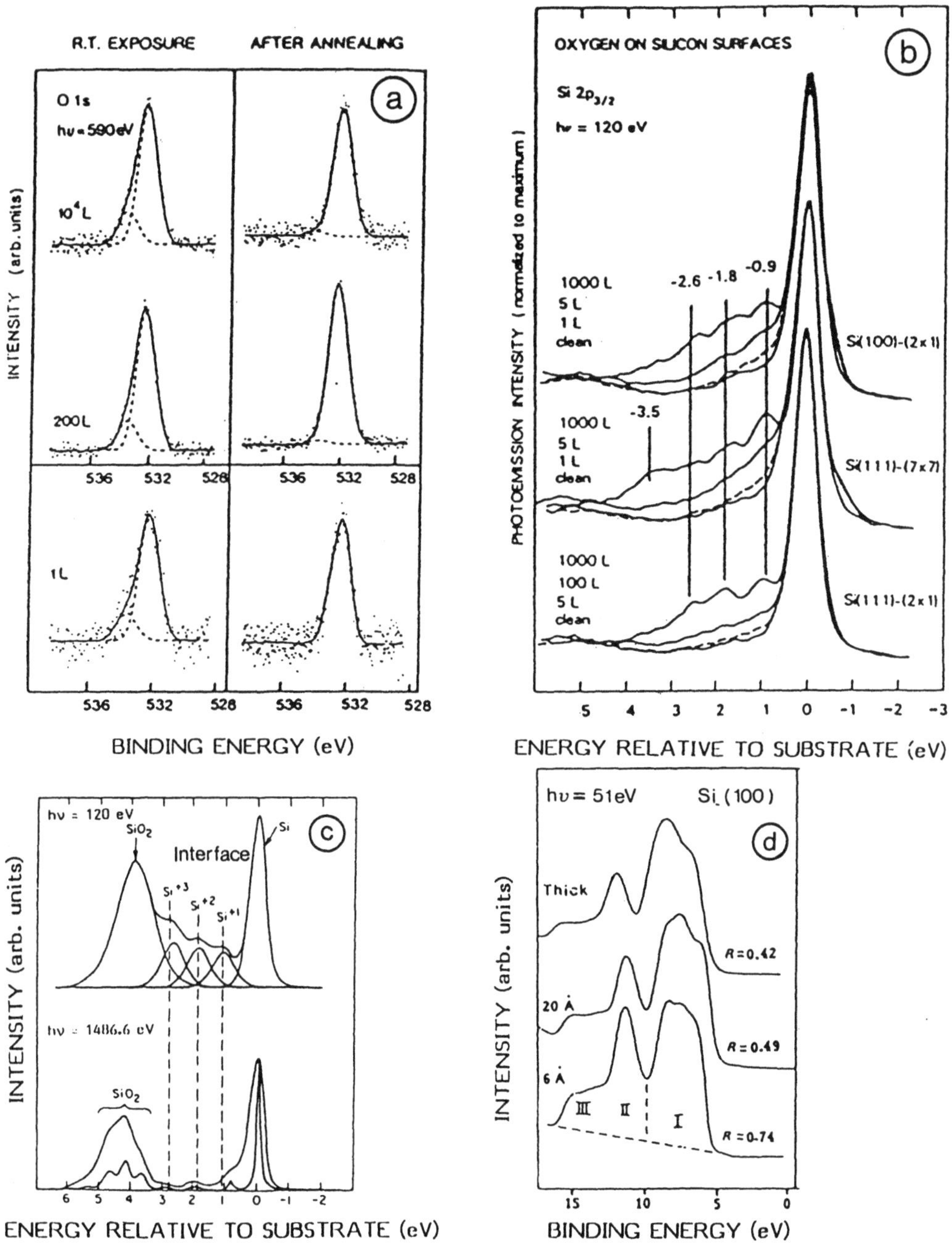

Fig. 16. Formation of the SiO_2-Si interface.

(a) O 1s spectra measured at hν = 590 eV, before and after annealing for oxygen exposed Si(111) surfaces (after ref. [40]).

(b) Multiple Si 2p core level shifts for oxygen adsorption in the monolayer range (after ref. [42]).

(c) The SiO_2-Si interface probed with Si $2p_{3/2}$ core levels using synchrotron radiation (ref. [43])and XPS (ref. [44]).

(d) SiO_2 valence band spectra for different oxide thicknesses. The variation of (III + II/I) band intensity ratio reveals the existence of threefold coordinated oxygens (after ref. [28]).

Surface science techniques including photoelectron spectroscopies have been widely used to investigate the early stages of silicon oxidation, the formation of the SiO_2-Si interface, the properties of buried interfaces and the nature of thin SiO_2 films.

In earlier studies many different bonding geometries have been proposed for oxygen adsorption ; they can be classified into atomic and molecular models. Synchrotron radiation photoemission studies of Si 2p, O 1s core levels and valence bands performed with high surface sensitivity and high energy resolution have clarified the situation. When oxygen is adsorbed at room temperature, oxidation of silicon occurs concurrently with oxygen chemisorption at coverages as low as 0.2 monolayer. This was established by detecting two components in O 1s spectra measured at hν = 590 eV (fig. 16a) [40].

A similar observation was done with XPS [40]. For one monolayer coverage, four silicon oxidation states are observed in Si 2p core level spectra measured at hν = 120 eV (fig. 16b). They correspond to tetrahedric Si atoms bonded to 1, 2, 3 or 4 oxygen atoms. These observations confirmed that oxygen atoms are inserted into the Si lattice with back bonds broken. Heating the sample leads to Si-O bond rearrangement and transformation of non-bridging oxygens into bridging oxygens. It is the first step of the formation of the SiO_2-Si interface [42] .

The transition layer at the SiO_2-Si interface was investigated with synchrotron radiation photoemission and XPS (fig. 16c). From the relative intensity of the Si^{+1}, Si^{+2}, Si^{+3} components in Si 2p spectra it was found that interfaces with thin SiO_2 films thermally grown at low temperature (~ 700°C) and low pressure (10^{-6}-10^{-4} torr) were not completely chemically abrupt. A width of 5 ± 1 Å was found for both (111) and (100) surfaces. As XPS probes a higher depth, intermediary oxidation states are more difficult to detect but GRUNTHANER et al. [44] were able to extract the distribution of the suboxides for SiO_2-Si interfaces grown with thick oxides in 1 atmosphere of oxygen at ~ 1000°C. The oxides were chemically thinned to allow XPS measurements : for such interfaces the transition layer was estimated to be about 1 monolayer thick [44]. Because the SiO_2-Si shift is higher in Auger processes Si KLL Auger spectra measured at grazing emission angle can also be used to detect intermediary states at the interface [20] (fig. 17).

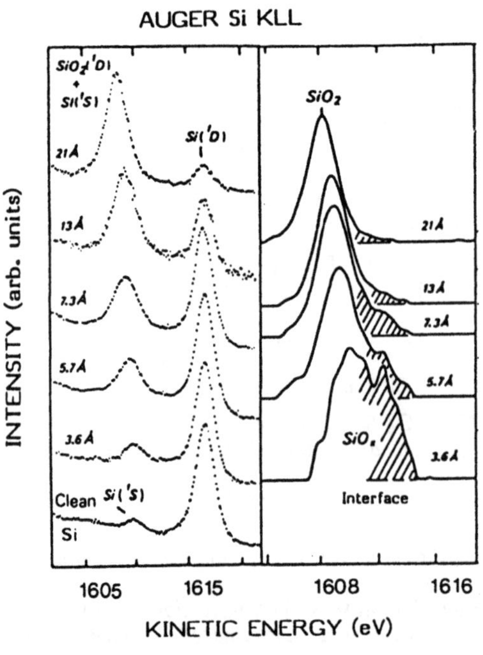

Fig. 17. Detection of interfacial intermediary oxidation states at the SiO_2-Si interface using Si KLL spectra. Left : raw data ; right : the Si substrate components have been subtracted. The spectra have been normalized to their maximum intensity (after ref. [20]).

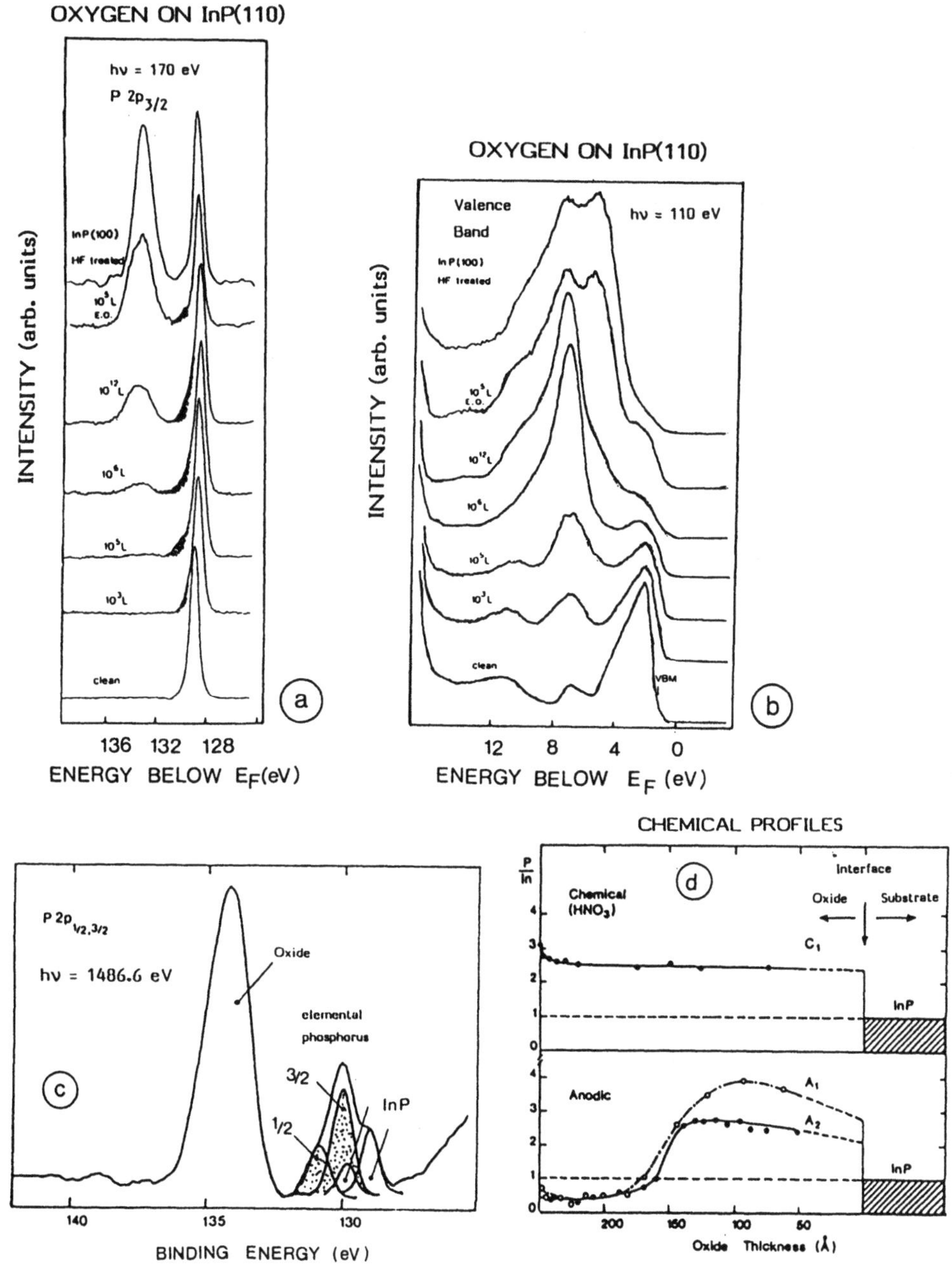

Fig. 18. Characterization of oxide-InP interfaces

(a) & (b) : P 2p core level and valence spectra for oxygen adsorption on InP(110) surfaces. 3 regimes were detected : chemisorption, nucleation, formation of a continuous layer (after ref. [45]).

(c) : Elemental phosphorus detected at the thermal oxide-InP interface (ref. [46]).

(d) : P/In chemical profiles measured for chemical and anodic oxides grown on InP(100) (after ref. [29]).

Within about 20 Å of the interface the properties of the SiO_2 phase appear to differ from those of bulk SiO_2. The origin of these anomalous properties is still the subject of much speculation. For abrupt interfaces a high density SiO_2 layer is expected to exist near the interface. GRUNTHANER et al. [44] suggested that the distribution of Si-O-Si angles in 4:2 coordinated SiO_2 changes as the interface is approached whereas a study of UPS valence band spectra measured at 51 eV for a series of thin and thick oxide layers (fig. 16d) allowed HOLLINGER et al. to propose the existence of threefold coordinated oxygen centers [28].

6.5 Oxide-InP Interfaces

Recently, considerable interest in the properties of oxide-InP interfaces has occurred because of their usefulness in MOSFET devices. The formation of the oxide-InP(110) interface in oxygen at room temperature was studied in situ using synchrotron radiation [44]. Changes observed in P 2p, In 4d and valence band spectra revealed that the oxidation of InP in the monolayer range is spatially inhomogeneous. Three stages were observed : a precursor chemisorption stage, a nucleation process and the formation of a uniform oxide layer (fig. 18).

Oxidation performed at higher temperatures (250°C - 550°C) in oxygen revealed that the oxide layer is made up of a microscopic mixture of In_2O_3 and $InPO_4$ as shown by XPS O 1s and valence band spectra. For temperatures higher than 300°C some elemental phosphorus is detected at the thermal oxide-InP interface [46]. This indicates that, in contrast to silicon oxidation, oxidation of InP occurs at the surface where In diffuses out preferentially.

The formation of chemical or electrochemical oxide-InP interfaces cannot be followed with UHV surface techniques but the chemical composition of the oxide layers can be estimated by sequential chemical thinning of the oxide layer. This technique was applied successfully to investigate the composition of anodic films on InP. $In(PO_3)_3$ glass-like condensed phosphates have been detected which explains the passivating properties of such films [29].

7. CONCLUSION

High-resolution monochromatic XPS and synchrotron radiation photoemission play a major role in research on semiconductor interfaces. Informations on chemical bonding and electronic states at the interface, Schottky barriers and heterojunction band discontinuities are obtained.

In the future the appearance of new high quality XPS systems and the availability of intense widely tunable radiations from synchrotrons, combined with advanced microscopies such as scanning tunneling microscopies and new electronic structure calculations will help us firstly to understand better the microscopic phenomena which control the properties of "perfect" interfaces, secondly to describe better real disordered interfaces.

REFERENCES

1. D.E. Eastman, J.J. Donelon, N.C. Hien, F.J. Himpsel: Nucl. Instr. Meth., 172, 327 (1980)
2. F.J. Himpsel: Adv. in Phys. 32, 1 (1983)
3. C.S. Fadley: In Progress in Surface Science, vol. 16, ed. S.G. Davison (Pergamon, 1984), p.275
4. W.E. Spicer: Phys. Rev. 112, 114 (1968)
5. M.P. Seah, W.A. Dench: Surf. Interface Anal. 1, 2 (1979)
6. R.M. Friedman, R.E. Watson, J. Hudis, M.L. Perlman: Phys. Rev. B8, 3569 (1973)
7. D.A. Shirley: In Photoemission in Solids, ed. by M. Cardona and L. Ley, Topics in Applied Physics, Vol.26 (Springer, Berlin 1978) p. 165

8. S.M. Goldberg, C.S. Fadley, S. Kono: J. Electron. Spectrosc. 21, 285 (1981)
9. G. Hollinger, Y. Jugnet, P. Pertosa, L. Porte, Tran Minh Duc: Proc. 7th Intern. Vacuum Congr. and 3rd Int. Conf. Solid Surfaces, Vienna (1977) p. 2229
10. G. Rossi: this volume
11. M.P. Seah: In Practical Surface Analysis, ed. by D. Briggs and M.P. Seah (Wiley, Chichester 1983) p. 181
12. K. Siegbahn: J. Electron. Spectrosc. 5, 3 (1974) and references therein
13. U. Gelius: Phys. Scripta 9, 133 (1974)
14. F.R. McFeely, J.F. Morar, N.D. Shinn, G. Landgren, F.J. Himpsel: Phys. Rev. B30, 764 (1984)
15. R. Holm, S. Storp: Appl. Phys. 9, 217 (1976)
16. M. Martin: Thesis, Lyon (1987)
17. P.J. Grunthaner, F.J. Grunthaner, A. Madhukar: J. Vacuum Sci. Techn. 19, 649 (1981)
18. D. Spanjaard, C. Guillot, M.C. Desjonqueres, G. Treglia, J. Lecante: Surf. Sci. Rep. 5, 1 (1985)
19. C. Wagner: Faraday Discuss. Chem. Soc. 60, 291 (1975)
20. G. Hollinger: Appl. Surf. Sci. 8, 318 (1981)
21. G. Kaindl, T.C. Chiang, T. Mandel: Phys. Rev. B28, 3612 (1983)
22. F.J. Himpsel, G. Hollinger, R.A. Pollak: Phys. Rev. B28, 7084 (1983)
23. J.J. Yeh, I. Lindau: Atomic Data and Nucl. Data Tables 32, 1 (1985)
24. E.P. O'Reilly, J. Robertson: Phys. Rev. B27, 3780 (1983)
25. G. Rossi, I. Lindau, L. Braicovich, I. Abbati: Phys. Rev. B28, 3031 (1983)
26. M.H. Brodsky, M. Cardona: J. Non-Crystall. Solids 31, 81 (1978)
27. C. Calandra, O. Bisi, G. Ottaviani: Surf. Sci. Reports 4, 271 (1984)
28. G. Hollinger, E. Bergignat, H. Chermette, F. Himpsel, D. Lohez, M. Lannoo: Phil. Mag. (in press) (1987)
29. G. Hollinger, J. Joseph, Y. Robach, E. Bergignat, B. Commere, D. Viktorovitch, M. Froment: J. Vacuum Sci. Technol. B (in press) (1987)
30. J.L. Brillson: Surf. Sci. Reports 2, 123 (1982)
31. R. Purtell, G. Hollinger, G.W. Rubloff, P.S. Ho: J. Vacuum Sci. Technol. A1, 566 (1983)
32. G.W. Rubloff: Surf. Sci. 132, 268 (1983)
33. R.A. Butera, M. Del Giudice, J.H. Weaver: Phys. Rev. B33, 5435 (1986)
34. G. Margaritondo: Solid-State Electr. 29, 123 (1986)
35. A.D. Katnani, G. Margaritondo: Phys. Rev. B28, 1944 (1983)
36. E.A. Kraut, R.W. Grant, J.R. Waldrop, S.P. Kowalczyk: Phys. Rev. Lett. 44, 1620 (1980)
37. R. Ludeke, R.M. King, E.H.C. Parker: In The Technology and Physics of Molecular Beam Epitaxy, ed. by E.H.C. Parker (Plenum, New-York 1985) p. 555
38. A.O. Katnani, P. Chiaradia, H.W. Sang, P. Zurcher, R.S. Bauer: Phys. Rev. B31, 2146 (1985)
39. Tran Minh Duc, C. Hsu, J.P. Faurie: J. Vacuum Sci. Technol. B (in press) (1987)
40. G. Hollinger, J.F. Morar, F.J. Himpsel, G. Hughes, J.L. Jordan: Surf. Sci. 168, 609 (1986)
41. P. Morgen, W. Wurth, E. Umbach: Surf. Sci. 151/152, 1086 (1985)
42. G. Hollinger, F.J. Himpsel: Phys. Rev. B28, 3651 (1983)
43. G. Hollinger, F.J. Himpsel: Appl. Phys. Lett. 44, 93 (1984)
44. F.J. Grunthaner, P.J. Grunthaner: Materials Sci. Reports 1, 65 (1986) and reference therein
45. G. Hollinger, G. Hughes, F.J. Himpsel, J.L. Jordan., J.F. Morar, F. Houzay: Surf. Sci. 168, 617 (1986)
46. E. Bergignat, G. Hollinger, Y. Robach (to be published in Surf. Sci.)

Two-Photon Photoemission in Semiconductors

J.M. Moison

Centre National d'Etudes des Télécommunications, Laboratoire de Bagneux, 196, Avenue Henri Ravéra, F-92220 Bagneux, France

Two-quantum photoemission is a process in which twice the photon energy is transferred to the emitted photoelectron. In two-photon photoemission, this transfer is obtained by successive photoexcitations. The transit of the electron in an intermediate state gives to its final state a "memory" of this normally-empty state. We review here the experimental problems involved and the information obtained this way on the empty states of semiconductors. The present state of the art includes the determination of static properties (density of states, dispersion,...) and dynamic ones (recombination or relaxation processes,...). Excited electrons can be probed along their relaxation path, from ballistic to relaxed behaviour by an adequate choice of experimental conditions. Finally, the surface and interface sensitivity of two-photon photoemission is evaluated.

1. Introduction

According to the first photoemission experiments, no electron could be emitted by a solid under irradiation, no matter how intense, unless its wavelength was lying under a value which was characteristic of the solid. This "photoelectric effect" which is one of the first demonstrations of the quantum nature of light, is now interpreted in terms of a threshold photon energy. The difference between the Fermi level - highest occupied level - and the vacuum level - lowest energy level at which an electron can leave the solid - is the work function. If the photon energy is smaller than this value, no optical transition can bring the electron above the vacuum level and no photoemission is observed, whatever the photon flux. Actually, this last statement holds only as long as this flux remains "weak". Under "strong" illumination, photoemission can occur at photon energies lying below the work function, because the light beam can transfer more than one quantum of energy to the electron by non-linear processes. Several mechanisms may lead to this transfer, but the most "useful" - and fortunately the most often observed - involves a two-stage process: an electron is promoted from its initial state to an intermediate state, where it acquires the extra energy which brings it above the vacuum level (two-quantum photoemission or 2QP). This is usually called a pump+probe process. The photoelectron observed in the vacuum may then keep a "memory" of the intermediate empty state by which it transited, and hence yield information about the properties of this normally-empty state. The ability of 2QP to probe empty states, like "negative electron affinity", partial yield, or inverse photoemission spectroscopies, has been put to several uses, like for instance the observation of image-potential states on metal surfaces [1]. However, we will focus here on studies dealing with empty states of semiconductors, including Te [2], Ge [3], PbI_2 and GaS [4], Si [5,6,7,8], GeSe [9], ZnTe and CdTe [10], and InP [11,12], after a review of the experimental problems involved.

2. Experimental

Two-quantum photoemission, like most non-linear optical processes, has a very low quantum yield and its observation requires very high light intensities, perilously close to the optical damage threshold. However, it has been observed on a variety of materials, including metals, alkali halides, or organic compounds (see a reference list in [13]). A few figures must be mentioned here. In a linear or one-quantum photoemission (1QP) process, the photoelectron flux is proportional to the photon flux. Their ratio, which is known as the 1QP quantum yield, lies around 10^{-3}. In 2QP, the photoelectron flux is proportional to the square of the photon flux, with a proportionality factor, or "two-quantum yield", lying in the 10^{-30} cm^2/s range. Therefore, under a common $1W/cm^2$ excitation (He-Ne laser), the yield of 2QP is 10^9 times smaller than the one of 1QP; they would become equal only at $1MW/cm^2$. It follows that 2QP can be observed clearly only 1) at photon energies below the work function, where 1QP vanishes, and 2) at light intensities where continuous-wave irradiation would induce optical damage. We have then to rely on pulsed visible/near-UV lasers, at least for the first photoexcitation, the second step being possibly taken up by other continuous-wave sources. For short pulses ($<10^{-6}$ s), the damage threshold is expressed not as a power threshold but as an energy/pulse threshold, lying around 0.1 J/cm^2. With nanosecond lasers, we may then go up to 100 MW/cm^2 without damaging the sample. Actually, 2QP can be observed at around 10 kW/cm^2 by using ultimate-sensitivity detectors and signal averaging. This is fortunate since besides genuine damage, many phenomena occur at high intensities which can bias 2QP measurements one way or the other. For instance, in the case of silicon [14], the occurrence of surface modifications [15], electron thermoemission [16], three-quantum photoemission [5,10], surface heating [17], non-linear optical absorption [18], photovoltage, band filling, etc..., and of course 1QP limit the observation range severely (see Fig. 1). The situation is significantly different for picosecond pulses. The energy/pulse

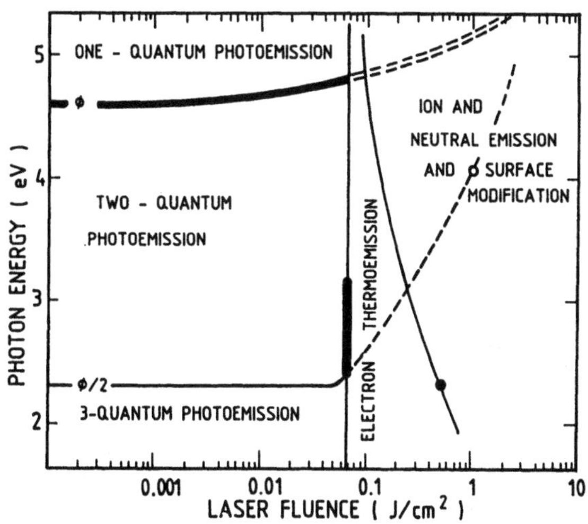

Figure 1: The various photoemission processes observed on Si(111) as a function of laser fluence and photon energy for nanosecond excitation. Heavy lines and dots represent actual measurements. Light lines are extrapolations. Dotted lines are virtual borders. ϕ is the work function.

needed is much smaller so that effects connected to this parameter, like heating, are less constraining. On the other hand, the instantaneous injection level is much higher, which enhances the problems associated with the high density of photogenerated carriers, like photovoltage.

These requirements are rather constraining but mandatory: constant checks of the observation of genuine 2QP must be made, for instance by the quadratic flux law. However, the allowed range of experimental parameters is quite sufficient to get as many data as currently obtained in 1QP. The quadratic flux law can be checked over six orders of magnitude of photoelectron flux by varying the beam energy and impact area [5]. Monitoring these parameters yields an absolute value of the 2QP yield, eventually versus the energy of one of the photons used (2QP spectroscopy). The energy [1,5-8,10-12] or angular [1,8,11,12] distribution of photoelectrons can be obtained with excellent resolution down to 50 meV and 2 degrees [1]. Of course, since 2QP which involves low-energy electrons is very sensitive to the state of the sample surface, reliable data require the usual environment of surface physics, both for surface preparation and monitoring. Furthermore, like most techniques, 2QP cannot stand alone, and for instance parallel 1QP experiments are most often a very useful complement.

3. The three-step model in two-quantum photoemission

As a first step in the interpretation of 2QP data, it may be observed that 2QP at a photon energy E and 1QP at a photon energy 2E transfer the same energy 2E to the electrons. The final states in the vacuum and the initial states in the solid involved in both processes are identical. For this reason, 2QP may be described by the classical "three-step model" [19] whose success in the field of 1QP is remarkable. This model divides the overall photoemission process into three successive steps: 1) photo-excitation of the electron from its initial state to its final state in the crystal, 2) transport to the surface, 3) escape into the vacuum (Fig. 2).

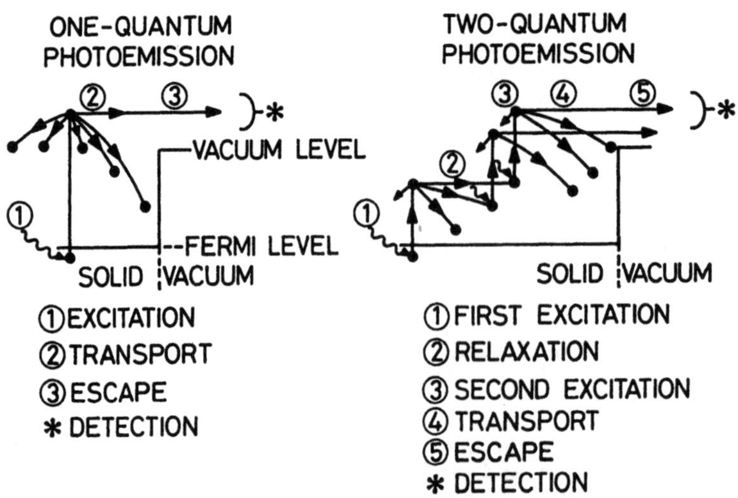

Figure 2: Multistep models of one-photon and two-photon photoemission

Little is known about the two final processes, but they are identical for 1QP and 2QP as the final states are identical. The transport is usually taken into account by an escape depth L: an electron at a depth z has a probability exp(-z/L) of reaching the surface without being scattered. As the escape depth lies around 10 Å for the low-energy final-state electrons (5-10 eV above the Fermi level), the probe step is very surface-sensitive, and nearly instantaneous, because the escape time is very short (10^{-14} s). The escape step is described by an escape probability function, taken equal to zero when the final state lies under the vacuum level, and to one otherwise. Actually, these two last steps are rather neutral, and the photoemission data roughly reflects the excitation step, which in turn depends on the band structure and on the excitation process.

4. Two-quantum excitation processes

In the case of 2QP, the excitation process is not a mere optical transition as in the case of 1QP, because it involves the transfer of two energy quanta. This is clearly demonstrated by many observations: 1) the relation between photon flux and photoelectron flux is quadratic, 2) no 2QP can be observed for photon energies below half the work function, 3) the upper level to which an electron can be brought is exactly the Fermi level plus twice the photon energy, and 4) photoelectron characteristic energies vary with the energy of one of the photons used, or as twice this energy if the same photons are used for both processes. Several processes which transfer two quanta can be put forward. Some of them involve a simultaneous transfer like second-harmonic generation followed by optical absorption, or two-photon absorption via a virtual intermediate state. Others involve successive transfers of a single quantum via a real intermediate state, like optical absorption followed by either free-carrier absorption or a non-radiative Auger process. If the first ones are operative, no influence of the intermediate state is expected, except possibly for the selection rules, and 2QP at a photon energy E and 1QP at energy 2xE are similar processes. However, the experiments show that both the related yield spectra and energy distributions are clearly different; the excitation process does involve a first excitation, a stay at the intermediate level and a final excitation. Within this class of 2QP processes, two main mechanisms for the second excitation can be invoked, 1) photoexcitation by a second photon or 2) energy pooling via collisions between two excited electrons promoting one of them and relaxing the other one (Auger two-particle process). Though this second mechanism may indeed be observed [1], it is most often concluded that the first one is operative [4,10,12]. In the following we refer only to this mechanism, which may then clearly be called two-photon photoemission (2PP).

5. Two-photon photoemission and the study of empty states

Its sensitivity to the intermediate level is a prime feature of 2PP, because it gives access to information about the normally unoccupied states located above the Fermi level. Such information, which is very important for both theoretical and applied purposes, has until now been obtained from calculations, from rather indirect transport measurements, or more recently from inverse photoemission (low-energy cathodoluminescence). The 2PP approach to this problem is to populate first the empty levels with electrons and to take them out into the vacuum before they are recombined. In this respect, 2PP can be described by a "five-step model" including: 1) optical excitation from the initial level to the intermediate level, 2) scattering, drift diffusion and recombination in the intermediate level, 3) excitation to the final level, 4) transport to the surface, 5) escape (Fig. 2). We have already noted that the final step 4) is rather neutral. Steps 1) and 3) are simple optical transitions; step 3), like the 1QP first step , conserves the energy and parallel momentum, which allows the derivation of the energy and momentum (and hence the dispersion [8,11]) in the intermediate state from the values in the final state. The remaining steps 2) and 5) determine to a large extent the information yielded by 2PP and will be discussed in more detail.

6. Ballistic and relaxed excited electrons

In principle, all kinds of electrons in the intermediate states, whatever their state of relaxation following the first excitation, i.e. be they ballistic or relaxed, should be observed simultaneously by 2QP, as long as they lie within the escape depth. However, photoemission possesses a built-in high-pass energy filter, the vacuum barrier. An electron in the intermediate state, which would have had just enough energy to leave the solid with the probe excitation, and which undergoes a slight energy relaxation, cannot contribute any more to photoemission. Therefore, the contribution of ballistic electrons is outstanding at low probe photon energies. The distribution in final states reflects the distribution of intermediate states as created by the first excitation. Information on the band structure can be obtained through adequate selection rules, as demonstrated in Si [5] and at low probe energy in InP [12] for high-energy conduction-band states.

On the other hand, at high probe energies, both ballistic and relaxed electrons can be extracted from their intermediate levels. As electrons that have relaxed down to critical points of the band structure have a longer transit time in these particular intermediate levels, the corresponding 2QP yield should be greater than the one corresponding to ballistic electrons: high probe energies enhance the contribution of electrons relaxed in long-lifetime levels. The distribution in final states then reflects the distribution in empty states after relaxation in particular points of the band structure. Such an accumulation of electrons in surface states of Si [8] and InP [11], in lateral valleys of the conduction band in InP [12] and possibly in ZnTe [10], and in the bottom of the conduction band in ZnTe have been reported. If very short pulses (<100 ps) are used, the relaxation of these populations can be followed by 2PP, with experiments involving a pump beam which populates them and a delayed probe beam [8,10,11]. With longer pulses, these populations may also be observed, by reducing the probe energy down to the point where lower-lying electrons cannot be emitted any more, whatever their overwhelming number.

2PP can then give information on both the static and dynamic properties of empty states, depending on the experimental conditions. For instance, 2PP in InP reveals either the ballistic or the relaxed distributions depending on the probe energy [12]. It may also be emphasized that it yields absolute energy levels, while purely optical experiments yield only transition energies.

7. Surface and interface sensitivity of 2PP

2PP is basically surface-sensitive: the escape length in the probe step is short, similar to the one involved in near-UV photoemission (~10 Å). The photoelectrons emitted originate from the surface or subsurface region; this is not the case for instance for negative-electron-activity photoemission where a long escape length is involved. However, the electrons may travel a long way at the intermediate level, all the longer as their lifetime in this level is longer, up to micrometers at the bottom of the conduction band. When they arrive at the surface, they may be "probed" by the second photon as such, or after trapping in surface states. This second situation where the bulk feeds the surface states [8,11] is favoured because many electrons can accumulate in surface states: when at most 10^{19} cm^{-3} electrons populate the conduction band, and hence only 10^{12} cm^{-2} can be emitted, up to 10^{15} cm^{-2} can populate a surface state and be emitted. Furthermore, possible final-state effects are minimized for surface states. In this respect, the information given by 2PP combines the static properties as can be given by inverse photoemission (densities of states, dipersion,...) but also the dynamic properties (trapping from bulk states to surface states). Indeed, for time- and energy-resolved experiments on Si and InP [8,10,11], the position of 2PP features is related to the surface states density and their time dependence to bulk transport parameters.

The surface may also act indirectly. When we monitor by 2PP a normally-empty bulk state, the corresponding yield depends on the population in this level near the surface, and hence in turn obviously on the bulk-surface exchanges. For instance, surface recombination greatly decreases the near-band-edge carrier density near the surface. Other mechanisms like intra-band relaxation can be promoted at the surface via surface states. This is shown in InP where introducing defects at the surface enhances these interactions and quenches the contribution of electrons relaxed in the lateral valley, leaving only a few ballistic ones [12].

Finally, 2PP is potentially sensitive to interfaces, because of the long drift path of the electrons in the intermediate level. An interface buried not too far from the surface (~ 50 Å ?) could then influence the 2PP, directly, by determining specific electron states, or indirectly by biasing the electron density through interface recombination or relaxation.

8. Conclusion

The need for information about the empty states of semiconductors and about the dynamics of injected carriers is increasing, for instance in microelectronics with the development of high-speed "ballistic" devices. In this field, the 2PP approach which is now proven to be realistic may prove fruitful. With two-beam experiments, electrons can be injected at given energy levels, and then probed with fair time resolution and excellent energy resolution along their relaxation path: 2QP may be described as a contact-less, energy-resolved, transport technique. It is hoped that 2QP can bridge the gap between experiments where ultra-short times are involved, like inverse photoemission, and transport experiments which deal with longer times and energy-integrated data.

REFERENCES

[1] K.Giesen, F.Hage, F.J.Himpsel, H.J.Riess, and W.Steinmann, Phys.Rev.Lett. 55,300(1985)

[2] M.Wautelet and L.D.Laude, Phys.Rev.Lett. 38,40(1977)

[3] L.D.Laude, M.Lovato, M.C.Martin, and M.Wautelet, Phys.Rev.Lett. 39,1565(1977)

[4] A.Kasuya, S.Togashi, T.Goto, and Y.Nishina, Phys.Stat.Sol. B89,K145(1978)

[5] M.Bensoussan, J.M.Moison, B.Stoesz, and C.Sebenne, Phys.Rev. B23,992(1981); J.M.Moison and M.Bensoussan, Sol.State Comm. 39,1213(1981); M.Bensoussan and J.M.Moison, Phys.Rev. B27,5192(1983)

[6] W.Eberhardt, R.Brickman, and A.Kaldor, Sol.State Comm. 42,169(1982)

[7] A.M.Malvezzi, J.M.Liu, and N.Bloembergen, Proc. Symp. Mat.Res.Soc., ed. by J.C.C.Fan and N.M.Johnson, 23,135, North-Holland, New-York

[8] J.Bokor, R.Storz, R.R.Freeman, and P.H.Bucksbaum, Phys.Rev.Lett. 57,881(1986)

[9] A.Kasuya and Y.Nishina, Jap.J.Appl.Phys. 20,L63(1981)

[10] R.T.Williams, J.C.Rife, T.R.Royt, and M.N.Kabler, J.Vac.Sci.Technol. 19,367(1981); R.T.Williams, T.R.Royt, J.C.Rife, J.P.Long, and M.N.Kabler, J.Vac.Sci.Technol. 19,367(1981)

[11] R.Haight, J.Bokor, J.Stark, R.H.Storz, R.R.Freeman, and P.H.Bucksbaum, Phys.Rev.Lett. 54,1302(1985); J.Bokor, R.Haight, R.H.Storz, J.Stark, R.R.Freeman, and P.H.Bucksbaum, Phys.Rev. B32,3669(1986)

[12] J.M.Moison and M.Bensoussan Scan.Elect.Micr. 3,1093(1985); J.M.Moison and M.Bensoussan, Phys.Rev. 35,914(1987); J.M.Moison, PhD Thesis, Orsay 1985, unpublished

[13] M.Bensoussan and J.M. Moison Appl.Phys. B28,95(1982)

[14] J.M.Moison and M.Bensoussan, in "Surface studies with lasers", edited by F.R.Aussenegg A.Leitner, and M.E.Lippitsch, Springer-Verlag, Berlin 1983

[15] J.M.Moison and M.Bensoussan, Surf.Sci.126,294(1983)

[16] M.Bensoussan and J.M.Moison, J.Phys. C7,149(1981)

[17] J.M.Moison, F.Barthe, and M.Bensoussan, Phys.Rev. B27,3611(1983)

[18] A.Lietola and J.F.Gibbons, J.Appl.Phys. 53,3207(1982)

[19] C.N.Berglund and W.E.Spicer, Phys.Rev. 136,1044(1964)

Formation and Electrical Properties of Metal-Semiconductor Contacts

L. Lassabatère

Laboratoire d'Etudes des Surfaces, Interfaces et Composants, UA/CNRS No. 040787, Université des Sciences et des Techniques du Languedoc Place Eugène Batillon, F-34060 Montpellier, France

I - INTRODUCTION

Metal-semiconductor contacts are at the same time an old and a new research topic. The earliest systematic investigation on these contacts is generally attributed to Braun, who in 1874 noted the dependence of the resistance of metal-lead sulfide contacts on the polarity of applied voltage. Afterwards, in 1904, point-contact rectifiers found practical applications and were used as radio-wave detectors. Since then numerous experimental and theoretical studies have been carried out, and now this structure plays a key role in many electronics devices. In the form of an ohmic contact, it is used in devices needing very low resistance contacts in order to get high performance and reliable operation. This is the case in lasers, light emitting diodes, Gunn diodes, MESFETs, and so on. In the form of rectifying contacts characterized by a high value of the interface barrier, it is found in Schottky diodes, in microwave mixer diodes, in metal-semiconductor field effect transistors and their associated integrated circuits, in photodiodes, in solar cells and so forth. Therefore, the need for reliable, well-controlled and easy to reproduce contacts is very high, and this explains why increasing research has been carried out in the last few years in order to obtain a better understanding and manufacturing processes of the contact. However, in spite of several decades of research, which really began in 1938 with Schottky's /1/ and Mott's /2/ work, our knowledge of the fundamental behaviour of this contact is still very far from complete. Its structure, made only of two components, is apparently simple, but in fact many parameters contribute to the real nature and to the final properties of the interface. The initial view of Schottky and Mott, who considered the role of a single phase of metal joined to a single phase of semiconductor, contributed to the understanding of this structure. Unfortunately, it corresponds to few metal-semiconductor couples. In most cases, this description was found not to apply and Bardeen /3/ proposed the presence of interface states as the explanation. These states localised at the interface and residing in the gap determine the barrier height which, contrary to the insight to be found in the theories of Mott and Schottky, no longer depends on the metal-semiconductor work function. Although satisfactory for the explanation of results relating to contacts achieved on real, i.e. oxide-covered surfaces, this model is no longer convenient when dealing, for example, with reactive contacts. Therefore, the insights of Bardeen, Schottky and Mott correspond to limiting cases of a more general and complex problem which has, for a holistic view, to take into account many parameters and mechanisms, and particularly the band structure, the interface states and the detailed microscopic behaviour of the contact which has to be analyzed by taking into account all the physical and chemical parameters which can contribute to the local properties, and consequently to the macroscopic features of the device.

The importance of microscopic chemical effects was first considered by Phillips /4/ and later by Andrew and Phillips /5/ and Brillson /6/ who emphasized the role of semiconductor stability and of interface reactivity through the interface heat of reaction, defined as the difference between the heats of formation of the semiconductor and of the most stable interface compound resulting from the interaction with the metal. Semiconductors exhibiting large heats of formation are relatively stable ; interface reactions, if present, are not very efficient and the interface barrier should then depend on the work function of the metal. This corresponds to the Schottky limit. However, in most cases, the reactivity cannot be neglected, and, as shown by Brillson, it can play a key role in determining the ϕ_B barrier in metal III-V or II-VI contacts.

In addition, interdiffusion of the elements at the interface can also affect this barrier. The role of the interdiffusion of the elements of the interface has also been emphasized. Other parameters, such as the semiconductor surface crystallographic structure, the electronic surface states, the stoichiometry of the surface, the defects resulting from the semiconductor surface preparation or from the metal deposition, can induce significant modification in the properties of the contact. Consequently, the exact nature of the interface is very complex and will have to be better known . This is why great effort has gone, over the years and especially in the last 15 years, into providing a systematic understanding of this seemingly simple but very difficult and important system. The corresponding approach, made possible by the rapid evolution of the techniques (for the vacuum, the surface analysis, etc ..) is based on the analysis, stage by stage, of the interface formation. Low energy electron diffraction (LEED), reflection high energy electron diffraction (RHEED), Auger electron spectroscopies (AES), low energy electron loss spectroscopy (LEELS), X-ray ultraviolet photoelectron spectroscopy (XPS, UPS), secondary ion mass spectroscopy (SIMS), work function measurements and so on, are frequently used in a correlative manner in order to link the different mechanisms which may occur during the interface formation and which can contribute to the final properties of the structure. Once formed, the contact is studied by electrical or photoelectrical methods. Current I versus applied voltage V curves give information on transport mechanisms, on the ϕ_B barrier, on the interface electronic states. Capacitance C versus V measurements provide information on the space charge in the semiconductor and also on the ϕ_B barrier. A good overview of the previous point can be found, in a detailed form, in /7-13/.

This paper deals with the most important of the previous points. We begin by introducing some notions on surface properties, band diagrams and elecronic devices concerned by the metal-semiconductor contact.

II - FROM THE SURFACE TO THE CONTACT AND TO THE DEVICE

There are two main ways of making a contact. We can take a metal point and press it against the semiconductor, we can deposit the metal by evaporation or by any other way and build up a thick metal layer. We will not discuss the first method which was used for the first radio detectors and we will only focus on the second one. We start with a given surface and we deposit metal. How is the contact made in these conditions ? How can we describe it with a band diagram ? What is its contribution in the limitation of component properties ?
In this section, we briefly recall the basic notions needed to understand the contact.

1 - Surface properties

The most frequently used surface in fundamental research is obtained by cleavage or by molecular beam epitaxy. It is a clean surface which is much more easily analysed than a surface prepared by chemical etching. However, as in electronic technology the contact is formed on a chemically etched surface, we will also describe their fundamental properties.

1.1. - Clean surface

By cleaving a semiconductor sample, we obtain, if the cleave is perfect, a surface which should be a perfect representation of the parallel atomic planes of the bulk. It is called a "bulk exposed plane". Such a surface shows the minimum disturbance of the solid arising from the formation of the surface. Even so, because many electronic properties of the bulk depend upon the three-dimensional periodicity of the potential inside the solid, the loss of periodicity in one dimension due to the existence of the surface will result in a change in the electronic states at the surface. Futhermore, when cleaving the crystal the bonds between the atoms of the planes between which the sample is cut are broken. Dangling bonds appear at the surface. As a result, the surface effect on electronic properties is interpreted by taking into account electronic surface states which are characterized by their density, N_t, their capture cross section K (K_p for holes, K_n for electrons), their electronic nature and their energy distribution $E_t(E)$. Surface states can also be associated with changes in the potential due to relaxation, reconstruction, structural imperfections, such as emerging dislocations, modifications in the surface stoichiometry, or adsorbed impurities. In the case of cleaved (111) Si surfaces, Fig. 1, depending on the temperature, different superstructures can be encountered (metastable (2x1) at low temperature, 7x7 at 350°K, 1x1 at high temperature). For this last surface, which is relaxed (0.33 Å inwards), half-filled dangling bond surface states appear in the gap and pin the Fermi level 0.3 eV above E_v /14/, while for the 2x1, superstructure /15/ this band splits into two parts with the same number of states. (110) surfaces are characterized by two dangling bonds per surface atom. The previous results have been recently discussed and new interpretations have been proposed (see G. Lelay in this school).

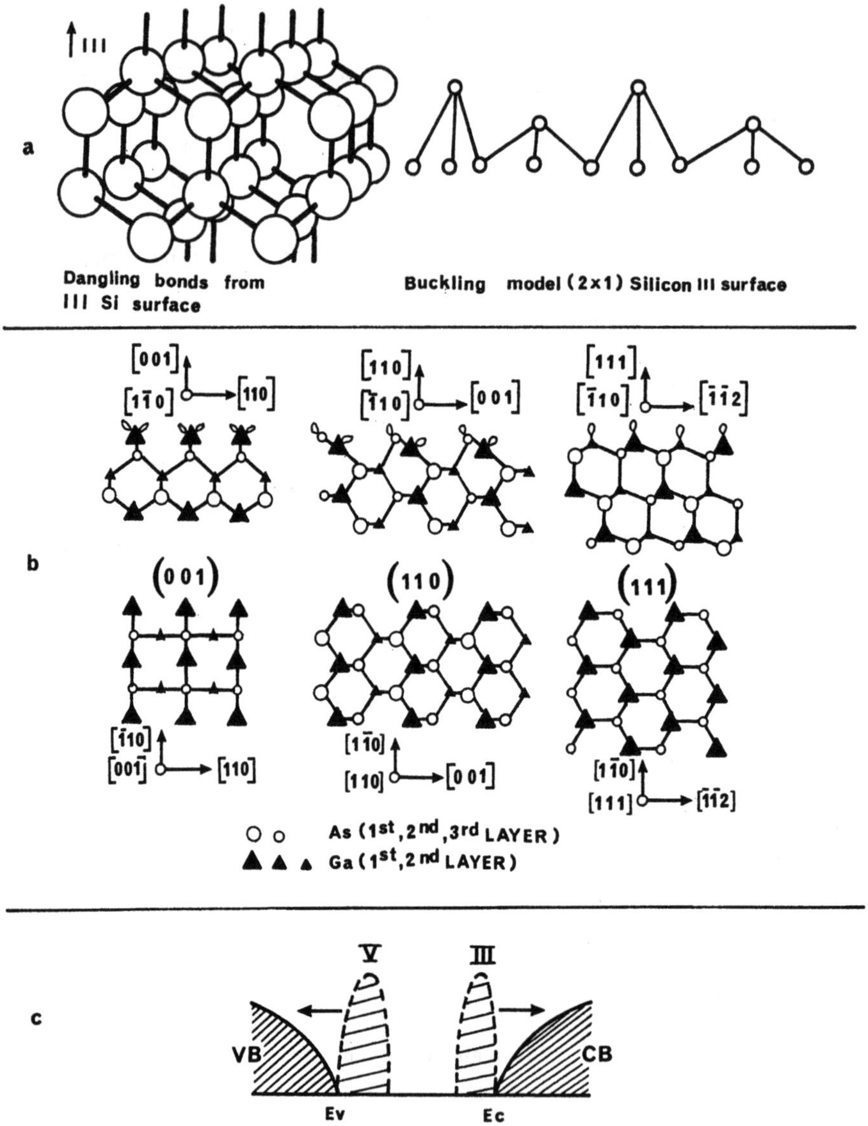

Fig. 1. (a) Atomic structure of Si (111) surface [15].
(b) Atomic structure of III-V compound surfaces [16].
(c) Electronic states corresponding to dangling bonds on (110) III-V compound surfaces [21].

The structure of (001) (110) (111) of a zinc blende compound semiconductor is given in Fig. 1 /16/. In the case of most III-V compounds, theoretical estimates /17, 18/ for unrelaxed (110) surfaces led to the conclusion that occupied and unoccupied states should exist in the band gap. Experimental results show /19, 20/ that no surface states can be detected on perfectly cleaved surfaces. As the symmetry of the surface mesh is not changed there is no reconstruction, and the displacement of the states outside of the band gap /21/ results from a surface relaxation which moves the anions outwards and the cations inwards /22/. (100) III-V compound surfaces are polar due an alternate arrangement of cations and anions. When obtained by molecular bean epitaxy they are clean, but they display numerous reconstructed forms. Thus, depending on the nature of the surface and also of the process used to prepare it, different structural and electronic states have to be envisaged. These electronic surface states behave as trapping or recombining centres. In the simplest case, each centre at the surface captures or releases only one electron, thereby introducing a single allowed energy level in the gap, but in general it can capture (or release) more than one electron and in each charge condition there can be several available energy levels (ground and excited states). Here we shall consider only the simplest case where only one electron takes part in the charge transfer process. If we call the energetic level corresponding to acceptor and donor states E_t (E_a, E_d), the probability, for such a centre, to be occupied by an electron is /23/

$$f_t = \left(1 + 1/g \exp \frac{(E_{to} - E_F)}{kT} \right)^{-1} = \left(1 + \exp \frac{(E_t - E_F)}{kT} \right)^{-1} ; \quad (1)$$

g is the ground state degeneracy factor and $E_t = E_{to} - kT \log g$. (2)

At room temperature kT log g is small and consequently it is often neglected.

This charge transfer, which occurs between the surface states and the semiconductor bands, results in the formation of a surface space charge. In a thin layer near the surface, the potential V (x) and the carrier densities n (x), p (x) are modified. When the majority carrier density increases, this layer is called an "accumulation layer". When this density decreases, it is a depletion layer. The inversion layer corresponds to a particular case for which there is a change in the nature of the majority carriers near the surface ($n \rightarrow p$ or $p \rightarrow n$).

Due to the space charge layer the work function of the semiconductor ϕ is modified and we have, cf. Fig. 2,

$$\phi = \phi_o - qV_S = X + (E_C - E_F)_o - qV_S. \quad (3)$$

In this relation ϕ_o is the work function of the semiconductor in flat band conditions, V_S is the surface barrier height and $(E_C - E_F)_o$ is the position of the Fermi bulk level relative to the conduction band.

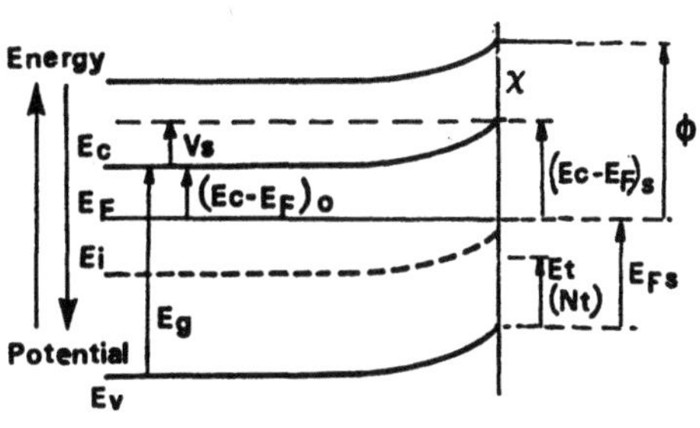

Fig. 2. Energy band diagram for a n-type semiconductor with acceptor surface states.

X = electron affinity
ϕ = semiconductor work function
V_s = surface barrier height
N_t^s = surface state density
E_t = surface state level
E_{Fs} = surface Fermi level position.

The states we have considered, resulted from the breaking of the crystalline lattice. If the surface structure is modified, these surface states may be changed. If defects (cleavage defects, vacancies, etc ...) appears at the surface they can induce new surface states. Thus the origins of the surface states can be numerous.

1.2. - Real surface

These surfaces, which are typically the surfaces used in microelectronics, are obtained by etching semiconductor wafers. The sample is first inserted in the suitable etchant. After etching, the remnants of the etch must be removed. A common practice is to remove the sample from the etchant and rinse it in deionized water. This procedure leaves a considerable residue on the surface, and the physical and chemical state of this surface depends to a large extent on the etchant employed. In the most common case, the surface is covered by a thin (a few x 10 Å) oxide film. Thus, we have to take into account the interface (semiconductor-oxide) states, the states located at the outer surface of the film and the states in the film itself ; the band diagram has to be changed in order to take into account a possible potential drop V_i in the film, cf. Fig. 3.

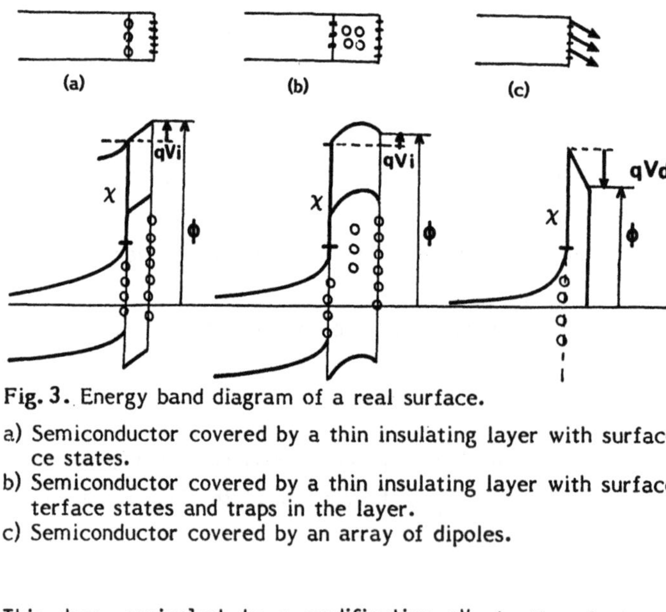

Fig. 3. Energy band diagram of a real surface.

a) Semiconductor covered by a thin insulating layer with surface and interface states.
b) Semiconductor covered by a thin insulating layer with surface states, interface states and traps in the layer.
c) Semiconductor covered by an array of dipoles.

This drop, equivalent to a modification qV_i in the electron affinity, can be as large as one eV or more. Similar modifications in X could be produced by adsorbed molecules having an electric dipole moment M, or by local charge transfer between adatoms and the surface resulting in the formation of bonds.

These modifications may be large if M is perpendicular to the surface. To the approximation that such a dipole array can be represented by a perfect double layer there will be no change in the electric field outside the layer and the barrier will remain unaltered. A sharp jump in potential will occur whose magnitude is given by

$$\left(\frac{G\,M}{\varepsilon}\right) = V_d \qquad (4)$$

with G = density of dipoles and ε = dielectric constant. Many et al. /23/ consider that $\varepsilon = \varepsilon_0$, the dielectric constant of free space. Mele and al. /24/ propose to use for clean surfaces ε values included between ε_s dielectric constant of the semiconductor and $\varepsilon_s \sqrt{2}$.

The corresponding modification of X for a charge transfer of 2 x 10^{-1} electron over 2.5 Å leads for $\varepsilon_s = \varepsilon_0$ to $\Delta X \simeq 0.4$ eV (for 1 ML).

In the case of (110) cleaved surfaces of III-V compounds characterized by a surface relaxation inducing a surface dipole modification for l = relaxation induced dipole length modification = 0.65 Å /25/ and for $\varepsilon = 8\varepsilon_0$, ΔX = 1.2 eV. Thus, variations in X values can be large and have to be seriously considered.

2 - Metal deposition

When a molecule (or an atom) impinges upon a surface, it may lose sufficient energy to become effectively bound to the surface. It is said to be adsorbed. This adsorption which is the first stage in the formation of an epitaxial film depends on many parameters and particularly on the affinity A and the ionization energy I of the adatom. Possible charge transfer between the substrate and the impinging atom depends on A, I and on the work function of the substrate. They can result in noticeable modification in ϕ and in V_S.

As more and more atoms or molecules arrive at the surface an adlayer begins to form. This formation is linked to the physical nature of the couple substrate - impinging atom. In the case of metal atoms impinging on a semiconductor surface three modes have to be considered. In the Van der Merwe mode, the metal film grows layer by layer ; in the Volmer-Weber mode, the deposit nucleates on the surface and forms a stable cluster ; in the Stranski-Krastanov mode, the deposit first forms an atomic monolayer and then nucleates to form islands on top of the monolayers. In the previous view no perturbations of the substrate surface were envisaged. In fact, it appears that, on III-V clean surfaces for example, the energy released in the formation of clusters might be quite sufficient to trigger an exchange reaction at the interface. This is the case of Al on GaAs /26/ and therefore reaction exchange can take place at the interface, modifying in this way its stoichiometry and its electrical properties. Futhermore, depending on the temperature of the substrate and of the rate of metal deposition, differences can be induced in the interface formation.

All these perturbations, unknown to Schottky when he proposed the first explanation of the metal semiconductor contact, are now at the center of the most recent research.

3 - Contribution of metal-semiconductor contacts to device properties

The first point we have to consider relates to the uniformity of the interface. To explain discrepancies in the results, non-abrupt boundary theories have been proposed, which suppose that considerable intermixing and disorder can be generated at the surface.

Following Freeouf and Woodall /27-29/ mixed phases may be present at the interface. In this hypothesis, the properties of the contact depend on the exact nature of the phases which are involved. Recent results on the Schottky barrier have been explained in this way.

The same conclusion applies to ohmic contacts and Braslau /30/ explains experimental results by non-uniform alloying of the contact. A.J. Valois and Robinson /31/ interpret the electrical characteristics of Au-based InP alloy contacts by a structure with Au_2P_3 clusters which penetrate into the semiconductor.

Electronic devices can be significantly affected in their behaviour by the properties of this contact /13-32/. Many of them depend for a good performance upon the attainment of contacts which do not restrict current flow nor introduce any non-linearity in the current they pass. For light-emitting diodes, lasers, and Gunn diodes values of the specific contact resistance in the range 10^{-2} - 10^{-5} $|cm^{-2}|$ have been found to be adequate. For small-signal FET devices any extraneous resistance will affect the gain.

In power semiconductor devices handling up to 10 A of current at 20 V bias, contacts must have a combined resistance of 0.1Ω if their effect is to be negligible. Schottky diodes showing rectifying properties are finding more and more applications in modern semiconductor technology. The main reason for this increasing usage is that they do not exhibit minority carrier effects, and thus can be operated as a majority carrier device with an inherently fast response and a very small charge storage. Many industrial applications exploit the inherent high speed advantage of the conduction of majority carriers.

Metal-semiconductor contacts can also be used as the drain and source contacts in an insulated gate field effect transistor, as a gate in FETs, CCDs and as a high efficiency photodetector. Microwave diodes and transistors, optical and particle detectors are important practical applications of Schottky diodes. Typical structures frequently to be found in devices are presented in Fig. 4.

III - CLASSICAL MODELS OF METAL-SEMICONDUCTOR CONTACTS

In this paragraph we briefly describe the classical models of Schottky, Bardeen, Cowley and Sze.

1 - Schottky model /1/

When a metal (work function ϕ_m) and a semiconductor (work function in flat band conditions ϕ_0) are brought into contact charge transfer between the metal and the semiconductor results in a dipole layer at the interface and an equilibration of Fermi energies. Schottky supposes that there are no modifications of ϕ_m and of X produced by the contact and that there are neither surface nor interface states.

Under these conditions, the formation of a surface barrier is needed to adjust ϕ_m and ϕ. At equilibrium $\phi = \phi_m$,

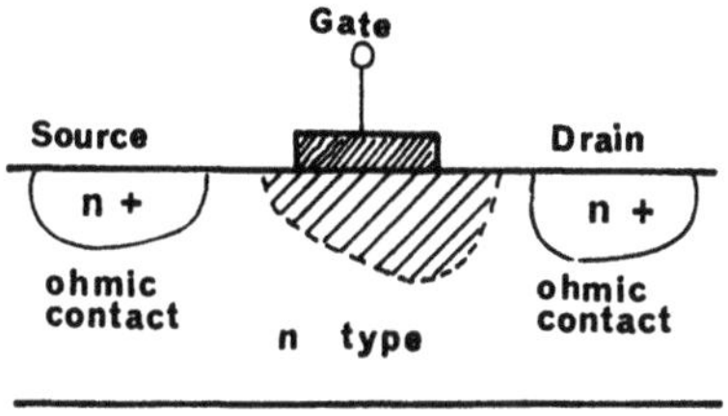

(a)

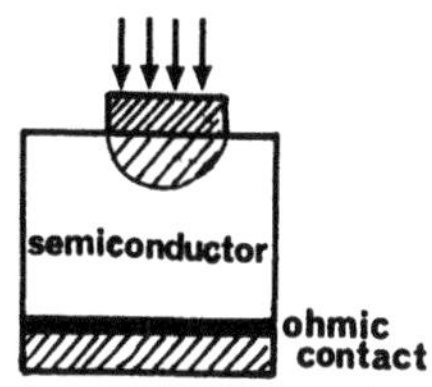

(b)

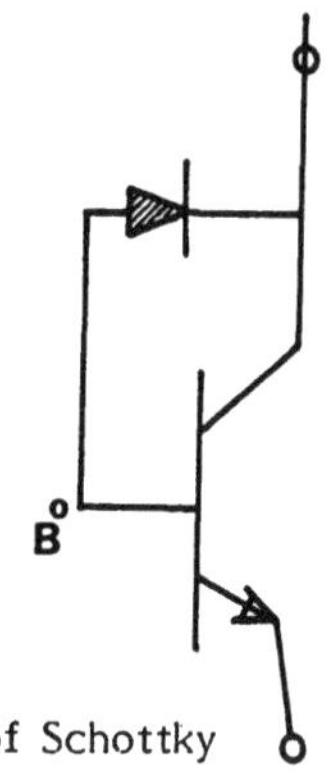

(c)

Fig. 4 . Example of the use of Schottky diodes in electronic devices.

a) Schottky gate structure

b) Schottky barrier photodiode

c) Clamped transistor. Schottky diode operates as a majority carrier device with a fast response. Incorporated between the base and the collector of the transistor it forms a clamped transistor with a very short saturation time constant.

d) Schottky barrier gate in a charge coupled device.

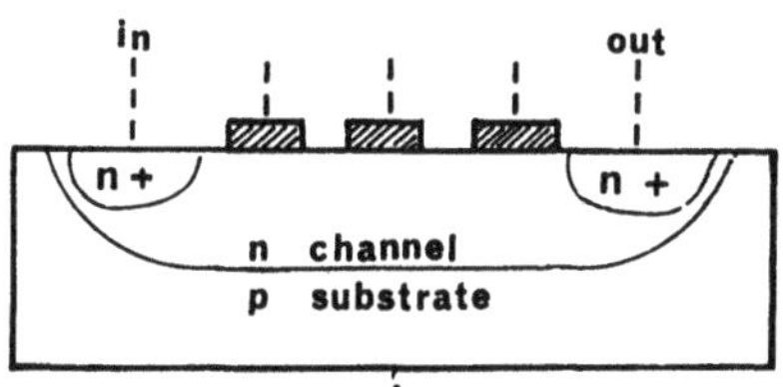

(d)

$$X + (E_C - E_F)_0 - qV_S = \Phi_m. \tag{5}$$

The barrier height efficient for the determination of the current passing through the contact is

$$\Phi_B = (E_C - E_F)_0 - qV_S = \Phi_m - X. \tag{6}$$

Following this model, the contact will be ohmic for accumulation surface layers and rectifying for depletion surface layers. In consequence by simply choosing a metal with an appropriate work function a rectifying diode or an ohmic contact could be obtained. However, experimental results show that strong dependence of barrier height on Φ_m is observed only in a predominantly ionic semiconductor. In many covalent semiconductors and in most III-V compounds, Φ_B is less sensitive to Φ_m. To explain this behaviour Bardeen proposed a model based on the action of surface states.

2 - Bardeen Model /3/

The assumption of the complete absence of surface states in the gap of semiconductors is not always applicable. As shown in the first part of this paper, surface or interface states may lead to the trapping of charges and to the existence of a space charge layer with a surface barrier even in the absence of a metal contact. When a metal is brought in contact with the semiconductor, the Fermi level in the semiconductor must change by an amount equal to the contact potential in order to reach the equilibrium conditions. If the density of states is sufficiently large, the charge transfer will mainly occur to or from the surfaces states and the interior of the semiconductor will be screened

from the metal ; the space charge will remain unaffected and ϕ_B will be independent of ϕ_m.
To emphasize the mechanism of charge transfer, Bardeen introduced the notion of a "neutral level". The position ϕ_0^* of this level is located in the gap in such a manner that when E_F and ϕ_0^* coincide, there is no charge on the surface and the bands are flat ; the states below ϕ_0^* are filled and donor-like, the states above ϕ_0^* are empty and acceptor-like. The surface density of charge Q_S becomes positive when E_F is below ϕ_0^*, negative when it is above. If the density of states is very large, a very small variation in the position of E_F around ϕ_0^* will result in a large variation of Q_S. Therefore, the surface Fermi level will be at ϕ_0^*. If we now assume that a thin insulating layer separates the metal from the semiconductor, or that the surface states are distributed in space inside the semiconductor, the necessary potential difference
$\phi_m - \phi = \phi_m - [X + E_g - \phi_0^*]$ will appear across the interface layer or across the thin layer inside which the states are distributed.
In these conditions, the barrier height is given by the equation (7), called the Bardeen limit :

$$\phi_B = E_g - \phi_0^* . \tag{7}$$

It is independent of the metal work function ; the surface states screen the semiconductor from the metal effect. The Fermi level is said to be "pinned".

3 - The Cowley and Sze approach /33/

The previous models correspond to limit cases of the more general model of Cowley and Sze /33/. This model, which has been recently reexamined by Rhoderick /12/ is a general analysis of metal-semiconductor contacts in the presence of surface states (density D_s cm^{-2} eV^{-1}) and an interfacial layer. It is based on the representation of Fig. 5 in which ε_i is the dielectric constant of the insulating layer and d its width.
This model supposes that the surface states, whose density is constant over the energy range from ϕ_0^* to the Fermi level, are independent of the metal.
The layer is assumed to have a thickness of a few angstroms and is therefore transparent to electrons. The main lines of the calculation are the following.
The surface charge density Q_S, the charge in the semiconductor Q_{SC} and the potential drop V_i across the layer are expressed in terms of D_S, E_g, ϕ_0^*, ϕ_{Bn}, ϕ_m, the temperature T and the doping of the semiconductor. From the relation

$$[Q_m = - Q_{SC} - Q_S] , \tag{8}$$

which results from the neutrality of the system, we deduce Q_m and the potential drop V_i across the layer. Then inspection of the energy band diagram allows the determination of ϕ_B. A complete analysis of the problem is very long and involves a lot of tedious algebra but it can be simplified by supposing that the junction is biased in such a way that the bands are flat (cf. Fig. 5). If the layer is thin enough (about 10 Å), the interface states are coupled by tunnelling to the metal conduction band. Their population is determined by the Fermi level in the metal and can be expressed as

$$Q_S = qD_s \times (\phi_{Bn} + \phi_0^* - E_g). \tag{9}$$

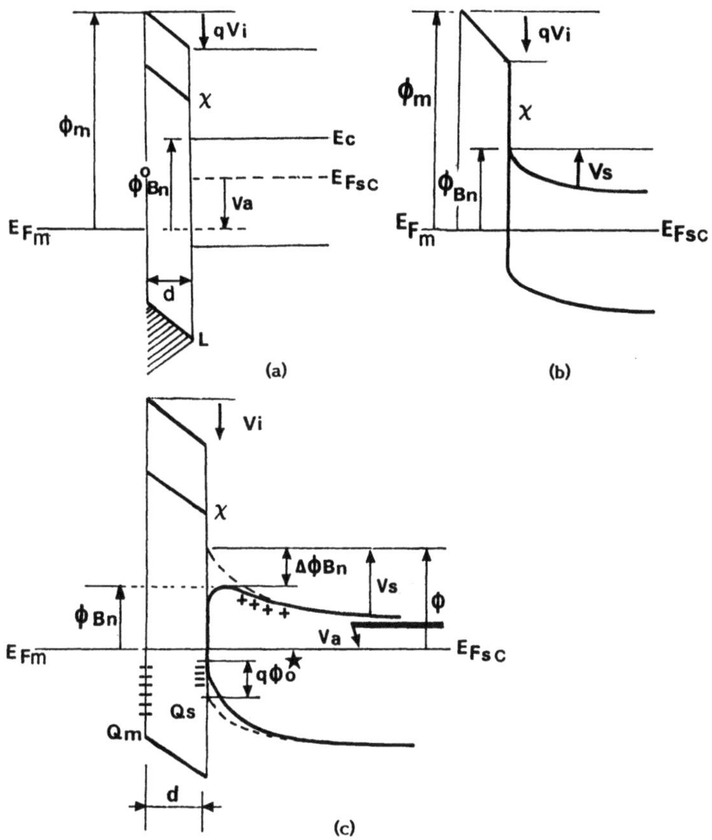

Fig. 5. Energy band diagram of a metal semiconductor contact with surface states and interfacial layer. (a) Semiconductor in flat band condition. (b) Contact with no applied bias. (c) General representation of a Schottky barrier. V_a = applied voltage.

From the field we deduce

$$V_i = \frac{dQs}{\varepsilon i} \tag{10}$$

Inspection of the energy band diagram then gives

$$\phi_{Bn}{}^{o} = \phi_m - qV_i - X, \tag{11}$$

$$\phi_{Bn}{}^{o} = A\,(\phi_m - X\,) + (1 - A)\,(E_g - \phi_o{}^{*}) \text{ for n type .} \tag{12a}$$

A similar calculation leads to

$$\phi_{BP}{}^{o} = A\,(E_g - \phi_m - X\,) + (1 - A)\,\phi_o{}^{*}, \tag{12b}$$

$$A = \frac{\varepsilon_i}{\varepsilon_i + q^2\, dD_s} \tag{13}$$

Therefore, for a given metal semiconductor system :

$$\phi_{Bn}{}^{o} + \phi_{Bp}{}^{o} = E_g\,. \tag{14}$$

Futhermore, A varies between 0 and 1 depending on the density of states. If

$q^2 d\, D_s \ll \varepsilon_i$, $A \simeq 1$

and $\phi_{Bn}{}^{o} = \phi_m - X \qquad \phi_{Bp}{}^{o} = E_g - \phi_m + X$.

This is the equation of Schottky.

If $q^2 d\, D_s \gg \varepsilon_i$, $A \ll 1$ and $\phi_{Bn}{}^{o} \simeq E_g - \phi_o{}^{*}$, $\phi_{Bp}{}^{o} \simeq \phi_o{}^{*}$

This is Bardeen's equation.

In this case, the barrier is independent of the metal work function. Comparison of these two results emphasizes the role of d and D_s. When D_s and d are small, the potential drop across the layer is negligible in comparison with the potential drop in the semiconductor. The contact potential difference is localised in the space charge. This is the Schottky contact. However, if D_s and d are large, the charge in the states and the potential drop across the layer can be large enough to screen the semiconductor from the electric field in the insulating film and to support the necessary potential difference between the metal and the semiconductor. This is the Bardeen contact.

Numerical calculations are useful in order to appreciate when the states have to be taken into account.

From (13), we deduce

$$D_s = \frac{\varepsilon_i (1 - A)}{q d A}. \qquad (15)$$

Taking d = 20 Å, which is an upper limit for good Schottky diodes, and $\varepsilon_i = 5\, \varepsilon_o$ a typical value for an insulator, leads to

$$D_s = 1.5 \times 10^{13} \times \frac{1 - A}{A} \ \mathrm{cm}^{-2}.$$

In order to obtain the Bardeen limit $A \ll 1$, D_s has to be higher than 10^{13} cm^{-2} ; when $D_s < 10^{12}$, we are in the Schottky limit; smaller d values imply higher interface state densities in order to screen the semiconductor from the metal.

An important experimental case corresponds to situations where the interface state density is not constant but is peaked about an energy E_t which differs from $\phi_o{}^{*}$. Let Q_t be the surface charge in the peak when the states are filled up to E_t, then the surface charge is given by (16), which is similar to (9),

$$Q_s = q D_s [\, \phi_{Bn}{}^{o} + E_t - E_g \,] + Q_t \qquad (16)$$

and

$$\phi_{Bn}{}^{o} = A\, [\, \phi_m - X - \frac{d\, Q_t}{\varepsilon_i} \,] + (1 - A)(E_g - E_t) \qquad (17)$$

for A = o $\quad \phi_{Bn}{}^{o} = (E_g - E_t)$. (18)

In the Bardeen limit, the Fermi level is close to E_t, that is to say, close to the peak in the interface state distribution.

3.1 - Contribution of charges in the insulator · /34/

When there are charges in the insulating layer, the previous analysis has to be modified. If Q_i is the charge per unit area, for n type semiconductors (12) has to be replaced by

$$\phi_{Bn}{}^{o} = A\,(\,\phi_m - X\,) + (\,1 - A)\,(\,E_g - \phi_o\,) - \frac{Ad\,Q_i}{\varepsilon_i}\,. \tag{19}$$

3.2 - Contribution of the electric field to the barrier

The potential difference between the metal and the semiconductor appears partly across the layer and partly across the semiconductor. Thus the barrier height depends on the field /33/ through the relation (20)

$$\phi_B = \phi_B{}^o - \alpha\,|\,\mathcal{E}_{max}|, \tag{20}$$

where $\mathcal{E}_{max}$ is the field at the top of the barrier and

$$\alpha = \frac{d\,\varepsilon_s}{\varepsilon_i + q^2\,d\,D_s} \tag{21}$$

When $q^2dDs \gg \varepsilon_i$ $\alpha \approx 0$ and ϕ_B is independent of the electric field. Conversely, if $q^2dDs \ll \varepsilon_i$, then ϕ_B will depend on $\mathcal{E}_{max}$.

We now have to determine the dependence of $\mathcal{E}_{max}$ on the applied voltage V_a. It can be found by solving Poisson's equation. Supposing the semiconductor is in the depletion condition, we obtain for n type

$$\mathcal{E}_{max} = \left(\frac{2q\;Nd}{\varepsilon_s}\right)^{1/2} \left(V_S - \frac{kT}{q}\right)^{1/2} \tag{22}$$

$$V_S = \phi_B - V_a - (\,E_C - E_F\,)_o\,/\,q, \tag{23}$$

$$\phi_{Bn} = A \times (\,\phi_m - X\,) + (\,1 - A)\,(\,E_g - \phi_o{}^*) - \alpha \times \left(\frac{2q\;N_d}{\varepsilon_s}\right)^{1/2} (V_S - kT/q)^{1/2} \tag{24}$$

that is to say,

$$\phi_{Bn} = \phi_{Bn}{}^{o} - \Delta\phi_{Bn}\,. \tag{25}$$

Numerical calculation shows that for moderate doping this lowering of the barrier should not exceed a few x 10 mV ; thus this correction is often ignored except in the case of heavier doping, of a thicker interfacial layer or under reverse bias because of the increase in V_S.

3.3 Image force lowering

An electron just inside the semiconductor induces a sheet of charge on the surface of the metal . The corresponding electric field attracts the electron towards the surface of the metal. Thus the energy necessary to remove the electron from the semiconductor is reduced by

$$\Delta\phi_B = \left(\frac{q\;\mathcal{E}_{max}\,)}{(4\,\pi\,\varepsilon_s{}^*)\;1/2}\right)^{1/2} \tag{26}$$

where $\mathcal{E}_{max}$ is the electric field at the interface resulting from the Schottky barrier. Because the velocity of the electron, when approaching the metal is very high (10^5 m/s = thermal velocity) the semiconductor is expected not to have enough time to become fully polarized. Thus, ε_s^* is the dynamic dielectric constant and it is usually smaller than ε_s. This reduction in the surface barrier height is taken into account in Fig. 5, which also shows the effect of the image lowering on the barrier for holes in p-type samples. In this case, the image force bends the valence band upwards. Numerical calculations show that for a given Schottky barrier field, $\Delta\phi_B$ increases when the doping increases and the position X_m of the top of the barrier in n type sample moves towards the interface. Typical values are in the range /12/ : $\Delta\phi_B \approx 10^{-2}$-$10^{-1}$ V, $X_m \approx 1$-20Å

IV - BRIEF REVIEW OF EXPERIMENTAL RESULTS

Over the last 30 years, many groups have attempted to develop a systematic study of the contact. By choosing a large set of semiconductors and metals they have tested the previous models and introduced new concepts. Before discussing the most recent approach in the explanation of the contact, it is useful briefly to review the most fundamental results presented in Fig. 6. Starting from (12), we can write the barrier height in the following linear form :

$$\phi_B = S(\phi_m - X) + C , \tag{27}$$

where C is a constant and S is characteristic of the interface behaviour, equal to 1 in the Schottky limit and to nought in the Bardeen limit. But, ϕ_m is not precisely known and the values reported in the literature often vary over a range of up to 0.5 eV. To overcome this difficulty, Aven and Mead /35/ have replaced ϕ_m by the electronegativity of the metal X_m, which is related to ϕ_m by the empirical relation /36/

$$\phi_m = 2.27\, X_m + 0.34. \tag{28}$$

They have plotted the variations versus X_m of the index of interface behavior X*, defined by

$$S^* = d\,\phi_B / d\,X_m . \tag{29}$$

The results, a recent compilation of which /37/ is presented in Fig. 6, show that $S^* \simeq 0.1$ for covalent semiconductors and the III-V compounds and $S^* \simeq 1$ for ionic materials (SnO_2, ZnS,...). In 1969, Kurtin et al. /38/ plotted the variation of S* versus ΔX, the semiconductor constituent electronegativity difference. This curve (Fig. 6) exhibits a sharp transition of S* at $\Delta X = 0.7$ separating the covalent ($S^* \simeq 0$) and the ionic materials ($S^* \simeq 1$).

Up to now, neither this sharp transition nor the saturation of S* for $\Delta X > 1$ have been well understood. Furthermore, a re-analysis of the data presented by Schluter /39/ challenges this result. In 1975, Andrew and Phillips /5/ published a plot of Schottky barrier height versus the heat of formation of silicides (Fig. 6) and analysed the barrier height

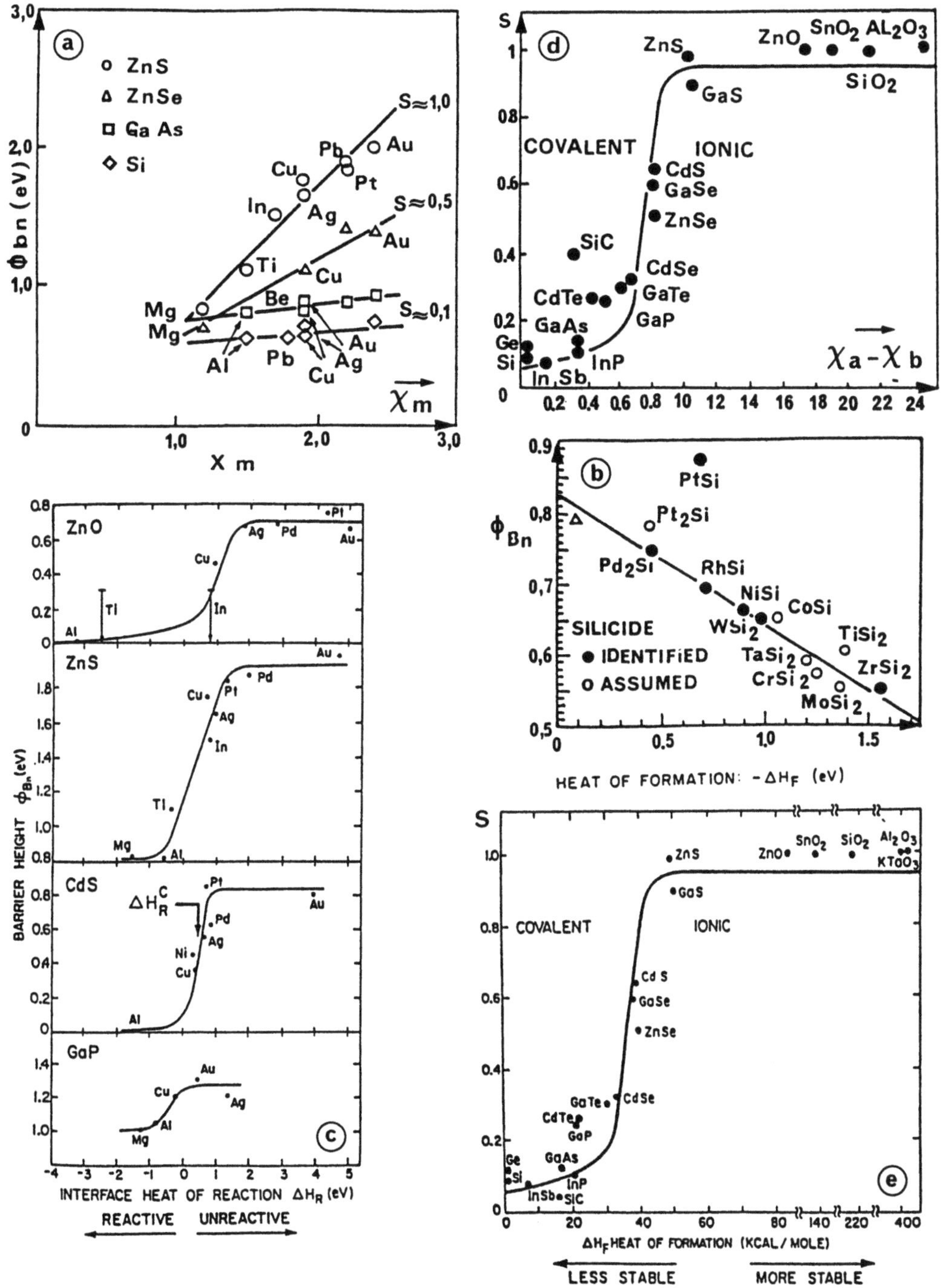

Fig. 6. **(a)** $\emptyset_{bn}$ versus metal electronegativity [37]. **(b)** $\emptyset_{bn}$ versus heats of formation of silicide [6]. **(c)** $\emptyset_{bn}$ versus interface heat of reaction. **(d)** Index of interface behaviour versus electronegativity difference $\chi_a - \chi_b$ for AB semiconductor [38]. **(e)** Index of interface behaviour versus heats of formation of the semiconductor [7].

on the basis of the nature of the chemical bond. The interaction metal-semiconductor is reactive and the reaction at the silicide-silicon interface moves the interface into the interior of the semiconductor and, in this way, screens the interface from the surface imperfections. Thus the reactive layer controls the interface and the barriers; chemical bonding across the interface leads to a large dipole which significantly changes the barrier : the larger the heat of formation of the silicide the lower the barrier. This approach has been recently developed by Brillson /7/ for III-V and II-VI compounds. His results, summarized in Fig. 6, emphasize the role of the heat of formation of the semiconductor and of the reactivity of the interface.

As a conclusion, we would argue that the previous results do not permit the deduction of a synthetical model capable of explaining all the experimental points. A large amount of data has yet to be provided and new studies are needed. Recent experiments are being carried out within this perspective.

V - INTERFACE FORMATION - RECENT STUDIES AND RESULTS

The role of the chemistry at the interface and of the interfacial layer was greatly emphasized by the results discussed above. The formation of the interface is highly dependent on the structure of the surface, the nature of the impinging metallic atoms, the deposition rate and the temperature. A better understanding of the contact implies an improved knowledge of the contribution of these different parameters. This requires a precise study of the successive stages of the interface formation. For the last 15 years, numerous studies have been devoted to this objective. Complementary sophisticated methods have been developed in order to obtain a microscopic view of the interface. Most of them will be described in detail during this school. Therefore, after a brief survey of the interface formation, we will concentrate on results obtained by the Kelvin method.

1 - Interface formation

If we want to follow the interface formation, we have first to prepare and characterize the initial surface, to deposit the metal layer by layer and then to form the diode. Detailed studies of these three stages are necessary. The surface, even when cleaved, can present structural defects like steps and vacancies capable of generating surface states. It can also be non-uniform, these defects being distributed in a random manner along the surface.

In the initial metal deposition, the density of surface atoms is low enough as to not interact. In this case one has to envisage a site-specific chemisorption. The energy released at this stage, such as the heat of condensation or reaction may dislodge surface atoms from their equilibrium sites. Therefore, the metal atom may interact with the semiconductor so as to yield new reaction products and dissociates the semiconductor. As a consequence, surface states of the clean semiconductor surface may be changed and new donor and acceptor states may be generated.

But the metal atoms may also move and in many cases, even at room temperature, they possess a large mobility. As the concentration grows clusters may be formed even if the coverage is less than 1/2 monolayer (this corresponds to what is sometimes called the "dilute limit").This stage plays an important role because the Fermi level pinning still appears at these coverages /40/ and the exchange reaction begins to occur /41, 42/. The following stage corresponds to the formation of the first monolayer. The structural aspects are strongly dependent upon the deposition temperature. In most systems, heat released as the metal nucleates induces a significant diffusion and the Fermi level is still completely pinned at this stage. Additional monolayers are needed to establish the metallic character of the layer; Kleinman et al. have found that three monolayers are generally required to converge to metallic-like properties /42, 43/.
They lead to complex interdiffusion and reactions on many III-V compounds.

2 - Work function measurements and Fermi level pinning

If work function measurements are not able to give microscopic information about the interface formation, they are very useful in acquiring information on the barrier and the electron affinity. The Kelvin method which is generally used to perform these measurements is very sensitive and non-disturbing.

2.1. Kelvin method /44/

A metallic probe is placed in front of the semiconductor surface. Due to the contact potential difference a potential drop $V_{CP} = \phi - \phi_m$ appears between the two plates of the capacitor. By vibrating one of these plates, we induce a current in the circuit. We insert an adjustable bias E in order to cancel the field inside the capacitor and the current. Then

$$\phi - \phi_m = qV_{cp} = qE. \tag{30}$$

Changes in ϕ result in changes in E and

$$\Delta\phi = q\Delta E. \text{ As } \phi = X - qV_S + (E_C - E_F)_o, \tag{31}$$

$$q\Delta E = \Delta X - q\Delta V_S. \tag{32}$$

If we suppose that we start with a semiconductor in flat band conditions and induce the formation of a barrier V_S from $\Delta\phi = q V_S$ and from the space charge relations /23/, we deduce the density of charged surface states. Depending on the doping densities as low as $10^8 - 10^9$ cm^{-2} can be detected /45/. This sensitivity is increased if doping is lowered.

Therefore, the method is very useful. However, it gives the sum : $\Delta X - q\Delta V_S$ or $X - qV_S$. If we want to try to separate X and V_S, we use the surface photovoltage SPV /23/, i.e. the change in V_S induced by photons. Depending on the energy of the photons, different possibilities have to be envisaged. When we increase hν the CPD shows a large change due to band-to-band transitions and creation of free electron-hole pairs which tends to flatten the band. SPV depends on the barrier height and on the photon flux. The flattening of the barrier is linked to the photon flux by a logarithmic law. Therefore, it is very difficult to achieve a total flattening.

Another application of the method is surface photovoltage spectroscopy. Photon-induced population or depopulation of states within the band gap will change the surface barrier and consequently the CPD. Lagowski and co-workers /46/ have explored a wide range of photovoltage phenomena and have shown that the SPV is most effective for wide band gap semiconductors and is capable of detecting charge densities orders of a magnitude underneath those typically observed by photoemission spectroscopy. However, the method is not very easy and the transitions between surface states are difficult to see. In spite of this, Brillson has, in recent years, strongly emphasized the possibilities of the method and has applied it to the study of many III-V and II-VI compounds.

Monch et al. /47/, Van Laar and Huijser /48/, Palau et al. /49/ and Ismail et al. /50/ have also successfully used the method, an interesting and supplementary asset of which consists in the possibility of plotting work function and photovoltage topographies. By moving the probe in front of the surface and keeping its distance constant from the surface, we get a very precise description of the electronic state and of its local variations /48 - 50/.

2.2. - Surface studies

The experiments we report were obtained in the following way : two III-V compounds samples, were cleaved and studied in the same run. We moved the probe successively in front of each of them and recorded the topographies of the CPD and SPV. Typical curves obtained are presented in Fig.7. They show a large variation of the work function and of the photovoltage. On well cleaved areas SPV is null and

$\Phi_p - \Phi_n \approx (E_{Fn} - E_{Pp})_o$. On badly cleaved areas SPV $\neq$ 0 and

$\Phi_p - \Phi_n \neq (E_{Fn} - E_{Fp})_o$. Furthermore, on the worse surfaces $\Phi_p - \Phi_n \simeq 0$: the surface Fermi level is pinned. From relation (4), we deduce

$$\Phi_p - \Phi_n = (E_{Fn} - E_{Fp})_o - qV_{Sp} + qV_{Sn}$$

and

$$q\,(V_{Sn} - V_{Sp}) = (\Phi_p - \Phi_n) - (E_{Fn} - E_{Fp})_o. \tag{33}$$

When

$$\Phi_p - \Phi_n = 0\,, \quad q\,(V_{Sn} - V_{Sp}) = (E_{Fp} - E_{Fn})_o, \tag{34}$$

when

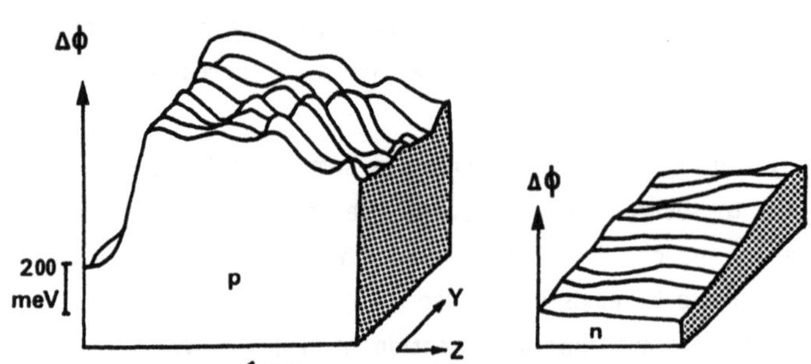

Fig. 7 . Work function topographies for n and p-type InP (110) cleaved surfaces.

$$\Phi_p - \Phi_n = (E_{Fn} - E_{Fp})_o \ , \qquad q\,(V_{Sn} - V_{sp}) = 0 \tag{35}$$

Measurement of SPV indicated that $V_{Sn} \lesssim 0$ and $V_{Sp} \gtrsim 0$. On perfectly cleaved areas $V_{Sn} = V_{sp} = 0$ and $\Phi_p - \Phi_n = (E_{Fn} - E_{Fp})_o$. On non-perfectly cleaved areas $V_{Sn} \neq 0$, $V_{Sp} \neq 0$ and on the worst areas $q|V_{Sn}| + qV_{Sp} = (E_{Fn} - E_{Fp})_o$.
Accordingly when moving from badly-cleaved to well-cleaved areas the barrier changes significantly and goes from nought to hundreds of millivolts. The corresponding SPV ($\Delta\Phi$) plots allow one to correlate SPV and V_S

2.3. - Interface formation

When we deposit a very small quantity of metal we rapidly obtain a flattening of the topographies. Typical variations of Φ_p and Φ_n versus coverage are given in Fig. 8.

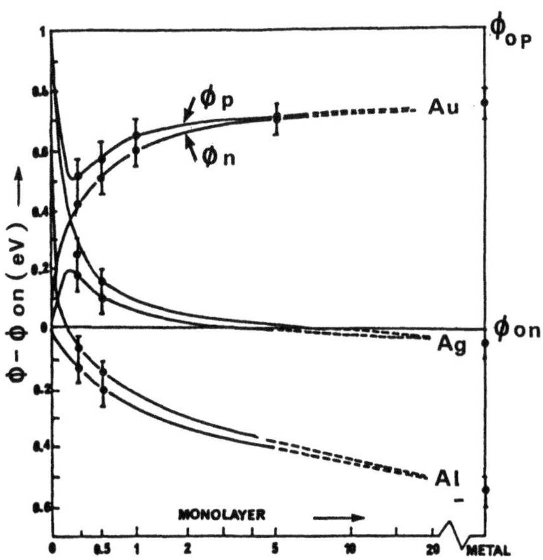

Fig. 8. Work function variations of n and p type InP (110) surfaces versus Au, Ag, Al coverage.

In all cases, a 1/4 ML deposit is enough to make $\Phi_p - \Phi_n$ as small as about 100 meV. As the electron affinity may be supposed similar for both types in the same deposit conditions, we deduce

$$\Phi_p - \Phi_n = (E_C - E_{Fp})_o - (E_C - E_{Fn})_o - q\,(V_{Sp} - V_{Sn}) \simeq 100\,\text{meV}.$$

The sum of the surface barriers is practically independent of the metal nature and whatever the metal is they vary in the same manner with the coverage.
Furthermore, information about X can be deduced by evaluating V_{Sn}, V_{Sp} from SPV. Measurements relative to the cleaved surface before metal deposition serve as a guide. From work function topographies, we directly obtain the barrier heights V_{Sn} and V_{Sp}. After metal deposition, we suppose that on n and p types, the surface photovoltage has been reduced in a similar manner for both samples. As $V_{Sp} - V_{Sn}$ is known, we use SPV curves for n and p types and determine V_{Sn} and V_{Sp} in order to obtain the same

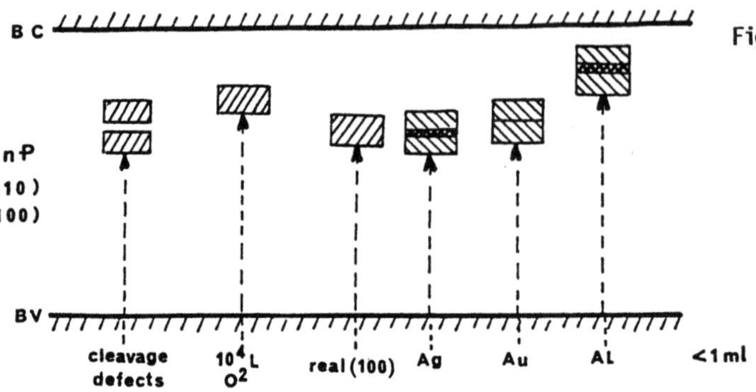

Fig. 9. Fermi-level pinning
$\theta < 1$

relative reduction in the photovoltage for n and p types. Results obtained in this way, in good agreement with results reached by other methods (UPS ...), are presented in Fig. 9. At submonolayer coverage, the Fermi level is pinned and this pinning position, located in the case of InP in the upper half of the gap, is characteristic of most III-V compounds.

VI - FERMI LEVEL PINNING

On III-V compounds but also on other semiconductors, Fermi level pinning often occurs and contributes to the definition of the electrical properties. Therefore, we will briefly discuss the mechanism of Fermi level pinning.

1 - Fermi level pinning by discrete surface states

1.1 - Qualitative analysis

We look at the case of a sample with acceptor surface states at an energy level E_t. The density n_t of electrons on the surface is given by (1)

$$n_t = N_t \left[1 + \exp \frac{(E_t - E_F)_o - qV_S}{kT} \right]^{-1} .$$

We start from a situation in which the states are unoccupied. The system is not at equilibrium. Electrons will move from the solid to the surface states. The barrier will develop and the states move towards the Fermi energy level. As the level nears and crosses the Fermi energy there is a very rapid drop in the required fractional occupancy. At some point, often when E_t is within a few kT of E_F, the Fermi condition becomes satisfied and the system is at equilibrium. This movement leads first to a limitation in V_S and to the pinning of E_F. Indeed V_S is limited to a value such that E_t has moved near to E_F. Therefore if N_t is very high, E_F will become pinned at the surface state level.

1.2. Quantitative analysis /51, 52/

We start from the relation

$$Q_S + Q_{SC} = 0 , \qquad (36)$$

where Q_{SC}, charge in the semiconductor, is given by

$$Q_{SC} = \frac{\varepsilon kT}{qL} F \,, \quad (37)$$

where F is a tabulated function which depends on V_S and L is the effective Debye length /23/.

Then we plot the variations of the equilibrium surface Fermi level position versus E_t for different values of the density of states. Figure 10 relates to acceptor states on an type semiconductor surface. When N_a is low, E_{FS} is practically independent of E_t. For $10^{11} < N_a < 10^{12}$ cm^{-2}, E_{FS} first does not move when E_t is down in the gap. The states are fully occupied and V_S is practically constant. When E_t approaches E_C, there is a modification in n_t and this modification will tend to change V_S in order to, as seen above, pin E_{FS}. Then for $N_t > 10^{12}$, E_{FS} always follows E_t. The density of states is high enough always to pin E_{FS}. Similar results are obtained for p type semiconductor and donor states.

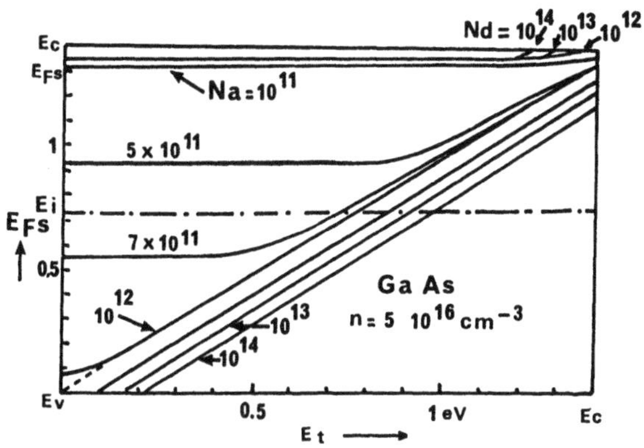

Fig. 10. E_{FS} variation versus E_t for acceptor surface states (density N_a cm^{-2}) or donor surface states (density N_d).

2 - Fermi level pinning induced by acceptor and donor states / 51 - 53/

In this case

$$Q_S = Q_{Sa} + Q_{Sd} = - Q_{SC} \,.$$

We suppose that the states are only once ionized, Q_{Sa} and Q_{Sd} are given by the relations 1 and Q_{SC} by the functions $F(V_s)$. Therefore, we can determine the position of E_{FS} at the surface. Curves given in Fig.**11** illustrate the variations of E_{FS} versus the position of a discrete acceptor state on a surface on which there is a fixed donor surface state. When $E_a > E_d$, in the three studied cases ($N_a > N_d$, $N_a = N_d$, $N_a < N_d$) on n type E_{FS} follows E_a. The donors are fully occupied and the electrical state of the surface is the same as if only the acceptor states were present. On p type, acceptors, when located above E_d are unoccupied and E_{FS} is pinned near E_d. Parallel results are obtained if we move E_d and keep E_a constant.

Following this approach, we can deduce the variations of E_{FS} versus N_a and N_d for a given position of the states. Figures 12 a-c illustrate the variations of E_{FS} versus

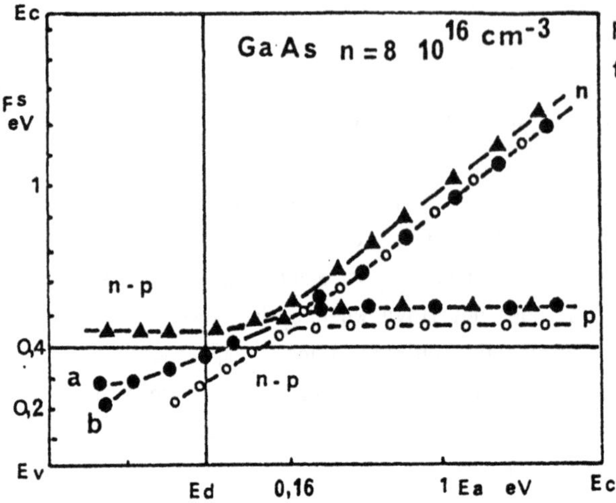

Fig. 11. E_{Fs} variations versus E_a for E_d = 0.4 eV.

● a : $Na = 10^{14} cm^{-2}$ Nd : $10^{14} cm^{-2}$

▲ b : $Na = 10^{13} cm^{-2}$ Nd : $10^{14} cm^{-2}$

○ c : $Na = 10^{14} cm^{-2}$ Nd : $10^{13} cm^{-2}$

N_a, the ratio N_a/N_d being kept constant. Three cases are considered corresponding to $N_a = 10N_d$, $N_a = 0.1N_d$, $N_a = N_d$. Figure 12a corresponds to $E_a > E_d$. Whatever the ratio N_a/N_d, the pinning is localized between the states and needs larger surface state densities than when $E_d > E_a$. This is due to the fact that the states have only a small charge, and to obtain the same Q_S we have to increase N_a N_d.

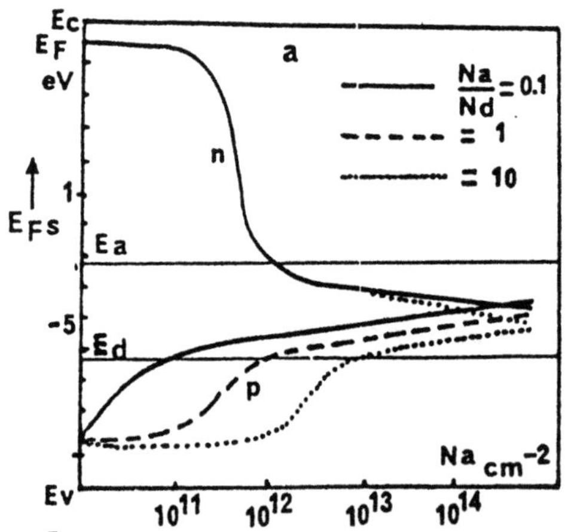

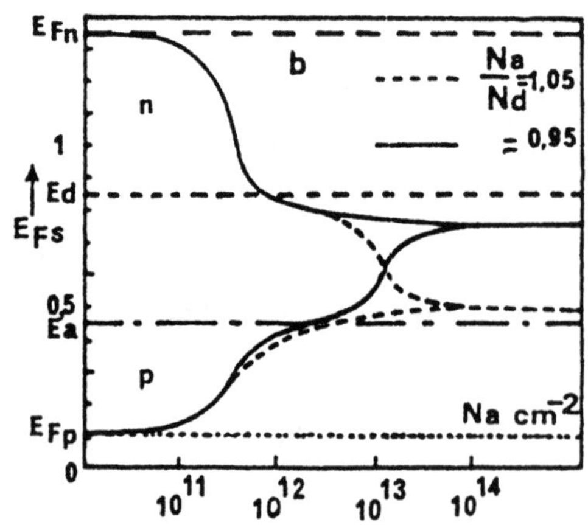

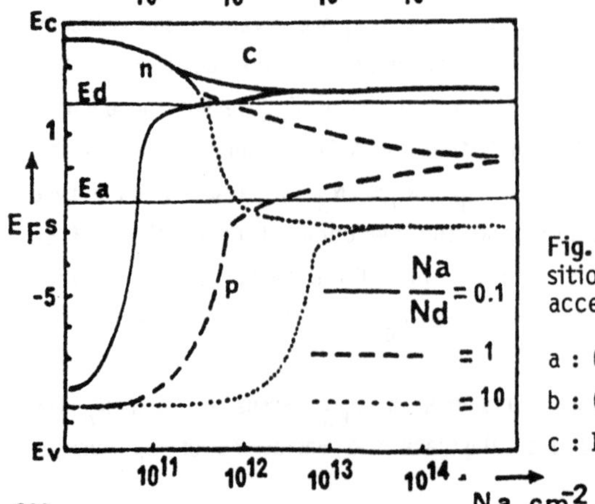

Fig. 12. GaAs and InP surface Fermi level position versus surface state density Na, Nd for acceptor (Ea) and donor (Ed) surface state.

a : GaAs $n = 5 \times 10^{16} cm^{-3}$ $P = 5 \times 10^{16} cm^{-3}$

b : GaAs $n = 5 \times 10^{16} cm^{-3}$ $p = 10^{17} cm^{-3}$

c : InP $n = 5 \times 10^{16} cm^{-3}$ $p = 5 \times 10^{16} cm^{-3}$

Fig 12b, c correspond to $E_d > E_a$. When $N_a = 10N_d$ the acceptors pick the electrons of the donors but as $N_d < N_A$, they are not fully occupied. Therefore the acceptors mainly determine the pinning of E_F. When $N_a = 0.1N_d$, the opposite occurs and the E_{FS} position is mainly determined by the donor.

If now we focus on states with similar densities the pinning will always occur near E_d or E_a but, as the compensation is higher than in the previous case, we have, to obtain the same barrier, to increase the density of states. Futhermore, the pinning is now located near, but between the states

An important conclusion results from these curves. In the case of $E_d > E_a$, when the densities are high enough to pin E_F small variations of N_a/N_d around 1 lead to E_F abruptly passing from E_a to E_d and conversely.

Now, we conclude with the following remarks :

- The pinning position has been determined by supposing a discrete distribution of states. In fact, these results can be extended to band states /51/ (Palau's thesis) within the limit that the band is not too wide.

-We have supposed that N_a/N_d was constant. Similar calculations can be used with respect to variations in N_a/N_d to explain certain results.

3 - Modelling

Experimental results give V_{Sn} and V_{Sp} for different coverages. Therefore, we can try to find the different combinations of N_t and E_t which can give the previous barriers. Starting from experimental V_{Sn} and V_{Sp}, we draw parametric curves /53/.

By analysing these curves and considering physical data it is possible to determine realistic distributions of states. The calculations used to establish the curves are tedious but the results are useful.

VII - CHARACTERIZATION OF DIODES

Capacitance-voltage , current - voltage and photocurrent - photon density measurements are the most frequently used in order to obtain ϕ_B.

1 - Capacitance - voltage measurements

The capacitance of the depletion layer at the interface is given by

$$C = A \; dQ_{SC} / d\,V_s \tag{38}$$

where A is the contact area. From (37), we deduce

$$C = A \times \varepsilon_s kT/qL \times dF/dV_s, \tag{39}$$

$$A^2/C^2 = 2[q\varepsilon_s(n_o + p_o)]^{-1} \times [|V_S| - kT/q] = 2(q\varepsilon_s N)^{-1}(|V_S| - \frac{kT}{q}) \tag{40}$$

for respectively n and p type samples $N = N_d$; $N = N_a$.

If the diode has no interfacial oxide and if we apply a reverse bias V_{ar}, (40) can be written

$$2 \times [\varepsilon_s q \ (n_o + p_o)]^{-1} \ (V_{So} + V_{ar} - kT/q) = C^{-2} A^2. \qquad (41)$$

A graph of C^{-2} as a function of V_{ar} should be a straight line with a slope $2 (\varepsilon_s qN)^{-1}$ and a negative intercept on the V_{ar} axes equal to $- V_{So} + kT/q$.

2 - Current voltage measurements

The current transfer from the metal to the semiconductor can result from different mechanisms which, depending on the experimental conditions, can occur simultaneously or separately. Schottky's model is based on the diffusion of majority carriers over the depletion barrier /54 , 13/ and Bethe's model on the thermionic emission of majority carriers over the barrier.

The expression for the current is in both cases

$$I = I_s \ \exp [qV_a / kT - 1] \qquad (42)$$

with

$$I_s = I_o \exp - q\phi_B/kT; \qquad (43)$$

I_o depends on the mechanism . For thermionic emission I_o depends on T^2; for diffusion I_o varies with the voltage but is less sensitive to temperature. A number of authors have combined these two theories and taken into account the recombination at the top of the barrier. I_o is then of the form /54/

$$I_o = qN_C \ V_R \times (1 + V_R / V_D)^{-1}. \qquad (44)$$

V_R is the recombination velocity at the top of the barrier and V_D the effective diffusion velocity. For high-mobility semiconductors the emission theory is justified.

Besides the diffusion and emission mechanisms, electrons can also be transferred by tunneling. This mechanism becomes important when the semiconductors are doped to degeneracy. Then the depletion region is very thin and electrons can tunnel from the semiconductor to the metal (field emission). Other electrons, at high temperatures, are able to rise high above E_F and can tunnel into the metal before reaching the top of the barrier.

Experimental studies of the diode show that the current voltage characteristic is of the form

$$I = I_s \ \exp \ \left[\frac{q \ V_a - R_S I - 1}{kT} \right] ; \qquad (45)$$

$R_S I$ is the potential drop across the series resistance of the sample and n is called the ideality factor. For ideal Schottky barriers n = 1 . In practical diode tunneling, intermediate layers, recombination or trapping, etc, contribute to n values different from 1.

The experimental determination of ϕ_B and n is achieved by drawing the curves Log I versus V_a

3 - Photoelectric measurements

Monochromatic light is incident on a metal contact. If $\phi_B < h\nu < E_g$ the light excites some electrons from the metal over the barrier. The resulting current is given by Fowler's theory /55/ as

$$I_{ph} \propto (h\nu - \phi_B)^2 .$$

From the plot of $\sqrt{I_{Ph}}$ versus hv, we deduce ϕ_B.

4 - Anomalies

Many deviations from the simplest law may be encountered. The first of them results from interface inhomogeneities inducing local variations in the Schottky barrier. I(V) curves are then characterized by modifications in the slope corresponding to the different regimes. The interface states' charge can be modified by V_a. This also changes the ideality factor. Tunneling can also occur.

As the barrier deduced from I(V) plots is a measure of the maximum height of the barrier as seen by an electron it will depend on all these perturbing effects.

The C(V) method measures the extrapolated barrier heights that would be found if the barrier remained parabolic right up to the metal. Anomalies on C(V) plots are also frequent. The barrier they give is noticeably different from the barrier deduced from I(V) curves. Futhermore, their slope is frequently different from the slope deduced from the doping. These anomalies can originate from the presence of a thin insulating layer at the interface /56, 57/.These anomalies and the possible errors in the results they provide have to be seriously taken into account when attempting to interpret the results in terms of physical effects./58, 59/

5 - Results

The barriers in the diodes are mainly deduced from I(V), C(V) curves. The numerical results frequently differ in a striking manner; ϕ_B deduced from I(V) is smaller than ϕ_B deduced from C(V). This results from the fact that ϕ_B I(V) corresponds to an activation energy for current flow and is therefore significantly dependent on the transition layer between the metal and the undisturbed bulk, while $\phi_B(V)$ deduced from relation (41), is representative of the space charge in the undisturbed semiconductor. Differences between the numerical values are then common and their magnitude is representative of the quality of the interface.

Results obtained by other methods also show discrepancies as emphasized by Freeouf in Fig. 13, and this is due to the fact that these methods do not measure exactly the same thing. Therefore these data are not of sufficiently high quality to allow theories seriously to attempt to explain differences as low as 0.100 eV as is the case for metal on clean semiconductors. However, they can be used to define tendencies. The following example emphasizes this view.

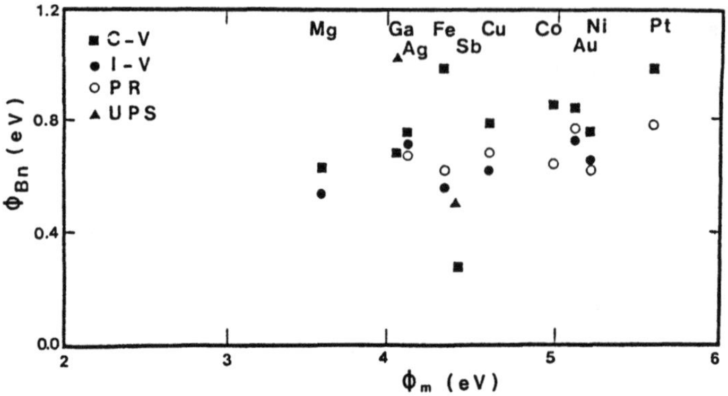

Fig 13 : Φ_{Bn} values obtained by different methods for several metal-Si contacts /58/

Data obtained on III-V compounds and relative to InP and GaAs show that, at submonolayer coverage, for GaAs the same pinning position is observed for many metals and for InP E_{FS} is located close to the conduction band edge for reactive metals and closer to the middle of the gap for Cu, Au, Ag. Similar results are obtained in the diodes, see Fig. 14. Futhermore, the sum of the barrier in diodes produced on n and p type

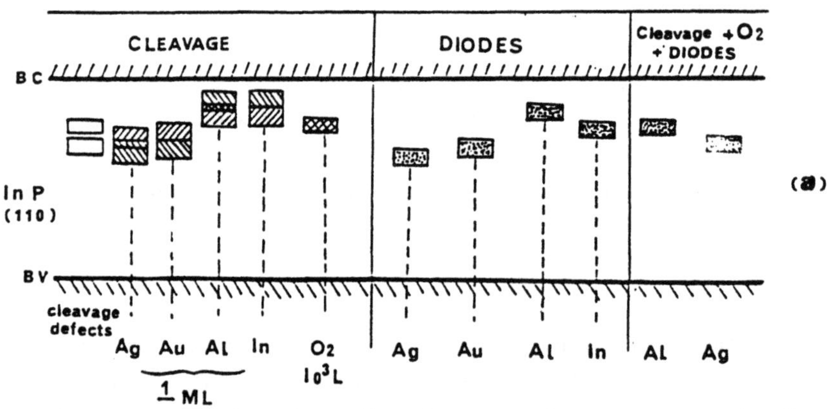

Fig. 14 a : Fermi level pinning position for submonolayer coverage and in the diodes.

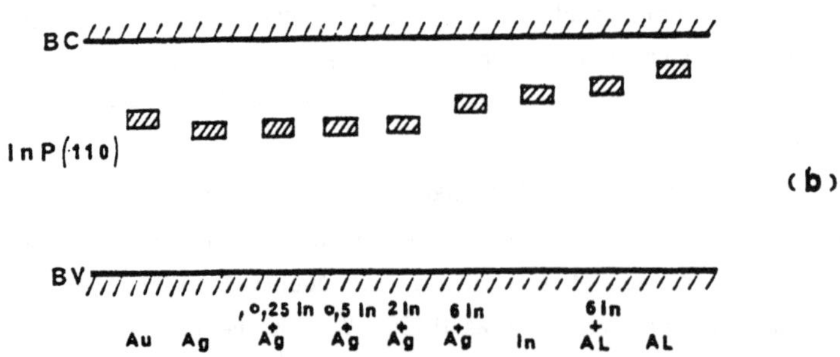

Fig. 14 b : Contribution of a thin interlayer to the pinning position in the diodes.

samples and the sum of the surface barriers at a monolayer coverage are practically identical. Therefore it seems that the pinning is similar at submonolayer coverage and in the diodes.

Other data relative to contact on vacuum-cleaved surfaces have been reported /8,10/. They show that the barrier depends upon the details of the band structure of the metal and they may in some cases decrease with ϕ_m. Schottky and Bardeen models are unable to explain all these data. New models have to be found.

VIII - RECENT MODELS

In spite of numerous data, the fundamental physical mechanisms are not well understood. Many questions have to be answered concerning the role of surface defects, of intrinsic surface states, of possible chemical reaction at the interface, and of the abruptness of this interface. Fermi level pinning, analysed above, results from the production of states in the band gap. The origin and the features of these states are one of the key issues in establishing a microscopic understanding of the barrier. This is why in recent years effort has been made and many new hypotheses and new models have been proposed.

1 - Intrinsic surface state model

Depending on the nature of the semiconductor and on its surface, interface states can be located in the band gap (in the case of silicon, for example) or outside the band gap (110 III-V compound surface). The question is : do these states play a role in Fermi level pinning? Even in the case of surface states lying outside the gap, we can envisage that the interaction with the metal and the resulting relaxation could pull these states into the gap. Detailed angle resolved photoemission measurements /60, 61/ seem to challenge this hypothesis, which, however, cannot be completely excluded

Therefore, an important problem is the evolution of these states when depositing metal. Heine /62/ has shown that the semiconductor derived states were appreciably modified by the metal and that the surface states derived from the clean surface states were transformed into an interface resonance.

Another approach has been proposed by Phillips /4/ who argues that interface bonding would determine the barrier height.

2 - Defect-induced model - Spicer et al. /62, 63-64/

To explain Fermi level pinning, Spicer et al. have proposed a new model which supposes that defects in the semiconductor at, or near, the surface are produced by metal deposition. These defects, in turn, produce surface states which pin E_F at the surface and in the diodes. If we remember the analysis of E_F pinning and if we suppose that every impinging atom creates one surface state, the surface state density will be high enough to pin E_F. Now the question is : What is the nature of these states? Spicer et al. supposed that, vacancies or more complex defects created by the submonolayer were responsible for these states and then for Schottky barrier formation. Because the pinning position was independent of the nature of the chemisorbed atom, they concluded that these defects were produced in an indirect way. They suggested that the heat of condensation of the metal atom can provide enough energy to initiate the formation of defects. In this case for the interaction of Au, Cs with CaAs, the heat of formation of GaAs is -17 kcal/mole while the heat of condensation of Au and Cs are respectively 89 kcal/mole and 80 kcal/mole. Therefore, semiconductor bonds are broken and vacancies are formed. These vacancies result in dangling bonds and in associated surface states which have been localized in the band gap and which act as donors (III atom vacancies) and as acceptors (V atom vacancies). Numerous experimental results and modelling, with a two state model are in good agreement with Spicer's model. This is why other groups have attempted to complete the previous hypothesis. So, Allen and Dow have first shown that, depending on its location, the same physical defect introduces different electronic states in the band gap /66-68/.

Then they calculated the energetic levels corresponding to vacancies stituated on the first, the second , etc .. surface planes, to antisite defects (anions on cations or cations on anions). Leading on from these results they have explained many experimental data and, precisely, by associating antisite to non-reactive metals and vacancies to reactive metals they have correctly interpreted the existence of two pinning levels for InP. Therefore, and according to their theoretical computations, Spicer's model and Brillson's model, which is based on the efficiency of the interface chemistry, could be linked.

3 - Brillson's model /6 , 7/

Even at submonolayer coverage, interdiffusion and exchange reaction can occur. In the case, for example, of Al on GaAs, if we compare the heat of formation of GaAs (- 17 kcal/mole) and of AlAs (-28 kcal/mole), we can envisage the replacement of Ga atoms by Al atoms. But the heat of condensation of individual adsorbate atoms does not seem to be sufficient to activate the process. Consequently, Zunger /26/ has suggested that metal atoms can form clusters at low coverages. The energy released in the formation of these clusters, which is equal to the product of the number of bonds between the atoms of the cluster and the average binding energy of nearest neighbors, would be high enough to promote surface chemical reaction locally. Increased temperature should promote a higher rate of cluster formation and a higher replacement reaction. The deposition of metal coverages in excess of a few layers also increases the interdiffusion and leads to complex reactions.

In these conditions, the formation of a reacted region with new dielectric properties and chemical composition and of an interdiffused region in the semiconductor correspond better to a real contact. For CA compounds, the interface widths are then dependent upon the reactivity of the interface through the heat of reaction

$$\Delta H_R = 1/x \,[\, H_F\,(CA) - H_F\,(M_x\,A)\,] \;, \qquad (47)$$

where H_F (CA) is the heat of formation of the semiconductor and H_F (MA) the heat of formation of the most stable MA product.

Brillson defined the reacted interface width "t_0" as the characteristic length over which

cation-anion bonding changes to metal metal-anion bonding and showed that "t_0" increases with ΔH_R; abrupt III-V compound semiconductor metal-interfaces correspond to systems with very strong metal-anion bonding.
The layer formed in this manner can then support a significant potential difference which can be assimilated when "t_0" is small to a modification in X.

4 - Metal induced gap states

Beyond the dilute phase, corresponding to submonolayer coverage, metal-induced states become important. Heine /61/ was the first to analyse the role of the metal Bloch wave functions. These wave functions exponentially decrease in the semiconductor and induce electronic states in the band gap. Electrons tunnel from the metal into the semiconductor and generate a negative space charge in the semiconductor and a positive charge in the metal. The form of the barrier is then distorted and this could explain differences between I (V) and C (V) data.
Later Inkson /69/ explained the pinning by a narrowing of the band gap of the semiconductor at the surface. All these explanations were insufficient but Heine's idea was elaborated on by Pellegrini /70/ and Louie et al. /37/. Using the jellium model previously used by Bennet and Duke /71/ to approximate the Al metal overlayer on top of Si they determined the total valence charge density as a function of the distance perpendicular to the surface and showed that new features appear in the density of states. Interface states, which are bulk-like in the Si, decay rapidly in the Al, interface states which are bulk-like in the Al decay rapidly in the Si and truly localized interface states decay away from the interface in both directions. The metal-like interface states which appear within the band gap and which exponentially decay in the semiconductor were called metal-induced gap states (MIGS).
This calculation has been generalized to GaAs, ZnSe and ZnS but unfortunately it has not been extended to other metals. More recently Tersoff /72, 73/ elaborated on the previous idea and, starting from the fact that the density of MIGS was high enough to screen any local modification of charge, defined an energy level E_B which was simply related to the semiconductor band structure. The Fermi level in the semiconductor is then pinned near, or at, this level. Therefore, the barriers in the diode are mainly determined by the semiconductor. The contribution of the metal and the local surface structure appear only as secondary factors.
Similar calculations have been carried out recently by Lannoo and co-workers /74/. Analysing the semiconductor surface dangling bonds they show that E_F would be pinned at a position E_D given by the relation

$$2 \times E_D = E_A + E_B \tag{48}$$

where the energies E_A and E_B are related to the anion and cation dangling bonds

5 - Freeouf and Woodall model /27 - 29/

Another approach to understanding the metal semiconductor contact is the effective work function model of Freeouf and Woodall. In this model, which corresponds to non-abrupt boundary conditions, the value of ϕ_B is not due to, or fixed, by surface defects or surface states but rather it is related to the work function of microclusters of the one or more microphases at the interface. These phases result from oxygen contamination, metal semiconductor reaction. Since each phase has its own work function , the Schottky equation is written in the modified form

$$\phi_{Bn} = \phi_{eff} - X$$

when ϕ_{eff} is the weighted average of the work function of the different phase. The effective work function model can be reasonably used to describe numerous experimental results.

6 - Interface Fermi level pinning

We have already investigated Fermi level pinning in the case of submonolayer coverage. Experimental results show that the Fermi level position at the interface between a bulk metal and a III-V semiconductor is frequently similar to its position in the submonolayer case. Hence, it was important to attempt to find a link between situations in which we have to deal with individual metal atoms or with a thick metallic layer. In the latter case, the density of MIGS is larger and these states can be charged. Consequently the question has to be asked : are the same states responsible for the pinning? Several papers based upon different hypotheses have been published recently. We will briefly summarize some of them. The first complete analysis has been presented by Zur et al./75/. In their model, they assume that defect induced interface states are localized in the semiconductor not more than a few angstroms away from the metal. They suppose that the metal terminates sharply at the interfaces and they use a jellium model to describe it and to determine the potential difference V_m between the jellium surface and the bulk. The band bending V_i in the interface region, in which the defects are located, is also determined. Poisson's equation is used to obtain the barrier in the semiconductor V_S (d). As the total potential difference across the interface $V_a = \phi_m - \phi_{SC}$ is the sum of $V_i + V_m + V_S$ they can deduce V_S as a function of $\phi_m + V_m^o - X = \beta$, where V_m^o is relative to the contact intrinsic semiconductor metal. Finally, they draw curves giving the Fermi level position at the interface versus β for several densities of states and several values of the metal response defined as $\partial V_m / \partial Q_m$. These curves emphasize the role of ϕ_m-X, N_t and of the metal response. The most general case is characterized by the same linear variation of V_S (d) for n and p type samples at low β values followed by a plateau and a new linear variation. This plateau, which corresponds to the pinning of E_F, which does not vary with ϕ_m, appears for defect densities $\simeq 10^{14}$ cm^{-2} , whereas for submonolayer coverage it appears at 10^{12} cm^{-2} . The difference between these results stems from the origin of the charge that balances the charge captured on surface defects. At low coverage, the source of these charges must be the depletion charge in the semiconductor and at high coverage the charges come from the metal.

This analysis of the pinning has been elaborated on by Spicer et al. /76/ who have made calculations using a metallic approximation and the model of Bardeen. In the model they proposed , the metal is separated by a distance d from the semiconductor . The defects are assumed to be located in a single plane parallel to the interface. These results are in good agreement with those of Zur et al.

Other alternative models based on very simple but realistic hypotheses have been published in the few last years. Starting from acceptor and donor defect-induced states, located in a thin layer at surface of semiconductor, Palau et al./77/ have shown, by calculating the potential drop in the undisturbed semiconductor and in the layer, that experimental data could be correctly explained. Futhermore, they have stated with precision the role of the layer thickness of the density of states and the shape of the barrier in the semiconductor.

More recently Lu /57/ using the MIGS hypothesis , has made very fine calculations. Assuming an exponentially decreasing density of MIGS and neglecting the potential drop in the metal, he has determined the shape and the height of the barrier and made the link between the Fermi level position and the metal work function more precise. Afterwards, he has analysed the behaviour of the interface in the diodes and explained, with this model, electrical anomalies in the diode.

IX - CONCLUSION

The main conclusion from the previous analysis is that the contact metal-semiconductor is not as simple as previously supposed. The numerous parameters which are capable of modifying the formation of the interface are in competition.

The properties of the bulk semiconductor , of its surface, the nature of the metal, the temperature and the deposition rate can modify in a competitive way the exact nature of the interface. Local inhomogeneities due to changes in the doping and in the structure can lead to microscopic modifications resulting in variations of the barrier height. Because of this complexity, there is no synthetical model to explain all the experimental results. Even if the electrical properties of the devices are well explained by the very general but macroscopic model of Cowley and Sze, the physical origin and nature of what is interpreted in terms of states, of the interfacial layer and of the barrier as well as the exact meaning of the experimental data are not very clear.

The model of Spicer et al., completed by the theory of Dow et al. is supported by numerous experimental results, but differences in the pinning position, linked to the metal nature, have been observed on the same surface. In some cases, the barrier is not built at submonolayer coverage; it changes when passing from submonolayer to thick coverage. According to the nature of the metal, chemical reactions are capable of modifying the barrier. Moreover, results relative to the electronic exhange at the interface, lead to time constants which are difficult to explain if we consider the position in the band gap and the location of the states.

MIGS can also explain the small dependence of the barrier on the nature of the metal but it is difficult to understand why the pinning positions are the same on GaAs-oxide and GaAs-metal, for example, why a monolayer of oxygen is able to change the barrier in the diode.

In this context, it is obvious that the physico-chemical approach as well as the approach based on the non-uniformity have to be associated.

In order to acquire a better knowledge of the interface it would be necessary to be able to extract from the global information the specific contribution of the different contributing mechanisms. This implies improving the study of the surface and of the metal deposit by carefully defining reproductive experimental conditions and also refining the methods used to deduce the barrier and the analysis of the data.

REFERENCES

1. W. Schottky : Naturwissenschaften 26 (1938), p. 843.

2. N.F. MOTT : Note on the contact between a metal and an insulator or semiconductor. Proc. Camb. Phil. Soc. 34.568-572, (1938).

3. J. Bardeen : Phys. Rev. 71, p.717-727 (1947).

4. J.C. Phillips : J. Vac. Sci. Technol. 11, p. 947-950 (1974).

5. J.M. Andrew, J.C. Phillips : Chemical bonding at metal-semiconductor interfaces. Phys. Rev. Lett. 35, p. 947-950 (1975).

6. L.J. Brillson : Transition in Schottky barrier formation with chemical reactivity. Phys. Rev. Lett. 40, p. 260-263 (1978).

7. L.J. Brillson : Surface Science Reports, Vol. 2, n° 2 (1982).

8. M.S. Tyagi : Physics of Schottky barriers junctions in "Metal Schottky barrier junctions and their applications", ed. by B.L. Sharma, Plenum, New-York, 1984, p. 1-60.

9. R.S. Bachrach : The same reference p. 61-112.

10. R.H. Williams : III-V semiconductor surface interaction in "Physics and Chemistry of III-V compound semiconductor interfaces", ed. by Carl W. Wilmsen, Plenum, New-York, (1985), p. 1-72.

11. G.Y. Robinson : Same reference, p. 73-162.

12. E.H. Rhoderick : Metal-semiconductor contacts. Ed. by Clarendon Press, Oxford, (1978).

13. V.I. Rideout : A review of the theory, technology and applications of metal-semiconductor rectifiers. Thin Solid Film, (1978), p. 48-261-291.

14. J.A. Appelbaum, D.R. Hamman : Rev. of Modern Phys., p. 48-479 (1976).

15. D.R. Haneman : Phys. Rev. 121, p. 1093 (1961).

16. W. Ranke, K. Jacobi : Progr. Surf. Sci. 10-1 (1981).

17. J.R. Chelikowski, M.L. Cohen : Phys. Rev. B, 19, p. 826-834 (1976).

18. J.D. Joannopoulos, M.L. Cohen : Phys. Rev. B, 10, p. 5073-5081 (1974).

19. A. Huijser, J. Van Laar : Surf. Sci. 52, p. 202-210 (1975).

20. J.M. Palau, E. Testemale, L. Lassabatère : J. Vac. Sci. Technol. 19 (2), July-August (1981).

21. M. Lannoo : Electronic structures of crystal defects and of disordered systems. Les Editions de Physique, (1981), p. 44.

22. D.J. Chadi : Surf. Sci. 99, p. 1-12 (1980).

23. A. Many, Y. Goldstein, N.B. Grover : Semiconductor surfaces, North-Holland (1965).

24. E.J. Mele, K.D. Joannopoulos : Phys. Rev. Lett. 341 (1978) Phys. Rev. B 18, 6998 (1978).

25. A. Kahn : Surf. Sci. Reports 3, p. 246 (1984).

26 A. Zunger : Phys. Rev. B 24, p. 4372-4391 (1981).

27. J.L. Freeouf, J.M. Woodall : Appl. Phys. Lett. 39, p. 727-729 (1981).

28. J.L. Freeouf, J.M. Woodall : J. Vac. Sci. Technol. 21, p. 570-573 (1984)

29. J.L. Freeouf : J. Vac. Sci. Technol. 18, (1981), p.910-916.

30. N. Braslau : J. Vac. Sci. Technol. 19, (1981), p. 803-807.

31. A.J. Valois, G.Y. Robinson : Solid State Electron. 25, 978 (1982).

32. D.L. Lile : J. Vac. Sci. Technol. B 2(3), p. 496-503.

33. A.M. Cowley, S.M. Sze : J. Appl. Phys. 36, (1965), p. 3212-3220.

34. O. Wada, A. Majerfield, P.N. Robson : Solid State Electron. 25 (1982).

35. M. Aven, C.A. Mead : Appl. Phys. Lett. 7, p. 8-10 (1965).

36. L. Pauling : The nature of the chemical bond. Ed. by Cornell University Press, Ithaca, New-York (1960).

37. S.G. Louie, J.R. Chelikowski, M.L. Cohen : Phys. Rev. B 15, 2154 (1977).

38. S.G. Kurtin, T.C. McGill, C.A. Mead : Phys. Rev. Lett. 22, (1969), p.1433-1436.

39. M. Schluter : Phys. Rev. B 17, (1978), p. 5044-5047.

40. I. Lindau, P.W. Chye, C.M. Garner, P. Pianetta, C.Y. Su, W.E. Spicer : J. Vac. Sci. Technol. 15, 1332 (1978).

41. R.Z. Bachrach, A. Bianconi : J. Vac. Sci. Technol. 15, 525 (1979).

42. K. Mednick, L. Kleinman : Phys. Rev. B 22, 5768 (1980).

43. E.B. Caruthers, L. Kleinman, G.P. Alldredge : Phys. Rev. B 9, 3330, 1974.

44. Lord Kelvin : Phil. Mag. 5, 46 (1978).

45. J. Bonnet, L. Soonckindt, L. Lassabatère : Proceedings of the 4th International Conference on Solid Surfaces, Cannes (1980), p. 1129.

46. M.C. Gatos, J. Lagowski : J. Vac. Sci. Technol. 10, 130 (1973).

47. W. Mönch, H.J. Clemens, S. Gorlich, R. Ennighorst, H. Gant : J. Vac. Sci. Technol. 19-3, 525 (1981).

48. J. Van Laar, A. Huijser : J. Vac. Sci. Technol. 13, 769 (1976).

49. J.M. Palau, E. Testemale, L. Lassabatère
J. Vac. Sci. Technol. 19 (2), 192 (1982).

50. A. Ismail, A. Ben Brahim, J.M. Palau, L. Lassabatère, I. Lindau : J. Vac. Sci. Tech nol, 36, (1986), p. 217-221.

51. J.M. Palau : Thesis, Montpellier (1982).

52. A. Ismail : Thesis, Montpellier (1985).

53. A. Ismail, J.M. Palau, L. Lassabatère : Journal de Physique 45 (1984), p. 1717-1723.

54. S.M. Sze : Physics of semiconductor devices, 2nd ed. John Wiley and Sons, New-York (1981).

55. R.H. Fowler : Phys. Rev. 38, p. 45-56 (1931).

56. E. Vieujot : Thesis, Montpellier (1986).

57. G.N. Lu : Thesis, Paris-Sud (1986).

58. J.L. Freeouf : Proc. 2nd Intern. Conf. on Surfaces and Interfaces, Trieste (1982).

59. A. Thanaïlakis : J. Phys. C 8.655-668 (1975).

60. A. Mc Kinley, G.J. Hugues, R.H. Williams : J. Phys. C. 10 (1977), p. 4545-4557.

61. W.E. Spicer, I. Lindau, P. Skeath, C.Y. Su : J. Vac. Sci. Technol. 17, (1980), p. 1019-1027.

62. V. Heine : Phys. Rev. 138, A 1689 (1965).

63. I. Lindau, PW Chye, PR Skeath, CY Su, WE Spicer - J. Vac Sc. Technol. 15 (1978) 1337

64. W.E. Spicer, I. Lindau, P. Skeath, C.Y. Su, P. Chye : Phys. Rev. Lett. 44, (1980), p. 420-423.

65. P. Skeath, C.Y. Su, I. Lindau, W.E. Spicer : J. Vac. Sci. Technol. 17, (1980), p. 874-879.

66. R.E. Allen, J.D. Dow : Phys. Rev. B. 25, 1423 (1982).

67. R.E. Allen, T.J. Humphreys, J.D. Dow, O.F. Jankey : J. Vac. Sci. Technol. B 2(3), (July-Sept. 1984), p. 449-452.

68. R.E. Allen, V.F. Jankey, J.D. Dow : Surf. Sci. 168,(1986), p. 376-385.

69. J.C. Inkson : J. Phys. C. Solid State Physics 6, p. 1350-1362 (1973).

70. B. Pellegrini : Phys. Rev. B 7, 5299 (1973).

71. A.J. Bennet, C.B. Duke : Phys. Rev. 160-541 (1967), 162-478 (1967).

72. J. Tersoff : Surf. Sci. 168 (1986).

73. J. Tersoff : J. Vac. Sci. Technol. B 3, 1157 (1985).

74. I. Lefevre, M. Lannoo, C. Priester, G. Allan : to be published.

75. A. Zur, T.C. McGild, D.L. Smith : Phys. Rev. B 28, 4.2060 (1983).

76. W.E. Spicer, S. Pan, D. Mo, N. Newman, P. Mahowald, T. Kendelewicz, S. Eglash : J. Vac. Sci. Technol. B 2 (3) (1984), 476-480.

77. J.M. Palau, A. Ismail, L. Lassabatère : Solid State Electron. 28.5 (1985), p. 499.

Deep Level Transient Spectroscopy for Semiconductor Surface and Interface Analysis

E. Rosencher

CNET-CNS, B.P. 98, F-38243 Meylan, France

1. INTRODUCTION

The purpose of the lecture is to describe the physical principles of the technique of Deep Level Transient Spectroscopy rather than to relate the countless applications of this technique /1/. Moreover, we will focus on the development of this method towards semiconductor surface or interface analysis. C.T. Sah and his coworkers were the first to study the capacitance transients of Schottky diodes in order to determine the concentration, the nature and the statistics of defects in semiconductor junctions /2/. D.V. Lang, by a proper signal processing of these transients /3/, allowed to extract spectroscopic informations from these relaxation signals. The method, called Deep Level Transient Spectroscopy (DLTS) led to a revolution in the physics of defects in semiconductors. More recently, DLTS was developed to study the insulator/semiconductor interface states /4,5/ and the metal/semiconductor near-interface states /6,7/.

In section 2, we will describe the mechanisms of capacitance transients in semiconductor junctions and thus explain the great sensitivity of DLTS. The principles of measurements are given in Section 3. In Section 4, we will explain how spectroscopic informations can be extracted from capacitance transcients. Finally, some applications of DLTS in semiconductor interface studies will be described in Section 5.

2. CAPACITANCE TRANSIENTS

Semiconductors are different from other materials (i.e, metals or insulators) in the fact that the surface concentration of free carriers can be easily modulated by applying an external potential at their surfaces. If one imposes a voltage drop V_a at the surface of a n doped semiconductor SC (doping level of N_D), the carriers will be repelled deeper in the SC, yielding a depletion layer where ionized donors screen the potential (Figure 1). The potential distribution is given by Poissons's equation

$$\frac{\partial^2 V}{\partial x^2} = \frac{q\,N\,(x)}{\varepsilon_s} \qquad (1)$$

where ε_s is the SC dielectric constant, q the elementary charge and N (x) the charge distribution. In the abrupt zone boundary model, these charges consist in the ionized donors only ($N\,(x) = N_D$). The depletion zone width W is given by

$$W = \left[\frac{2\,\varepsilon_s\,V_a}{q\,N_D}\right]^{\frac{1}{2}} \qquad (2)$$

yielding a unit area capacitance of the junction

$$C = \varepsilon_s\,\frac{1}{W}\,. \qquad (3)$$

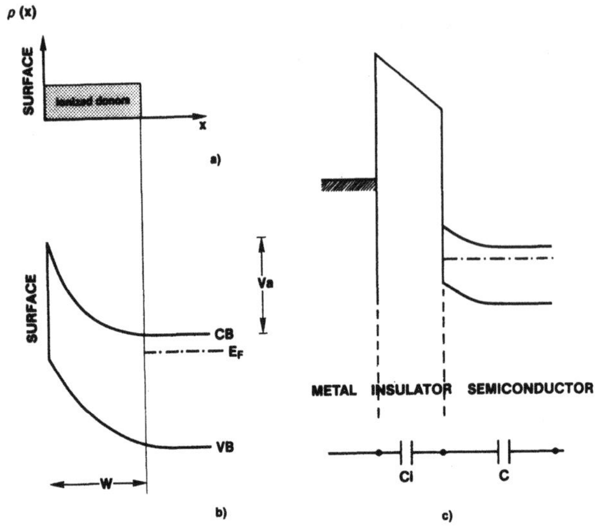

Fig. 1 : Charge distribution (a) and energy band diagram in a semiconductor junction (b). Band diagram of a M.I.S. structure (c) : equivalent capacitances are indicated.

Let us imagine a uniform variation ΔN of the charge distribution in the semiconductor (we shall see how and why later). This will lead to a variation of depletion length W via Eq. 2 and, consequently, to a change of the junction capacitance given by

$$\frac{\Delta C}{C} = \frac{1}{2}\,\frac{\Delta N}{N_D}. \qquad (4)$$

Relative variations of capacitance ΔC/C as small as 10^{-5} can be detected using home-made or commercial capacitance bridges. For a doping concentration of $10^{14}\,cm^{-3}$, defect concentration of 10^{9} impurities/cm³ may thus be detected, i.e, less than a picomole ! The fundamental reasons for this sensitivity rely on the overall small number of charges which are at stake in a semiconductor junction.

If the defects are concentrated at the surface of the semiconductor, the capacitance variation depends on the capacitance of the structure outside the SC. For instance, for a Metal-Insulator-Semiconductor (MIS) structure, the capacitance variation is given by (see Figure 1c)

$$\frac{\Delta C}{C} = \frac{C^2}{\varepsilon_s C_i}\,\frac{\Delta N}{N_D} \qquad (5)$$

where C and C_i are the capacitances of the MIS device and insulator per unit area, respectively. For a relative variation ΔC/C of 10^{-5}, a variation of 10^{8} charges/cm² may be detected i.e 10^{-7} of an atomic monolayer.

3. PRINCIPLES OF MEASUREMENTS

An electrical defect in a SC is an allowed state in the otherwise forbidden gap of the SC. Its charge state can be changed, whether an electron is present in this level or not. For a donor state, the level is empty above the Fermi level and/or full below the Fermi Level (see Figure 2).

In DLTS experiments, these levels are successively filled and emptied by applying a cycle of external biases. In a first part of the cycle, the SC junction is for-

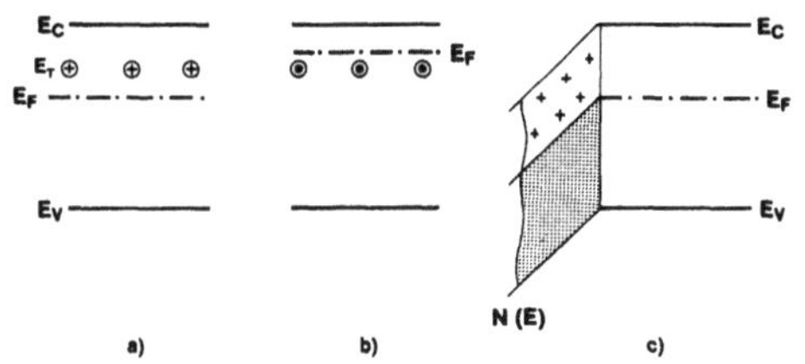

Fig. 2 : Donor levels in the SC forbidden gap : the level is empty (ionized) above the Fermi level E_F (a), full (neutral) below E_F (b). Figure c indicates the charge state for a continuum of donor levels.

ward biased. The potential drop at the SC surface is decreased, the concentration of carriers is dramatically enhanced at the surface, the impurity levels situated below the Fermi level get filled by the electrons of the conduction band (see Figure 3) : this is the "filling pulse" part of the cycle. Then, the junction is reverse biased : the impurity levels in the depletion zone are above the bulk SC Fermi level. The system relaxes towards equilibrium by emission of the trapped carriers above E_F into the SC conduction band. The variation of charge in the depletion layer thus leads to a variation of the junction capacitance, as explained in Section 2 /8/.

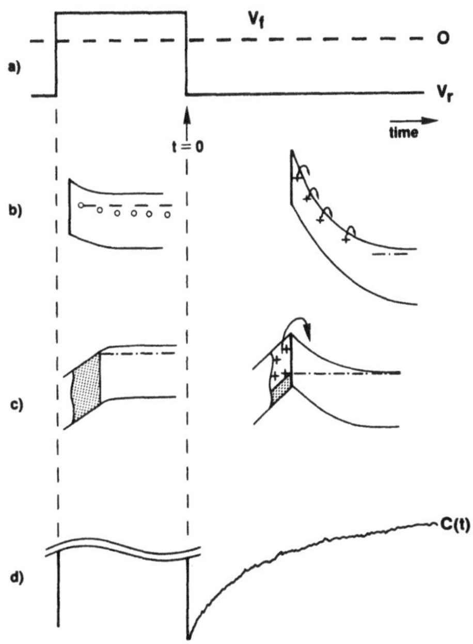

Fig. 3 : Pulse configuration in DLTS (a) ; band diagram during cycle for a discrete level (b), for a continuum of surface states (c) ; capacitance transient due to the emptying of trapped carriers (d).

Let us introduce the occupation factor f (E, T) of a level situated at $E = E_C - E_T$ below the conduction band. After been filled (f = 1) at time t = 0, the occupation factor is given by /9/

$$f\ (E,\ t) = e^{-e_n \cdot t} \qquad (6)$$

where e_n is the emission time constant from the level. We give below a heuristic derivation of e_n from a phenomenological point of view. A more involved derivation can be found in Ref. 9.

This frequency of emission e_n is proportional to i) the free place available for the electrons (i.e, the density of states in the conduction band N_c) ii) the frequency of trials to escape (i.e, the thermal velocity V_{th}), iii) the proportion of successful trials (i.e, the Boltzmann factor exp - E/kT) and, finally, a factor describing the innate geometry of the defect, i.e the capture cross-section σ_n:

$$e_n = \sigma_n \cdot V_{th} \cdot N_c \cdot \exp - \frac{E_c - E_t}{kT} \; . \qquad (7)$$

This extremely important formula (sometimes called the "detailed balance" formula) is the keypoint of Deep Level Transient Spectroscopy (also named by the way Emission Time Spectroscopy). The capacitance transient due to the emptying of a single level (concentration N_t) is thus given by Eq. 4, 6 and 7:

$$\Delta C = \frac{1}{2} C \frac{N_t}{N_D} e^{-e_n \cdot t} \; . \qquad (8)$$

If we are dealing with a continuum of levels N_t (E) in the SC gap at the surface of the SC, the capacitance transient is given by

$$\Delta C = A \int_{E_v}^{E_c} N_t (E) . e^{-e_n(E).t} \, dE \qquad (9)$$

with A given by Eq. 5 and neglecting the exchange of carriers with the SC valence band. We shall see in the next section how a spectroscopic information can be extracted from Eq. 9.

4. EMISSION TIME SPECTROSCOPY

The following derivation is successively due to Simmons and Wei /10/, Schultz and Johnson /4/, Nicollian and Brews /11/ and Rosencher et al /12/. Figure 4 shows the occupation factor f(E, t) of a continuum of levels in the SC gap at different times and for two different temperatures : these curves are calculated from Eq. 6 and 7 neglecting any energy dependence of capture cross section σ_n.

Let us imagine that we design an experimental set up which is sensitive only to transient signals in a definite range of time constants : an oscilloscope on time base $t_o \simeq 10$ µs for instance (see Figure 4). It is clear, from Figure 4, that the variations of f with temperature T and energy E are extremely steep (an exponential of an exponential). In other terms, only an extremely small fraction of states in the energy continuum participates in the transient signal for a time constant t_o at a temperature T. The position of this fraction of levels relative to the conduction band is in the region where f = 1/2 (see Figure 4), which is given by /12/

$$E_{\frac{1}{2}} = kT \ln \left[\frac{e_{no} \, t_o}{\ln 2} \right] \qquad (10)$$

where $e_{no} = \sigma_n V_{th} N_c$. When the temperature is varied, the position of this level scans the SC gap leading to a spectroscopic information.

Let us apply some mathematics. In Equation 9, we make following change variable:

$$u = e^{-E/kT} \; . \qquad (11)$$

Then /4/

$$\Delta C = A \, kT \int_0^1 N_t (E) \frac{1}{u} e^{-e_{no} ut} \, du \; . \qquad (12)$$

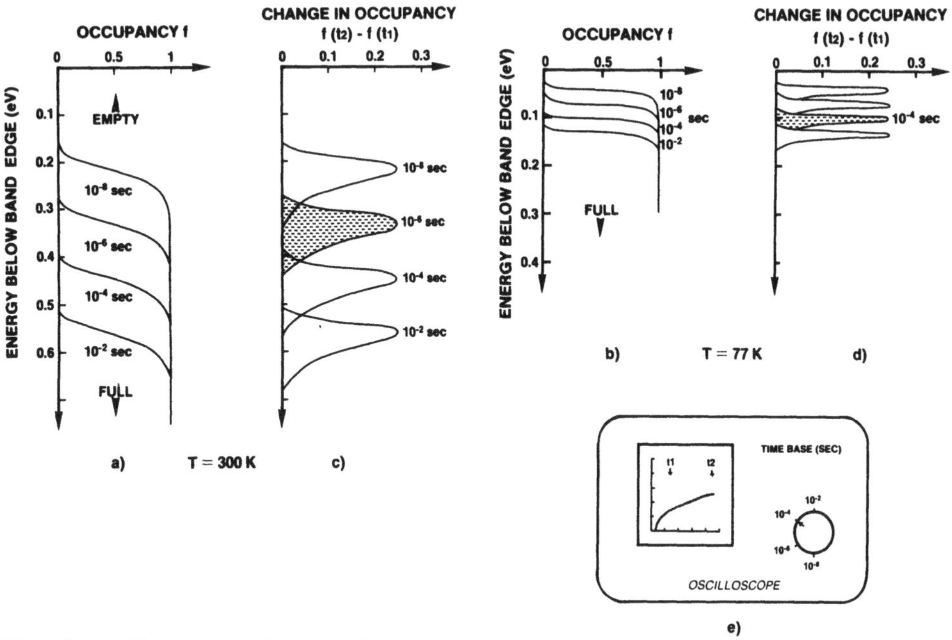

Fig. 4 : Emptying of interface traps with time after a pulse into depletion : occupancy f (E, t) from Eq. 6 for various times at 300 K (a) and 77 K (b) ; change in occupancy f ($2t_1$) - f (t_1) for different time scale t_1 at 300 K (c) and 77 K (d). The oscilloscope (e) indicates an arbitrary time constant (t_o= 10µs) : only the states in the energy window indicated in the shaded area are participating in the transient signal observed on the oscilloscope.

We know, from the preceding discussion, that if we observe the transient signal in a definite time interval, only a very small fraction of levels N_t (E) will be at stake, so that N_t (E) can be taken out from the integral in Eq 12. In 1974, D.V. Lang proposed, in order to define this time base, to use a box car sampling the signal at time t_1 and t_2 and introduced the DLTS signal S (T) /3/:

$$S(T) = C(t_2) - C(t_1), \tag{13}$$

i.e.

$$S(T) = A\,kT\,N_t(E_w) \int_0^1 \frac{1}{u} \left[e^{-e_{no} u t_2} - e^{e_{no} u t_1} \right] du . \tag{14}$$

Figure 4 shows the value of the integrand function (i.e f (t_2) - f (t_1)) and the position where this function is sharply peaked defines the energy window E_w :

$$E_w = kT \ln \left(e_{no} \cdot \frac{t_2 - t_1}{\ln (t_2/t_1)} \right) . \tag{15}$$

The value of the DLTS signal is thus given by

$$S(T) = A\,kT\,N_t(E_w) \ln (t_2/t_1) . \tag{16}$$

A graphic interpretation of Eq 15 and 16 is shown in Fig. 5. When the measurement temperature T is increased, the energy window E_w scans the SC energy gap. Only the levels within few kT of this energy window give rise to the transient signal S (T) : this is the basis of time constant spectroscopy.

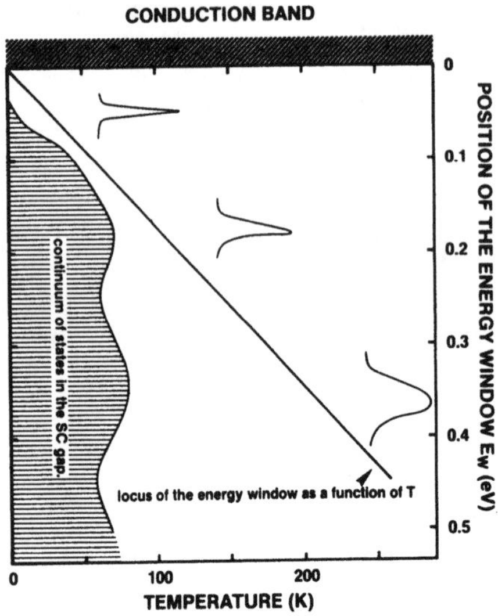

Fig.5 : As the temperature is increased, the time constant energy window E_w (Eq. 14) scans the SC forbidden gap. The width of the energy window is increasing with T.

Fig. 6 : Schematic experimental set-up for DLTS.

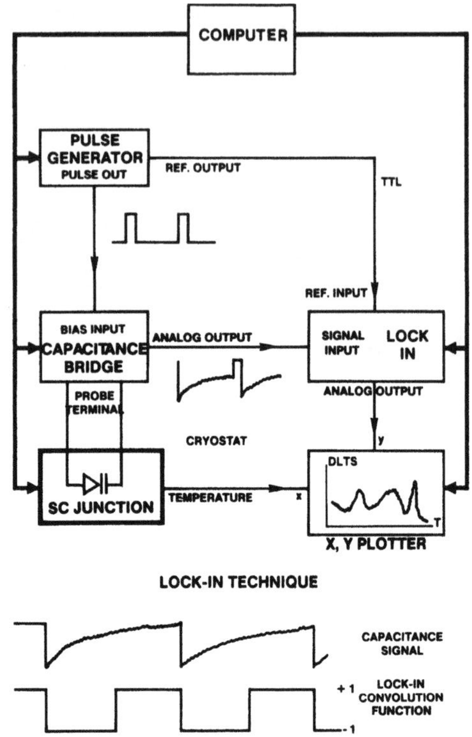

Other sampling functions have been used by various authors. One of the most practical ones is the one obtained by the lock-in technique :

$$S(T) = \int_0^{T/2} C(t)\,dt - \int_{T/2}^{T} C(t)\,dt \qquad (17)$$

which defines another energy window /13/

$$E_w = kT \ln (e_{no}/e_n) \qquad (18)$$

where the emission rate e_n is given by the implicit equation

$$e_n\, T/2 = \ln (1 + e_n\, T/2). \qquad (19)$$

A symbolic experimental set-up is shown in Fig. 6.

5. APPLICATIONS TO INTERFACE STATE STUDIES

Deep Level Transient Spectroscopy has been extensively applied to the study of discrete levels in SC. A review can be found in Ref. 1. As far as SC interfaces are concerned, the main applications have been : the Si/SiO_2 interface /13-18/, the Si/Silicide interface /6-7/, and, very recently, the GaAs surface covered by few monolayers of gold /24/.

Fig. 7

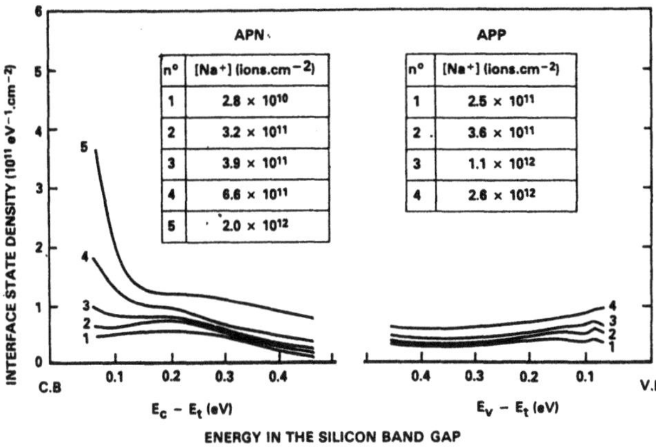

Fig. 8

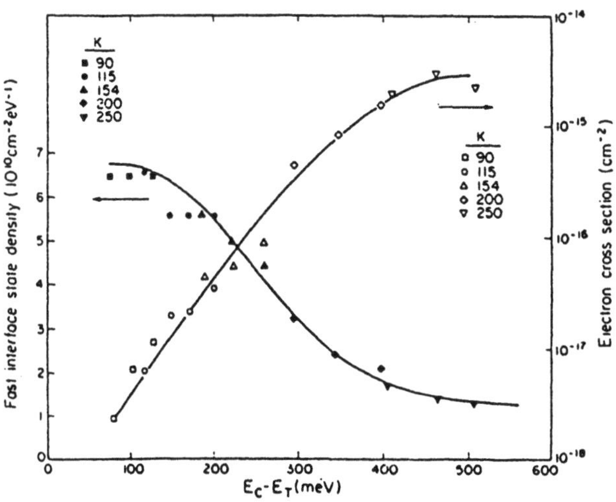

Fig. 7 : Si/SiO_2 interface state density for increasing Na^+ ion concentration at the Si surface. The Na^+ concentration is indicated in the inset /15/.

Fig. 8 : Evidence of the decrease of Si/SiO_2 interface state capture cross section near the Si conduction band edge /17/.

Using DLTS, SIMS and ESR measurements, Johnson et al /14/ have shown that Si/SiO_2 interface states could be annealed by the hydrogenation of dangling bonds at the Si surface. Rosencher and Coppard /13/ have shown that sodium ions at the Si/SiO_2 interface induce a coulombic impurity band tail near the Si conduction band but no change in the rest of Si band gap (see Figure 7). Using a small DLTS pulse configuration ("Energy - Resolved DLTS" /16/), Tredwell and Wiswanathan /17/ have shown that the capture cross sections σ_n of interface traps were dramatically decreasing near the Si conduction band (see Figure 8). This result, confirmed by many authors /16-18/, has still not been given a satisfactory explanation.

More recently, Rosencher et al /12/ have demonstrated the donor nature of residual Si/SiO_2 interface states in the upper gap of silicon, the level being positively charged when empty. Using field-enhanced emission from those traps, these authors have been able to partially determine the impurity potential well at the Si/SiO_2 interface. The results, shown in Figure 9, demonstrate the coulombic nature of the impurity potentials /12/. Let us note finally that, since commercial setups are becoming available, the DLTS technique is becoming a widespread characterisation method for MOS capacitors in microelectronics R and D centers.

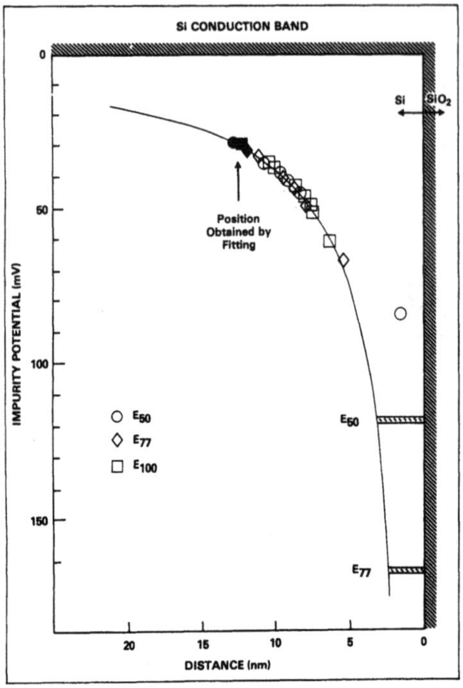

Fig. 9 : Potential of the electron traps as a function of distance from the Si/SiO_2 interface. The different points indicate levels in the Si gap investigated at different DLTS temperatures /12/.

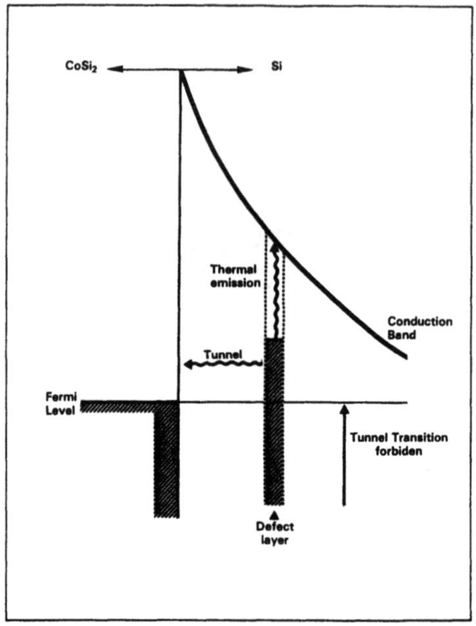

Fig. 10 : A model for the emission mechanism from $CoSi_2/Si$ "near interface" states /6/.

The use of DLTS for the study of metal/semiconductor interface states is more problematic. Interface states are supposed to control the metal/semiconductor (M/SC) barrier heights. These states may originate from a very thin insulator layer between the metal and the SC (Bardeen states) /19/ or from a localized combination of metal and SC wavefunctions /20/. No electrical measurements have been able to determine the nature of those interface traps. Recent results, using conductance techniques, are still very controversial /21-23/.

Indeed, the M/SC interface states exchange carriers with both the metal and the SC. There are no potential barriers between those levels and the metal : those levels are thus in resonance with the metal conduction band, with extremely short time constants which cannot be observed in time-resolved measurements (DLTS or conductance). Only when those states are located deeper in the SC, may they be observable. Such levels (called "near-interface states") have been observed by Rosencher et al /6/ at $Si/CoSi_2$ epitaxial interfaces : carbon contamination on the Si surface prior to $CoSi_2$ growth induces defects in silicon, at a tunneling distance from the $Si/CoSi_2$ interface (see Fig. 10). This results in a fluctuation of the C-V barrier height and DLTS signals, the peculiar behaviour of which may be found in Réf. 6. Similar results have been obtained by Werner et al using admittance measurements at GaAs/Au interface /23/.

In conclusion, the interface states responsible for the intrinsic I-V barrier heights, i.e really at the M/SC interface, cannot possibly be detected by time resolved techniques such as DLTS but also by conductance techniques. This has been recently confirmed by Chantre et al /7/. These authors have studied the epitaxial

$NiSi_2$/Si system, in which the I-V barrier height is dependent on the orientation of $NiSi_2$ crystal relativel to the Si one. No DLTS signal has been detected in these diodes, confirming that the intrinsic interface states at the origin of barrier heights cannot be detected by time-resolved experiments.

1. G.L. Miller, D.V. Lang, L.C. Kimerling : Ann. Rev. Mater Sci. 1977, 377-448
2. C.T. Sah, L. Forbes, L.I. Rosier, A.F. Tasch, Jr : Solid State Electron. 13, 759 (1970)
3. D.V. Lang, J. Appl. Phys. 45, 3023 (1974)
4. M. Schulz, N.M. Johnson, Solid State Comm. 25, 481 (1978)
5. K.L. Wang, A.O. Evwaraye, J. Appl. Phys. 47, 4574 (1976)
6. E. Rosencher, S. Delage, F. Arnaud d'Avitaya : J. Vac. Sci. Technol. B3, 762 (1985)
7. A. Chantre, A.F.J. Levi, R.T. Tung, W.C. Dautremont. Smith and M. Anzlowar, Phys. Rev. B 34, 4415 (1986)
8. There is also a transient current, associated to the detrapping of carriers, which can be used for DLTS
9. W. Shockley, W.T. Read, Jr : Phys. Rev. 87, 835 (1952)
10. J.G. Simmons, L.S. Wei : Solid State Electron 16, 53 (1973)
11. E.H. Nicollian and J.R. Brews : In MOS Physics and Technology (Wiley, New York 1982), p. 362
12. E. Rosencher, R. Coppard, D. Bois : J. Appl. Phys. 57, 2823 (1985)
13. E. Rosencher, R. Coppard : J. Appl. Phys. 55, 971 (1984)
14. N.M. Johnson, D.K. Biegelsen and M.D. Moyer : In Proceedings at the International Conference on the Physics of MOS Insulators, Ed. S.T. Pantelides
15. E. Rosencher, R. Coppard : In Proceeding of the 17th International Conference on the Physics of Semiconductors, Ed. J.D. Chadi and W.A. Harrison (Springer-Verlag, New York, 1985) p. 237
16. N.M. Johnson : Appl. Phys. Lett. 34, 802 (1979)
17. T.J. Tredwell, C.R. Wiswanathan : Appl. Phys. Lett. 36, 462 (1980)
18. M. Schultz : Surface Science 132, 422 (1983)
19. J. Bardeen : Phys. Rev. 71, 717 (1947)
20. V. Heine : Phys. Rev. 138, 1689 (1965)
21. C. Barret, P. Muret : Appl. Phys. Lett. 42, 890 (1983)
22. P.S. Ho, E.S. Yang, H.L. Evans, Xu Wu : Phys. Rev. Lett. 56, 177 (1986)
23. J. Werner, K. Ploog, H.J. Queisser : Phys. Rev. Lett. 57, 1080 (1986)
24 G. Marrakchi, E. Rosencher, M. Gavand, G. Guillot, A. Nouailhat : (to be published in Revue de Physique Appliquée).

Admittance Spectroscopy of Interface States in Metal/Semiconductor Contacts

P. Muret

Centre National de la Recherche Scientifique, Laboratoire d'Etudes des Propriétés Electroniques des Solides, associated with Université Scientifique, Technologique et Médicale de Grenoble, B.P. 166, F-38042 Grenoble Cedex, France

1. INTRODUCTION

The identification of interface states at metal/semiconductor (M/SC) contacts have long been key issues in understanding electronic properties of such junctions. Considerable spectroscopic evidence got from studies of a few monolayers deposited on semiconductor surfaces suggests that various effects (chemical formation of a compound, interdiffusion, defects creation...) take place which all can promote localized charge formation /1/. But there is also an interest for thicker layers both because they are really used in technological domains and electrical characterizations can easily apply. A useful tool which allows the detection of charges localized near the interface appears to be capacitance and conductance measurements. If an external force like illumination, electron beam or, like here, voltage difference is applied to the samples, variation of the population of carriers trapped within interface states can take place. This charge variation involves a capacitance and also a conductance (or dielectric losses) because a new steady-state is reached only after some delay corresponding to the time constant of the processes for carriers exchanges between interface states and carrier reservoirs. These phenomena will be analyzed in Section 2 taking the discrete monovalent trap at M/SC interface as a basic example. A brief review of experimental methods will be given in Section 3 and some illustrative results will be shown. Finally, a more detailed theory will be outlined in Section 4 for describing how the interface states density controls capacitance and conductance of a M/SC diode.

2. INTERFACE STATE KINETICS

For a given interface center C^s with s positive elementary charges, transfer of charge results from captures and emissions of electrons e^- and holes h^+ as described by the following reactions :

$$C^s + e^- \underset{e_n}{\overset{k_n}{\rightleftharpoons}} C^{s-1}\ , \tag{1}$$

$$C^s + e^- \underset{e_m}{\overset{k_m}{\rightleftharpoons}} C^{s-1}\ , \tag{2}$$

$$C^{s-1} + h^+ \underset{e_p}{\overset{k_p}{\rightleftharpoons}} C^s\ , \tag{3}$$

in which k and e represent respectively capture and emission probabilities per unit time. Subscripts n, m or p appear respectively for the SC and metal conduction band and for the SC valence band. For the SC cases, reaction rates are respectively proportional to the product of e by the number of occupied centers (occupied by e^- for (1) and by h^+ for (3)) for emission processes and to the product of k by the number of unoccupied centers for capture processes. The capture probability /2/ is given by the product of a capture cross section (σ_n or σ_p), a thermal velocity ($<v_{tn}>$ or $<v_{tp}>$) and the corresponding free carriers concentration at interface (n or p). Reaction (2) corresponds to tunnelling of electrons from interface states to metallic states or the reverse. Its rate depends on the difference of the electronic occupancy function at equilibrium f_s^o and out of equilibrium f_s /3/.

At equilibrium, detailed balancing implies that for each reaction, the rate of change from the left side to the right side equals the reverse one. Using a Fermi-Dirac occupancy function, this rule yields

$$e_n = \sigma_n <v_{tn}> N_c \exp\left\{-\frac{E_c - E_s}{kT}\right\}, \tag{4}$$

$$e_p = \sigma_p <v_{tp}> N_v \exp\left\{-\frac{E_s - E_v}{kT}\right\} . \quad (5)$$

where N_c (N_v) and E_c (E_v) are respectively the effective states density and band edge at interface for the SC conduction (valence) band. E_s is the energy position of the center inside the band gap. The differences $E_c - E_s$ and $E_s - E_v$ actually represent Gibbs free energy changes $\Delta G = \Delta H - T.\Delta S$. Therefore if an Arrhenius plot is made from experimental data of e_n (e_p), an enthalpy would be deduced.

Out of equilibrium, the rate of change of the charge per unit surface trapped in interface centers of density N_s is

$$qN_s \frac{\partial f_s}{\partial t} = qN_s \left[c_n n(1-f_s)-e_n f_s+e_p(1-f_s)-c_p p f_s+e_m(f_s^0-f_s)\right] \quad (6)$$

with $c_n = \sigma_n <v_{tn}>$ and $c_p = \sigma_p <v_{tp}>$.

At steady-state $\partial f_s/\partial t$ vanishes and the occupancy function can be obtained :

$$f_s = \frac{c_n n + e_p + e_m f_s^0}{c_n n + e_n + e_p + c_p p + e_m} . \quad (7)$$

The time relaxation of a perturbation δfs induced by δn and δp deviations of steady-state carriers concentrations n and p at interface can be evaluated from an expansion of (6) at first order. These driving forces δn, δp and the resulting shift δf_s can be taken into account conveniently by introducing respective electrochemical potentials V_n, V_p and V_s (or corresponding quasi-Fermi levels F_n, F_p, F_s) such as

$$\delta n = -\frac{q}{kT} n\, \delta(V_n - V_i) , \quad (8)$$

$$\delta p = \frac{q}{kT} p\, \delta(V_p - V_i) , \quad (9)$$

$$\delta f_s = -\frac{q}{kT} f_s(1 - f_s)\, \delta(V_s - V_i) , \quad (10)$$

coming from Boltzmann statistics for (8) and (9) and from Fermi-Dirac statistics for (10) ; V_i being the intrinsic potential. Then, (6) turns out to be a Kirchhoff equation which can be also represented by the electrical equivalent circuit /4/, shown in Figure 1. The left member of (6) corresponds to a capacitive current where the capacitance

$$C_s = \frac{q^2}{kT} N_s f_s (1 - f_s) \quad (11)$$

is maximum when $f_s = 1/2$.

We now examine which conditions apply for getting $f_s = 1/2$, assuming a n-type SC. This is achieved when the capture probability of majority carriers $c_n n$ equals the dominant reemission probability. If the latter is e_n, the maximum takes place when $F_n = E_s$ (or $c_n n = e_n$) which can be simply monitored from the forward bias voltage, the quasi-Fermi level gradient between the SC terminal and the interface being negligible in a M/SC diode /5/. In other cases, the maximum takes place at $F_n = E_{dn}$ or E_{dm} (demarcation levels) :

$$E_{dn} = E_v + q\phi_{Bn} + kT \ln\left(\frac{c_p N_v}{c_n N_c}\right) \quad (12)$$

if capture of minority carrier dominates (ϕ_{Bn} is the barrier height) or by

$$E_{dm} = E_c - kT \ln\left(\frac{c_n N_c}{e_m}\right) \quad (13)$$

if the tunnelling emission probability e_m dominates /6,7/. In the last case, if e_m approaches or exceeds $c_n N_c$, the condition $e_m = c_n n$ will never be obtained. No capacitance peak will occur because interface states cannot be filled enough, turning too rapidly to equilibrium with the metal.

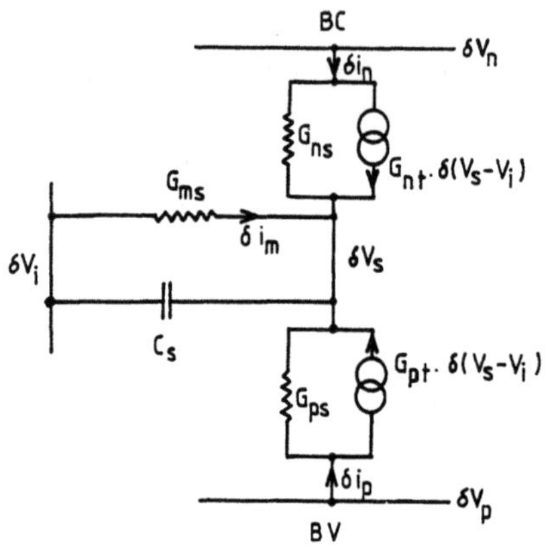

Fig. 1 : Equivalent circuit for a discrete interface state. Conductances and transconductances are defined below, following Sah /4/ :

$$C_s = \frac{q^2 N_s}{kT} f_s (1 - f_s) \qquad G_{ns} = \frac{q^2 N_s}{kT} c_n n (1 - f_s) \qquad G_{ps} = \frac{q^2 N_s}{kT} c_p p f_s$$

$$G_{ms} = \frac{q^2 N_s}{kT} e_m f_s (1 - f_s) \qquad G_{nt} = \frac{q^2 N_s}{kT} (1 - f_s)[c_n n (1 - f_s) - e_n f_s]$$

$$G_{pt} = \frac{q^2 N_s}{kT} f_s [c_p p f_s - e_p (1 - f_s)]$$

3. EXPERIMENTAL MEASUREMENTS

From the preceding analysis, it can be seen that a spectroscopy of interface states becomes possible by sweeping the quasi-Fermi level of majority carriers (here F_n) across the band gap. The maximum of the response will be obtained every time F_n will cross an interface state energy E_s or a demarcation level. In the former case, the true energy location of the center will be deduced. Such a sweep can be achieved in two ways : either the emission probabilities e_n are kept constant by working at constant temperature and $c_n n$ varies exponentially as a function of bias voltage or e_n is changed by a temperature scan while

$$c_n n = \sigma_n \langle v_{th} \rangle N_c \exp\left\{ - q \frac{\phi_{Bn} - |V_n|}{kT} \right\} \quad (14)$$

is maintained at a constant value. Assuming a negligible variation of pre-exponential factors, this last condition is simply achieved by monitoring a constant conductance of the M/SC diode because it is also proportional to n.

Two types of detection methods can be considered. First, the C_s variations are not detected directly but rather the capacitance of the depletion zone is used as a probe of the population variation $N_s\,\delta f_s$ which reacts on the depletion depth in order to maintain overall neutrality during a transient decay. Secondly, the C_s variations are directly measured when the sample is submitted to the superposition of both a steady-state voltage and a small ac component at frequency $\omega/2\pi$. In this last case, the maximum capacitance no longer appears at $c_n n = e_n$ when ω exceeds the reciprocal time constant τ^{-1} of the interface states but rather at $\omega_m = c_n n$ if e_n is constant and at $\omega_m = e_n$ if $c_n n$ is constant. This happens because the measured capacitance is not C_s but its parallel equivalent $C_{is} = C_s/(1 + \omega^2\tau^2)$.

The second method is used and Figure 2 illustrates the constant conductance case where the logarithmic plot of $\omega_m/2\pi\, v_{tp}N_v$ allows an independent determination of the energy E_s, in a sample made of nickel deposited on a p Si(111) substrate which has been cleaved in UHV /8/.

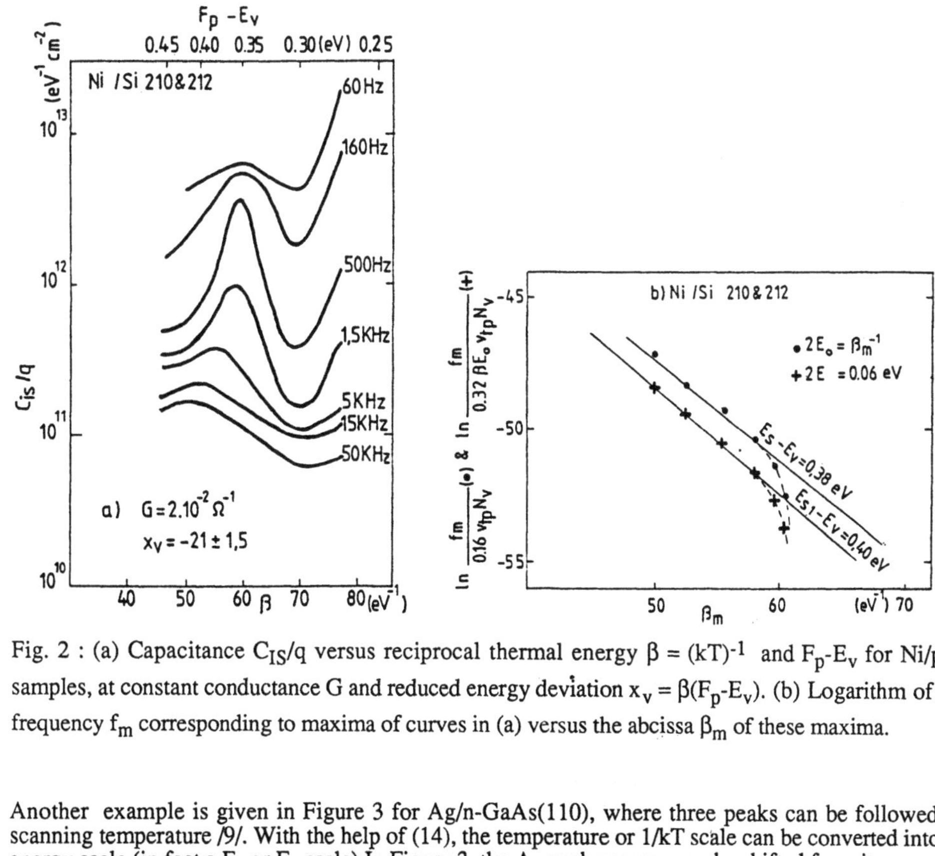

Fig. 2 : (a) Capacitance C_{IS}/q versus reciprocal thermal energy $\beta = (kT)^{-1}$ and F_p-E_v for Ni/p-Si samples, at constant conductance G and reduced energy deviation $x_v = \beta(F_p-E_v)$. (b) Logarithm of the frequency f_m corresponding to maxima of curves in (a) versus the abcissa β_m of these maxima.

Another example is given in Figure 3 for Ag/n-GaAs(110), where three peaks can be followed by scanning temperature /9/. With the help of (14), the temperature or 1/kT scale can be converted into an energy scale (in fact a F_p or F_n scale).In Figure 3, the A_n peak energy may be shifted from its apparent location (E_c - 0.7 eV) by the influence of minority carriers recombination. Such an effect would lead to an actual energy nearer the metal Fermi level (E_c - 0.95 eV) /9/.

From the systematic study of M/SC interfaces (n and p semiconductor, several metals) some conclusions can be inferred. First, featureless spectra of capacitance are seldom, even in so-called abrupt interfaces. This means that tunnelling probabilities are not so high as to delete all interface charge variations. Then the distance between the metallic layer edge and the centers must be larger than 10 Å or more in thick layer M/SC systems. Secondly, the spectra depend both on metal and SC but in different ways : they are completely different for Ag, Au or Ni/Si /8/ whereas true similarities exist for Ag, Au, Al/GaAs /9/. This shows that the reactivity of the SC plays a role in the formation of interface states.

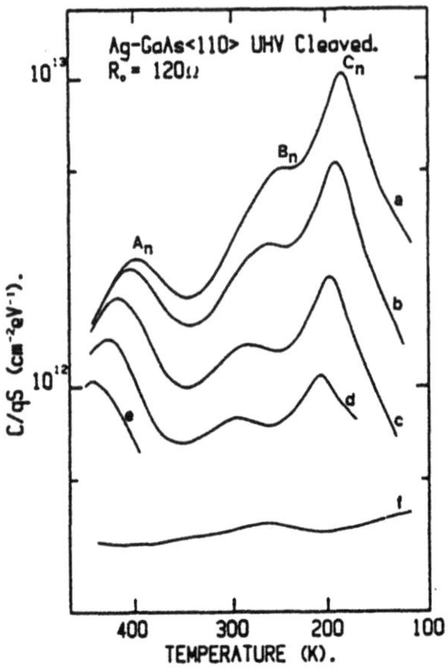

Fig. 3 : Variation of the capacitance of an Ag-n GaAs (110) UHV cleaved diode as a function of temperature for different values of the frequency : (a) 120 Hz ; (b) 220 Hz ; (c) 500 Hz ; (d) 1 kHz ; (e) 2 kHz ; (f) 100 kHz. Measurements are done at constant conductance $G = 1/120\ \Omega^{-1}$.

4. EFFECTS OF THE INTERFACE STATE DENSITY UPON THE CONDUCTANCE AND CAPACITANCE OF M/SC CONTACTS UNDER FORWARD BIAS

In the previous sections, only energy location of interface states have been discussed. If the quantitative influence upon electrical properties has to be considered, a more detailed theory is needed. A key point is given by the determination of the electrostatic potential V_i at the interface which occurs in (8), (9) and (10). It depends directly on the dipole strength of the interface charge $q\ N_s\ f_s$ and it determines also the barrier height (ϕ_{Bn} in (14)). Thus the conductance of the diode is now governed both by the applied voltage and by the barrier height modulation resulting from the charge stored inside interface states.

This effect has already been seen for conductance /10/ but also for capacitance /11/ because, as the completion of the charge $q\ N_s\ f_s$ is delayed by the time constant τ, the conductance becomes actually a complex admittance. The corresponding capacitance generally overcomes the direct capacitance C_{is} by a factor $q^2d/kT\ \sigma_n\ \varepsilon_i$ where d is the thickness of the interface layer and ε_i its permittivity /12/. The measured capacitance is then proportional to the dipole strength of the interface states which actually governs the Fermi level pinning (see LASSABATERE in this issue).

Moreover, since the free carrier concentrations at the interface depend now on the charge stored within interface states, this is true for capture rates too. Such a self-consistent treatment leads to a non-linear relationship between the capacitance and the interface states density /12/.

As a conclusion, basic mechanisms which govern the admittance of M/SC contacts have been outlined. The purpose of such an analysis is twofold. First, it provides techniques of characterization of interface states, which is of fundamental interest in understanding of M/SC interactions. Secondly, it gives a better knowledge of the electrical properties of M/SC diodes and shows what parameters must be controlled to improve the performances of such devices.

1. L.J. Brillson, Surface Science Reports 2, 123 (1982).
2. W. Shockley, W.T. Read, Phys. Rev. 87, 835 (1952).
3. L.B. Freeman, W.E. Dahlke, Solid State Electronics 13, 1483 (1970).
4. C.T. Sah, Proc. I.E.E.E. 55, 654 (1967).
5. C.R. Crowell, M. Beguwala, Solid State Electronics 14, 1149 (1971).

6. P. Muret, A. Deneuville, J. Appl. Phys. 53, 6289 (1982).
7. C. Barret, F. Chekir, A. Vapaille, J. Phys. C16, 2421 (1983).
8. P. Muret, A. Deneuville, Surface Science 162, 640 (1985).
9. F. Chekir, C. Barret, Surface Science 168, 838 (1986).
10. C. Barret, P. Muret, Appl. Phys. Letters 42, 890 (1983).
11. J. Werner, K. Ploog, H.J. Queisser, Phys. Rev. Letters 57, 1080 (1986).
12. P. Muret, (to be published).

Part V

Optical and Vibrational Properties of Interfaces

Optical Properties of Surfaces and Interfaces

P. Chiaradia

Istituto Struttura Materia CNR, Via E. Fermi 38,
I-00044 Frascati, Italy

1. Introduction

Light reflected from a crystal surface carries information both on bulk and surface optical properties. In this context the surface is defined as the transition region where the value of the dielectric function is different from both the ambient and the bulk values. The width of this region is tipically of the order of 5-10 Å. The main problem encountered in surface optical spectroscopies is to separate the surface term from the bulk term, which is usually much larger. There are two reasons why the surface dielectric function (DF) is distinct from the usual bulk-like DF, namely the peculiar electronic structure characteristic of the surface (surface states) and the fact that bulk-like optical transitions are modified by the surface, for example because of the loss of translational symmetry in the direction perpendicular to the surface. In principle, both the abovementioned aspects contribute to the surface optical properties. Consequently great care must be taken in order to disentangle them and - for instance - extract the information regarding surface states alone from the experimental data.

These notes are organized as follows: section 2.1 presents the basic ideas underlying surface differential reflectivity and gives examples of important results obtained with this technique. In section 2.2 the relevant theoretical aspects of the Surface Dielectric Response are presented. Finally sections 3. and 4. give an introductory account of principles and applications to surface science of Second-Harmonic Generation and Photothermal Displacement Spectroscopy, respectively. Throughout the paper extensive reference is made to the case of semiconductors, while only a few results obtained on metal surfaces are quoted.

2. Surface Differential Reflectivity

2.1 Introduction and Experimental Results

In this section a sketchy presentation of the Surface Differential Reflectivity (SDR) technique will be given. The reader is referred to the abundant literature available in this field, including a recent review /1/.

In SDR the separation of bulk and surface contributions to reflectivity is based on the fact that when the surface condition is changed (usually upon contamination of an atomically clean surface) surface related terms change accordingly, while the bulk term is not affected. Under simplifying hypotheses it can be shown that the fractional change ΔR/R contains only the surface contribution to the overall reflectivity. In fact the analysis developed by McIntyre and Aspnes /2/, based on a linearized treatment of the classical problem of Maxwell equations in a three-media system /3/, shows that ΔR/R at normal incidence is given by

$$\frac{\Delta R}{R} \cong - \mathrm{Im}\left(\frac{\hat{\varepsilon}_1 - \hat{\varepsilon}_2}{\hat{\varepsilon}_1 - \hat{\varepsilon}_3}\right) \qquad (2.1.1)$$

where $\hat{\varepsilon}_j = \varepsilon_j' + \varepsilon_j''$ (j=1,2,3) is the dielectric function of ambient, surface and substrate, respectively, while d is the thickness of the surface layer. For the usual case of vacuum as first medium ($\hat{\varepsilon}_1$=1) and non-absorbing substrate ($\hat{\varepsilon}_3 = \varepsilon_b$, real) the result is

$$\frac{\Delta R}{R} \cong - \frac{4 n_1}{n_1^2 - n_3^2} \frac{\omega}{c} (\varepsilon_2'' d) \qquad (2.1.2)$$

showing that ΔR/R is simply proportional to the imaginary part of the surface dielectric function through a structureless coefficient. For the above analysis to be valid it is necessary that the oxide contribution to reflectivity is negligible in the energy range of interest. For instance in Si(111)2x1 this is certainly true below 3.5 eV, as demonstrated by investigations of the oxide optical properties performed with Electron Energy Loss Spectroscopy /4/.

At the outset SDR has been used in the (multiple) internal total reflection (MITR) configuration and the results have been given in terms of the so-called surface absorption constant /5,6/. The latter is simply related to the reflectivity change ΔR/R previously discussed /7/

Figure 2.1.1,a shows two absorption spectra obtained by MITR in Ge(111)2x1 before and after oxidation. They differ because of the

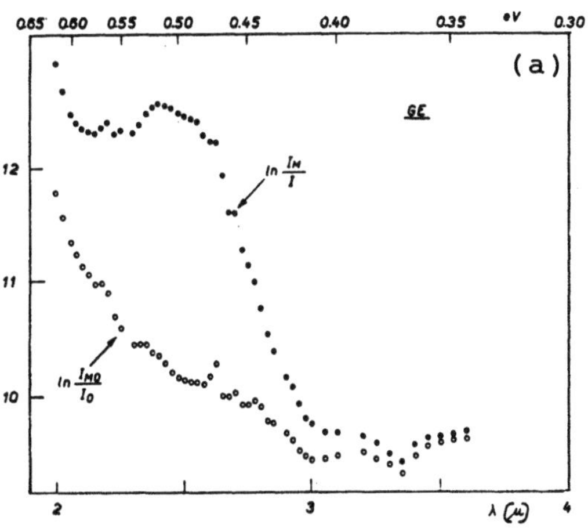

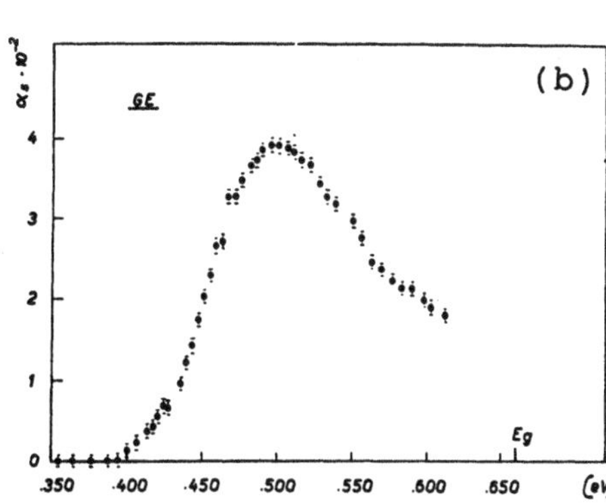

Fig.2.1.1 Natural logarithm of the ratio I_0/I as a function of wavelength for a cleaved surface of Ge and for the same surface after oxidation (a); absorption constant vs photon energy for a surface of Ge cleaved in UHV (b) (from reference 5).

change in surface dielectric function brought about by the saturation of dangling bonds. The difference is reported in fig.2.1.1,b and has been interpreted as a surface state to surface state (of dangling bond type) optical transition /5/. Similar results have been obtained in Si(1111)2x1, showing that in both cases an optical gap of the order of 0.5 eV exists between filled and empty surface states. In GaAs(110) however no such optical transitions have been detected with this method, indicating that the energy separation of surface states in GaAs(110) is larger than the fundamental gap /8/. This result also points out an intrinsic limitation of MITR, since this technique is obviously feasible only in the range of transparency of the material. This restraint has been successfully overcome by using external reflectivity configuration. After checking the equivalence of the two methods at photon energies smaller than the bandgap /7/, external reflectivity has been applied to the study of surface state transitions above the energy gap of covalent as well as 3-5 semiconductors /9,10/. Another advantage of external vs internal reflection is the viability of polarized light (in MITR this possibility is hampered by depolarization of the evanescent wave). The SDR spectra of single domain Si(111)2x1 both below and above the bandgap as obtained at normal (external) incidence with polarized light are shown in fig.2.1.2 /11/.

Surface atoms in Si(111)2x1 are arranged in a chain-like structure, according to the π-bonded chain model developed by Pandey /12/. The strong anisotropy of the 0.45 eV peak in fig.2.1.2 is related to the almost one-dimensional nature of surface (dangling-bond) bands /13,14/. This finding constitutes one of the soundest experimental evidence of Pandey's model and it is an example of structural information inferred from optical techniques, using symmetry arguments /15/.

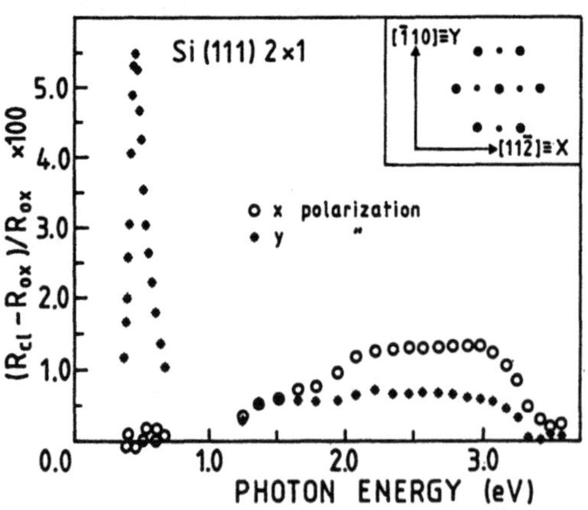

Fig.2.1.2 Differential reflectivity spectra of a Si(111)2x1 single domain surface. Light was polarized along directions parallel (asterisks) and perpendicular (open circles) to the chains of Pandey's model (from reference 11).

Recent progress in theoretical calculations of surface contribution to reflectivity has shown that the interpretation of SDR spectra above the bandgap of a semiconductor, where the substrate is absorbing, is fairly intriguing /16,17,18/. For instance, the broad band at photon energies higher than 1 eV in fig.2.1.2 is attributable not only to transitions between surface states but also to bulk-like transitions

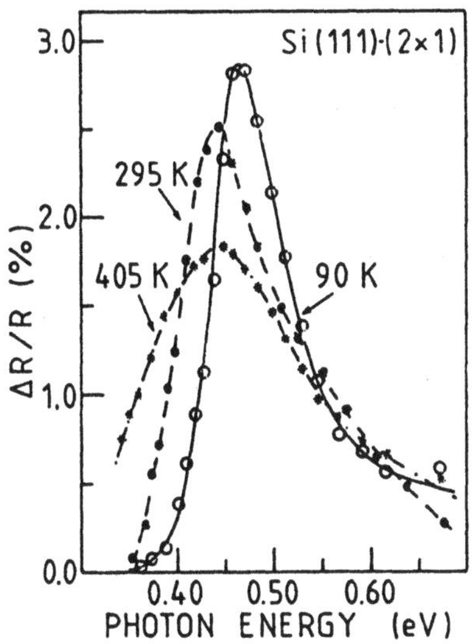

Fig.2.1.3 Differential reflectivity as a function of the photon energy in Si(111)2x1 at three different sample temperatures (from reference 20).

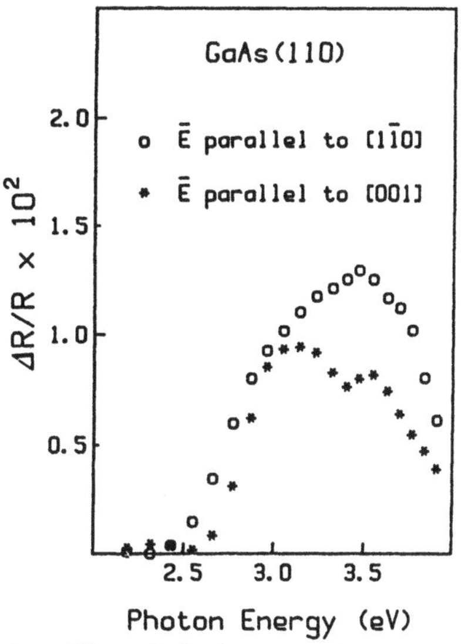

Fig 2.1.4 Surface Differential Reflectivity vs photon energy for a cleaved GaAs(110) surface. Light is polarized along [001] (asterisks) and [1$\bar{1}$0] (circles) directions (from reference 21).

"modulated" by the surface /19/. These effects are beyond the simple three-media classical analysis previously outlined and will be further discussed in the following section. The temperature-dependence of the SDR peak at 0.45 eV in Si(111)2x1 has been recently studied /20/. The result, shown in fig.2.1.3, indicates that the temperature-dependence of these optical transitions between surface states bears much analogy with the behaviour of localized centers in solids, thereby suggesting a strong electron - phonon interaction for dangling-bonds in Si(111)2x1 /20/.

Experiments of SDR above the bandgaps of GaAs(110) and GaP(110) have shown two broad bands at about 3 and 3.5 eV, respectively, that are certainly related to surface state transitions /10/. Unlike the case of Si(111)2x1, the polarization-dependence of these bands is weak /21/, as shown in fig.2.1.4 for GaAs. The origin of the small anisotropy is still uncertain. Recently a similar anisotropy has been observed in chemically etched GaAs(110) surfaces by Aspnes /22/ and it has been attributed to surface-induced bulk-like optical processes. On the other hand Safarov and coworkers have interpreted two peaks at 2.8 and 3.1 eV observed in their experiments on GaAs(110) as due to surface states and bulk-like transitions respectively /23/. Both Aspnes and Safarov make use of an experimental method different from SDR, as previously described, in the sense that they obtain ΔR by changing the polarization vector in the surface plane. No similar data are available for GaP(110) yet.

As already mentioned, the interpretation of SDR with absorbing substrates is complicated by the possible occurrence of a number of surface-related effects other than transitions between surface states.

This is the main reason why so far SDR has been successful with semiconductors and much less so with metals /24/. However the situation is rapidly changing owing to new advances in theoretical analysis and computational methods /17,19,25/.

2.2 Surface Dielectric Response: Theoretical Aspects

The dielectric function of an infinite crystal can be calculated starting from a knowledge of the electronic bands, with some approximate method, for instance the Random Phase Approximation /26/. Local field effects /27/ might also be accounted for within this approximation /28/. Usually exchange annd correlation effects are neglected. In spite of this approximation such a kind of calculation is indeed much time-consuming. Therefore some drastic approximations are often introduced, like for instance considering a homogeneous electron gas (jellium) or neglecting local field effects. The presence of the surface renders this computation a formidable task, because in this case the usual approximations are not justified, generally speaking. Actually the termination of the crystal brings about many conceptual difficulties, namely:

i) non-local response (spatial dispersion).

If we refer to the equation defining the dielectric tensor (within linear response theory and after time Fourier-transformation)

$$D_{\omega}(r) = \int d r' \, \varepsilon_{\omega}(r,r') E_{\omega}(r'), \qquad (2.2.1)$$

local response means that $\varepsilon(r,r')$ can be written as $\varepsilon(r')\delta(r-r')$, implying that $E(r')$ varies slowly in the region where $\varepsilon(r,r')\neq 0$. This is by no means the case for the normal component of E, which is classically discontinuous at the surface. In a crystal, nonlocality (in r-space) is equivalent to spatial dispersion (in k-space). A crystal being translationally invariant, $\varepsilon(r,r')$ takes the form $\varepsilon(r-r')$ and the convolution theorem ensures that

$$D_{\omega}(k) = \varepsilon_{\omega}(k) \; E_{\omega}(k). \qquad (2.2.2)$$

Spatial dispersion means that the dielectric function is a function of the wave vector k /29/.

ii) inhomogeneity (local field effects).

The previous condition $\varepsilon(r,r') = \varepsilon(r-r')$ does not apply to the z component of the dielectric function near the surface (z being the direction perpendicular to the surface), i.e.

$$\varepsilon(z,z') \neq \varepsilon(z-z') \qquad (2.2.3)$$

because of the rapid variations of the induced electric field ($E_{ind} = E - E_{ext}$) at the surface, even on the scale of the unit cell /30/. Consequently the great advantage of the convolution theorem (an algebraic instead of an integral relationship between D and E) cannot be exploited and the dependence of ε upon z and z' must be retained. In-homogeneity implies that local-field effects (LFE) are important at the surface. LFE's arise from the fact that the microscopic electric field differs from the macroscopic field (average of the microscopic field) /27/. This is due to the inhomogeneity of the electron distribution, which is expected to be larger at the surface than in the bulk /31/.

iii) anisotropy.
Even when the ε tensor is diagonal in the bulk, we do expect the off-diagonal terms to be nonzero near the surface, in particular those ones coupling the z direction to the directions parallel to the surface.

The combined effect of non-locality, in-homogeneity and anisotropy is such that the equation of light propagation for a finite crystal (obtained from Maxwell equations) is actually a set of three coupled integro-differential equations /32/. Every theory of the surface dielectric response can be looked at as a particular approximate solution of these equations. For instance the Fresnel solution represents the simplest case:

$$\varepsilon(r,r') = \delta_{\alpha\beta}\delta(r-r')\left[1-\theta(z)(\varepsilon_b-1)\right] \tag{2.2.4}$$

($\alpha,\beta=x,y,z$) where $\theta(z)$ is the step function and ε_b is the bulk dielectric function. The latter is clearly assumed to be local, homogeneous, isotropic and terminated in a steplike way at the surface. In this case the equations can be solved and the Fresnel coefficient of reflectivity is obtained. However the presence of the surface is not accounted for at all.

The model of McIntyre and Aspnes (MA) /2/ represents a step forward. By considering three classical Fresnel-like media (instead of two) within linear approximation, they obtain formulas for differential reflectivity having a simple physical interpretation. These formulas qualitatively account for some experimental results /7/. Like Fresnel theory, this model avoids all complications brought about by the surface.

Subsequently Nakayama has developed a theory of the surface dielectric response including all surface peculiarities /33/. In this case the solution is obtained only in the perturbative limit, which is not justified in the case of p-polarization. For the s-wave Nakayama's results are correct and confirm the MA approach.

Feibelman has found an exact solution of the surface optical response in the case of jellium /34/. He has confirmed the validity of MA results for the s-wave, while reaching no conclusion for the p-wave.

More recently Bagchi and coworkers have elegantly solved the integro-differential equations by using E_x, E_y and D_z (instead of E_z) as variables /35/. D_z is continuous at the surface and this allows one to put the equations in local form and use the convolution theorem. Then the solution is straightforward and the result justifies the structure of MA formulas both for s and p polarizations /35/.

The MA model is useful because it often allows one to physically interpret the results of SDR experiments; however it provides no help in realistic calculations of the surface optical response. The latter is certainly a difficult task: first of all one needs a solution of the electronic problem in the presence of a surface (surface electronic structure) and then use it in time-consuming calculations.

A great computational effort is now being developed, yielding increasingly sophisticated theories. However, it does not seem possible yet to fully account for all aspects of the surface dielectric response. For instance, most recent calculations either neglect LFE /19,36/ or treat it in a simplified way, assuming that the atomic polarizability at the surface is the same as in the bulk /37/. Nevertheless these calculations have improved our understanding of the surface optical properties significantly.

Finally, an interesting implication of the theory of surface dielectric response regards the surface photoelectric effect /30/ and in general the consequences of spatial dispersion in metal optics near the plasma frequency /38/. This issue, while being conceptually relevant especially in photoemission spectroscopy, is beyond the scope of the present notes.

3. Second-Harmonic Generation

Nonlinear optical effects, in particular Second-Harmonic Generation (SHG) have been used recently as a new probe of surface properties /39/. Generally speaking, SHG is ascribed to the second order nonlinear electrical susceptibility χ_2, defined as

$$P(r,t) = \text{first order term} + \int \chi_2(r-r_1,t-t_1;r-r_2,t-t_2): E(r_1,t_1) \times E(r_2,t_2)dr_1dr_2dt_1dt_2 + \text{higher order terms} \quad (3.1)$$

where P is the polarization vector and E is the macroscopic electric vector appearing in Maxwell equations.

The viability of SHG as a surface probe is based on the following two considerations. Firstly, the necessary surface to bulk contrast is easily achieved in media possessing inversion symmetry since in this case (considering only electric dipole processes) χ_2 vanishes everywhere but in the proximity of the surface, which is obviously not centrosymmetric /40/. The second reason is that submonolayer surface sensitivity can be obtained in SHG by using pulsed lasers /41/.

SHG shows promise as a tool for studying interfaces, as well as free surfaces both clean and covered with adsorbates. A sketch of a SHG process is shown in fig.3.1. The thickness d of the surface layer

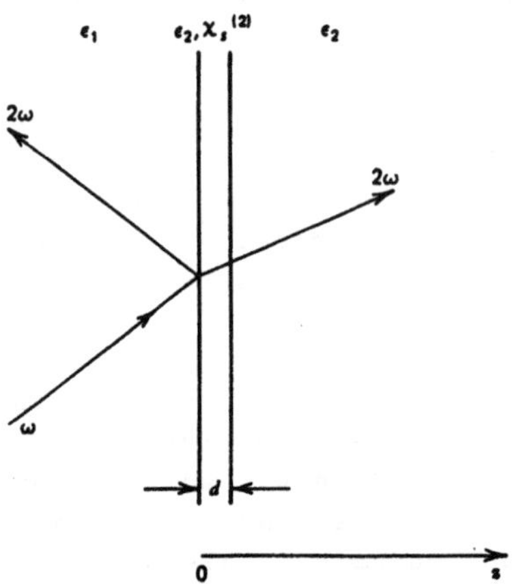

Fig.3.1 Sketch of second-harmonic generation from an interface between two isotropic media. The interfacial layer is specified by a linear dielectric constant ε_2 and a second order surface nonlinear susceptibility χ_2 (from reference 39).

with nonlinear susceptibility χ_2 is of the order of the interatomic distance. Provided very short laser pulses are used, SHG can yield information about dynamical processes, for instance vibrational transitions of adsorbates /39/. An example of SHG applied to surface studies is shown in fig.3.2, in which signals measured from Si(111) surfaces exhibiting 2x1 and 7x7 reconstructions are compared /42/. The existence of a mirror plane in the former surface and the threefold symmetry of the latter surface have been accurately confirmed /42/.

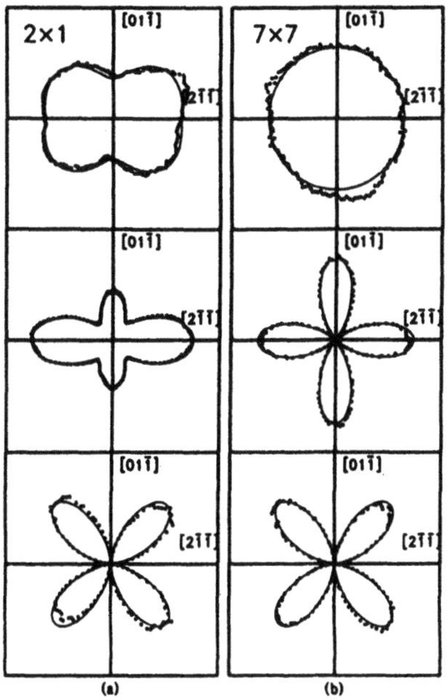

Fig.3.2 Second-harmonic intensity from (a) Si(111)2x1 and (b) Si(111)7x7 surfaces as a function of the polarization of the normally incident pump beam. The top panels display the total SHG signal; the middle and lower panels show, respectively, the SHG signal polarized along the $[2\bar{1}\bar{1}]$ and $[01\bar{1}]$ directions. The dotted curves represent the experimental data, while the solid curves show the result of a fit (from reference 42).

4. Photothermal Displacement Spectroscopy

Photothermal Displacement Spectroscopy (PDS) is a non-conventional way of measuring absorption coefficients in solids /43/. This relatively new technique exploits the thermal expansion of a sample heated upon light absorption. Clearly the theory of PDS involves the solutions of both the heat equation and light propagation in an inhomogeneous medium /43/. Therefore PDS yields information on optical properties in an indirect way. In principle both surface and bulk properties are explored, however under special circumstances the surface contribution can be enhanced /44/. An advantage of PDS with respect to SDR is the possibility of obtaining information on the surface electronic structure just from a measurement on the clean surface, without resorting to contamination.

A sketch of a typical set-up for PDS measurement is reported in fig.4.1. Usually two laser beams are employed: the first one creates a "bump" on the sample (pump beam) while the second one is deflected by the bump (probe beam).

One of the most important applications of PDS to surface science has been the study of Si(111)2x1 and Ge(111)2x1 with polarized light at photon energies below the gap. In Si(111)2x1 /45/ the results are

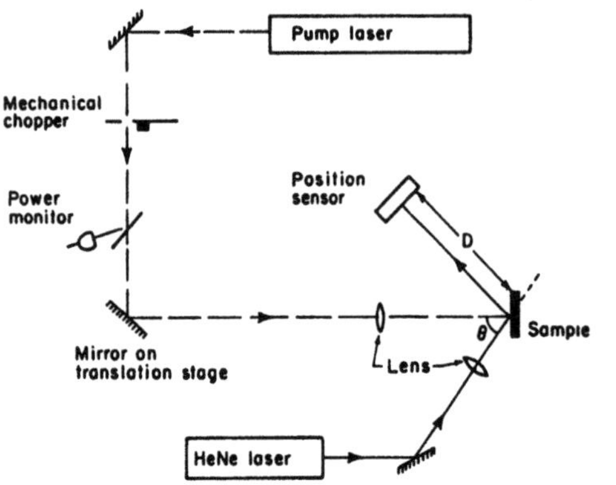

Fig.4.1 Experimental configuration of a photothermal displacement apparatus (from reference 44).

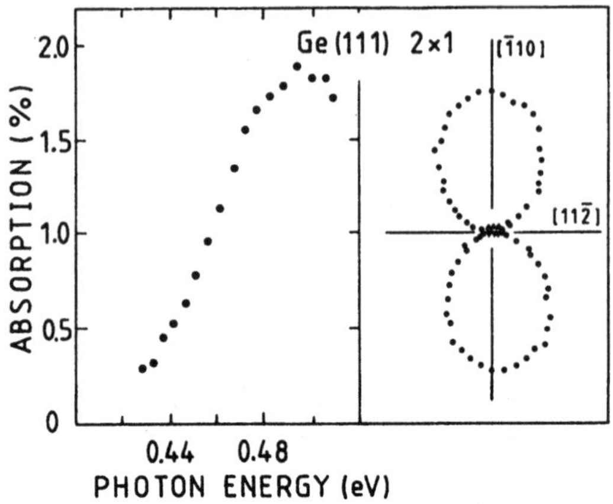

Fig.4.2 Surface state absorption spectrum of Ge(111)2x1 with incident light polarized along the [1̄10] direction (left). Polar plot of the polarization dependence at 0.496 eV (right). Data are taken from reference 46.

practically coincident with those obtained with SDR /13/ thereby strengthening the conclusions drawn from these experiments /15/. Similar results have been obtained with PDS also in Ge(111)2x1 /46/. They are reported in fig.4.2 and clearly show the strong anisotropy of the 0.5 eV peak in surface absorption.

5. Acknowledgements

The experimental work forming the basis of the present notes has been done in the group of Prof.Chiarotti in Frascati. Useful conversations with R.Del Sole are gratefully acknowledged.

6. References

/1/ S.Selci, F.Ciccacci, G.Chiarotti, P.Chiaradia A.Cricenti: J.Vac.Sci.Technol., in press

/2/ J.D.E.McIntyre, D.E.Aspnes: Surf.Sci. 24, 417 (1971)

/3/ P.Drude: In The Theory of Optics (Longmans, Green and Co, London, 1920); O.S.Heavens: In Optical Properties of Thin Films (Butterworths, London, 1955)

/4/ H.Ibach, J.E.Rowe: Phys.Rev. B9, 1951 (1974)

/5/ G.Chiarotti, S.Nannarone, R.Pastore, P.Chiaradia: Phys.Rev. B4, 3398 (1971)

/6/ I.Bartos: Phys.Stat.Sol. 33, 779 (1979)

/7/ P.Chiaradia, G.Chiarotti, S.Nannarone, P.Sassaroli: Sol.St.Comm. 26, 813 (1978)

/8/ P.Chiaradia, G.Chiarotti, I.Davoli, S.Nannarone, P.Sassaroli: Proc. 14th Intern. Conf. on the Phys. of Semic., Edinburgh, 1978, in Physics Conf. Series No 43 (1979) p.195

/9/ S.Nannarone, P.Chiaradia, F.Ciccacci, R.Memeo, P.Sassaroli, S.Selci, G.Chiarotti: Sol.St.Comm. 33, 593 (1980)
/10/ P.Chiaradia, G.Chiarotti, F.Ciccacci, R.Memeo, S.Nannarone, P.Sassaroli, S.Selci: Surf.Sci 99, 70 (1980)
/11/ S.Selci, P.Chiaradia, F.Ciccacci, A.Cricenti, N.Sparvieri, G.Chiarotti: Phys.Rev. B31, 4096 (1985)
/12/ K.C.Pandey: Phys.Rev.Lett. 47, 1913 (1981)
/13/ P.Chiaradia, A.Cricenti, S.Selci, G.Chiarotti: Phys.Rev.Lett. 52,1145 (1984)
/14/ R.Del Sole, A.Selloni: Sol.St.Comm. 50, 825 (1984)
/15/ P.Chiaradia, A.Cricenti, G.Chiarotti, F.Ciccacci, S.Selci: In The Structure of Solids, ed. by M.A.Van Hove and S.Y.Tong, Springer Series in Surface Sciences 2 (Springer, 1985)
/16/ R.Del Sole: J.Phys.C, 8, 2971 (1975)
/17/ L.Mochan, R.G.Barrera: Phys.Rev. B23, 5707 (1981) and B32, 4984 (1985)
/18/ D.E.Aspnes, A.Studna: Phys.Rev.Lett., 54, 1956 (1985); D.E.Aspnes: J.Vac.Sci.Technol. B3, 1498 (1985)
/19/ A.Selloni, P.Marsella, R.Del Sole: Phys.Rev. B33, 8885 (1986)
/20/ F.Ciccacci, S.Selci, G.Chiarotti, P.Chiaradia: Phys.Rev.Lett. 56, 2411 (1986)
/21/ S.Selci, F.Ciccacci, A.Cricenti, A.C.Felici, C.Goletti, P.Chiaradia: Sol.St.Comm., in press
/22/ D.E.Aspnes: J.Vac.Sci.Technol., in press
/23/ V.L.Berkovits, I.V.Makarenko, T.A.Minashvili, V.I.Safarov: Sol.St.Comm. 56, 449 (1985)
/24/ R.G.Greenler: J.Chem.Phys. 44, 310 (1966); G.W.Rubloff, J.L.Freeouf: Phys.Rev. B17, 4680 (1978); J.B.Restorff, H.D.Drew: Surf.Sci. 88, 399 (1979)
/25/ W.L.Mochan, R.G.Barrera: Phys.Rev. B32, 4989 (1985)
/26/ H.Ehrenreich, M.H.Cohen: Phys.Rev. 115, 786 (1959)
/27/ C.Kittel: Introduction to Solid State Physics, 3rd ed. (J.Wiley, New York, 1968), ch.12
/28/ S.Adler: Phys.Rev. 126,413 (1962); N.Wiser: Phys.Rev. 129, 62 (1963)
/29/ V.M.Agranovich, V.L.Ginzburg: Spatial Dispersion in Crystal Optics and the Theory of Excitons, ed. by R.E.Marshak (J.Wiley, New York, 1966)
/30/ P.J.Feibelman: Prog.Surf.Sci. 12, 287 (1982)
/31/ R.Del Sole, E.Fiorino: Phys.Rev. B29, 4631 (1984)
/32/ R.Del Sole: J.Phys C 8, 2971 (1975)
/33/ M.Nakayama: J.Phys.Soc.Japan 39, 265 (1975)
/34/ P.J.Feibelman: Phys.Rev. B14, 762 (1976)
/35/ A.Bagchi, A.K.Rajagopal: Sol.St.Comm. 31, 127 (1979) and A.Bagchi, R.G.Barrera, A.K.Rajagopal: Phys.Rev. B20, 4824 (1979)
/36/ A.Selloni, R.Del Sole: Surf.Sci. 168, 35 (1986)
/37/ W.L.Mochan, R.G.Barrera: Phys.Rev.Lett. 56, 2221 (1986)
/38/ F.Forstmann, R.R.Gerhardts: Festkorperprobleme 22 291 (1982)
/39/ Y.R.Shen: In The Principles of Nonlinear Optics (J.Wiley, New York, 1984) ch.25
/40/ reference 39, p.28
/41/ Y.R.Shen: J.Vac.Sci.Technol. B3, 1464 (1985)

/42/ T.F.Heinz, M.M.T.Loy, W.A.Thompson: Phys.Rev.Lett. 54, 63 (1985)
/43/ W.B.Jackson, N.M.Amer, A.C.Boccara, D.Fournier: Appl.Optics, 20, 1333 (1981)
/44/ M.A.Olmstead, N.M.Amer, S.Kohn, D.Fournier, A.C.Boccara: Appl.Phys. A32, 141 (1983) and N.M.Amer, M.A.Olmstead: J.Vac.Sci.Technol. B1, 751 (1983)
/45/ M.A.Olmstead, N.M.Amer: Phys.Rev.Lett. 52, 1148 (1984)
/46/ M.A.Olmstead, N.M.Amer: Phys.Rev. B29, 7048 (1984)

Vibrational Properties at Semiconductor Surfaces and Interfaces

Y.J. Chabal

AT&T Bell Laboratories, Murray Hill, NJ 07974, USA

1. INTRODUCTION AND MOTIVATION

The understanding and control of the formation and properties of semiconductor interfaces require the knowledge of surface *chemistry*, role of defects, adsorbate *geometry*, *dynamics* and *kinetics*. Vibrational spectroscopy, a direct probe of the chemical nature of the surface, can also give valuable information on the geometry and dynamics of adsorbates. Furthermore, it can indirectly probe defects "decorated" by probe atoms such as hydrogen. Finally, recent time-resolved vibrational data [1-3] show that kinetic phenomena can also be studied by this spectroscopy. Therefore, vibrational spectroscopy is well suited to address the fundamental questions of semiconductor interfaces.

Among the available techniques such as Electron Energy Loss Spectroscopy (EELS), Inelastic Atom Beam Scattering (IABS) and Raman scattering, surface infrared spectroscopy (SIRS) is particularly powerful because of its *high resolution, polarization* properties and sensitivity to high frequency modes ($>1000cm^{-1}$). At semiconductor interfaces, the vibrational lines are very sharp ($\sim 1cm^{-1}$) and cannot be resolved by EELS. For light adsorbates, the modes are too high in frequency to be probed by IABS and too weak Raman scatterers to be detected by Raman spectroscopy. Furthermore, the advances of fast scanning interferometers, IR detectors and computers have made SIRS a relatively straightforward technique. As a result, a number of groups are now actively using SIRS to probe semiconductor interfaces. The basic design of experimental set ups is simple and has been well described elsewhere [4,5].

The purpose of this lecture is to give the basics of SIRS as it applies to semiconductor interfaces, rather than an exhaustive review of all the work done in the field [6]. We focus on the study of semiconductor/vacuum interfaces with monolayer or submonolayer coverage of simple adsorbates. In section 2, explicit expressions are derived in terms of the macroscopic dielectric response of the various media to calculate the electric fields and the optical absorption of monolayers. The absorption is then related to the microscopic quantities of the adsorbate monolayer and various simplifying assumptions are discussed. The purpose of this exercise is to give a quantitative basis for the choice of experimental geometries and parameters as presented in section 3, and to provide a framework for quantitative analysis of the data as discussed in section 4. Section 3 spells out, for instance, various advantages of multiple internal reflection (MIR), a technique pioneered by HARRICK in the sixties [7,8]. Among the less obvious attributes of MIR is its ability to probe vibrational modes polarized parallel to the surface with as much sensitivity as modes polarized perpendicular to the surface. Section 4 presents examples selected mostly for their tutorial aspect. The first deals with an experimentally very well-defined system, i.e. the most stable hydride phase on Si(100), for which a rigorous and quantitative analysis of the data can be performed. Based on the knowledge extracted about this phase, semiquantitative information can then be obtained on all of the hydride phases on Si(100) and Ge(100), and interesting dynamics inferred about this stable phase. Finally, water decomposition on Si(100) is summarized as an instructive example of simple chemistry at surfaces. From these illustrations, a quantitative assessment of the technique should result. Section 5 deals more directly with the limitations of the technique and addresses the areas of expected growth such as *kinetic, in situ* measurements in conventional and laser-induced chemical vapor deposition (CVD) processes.

2. PHOTON INTERACTION WITH SEMICONDUCTOR INTERFACES

For a quantitative analysis of optical reflection spectra of adsorbates on a surface, a microscopic theory of the surface optical response must be worked out. Since a first principle treatment is generally not possible, a simple three-layer model with step function changes of dielectric constant has been used to approximate the vacuum/adsorbate layer/substrate system and to deduce the field strengths at the adsorbate position. In this section, the results for a thin adsorbate layer with anisotropic response are first presented. A simple Lorentz oscillator model is then used to relate the measured reflectance to microscopic quantities such as adsorbate oscillator effective charges. Finally, the assumption of sharp boundary conditions (step function change of dielectric constants) is discussed, particularly as it relates to non-absorbing semiconductor/vacuum interfaces.

2.1 Macroscopic Theory (3 layer model)

The model used is shown in Fig. 1. To make it more concrete and relevant to the experimental geometries used most commonly, a beam traveling in a semiconductor substrate is taken incident in the x-z plane onto the substrate/vacuum interface separated by a thin ($d << \lambda$) adsorbate layer. Both the substrate and vacuum are characterized by real dielectric constants, ϵ^s and ϵ^v respectively. The active adsorbate layer is characterized by a complex dielectric constant $\tilde{\epsilon}$. The present derivation follows the work of McINTYRE and ASPNES [9,10]. Since we are interested in submonolayer coverages of atoms or molecules oriented on the semiconductor surface, the active adsorbate layer is taken to be anisotropic but with the principal dielectric axes along x,y,z. The dielectric tensor is therefore diagonal with principal values $\tilde{\epsilon}_x,\tilde{\epsilon}_y,\tilde{\epsilon}_z$. For non-magnetic media, the relationship between the electric and magnetic fields and induction is [11]:

$$D_i = \tilde{\epsilon}_i E_i \quad \text{where} \quad i = x,y,z \tag{2.1}$$

$$\vec{B} = \vec{H} \tag{2.2}$$

From Maxwell's equations for media with no free charges:

$$\vec{\nabla} . \vec{D} = 0 \tag{2.3}$$

$$\vec{\nabla} . \vec{B} = 0 \tag{2.4}$$

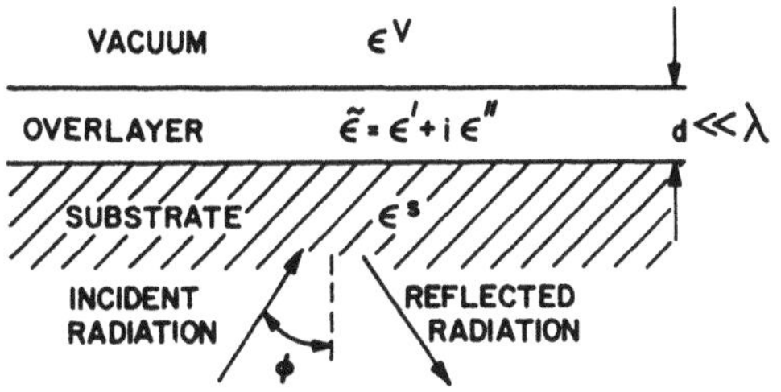

Fig. 1: Model used to describe the response of a thin adsorbate layer at a semiconductor/vacuum interface. The radiation is shown to be incident from the substrate side as is the case for total *internal* reflection spectroscopy. Note that to describe an *external* reflection geometry, the substrate is interchanged with vacuum (ϵ^s=1) and the vacuum interchanged with substrate.

$$\vec{\nabla} \times \vec{E} + \frac{1}{c}\frac{\partial \vec{B}}{\partial t} = 0 \tag{2.5}$$

$$\vec{\nabla} \times \vec{H} - \frac{1}{c}\frac{\partial \vec{D}}{\partial t} = 0 \tag{2.6}$$

the wave equation is obtained, a solution of which can be written:

$$\vec{E}(\vec{r},t) = \vec{E}^o e^{i\vec{k}\cdot\vec{r} - i\omega t} \tag{2.7}$$

From eqs. (2.5) and (2.6) we have:

$$(\frac{\omega}{c})\vec{B} = \vec{k} \times \vec{E} \quad \text{and} \quad (\frac{\omega}{c})\vec{D} = -\vec{k} \times \vec{H} \tag{2.8}$$

Hence, $\vec{k}$, $\vec{D}$ and $\vec{H}$ (or $\vec{B}$) are mutually orthogonal in all media. But $\vec{E}$ is not orthogonal to k and H in general. We define the vector

$$\vec{n} \equiv (\frac{c}{\omega})\,\vec{k} \tag{2.9}$$

the magnitude of which depends on the direction of propagation and on the polarization.

For the calculation, we consider s-polarized radiation characterized by $\vec{E} = E\hat{y}$ and p-polarized radiation characterized by $H = H\hat{y}$. For these two cases, the wave equations are, using eqs. (2.1), (2.2), (2.5) and (2.6):

$$(k_x^{j\,2} + k_z^{j\,2})E_y^j = (\frac{\omega}{c})^2 \epsilon_y^j E_y^j \qquad \text{for s-polarization} \tag{2.10}$$

and

$$(\frac{k_x^{j\,2}}{\epsilon_z} + \frac{k_z^{j\,2}}{\epsilon_x})H_y^j = (\frac{\omega}{c})^2 H_y^j \qquad \text{for p-polarization} \tag{2.11}$$

where j refers to the medium j = s, v or overlayer (no superscript). The existence of boundary conditions at each interface requires that the component of the wave vector parallel to the surface, k_x^j, be conserved, i.e. $k_x^s = k_x^v = k_x$. From our definition of ϕ (see Fig. 2.1) and from eq. (2.9), we have:

$$k_x \equiv (\frac{\omega}{c})n_x = (\frac{\omega}{c})\sqrt{\epsilon^s}\sin\phi \tag{2.12}$$

We are now ready to consider the cases of s-polarization and p-polarization.

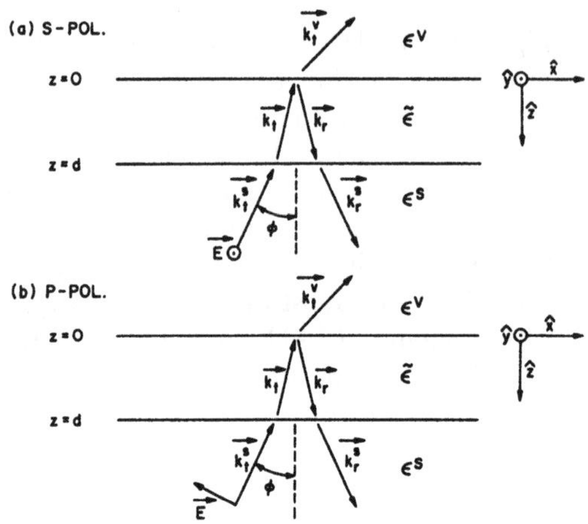

Fig. 2: Schematic representation of the electric field polarization and of the propagation vectors in the various layers for (a) s-polarization, and (b) p-polarization. The cartesian axes are defined with z=0 at the vacuum/overlayer interface and (xz) the plane of incidence. The subscripts t and r refer to transmitted and reflected beams. $\hat{x}$, $\hat{y}$ and $\hat{z}$ are unit vectors.

2.1.1 S-Polarization

Fig. 2(a) shows the geometry for s-polarization. In each layer j, the perpendicular component of the wave vector is obtained from eqs. (2.10) and (2.12):

$$\left(\frac{c}{\omega}\right)k_z^j \equiv n_z^j = (\epsilon_y^j - \epsilon^s \sin^2\phi)^{1/2} \tag{2.13}$$

We can proceed simply without resorting to matrix notation [12,13] by starting in the last medium, vacuum, where there is no reflected beam r. The wave vector of the transmitted beam is

$$\vec{k}_t^v = -k_z^v\hat{z} + k_x\hat{x}$$

In the second medium, the active layer, the wave vectors are:

$$\vec{k}_t = -k_z\hat{z} + k_x\hat{x}.$$

$$\vec{k}_r = k_z\hat{z} + k_x\hat{x}$$

and the fields can be written using eq. (2.8) to calculate H:

$$\vec{E}^v(\vec{r}) = \left[E_t^v e^{-ik_z^v z + ik_x x}\right]\hat{y}$$

$$\vec{H}^v(\vec{r}) = E_t^v(n_z^v\hat{x} + n_x\hat{z})e^{-ik_z^v z + ik_x x}$$

in vacuum, and:

$$\vec{E}(\vec{r}) = \left[E_t e^{-ik_z z + ik_x x} + E_r e^{ik_z z + ik_x x}\right]\hat{y}$$

$$\vec{H}(\vec{r}) = E_t(n_z\hat{x} + n_x\hat{z})e^{-ik_z z + ik_x x} + E_r(-n_z\hat{x} + n_x\hat{z})e^{ik_z z + ik_x x}$$

in the overlayer. Note that we use no superscripts for the overlayer for simplicity. The boundary conditions at z = 0 (vacuum/overlayer interface) are:

tang. E: $E_t^v = E_t + E_r$ (2.14a)

tang. H: $n_z^v E_t^v = n_z(E_t - E_r)$ (2.14b)

normal D: $0 = 0$ (2.14c)

normal B: $E_t^v = E_t + E_r$ (2.14d)

From eqs. (2.14) we get:

$$E_t = \frac{1}{2}(1 + n_z^v/n_z)E_t^v \tag{2.15a}$$

$$E_r = \frac{1}{2}(1 - n_z^v/n_z)E_t^v \tag{2.15b}$$

We now propagate the fields in the overlayer from $z=0$ to $z=d$ (overlayer/substrate interface) and make use of the fact that $d << \lambda$, i.e. $e^{\pm ik_z d} \approx 1 \pm ik_z d$. The fields can be written, using eqs. (2.7), (2.8) and (2.15):

$$\vec{E}(x\hat{x}+d\hat{z}) = E_t^v\left[\frac{1}{2}(1+n_z^v/n_z)(1-ik_z d)e^{ik_x x} + \frac{1}{2}(1-n_z^v/n_z)(1+ik_z d)e^{ik_x x}\right]\hat{y}$$

$$= E_t^v(1 - ik_z d n_z^v/n_z)e^{ik_x x}\hat{y}$$

$$\vec{H}(x\hat{x}+d\hat{z}) = \frac{1}{2}E_t^v\left[(1+n_z^v/n_z)(n_z\hat{x}+n_x\hat{z})(1-ik_z d) + (1-n_z^v/n_z)(-n_z\hat{x}+n_x\hat{z})(1+ik_z d)\right]e^{ik_x x}$$

$$= E_t^v\left[n_z(n_z^v/n_z - ik_z d)\hat{x} + n_x(1 - ik_z d n_z^v/n_z)\hat{y}\right]e^{ik_x x}$$

in the overlayer and:

$$\vec{E}^s(\vec{r}) = \left[E_t^s e^{-ik_z^s(z-d)+ik_x x} + E_r^s e^{ik_z^s(z-d)+ik_x x}\right]\hat{y}$$

$$\vec{H}^s(\vec{r}) = E_t^s(n_z^s\hat{x}+n_x\hat{z})e^{-ik_z^s(z-d)+ik_x x} + E_r^s(-n_z^s\hat{x}+n_x\hat{z})e^{ik_z^s(z-d)+ik_x x}$$

in the substrate. The boundary conditions at $z=d$ (overlayer/substrate interface) are:

tang. E: $E_t^v(1 - ik_z d n_z^v/n_z) = E_t^s + E_r^s$ (2.16a)

tang. H: $E_t^v n_z(n_z^v/n_z - ik_z d) = n_z^s(E_t^s - E_r^s)$ (2.16b)

norm. D: $0 = 0$ (2.16c)

norm. B: $E_t^v n_x(1 - ik_z d n_z^v/n_z) = n_x(E_t^s + E_r^s)$ (2.16d)

From eqs. (2.16) we obtain:

$$E_t^s = \frac{1}{2} E_t^v \left[1 + \frac{n_z^v}{n_z^s} - ik_z d \left(\frac{n_z^v}{n_z} + \frac{n_z}{n_z^s} \right) \right]$$

$$E_r^s = \frac{1}{2} E_t^v \left[1 - \frac{n_z^v}{n_z^s} - ik_z d \left(\frac{n_z^v}{n_z} - \frac{n_z}{n_z^s} \right) \right]$$

After some algebra, the ratio $\tilde{r} \equiv E_r^s / E_t^s$ can be obtained, again assuming $d << \lambda$:

$$\tilde{r} \equiv \frac{E_r^s}{E_t^s} = \left(\frac{n_z^s - n_z^v}{n_z^s + n_z^v} \right) \left[1 + 2ik_z d \frac{n_z^s (n_z^{v2} - n_z^2)}{n_z (n_z^{v2} - n_z^{s2})} \right] \tag{2.17}$$

Using eqs. (2.12) and (2.13), the above ratio becomes:

$$\tilde{r} = \tilde{r}^o \left(1 + 2i \frac{\omega}{c} d \sqrt{\epsilon^s} \cos\phi \frac{\epsilon_y^v - \tilde{\epsilon}_y}{\epsilon_y^v - \epsilon^s} \right) \tag{2.18}$$

where

$$\tilde{r}^o = \frac{\sqrt{\epsilon^s} \cos\phi - \sqrt{\epsilon^v - \epsilon^s \sin^2\phi}}{\sqrt{\epsilon^s} \cos\phi + \sqrt{\epsilon^v - \epsilon^s \sin^2\phi}} \tag{2.18a}$$

is the field ratio for the vacuum/substrate interface. Note that in (2.18) the y subscript of ϵ^s has been dropped since the substrate is assumed isotropic. The measured change in reflectance is:

$$\frac{\Delta R}{R}\Big|_{S-pol} \equiv 1 - \left| \frac{\tilde{r}}{\tilde{r}^o} \right|^2 = 4 \frac{\omega}{c} d \sqrt{\epsilon^s} \cos\phi \ \mathrm{Im} \left[\frac{\epsilon_y^v - \tilde{\epsilon}_y}{\epsilon_y^v - \epsilon^s} \right] \tag{2.19}$$

Using $\tilde{\epsilon}_y \equiv \epsilon_y' + i\epsilon_y''$ and $\omega = 2\pi c\tilde{\nu}$, eq. (2.19) becomes:

$$\frac{\Delta R}{R}\Big|_{S-pol} = 8\pi\tilde{\nu}\sqrt{\epsilon^s} d \cos\phi \left(\frac{\epsilon_y''}{\epsilon^s - \epsilon_y^v} \right) \tag{2.20}$$

2.1.2 P-Polarization

Fig. 2(b) shows the geometry for p-polarization. In each layer, j, the perpendicular component of the wave vector is obtained from eqs. (2.9), (2.11) and (2.12):

$$\left(\frac{c}{\omega} \right) k_z^j \equiv n_z^j = \left[\epsilon_x^j \left(1 - \frac{\epsilon^s}{\epsilon_z^j} \sin^2\phi \right) \right]^{\frac{1}{2}} \tag{2.21}$$

Since the wavevectors can be written out in a similar way as in the s-polarization case, the fields in vacuum and in the overlayer are, starting now with the more convenient H field and using eq. (2.8) to calculate the D fields:

$$\vec{H}^v(\vec{r}) = \left[H_t^v e^{-ik_z^v z + ik_x x}\right]\hat{y}$$

$$\vec{D}^v(\vec{r}) = H_t^v(-n_z^v\hat{x} - n_x\hat{z})e^{-ik_z^v z + ik_x x}$$

in vacuum, and:

$$\vec{H}(\vec{r}) = \left[H_t e^{-ik_z z + ik_x x} + H_r e^{ik_z z + ik_x x}\right]\hat{y}$$

$$\vec{D}(\vec{r}) = H_t(-n_z\hat{x} - n_x\hat{z})e^{-ik_z z + ik_x x} + H_r(n_z\hat{x} - n_x\hat{z})e^{ik_z z + ik_x x}$$

in the overlayer. Note that the k_z^j's are defined by eq. (2.21) and not by eq. (2.13), i.e. the k_z^j's are different for the two polarizations. The boundary conditions at $z = 0$ (vacuum/overlayer interface) are:

tang. E:	$(n_z^v/\epsilon_x^v)(-H_t^v) = (n_z/\epsilon_x)(-H_t+H_r)$	(2.22a)
tang. H:	$H_t^v = H_t + H_r$	(2.22b)
norm. D:	$-n_x H_t^v = -n_x(H_t+H_r)$	(2.22c)
norm. B:	$0=0$	(2.22d)

From eqs. (2.22), we get:

$$H_t = \frac{1}{2}H_t^v\left(1 + \frac{n_z^v\epsilon_x}{n_z\epsilon_x^v}\right) \tag{2.23a}$$

$$H_r = \frac{1}{2}H_t^v\left(1 - \frac{n_z^v\epsilon_x}{n_z\epsilon_x^v}\right) \tag{2.23b}$$

After propagation from $z = 0$ to $z = d$ (overlayer/substrate interface) and approximation based on $d << \lambda$, we obtain the following expressions for the fields, using eqs. (2.7), (2.8) and (2.23):

$$\vec{H}(x\hat{x}+d\hat{z}) = H_t^v\left(1 - ik_z d\frac{n_z^v\epsilon_x}{n_z\epsilon_x^v}\right)e^{ik_x x}\hat{y}$$

$$\vec{D}(x\hat{x}+d\hat{z}) = H_t^v\left[n_z\left(-\frac{n_z^v\epsilon_x}{n_z\epsilon_x^v} + ik_z d\right)\hat{x} - n_x\left(1 - ik_z d\frac{n_z^v\epsilon_x}{n_z\epsilon_x^v}\right)\hat{z}\right]e^{ik_x x}$$

in the overlayer, and:

$$\vec{H}(\vec{r}) = \left[H_t^s e^{ik_z^s(z-d) + ik_x s} + H_r^s e^{ik_z^s(z-d) + ik_x x}\right]\hat{y}$$

$$\vec{D}(\vec{r}) = H_t^s(-n_z^s\hat{x} - n_x\hat{z})e^{-ik_z^s(z-d) + ik_x x} + H_r^s(n_z^s\hat{x} - n_x\hat{z})e^{ik_z^s(z-d) + ik_x x}$$

in the substrate.

The boundary conditions at z=d (overlayer/substrate interface) are:

tang. E: $$(H_t^v/\epsilon_z)(-\frac{n_z^v\epsilon_x}{\epsilon_x^v}+in_zk_zd)=(n_z^s/\epsilon^s)(-H_t^s+H_r^s) \tag{2.24a}$$

tang. H: $$H_t^v(1-ik_zd\frac{n_z^v\epsilon_x}{n_z\epsilon_x^v})=H_t^s+H_r^s \tag{2.24b}$$

norm. D: $$-n_xH_t^v(1-ik_zd\frac{n_z^v\epsilon_x}{n_z\epsilon_x^v})=-n_x(H_t^s+H_r^s) \tag{2.24c}$$

norm. B: $$0=0 \tag{2.24d}$$

From eq. (2.24) and after some algebra assuming $d<<\lambda$, the ratio $\tilde{r}\equiv H_r^s/H_t^s$ can be obtained:

$$\tilde{r}=(\frac{n_z^s\epsilon_x^v-n_z^v\epsilon^s}{n_z^s\epsilon_x^v+n_z^v\epsilon^s})\left[1+2ik_zd(\frac{n_z^s\epsilon^s}{n_z\epsilon_x})(\frac{n_z^2\epsilon_x^{v2}-n_z^{v2}\epsilon_x^2}{n_z^{s2}\epsilon_x^{v2}-n_z^{v2}\epsilon^{s2}})\right] \tag{2.25}$$

Using eqs. (2.12) and (2.21), the above ratio becomes:

$$\tilde{r}=\tilde{r}^o\left[1+2i(\frac{\omega}{c})d\sqrt{\epsilon^s}\cos\phi\frac{(\epsilon_x^v-\epsilon_x)-(\frac{\epsilon_x^v}{\epsilon_z}-\frac{\epsilon_x}{\epsilon_z^v})\epsilon^s\sin^2\phi}{(\epsilon_x^v-\epsilon^s)-(\frac{\epsilon_x^v}{\epsilon^s}-\frac{\epsilon^s}{\epsilon_z^v})\epsilon^s\sin^2\phi}\right] \tag{2.26}$$

where

$$\tilde{r}^o=\frac{\epsilon_x^v\sqrt{\epsilon^s}\cos\phi-\epsilon^s\left[\epsilon_x^v(1-\frac{\epsilon^s}{\epsilon_z^v}\sin^2\phi)\right]^{\frac{1}{2}}}{\epsilon_x^v\sqrt{\epsilon^s}\cos\phi+\epsilon^s\left[\epsilon_x^v(1-\frac{\epsilon^s}{\epsilon_z^v}\sin^2\phi)\right]^{\frac{1}{2}}} \tag{2.26a}$$

is the field ratio for the vacuum/ substrate interface. Again, note that the x and z subscripts of ϵ^s have been dropped since the substrate is assumed isotropic. The measured change in reflectance is, recalling that $\omega=2\pi c\tilde{\nu}$:

$$\frac{\Delta R}{R}\Big|_{p-pol.}\equiv1-\left|\frac{\tilde{r}}{\tilde{r}^o}\right|^2=8\pi\tilde{\nu}\sqrt{\epsilon^s}d\cos\phi\ \mathrm{Im}\left[\frac{(\epsilon_x^v-\tilde{\epsilon}_x)-(\frac{\epsilon_x^v}{\tilde{\epsilon}_z}-\frac{\tilde{\epsilon}_x}{\epsilon_z^v})\epsilon^s\sin^2\phi}{(\epsilon_x^v-\epsilon^s)-(\frac{\epsilon_x^v}{\epsilon^s}-\frac{\epsilon^s}{\epsilon_z^v})\epsilon^s\sin^2\phi}\right] \tag{2.27}$$

where the ~ sign was reintroduced to emphasize that only $\tilde{\epsilon}\equiv\epsilon'+i\epsilon''$ will be kept imaginary. Note that the subscripts of ϵ^v have been kept so that eq. (2.27) can be used for *external* reflection ($\epsilon^s=1$) on a non-isotropic substrate. In order to simplify the expression and cast it in a form similar to eq. (2.20) for s-polarization, we now relax the assumption that the last medium be anisotropic. This is a justified assumption since, for *internal* reflection, this medium is vacuum and is therefore isotropic.

Thus, setting $\epsilon_x^v=\epsilon_z^v=\epsilon^v$ and $\epsilon_x=\epsilon_x'+i\epsilon_x''$, $\epsilon_z=\epsilon_z'+i\epsilon_z''$, we can rewrite eq. (2.27):

$$\frac{\Delta R}{R}\Big|_{p-pol.} = 8\pi\tilde{\nu}\sqrt{\epsilon^s}\, d\cos\phi\frac{(\frac{\epsilon^s}{\epsilon^v}\sin^2\phi-1)\epsilon_x''+(\epsilon^s\epsilon^v\sin^2\phi)\frac{\epsilon_z''}{\epsilon_z'^2+\epsilon_z''^2}}{(\epsilon^s-\epsilon^v)\left[\frac{\epsilon^s+\epsilon^v}{\epsilon^v}\sin^2\phi-1\right]} \tag{2.28}$$

The point of writing out eqs. (2.20) and (2.28) explicitly is to gain insight into the relevant experimental parameters. These are the relative strengths of the electric fields responsible for the absorptions and the role of the various components of $\tilde{\epsilon}$ on the spectra.

If the spectrometer source is completely unpolarized and rotation of a linear polarizer is used to provide s-and p-polarized radiation with unit strength, then the three relevant normalized field intensities are the coefficients of ϵ_y'', $\frac{\epsilon^{v2}}{\epsilon_z'^2+\epsilon_z''^2}\epsilon_z''$ and ϵ_x'' in eqs. (2.20) and (2.28):

$$\text{for s-pol.:}\qquad I_y = \frac{4\epsilon^s\cos^2\phi}{\epsilon^s-\epsilon^v} \tag{2.29}$$

$$\text{for p-pol.:}\qquad I_z = 4\epsilon^s\cos^2\phi\frac{(\epsilon^s/\epsilon^v)\sin^2\phi}{(\epsilon^s-\epsilon^v)\left[\frac{\epsilon^s+\epsilon^v}{\epsilon^v}\sin^2\phi-1\right]} \tag{2.30}$$

$$I_x = 4\epsilon^s\cos^2\phi\frac{(\frac{\epsilon^s}{\epsilon^v}\sin^2\phi-1)}{(\epsilon^s-\epsilon^v)\left[\frac{\epsilon^s+\epsilon^v}{\epsilon^v}\sin^2\phi-1\right]} \tag{2.31}$$

which are identical to equations (17-19) derived in ref. [7] for a vacuum/substrate interface [14]. For small absorptions typical of monolayer films at a semiconductor/vacuum interface: $\epsilon_z'^2+\epsilon_z''^2\approx\epsilon^v=1.0$ and $\epsilon^s>>\epsilon^v$. Eqs. (2.29) - (2.31) therefore show that all three intensities are roughly equal to $4\cos^2\phi$. For a typical and practical incidence angle $\phi=45°$, the normalized field intensities are roughly 2.

The second aspect deals with the active layer absorption itself and its spectral signature and will be dealt with in some detail in the next section. For now, we rewrite eqs. (2.20) and (2.28) as:

$$\frac{\Delta R}{R}\Big|_{s-pol.} = \frac{2\pi\tilde{\nu}}{\sqrt{\epsilon^s}\cos\phi}(I_y)(\epsilon_y''d) \tag{2.32}$$

$$\frac{\Delta R}{R}\Big|_{p\text{-pol.}} = \frac{2\pi\tilde{\nu}}{\sqrt{\epsilon^s}\cos\phi}\left[(I_x)(\epsilon_x''d)+I_z\left(\frac{\epsilon^{v2}}{\epsilon_z'^2+\epsilon_z''^2}\right)(\epsilon_z''d)\right] \qquad (2.33)$$

Note that the $1/\cos\phi$ factor arises from the sampling area and the $1/\sqrt{\epsilon^s}$ factor from the index matching with the active layer as discussed in detail by HARRICK [15]. Eqs. (2.32) and (2.33) show that, while the "parallel" spectra (x and y components) depend only on the absorption ($\mathrm{Im}\tilde{\epsilon}$) of the layer, the perpendicular spectrum (z-component) depends on $\mathrm{Im}(1/\tilde{\epsilon})$ and will therefore be affected both by the real and imaginary parts of $\tilde{\epsilon}$. As a result, a spectral *shift* will occur for resonant absorption, the magnitude of which depends on the optical response of the layer (see section 2.2).

2.2 Microscopic model of the active layer

The previous macroscopic considerations are useful only if $\tilde{\epsilon}$ can be expressed in terms of *relevant* microscopic quantities such as the dynamic charge of the adsorbate mode and the electronic screening within the adsorbate layer. Since ab initio calculations are in general difficult, we present here the parametrized Lorentz oscillator response of the layer [16], which can be directly related to ab initio calculations only if every parameter can be explicitly calculated. In this model, assuming a layer formed of identical oscillators, the dielectric function is:

$$\tilde{\epsilon}_i = \epsilon_{\infty i} + \frac{\tilde{\nu}_{pi}^2}{\tilde{\nu}_{oi}^2 - \tilde{\nu}^2 - i\Gamma_i\tilde{\nu}} \qquad i = x,y,z \qquad (2.34)$$

where ϵ_∞ is the electronic screening of the layer, $\tilde{\nu}_0$ the resonant frequency, Γ the natural linewidth and $\tilde{\nu}_p$ the plasma frequency (in cm^{-1}) of the oscillator. The last is related to the effective charge e^* and effective mass m^* of the oscillator by:

$$\tilde{\nu}_{pi}^2 = \frac{N_i}{\pi}\frac{e_i^{*2}}{m_i^* c^2} \qquad (2.35)$$

where N_i is the number of oscillators per unit volume which contribute to the i-th component of the response. A useful definition of the coverage N_{si}, the number of oscillators per unit area contributing to the i-th component of the absorption, is [17]:

$$N_{si} = N_i d \quad \text{so that} \qquad (2.36)$$

$$\tilde{\nu}_{pi}^2 d = \frac{N_{si}}{\pi}\frac{e_i^{*2}}{m_i^* c^2} \qquad (2.37)$$

From eqs. (2.32), (2.33), (2.34), and (2.37) it is clear that at least for the parallel components x and y, $\frac{\Delta R}{R}$ does not depend on d explicitly. However, for the z component, the quantity $\epsilon_z'^2+\epsilon_z''^2$ in the denominator requires the knowledge of the effective thickness. Following MAHAN and LUCAS [18], and as pointed out by others [19], we can write:

$$d = 4\pi N_s / \sum_{i \neq j} (1/r_{ij})^3 \tag{2.38}$$

where the denominator is the dipole sum which depends on the specific arrangement of the adsorbate at the surface. In general, it can be written $\sum = C N_s^{3/2}$ where C = 9.0336 for a square lattice [20].

We can now work out estimates for the spectral shift occurring for the z-component of p-polarized radiation mentioned at the end of section 2.1.2. That is, the resonance of $\mathrm{Im}(1/\tilde{\epsilon}_z) = \epsilon_z''/(\epsilon_z'^2 + \epsilon_z''^2)$ occurs, using eq. (2.34), at $\tilde{\nu}^2 = \tilde{\nu}_0^2 + \tilde{\nu}_p^2/\epsilon_{\infty z}$. For small shifts, the magnitude of the shift from $\tilde{\nu}_0$ is therefore $\Delta\tilde{\nu} \sim \tilde{\nu}_p^2/(2\epsilon_{\infty z}\tilde{\nu}_0)$.

If the coverage, effective charge and mass of an adsorbate are such that $\Delta\tilde{\nu}$ is larger than the linewidths, then one must be careful to remove that shift before comparing the data to calculated resonant frequencies of ϵ_x, ϵ_y and ϵ_z. Such data analysis will be relevant to the analysis of H on Si(100) presented in section 4.1.

2.3 Discussion of the assumption of sharp boundaries

As pointed out clearly by FEIBELMAN [21], the previous model gives an unphysical description of the nature of induced surface charges on a *microscopic* scale. That is, the electric field relevant to the vibrational excitation of the adsorbate modes is not well described by a step function discontinuity. It is for instance not clear that, for substrate/adsorbate modes, the relevant electric fields are the same as those on the vacuum side of the interface.

For electric fields polarized in the plane of the surface (x and y components), there is no surface charge induced and microscopic effects can be ignored. That is, since the parallel component of the electric field is continuous throughout the interface, the value of the field at the adsorbate can be accurately calculated from macroscopic quantities, ϵ^s and ϵ^v, even if the precise position of the absorbing dipole is not known.

The problem arises for fields polarized perpendicular to the surface plane (z component). For infrared frequencies (i.e. below the substrate plasma frequency), FEIBELMAN defined the *surface* response function:

$$d_\perp(\omega) \equiv \frac{1}{(1-\epsilon^s/\epsilon^v)} \int_{z^s}^{z^v} dz' z' \frac{d}{dz'}(\epsilon_z) \tag{2.39}$$

where $\epsilon_z = E_z^s/E_z^v$. The values of z^s and z^v are chosen such that *all* the surface absorption occurs between these two values. For instance, if we make the same assumptions as in the 3-layer model, i.e. all the absorption occurs between z=0 and z=d, then eq. (2.39) becomes, making use of the continuous nature of D_z:

$$d_\perp(\omega) = d \frac{1/\tilde{\epsilon} - 1/\epsilon^v}{1/\epsilon^s - 1/\epsilon^v} \tag{2.40}$$

showing clearly that $d_\perp(\omega) = 0$ if $\tilde{\epsilon} = \epsilon^v$. The physical meaning of $d_\perp(\omega)$ is discussed in ref. [21] where FEIBELMAN shows that Re $[d_\perp(\omega)]$ is the centroid of the surface position, i.e. the centroid of the induced charge which is responsible for the surface absorption. Moreover, in the case of semiconductor/vacuum interfaces for which there is *no bulk absorption,* $\mathrm{Im}[d_\perp(\omega)]$ is the surface power absorption.

When surface photoeffects are small and for frequency well below the substrate plasma frequency, FEIBELMAN concludes that the assumption of step function dielectric constants is good as far as computation of the fields is concerned. The problem arises, within the 3-layer model, with the computation of the effective thickness d when one is dealing with submonolayer coverages.

In conclusion, for the particular study of weakly absorbing layers at semiconductor/vacuum interfaces, the 3-layer model is well justified to calculate the electric fields. Calculations of the power absorption as they relate to geometry optimization are again well justified within the 3-layer model. The aspect of the model which should be dealt with care is the determination of the effective thickness relevant for the layer response to perpendicular fields (z-components). Fortunately, one can see from eqs. (2.33), (2.34) and (2.37) that, for weakly absorbing layers $\epsilon_z'' \ll \epsilon_z' \approx 1$, an error in d only affects the spectral shift calculation and is negligible as far as *absorption intensities* are concerned.

3. EXPERIMENTAL GEOMETRIES FOR INFRARED SPECTROSCOPY OF SEMICONDUCTOR SURFACES

To design an experiment where the vibrational spectrum of submonolayer coverage of molecules or atoms at semiconductor surfaces can be probed sensitively, we separate the problem into two cases. The first involves the study of modes with a resonant frequency within the optical gap of the substrate, i.e. higher than any bulk phonon absorption and lower than bulk electronic absorption. Note that samples of high enough resistivity are used so that the free carrier absorption is negligible in the region of interest. For this first case, the formalism developed in section 2 is appropriate where ϵ^v and ϵ^s are real and isotropic. The second case involves the study of modes with a resonant frequency in a region where the substrate absorbs sufficiently to prevent the use of internal reflection. In this case, the formalism developed in section 2 cannot be used as such but it can be modified to give semi-quantitative information.

3.1 Vibrational modes in substrate optical gap

For conventional sources (e.g. globar), the surface infrared spectroscopist is usually in the regime where the signal-to-noise ratio, S/N, is proportional to the total radiation intensity incident on the detector, I. Although the limitation for *transmission* spectroscopy with broadband interferometers in the 2000-3000cm^{-1} region can be the A/D limit, this is not true of *surface* spectroscopy for which some radiation is lost due to the requirements of small solid angle and spot size at the sample. Furthermore, considering that the capabilities of A/D's will improve in the near future beyond the 16 bit and 120KHz specifications, it is appropriate to assume that S/N is proportional to I. Based on this assumption, GREENLER [22] pointed out that the quantity $\Delta R = R(0) - R(d)$ had to be optimized rather than $\Delta R/R = 1 - R(d)/R(0)$.

To decide between external and internal reflection geometries to study a thin active layer at a semiconductor/vacuum interface, we first consider the case of a *single* reflection. The results of section 2 are applicable to *external* reflection simply by setting $\epsilon^s=1$ (vacuum) and $\epsilon^v=$ substrate dielectric constant, and vice versa for *internal* reflection.

For s-polarization, we have:

$$\Delta R|_{s-pol.}=\frac{2\pi\tilde{\nu}}{\sqrt{\epsilon^s}\cos\phi}|\tilde{r}^o|^2 I_y(\epsilon_y''d) \tag{3.1}$$

where $\tilde{r}^o$ is defined in eq. (2.18a), I_y in eq. (2.29) and $\epsilon_y''d$ in eqs. (2.34) and (2.37).

For p-polarization, we have:

$$\Delta R|_{p-pol.}=\frac{2\pi\tilde{\nu}}{\sqrt{\epsilon^s}\cos\phi}|\tilde{r}^o|^2\left[I_x(\epsilon_x''d)+I_z\left(\frac{\epsilon^{v2}}{\epsilon_z'^2+\epsilon_z''^2}\right)(\epsilon_z''d)\right] \tag{3.2}$$

where $\tilde{r}^o$ is defined in eq. (2.26a), I_x and I_z in eqs. (2.30) and (2.31), and $\tilde{\epsilon}_z$ in eqs. (2.34), (2.37) and (2.38).

Note that $|\tilde{r}^o|^2=1$ for both polarizations in the case of total *internal* reflection, $\phi\geqslant\sin^{-1}(\sqrt{\epsilon^v/\epsilon^s})$. In Figs. 3 (a-b), we show ΔR for both polarizations in the case of a single external and internal reflection using: $N_s=5\times10^{14}cm^{-2}$, $\epsilon_{\infty z}=1.3$, $\frac{(e^*/e)}{(m^*/m)^{1/2}}=0.05$, $\Gamma=5cm^{-1}$ and $\tilde{\nu}_o=2100cm^{-1}$ which represents the case of H on Si surfaces within factors of 2 to 3 (see section 4.1). For an estimate of the sensitivity, we assume the active layer to be isotropic. The above case is typical of a very weakly absorbing monolayer. Figs. 4 (a-b) show the more general case where $\frac{(e^*/e)}{(m^*/m)^{1/2}}=0.5$, the other parameters remaining the same for the sake of argument.

Figs. 3 and 4 first show that for a *single* reflection, internal reflection gives a factor of 2 to 10 better S/N depending on polarization and strength of the active layer absorption. Second, they show that the external reflection spectrum is more complex due to interference between the negative x and positive z components in p-polarized spectra. The physical interpretation of these changes of sign is that, for external reflection, the thin active layer can act as a reflection or antireflection coating depending on the polarization, angle of incidence and layer optical response. Third, the frequency shift of the z-component, negligible for weak absorbers such as Si-H, is apparent in Fig. 4, i.e. for a strongly absorbing monolayer. We note that the shift is given by the expression derived at the end of section 2.2: $\Delta\tilde{\nu}\sim\tilde{\nu}_p^2/(2\epsilon_{\infty z}\tilde{\nu}_o)=18cm^{-1}$ for the strongly absorbing monolayer.

The above results, along with similar calculations of ΔR for a range of angles of incidence show that (a) single internal reflection always gives a better S/N than a single external reflection by a factor of 2 to over 10; (b) The spectrum obtained by internal reflection is simple (simple addition of the positive components of the field intensities, I_x,I_y and I_z); and (c) I_x,I_y,I_z are very close in magnitude for a wide range

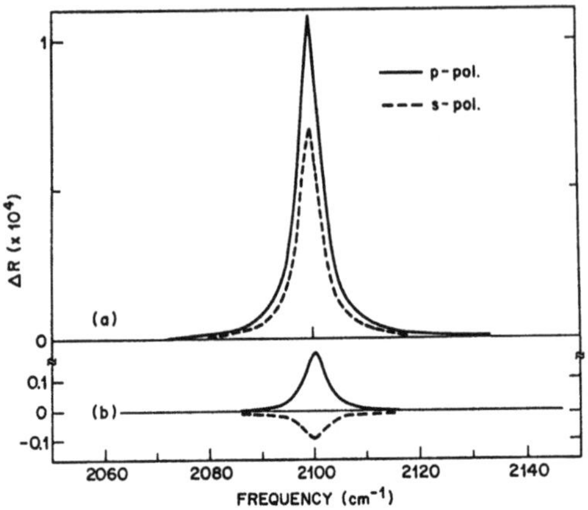

Fig. 3: Model calculations of $\Delta R \equiv R(0)-R(d)$ in the case of (a) *internal* reflection, and (b) *external* reflection for both p- and s-polarizations. The parameters used are $\epsilon^v=1$, $\epsilon^s=11.7$, $\tilde{\nu}_{ox}=\tilde{\nu}_{oy}=\tilde{\nu}_{oz}=2100\text{cm}^{-1}$, $\Gamma_x=\Gamma_y=\Gamma_z=5\text{cm}^{-1}$, $N_{sx}=N_{sy}=N_{sz}=5\text{x}10^{14}\text{cm}^{-2}$, $m_x^*=m_y^*=m_z^*=1$a.m.u. $(e^*/e)_x=(e^*/e)_y=(e^*/e)_z=0.05, \epsilon_{\infty z}=1.3$, $\phi=45°$. This represents the case of a weakly absorbing, isotropic monolayer.

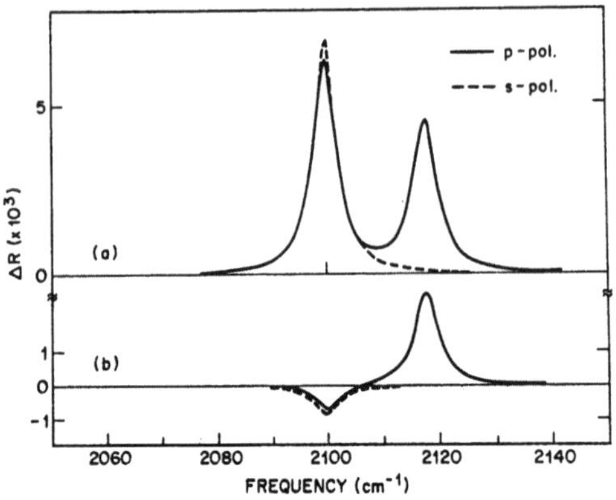

Fig 4: Model calculation of ΔR in the case of (a) *internal* reflection, and (b) *external* reflection for both polarizations. The parameters used are identical to those used for Fig. 4 except that $(e^*/e)_x=(e^*/e)_y=(e^*/e)_z = 0.5$. This represents the case of a strongly absorbing isotropic monolayer.

of angles so that analysis of the data can be done accurately even for highly diverging radiation and poor knowledge of the angle of incidence. This is not true of external reflection for which positive or negative contributions can occur and where the parallel components are *always* small.

For *multiple* internal reflection, a thin (~1mm) plate with bevelled edges is used. In practice, only a fraction of the incident radiation, f, couples because the spot size at the focus is usually larger than the coupling edge. In addition, there are reflection losses at the input and output edges [7], so that only a fraction, T_0, of the radiation reaches the detector. As a result, the measured ΔR for N reflections (on one side of the plate only) is:

$$\Delta R(N)=f\,T_0\,N\,\Delta R(1) \qquad (3.3)$$

where ΔR (1) is given by eqs. (2.41) or (2.42). For a 5mm source size and a Si plate (t =0.05cm,ℓ=5cm), ΔR(50refl.)=6.25ΔR(1). Since ΔR(1) is between a factor of two and over an order of magnitude larger than that for external reflection, multiple internal reflection practically gives one to two orders of magnitude better S/N than external reflection. Having established this point, we refer the reader to ref. [7] for a review of experimental MIR geometries.

3.2 Vibrational modes in substrate absorption region

Since it is not possible in this case to use multiple *internal* reflection because of the substrate absorption, the alternatives are external reflection or transmission through a very thin sample. The case in which the semiconductor absorption is present but still small compared to metals is complex and is discussed by P. CHIARADIA in the next lecture [23]. It is straightforward to calculate the expected $\Delta R/R$ from eqs. (2.19) and (2.27) by letting $\epsilon_y^v, \epsilon_x^v, \epsilon_z^v$ be complex. Unfortunately, there are no simple explicit expressions corresponding to eqs. (2.20) and (2.28). The optimum parameters (e.g. angle of incidence) depend strongly on the substrate medium and the nature of the overlayer.

What we can do here to gain some physical intuition is to work out the two limits of very weakly and very strongly absorbing substrates. For the weakly absorbing substrates, the equations derived in section 2 are approximately valid and it is found that the largest $\Delta R|_{p-pol.}$ is obtained for grazing incidence (e.g. $80° < \phi < 87°$ for a silicon substrate). Note however, that ΔR is negative even for the z-component and that the x and y components are small. The latter components are best probed near normal incidence, either by reflection or transmission through a thin sample.

The other limit implies $|\tilde{\epsilon}^v| >> |\tilde{\epsilon}| \approx \epsilon^s = 1$ (for vacuum). This is the same assumption made for metal substrates, for which the following expressions are obtained [17]:

$$\Delta R|_{s-pol.} = 8\pi\tilde{\nu}\cos\phi|\tilde{r}^o|^2 \mathrm{Im}\left[\frac{\tilde{\epsilon}_y - \tilde{\epsilon}_y^v}{1-\tilde{\epsilon}_y^v}\right] \tag{3.4}$$

and

$$\Delta R|_{p-pol.} \approx 8\pi\tilde{\nu}|\tilde{r}^o|^2\sin^2\phi\left[\cos\phi\left(1+\frac{1}{|\tilde{\epsilon}^v|^2}\frac{\sin^2\phi}{\cos^4\phi}\right)\right]^{-1}\mathrm{Im}\left(\frac{-1}{\tilde{\epsilon}_z}\right) \tag{3.5}$$

Eqs. (2.18a) and (2.26a) show that $|\tilde{r}^o| \approx 1$ for both s- and p-polarizations. However, eqs. (2.29 - 2.31) emphasize that the y and x components of the fields are very small. Therefore, as apparent in eqs. (3.4) and (3.5), only the z-component of the active layer dielectric constant can be probed with grazing incidence reflection. The optimum angle (in radian) is:

$$\phi = \pi/2 - (3^{\frac{1}{4}}/|\epsilon^v|^{\frac{1}{2}})$$

The general conclusion of this section is that, for absorbing substrates, external reflection must be used for which the sensitivity to modes parallel to the surface is substantially reduced and for which the maximum sensitivity to modes perpendicular to the surface is achieved by grazing incidence. Transmission is best for modes parallel to the surface on a weakly absorbing substrate.

As a result, one may want to investigate alternatives depending on the particular system under study. For instance, if the Si-O modes are of interest on a silicon substrate, standard multiple internal reflection on a Si plate is ill advised since the modes under question occur at a frequency ($\sim$900cm^{-1}) where the Si substrate absorbs.

However, epitaxial Si can be grown on Ge single crystals which are transparent above 830cm^{-1}. Multiple internal reflection using a Ge plate with a thin (~20Å) Si layer is particularly attractive if modes parallel to the surface need to be studied. Alternatively, a single internal reflection can be used through a very thin Si prism. The small path length through the bulk Si gives an easily detectable reflected intensity and the internal reflection gives a better sensitivity to the parallel modes.

It is clear that, depending on the particular system under study, a variety of approaches are possible. A quantitative estimate of the sensitivity of the various approaches can be done by means of eqs. (3.1), (3.2) and (3.3) for transparent or weakly absorbing substrates or of eqs. (2.19), (2.27) and (3.3) in general.

4. SELECTED EXAMPLES

In this section, selected examples are presented which do not represent an exhaustive list of all vibrational spectroscopy performed at semiconductor interfaces [8,24]. Rather, they have been chosen for their tutorial aspect. First, the detailed geometry of the Si(100)-(2x1)H system is deduced from surface infrared spectroscopy and ab initio cluster calculations. In this case, the results derived in section 2 can be used fully to make connection with first principle calculations. Second, the nature of the H-saturated Si(100) and Ge(100) surfaces is inferred from infrared spectroscopy on a semi-quantitative basis. Some dynamics are inferred from the IR data and molecular dynamics simulations for the well-characterized Si(100)-(2x1)H system. Finally, semi-quantitative information is also obtained for water decomposition on Si(100).

4.1 Structure of the Si(100)-(2x1) H System:

The exposure of the Si(100)-(2x1) surface to atomic hydrogen first leads to the development of a sharp, well-defined (2x1) LEED pattern before becoming a (1x1) pattern at saturation coverage. The general understanding of this phenomenon is that (a) the clean (100) surface reconstructs by forming dimers and (b) the addition of 1 monolayer of H saturates the dangling bonds of each dimer unit as shown in Fig. 5. The unit cell is therefore (2x1) and contains 2H. There are therefore 6 vibrational

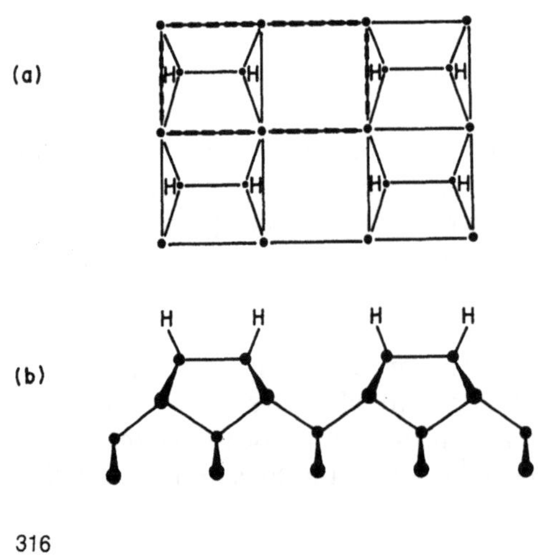

Fig. 5: (a) top view and (b) side view of a portion of a uniform monohydride phase on Si(100) or Ge(100). The resulting 2x1 unit cell is outlined in dashed lines. Dimerization of the top Si or Ge layer is depicted by reduction of the Si-Si or Ge-Ge distance.

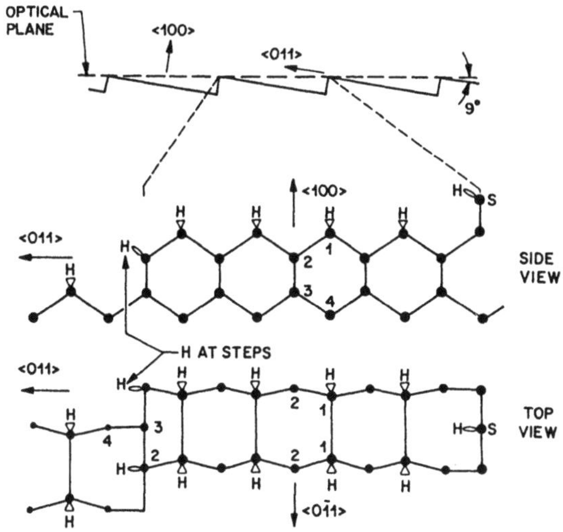

Fig. 6: Schematic representation of a vicinal surface cut at 9° from the (100) plane about the $<0\bar{1}1>$ axis, i.e. the optical plane with respect to which the probing electric field vectors (E_x, E_y or E_z) are described makes a 9° angle with the (100) terraces. The atomic representation of the vicinal surface including one terrace and two steps is given below. The projection of H adsorbed at all available dangling bonds of the clean surface (assuming that the dimers are not broken) is also shown schematically.

modes associated with this structure, two of which involve the stretch of the Si-H bonds. Since the other 4, involving the bend and frustrated translation of the H are too low in frequency to be detected with conventional surface IR, we focus on the two stretch modes occurring around 2100cm^{-1}, from which accurate *structural* information can be learned [25].

Since a nominally flat Si(100) surface displays two domains rotated by 90° with (2x1) symmetry, surfaces cut a few degrees off the (100) plane around the $<0\bar{1}1>$ axis were used for which only one domain of (2x1) symmetry was formed [25,26] as shown in Fig. 6. The importance of this geometry is that the dielectric constants of the Si-H layer can be associated with distinct modes: contributions to $\epsilon_{<011>}$ come only from H adsorbed at steps, while contributions to $\epsilon_{<100>}$ and $\epsilon_{<0\bar{1}1>}$ primarily come from ⊥ and // components of H adsorbed on terraces (see Fig. 6). Fig. 7(a) shows that these modes are the symmetric (⊥ to surface) and antisymmetric (// to surface) stretches of the "monohydride". Note that, unless the small ⊥-component of the mode associated with H at steps has a resonance at the same frequency as the symmetric stretch of the monohydride, its contribution to $\epsilon_{<100>}$ can be sorted out. As shown in Fig. 8, the use of two samples oriented in orthogonal directions with respect to the probing radiation (cases I and II) makes it possible for s- and p-polarizations to probe (I) S_I:$\epsilon_{<011>}$ and P_I: $\epsilon_{<0\bar{1}1>}$ and $\epsilon_{<100>}$, and (II) S_{II} : $\epsilon_{<0\bar{1}1>}$ and P_{II} : $\epsilon_{<011>}$ and $\epsilon_{<100>}$. As a result, the three components of ϵ can be measured unambiguously.

Results of such measurements showed that the monohydride modes exhibit a resonance in $\epsilon_{<0\bar{1}1>}$ at 2087cm^{-1} and in $\epsilon_{<100>}$ at 2099cm^{-1} at 300K, while $\epsilon_{<011>}$ is dominated by a strong resonance at 2087cm^{-1} and much weaker contributions elsewhere in the spectrum [25]. Based on this knowledge, it was straightforward to notice that, upon annealing to 625K, *only* the resonances associated with the monohydride on the terraces remain, as shown in Fig. 9, i.e. all the resonances associated with H at steps disappear. To the extent that the monohydride, characterized only by two sharp resonances with distinct frequencies and polarizations, is the most stable hydride on Si(100), it represents a *model* system which could be experimentally prepared and for which quantitative information could be obtained, as explained below.

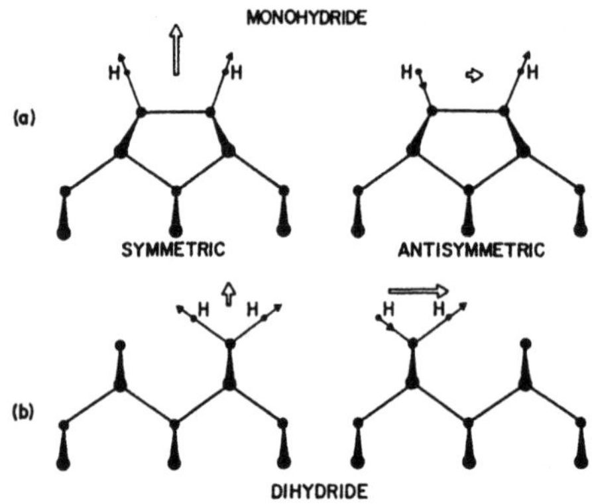

Fig. 7: Scale drawing of (a) the monohydride, and (b) the dihydride structures. The arrows (not to scale) represent the direction of H displacements (solid arrows) and the *polarization* (double arrows) of the two normal modes involving the stretching of SiH bonds. The different magnitude of the double arrows schematically indicates that the net dipole associated with each normal modes are different for the two structures, with $\mu_{//} < \mu_{\perp}$ for the monohydride and $\mu_{//} > \mu_{\perp}$ for the dihydride.

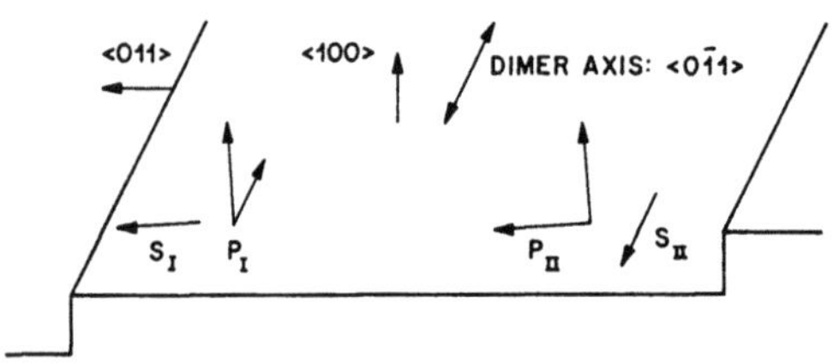

Fig. 8: Schematic representation of the vicinal surface with a (100) terrace. Note that the relevant electric field directions, labelled S and P are not all exactly collinear with the <011>, <100> and <01̄1> axes due to the angle made between the optical plane and the (100) terraces. Straightforward projection of the field vectors along the crystal axes is performed for quantitative data analysis.

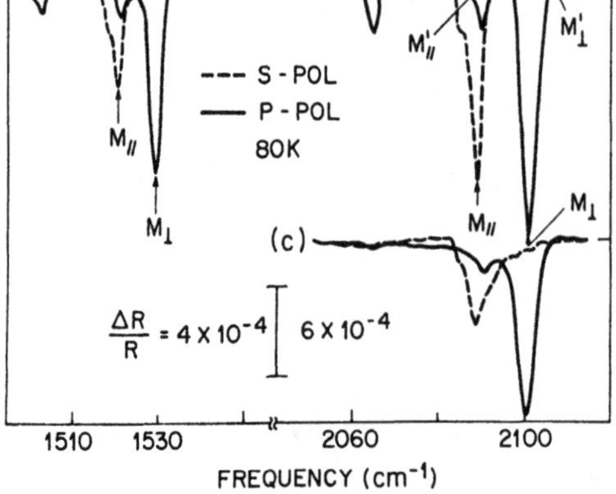

Fig. 9: Surface infrared spectra associated with (a) the (3x1) phase prepared by H-saturation at 375K, (b) the (2x1) phase obtained upon a 475K anneal, and (c) the (2x1) phase obtained upon a 625K anneal. On the left-hand side of the figure, spectra resulting from pure D exposures are shown. On the right-hand side of the figure, spectra resulting from pure H exposures are shown. The sample is cut 5° off the (100) plane. The s-polarization (dashed lines) has an electric field exactly along the <01̄1> axis. The p-polarization (solid lines) has components 5° off the <100> and <011> directions. The data are taken at 80K resulting in narrower lines and a small ($\sim 3cm^{-1}$) blue shift from the room temperature frequencies (see section 4.3). The resolution is $0.5cm^{-1}$ for the SiH region and $1cm^{-1}$ for the SiD region.

First, it is important to find the origin of the frequency shift of the resonance in $\epsilon_z = \epsilon_{<100>}$. As mentioned in sections 2.1.2 and 2.2, it can arise from extended dynamical effects, i.e. inter-unit cell dynamical coupling. The resulting shift, $\Delta\tilde{\nu} \approx \tilde{\nu}_p^2/2\tilde{\nu}_0\epsilon_{\infty z}$ will be detected if $\Delta\tilde{\nu} > 0.5\text{cm}^{-1}$. A shift between the antisymmetric (// modes) and symmetric (⊥ mode) can also arise because of *local* dynamical effects. The latter is sometimes referred to as "chemical" shift because it involves the coupling of normal modes *within* the unit cell and is therefore a signature of the particular chemical bonding within the cell. The two effects must be distinguished if, as is the case for this system, the data are compared to cluster calculations performed on a *single* unit cell.

Experimentally, the inter-unit cell dynamical coupling (extended) can be measured by means of isotopic mixture experiments. As more D is substituted for H at constant coverage, the Si-H normal modes will approach their "isolated frequency" value, i.e. the value devoid of dynamical coupling. If all the splitting is due to extended coupling, then it should decrease to zero as the D concentration increases. If, on the other hand, it is due to local effect (intra unit cell), then the splitting will remain constant as long as enough unit cells still have only H, then the two modes will disappear and a new mode will appear (usually in between the two normal modes) corresponding to only 1H in the unit cell next to a D. These experiments were performed and it was found [25] that the extended dynamical coupling accounts only for $\sim 3\text{cm}^{-1}$ for H and $\sim 2\text{cm}^{-1}$ for D. The rest of the splitting, 9cm^{-1} for H and 7.5cm^{-1} for D, is due to local effects as confirmed by the appearance of a new isolated frequency for very low H concentration at $\sim 2092\text{cm}^{-1}$ (i.e. in between the // and ⊥ components).

Since the number of terrace sites ($n_s = 6\text{x}10^{14}\text{cm}^{-2}$ for the 9° bevel), the hydrogen effective mass ($m^* = 1.67\text{x}10^{-24}\text{gm}$) and the resonant frequencies ($\tilde{\nu}_{0//} = 2087\text{cm}^{-1}, \tilde{\nu}_{0z} = 2096\text{cm}^{-1}$) are known, we are left with 3 unknowns: $e_{//}^*, e_\perp^*$ and $\epsilon_{\infty z}$.
The data provides us with 3 independently determined quantities:

$$\int \epsilon_{<011>} d\tilde{\nu}$$

from s-polarization in configuration II,

$$\int \frac{\epsilon^{v2}}{\epsilon'^2_{<100>} + \epsilon''^2_{<100>}} \epsilon''_{<100>} d\tilde{\nu} \approx \frac{1}{\epsilon^2_{\infty z}} \int \epsilon''_{<100>} d\tilde{\nu}$$

from p-polarization in configuration I,

and $\Delta\tilde{\nu} \approx \nu_p^2/2\nu_0\epsilon_{\infty z}$ from isotopic mixture experiments.

Fitting the data with $\Delta\tilde{\nu} = 3\text{cm}^{-1}$ yields $(e^*_{//}/e) \approx 0.03$, $(e_z^*/e) \approx 0.05$ and $\epsilon_{\infty z} = 1.4$. The unscreened ratio

$$r \equiv \int \epsilon''_{<01\bar{1}>} d\tilde{\nu} / \int \epsilon''_{<100>} d\tilde{\nu}$$

can then be obtained, $r \approx 0.35 \pm 0.1$. We note that the large error in r is due mostly to the uncertainty in $\epsilon_{\infty z}$. Here $\Delta\tilde{\nu}$ is used to determine $\epsilon_{\infty z}$. Because it is so small and known only within 0.5cm^{-1}, it leads to a ±30% error in r. In summary, the local (or "chemical") splitting is $+9\text{cm}^{-1}$ for H, $+7.5\text{cm}^{-1}$ for D and the unscreened intensity ratio is $r=0.35\pm0.1$.

In the cluster calculations, a large cluster (Si_9H_{14}) was used to calculate the geometry by minimizing the cluster total energy with respect to bond lengths and bond angles. From this geometry, the vibrational frequencies and intensities could be calculated using a smaller cluster (Si_6H_{14}). Although the absolute value of vibrational frequencies is too high by about 10%, it was found that the splitting between similar normal modes (e.g. symmetric and antisymmetric stretch modes) could be calculated accurately (within a few cm^{-1}). The results for the monohydride and monodeuteride gave a symmetric stretch frequency 11cm^{-1} and 9cm^{-1}, respectively, higher than the antisymmetric stretch frequency, in good agreement with the data. The calculated intensities associated with each normal mode gave the ratio:

$$\mu_{//}/\mu_{\perp}=0.59, \text{ i.e. } r=(0.59)^2=0.35,$$

again in good agreement with the data.

It is important to stress here the limitations of surface IR spectroscopy without theoretical input. For instance, if $\Delta\tilde{\nu}$ is too small to be observable, then $\epsilon_{\infty z}$ cannot be determined unless $\mu_{//}$ or $\mu_{\perp}$ is known. The bare dipole moments are in fact rarely accessible experimentally. As a result, the geometry (bond length and angle) cannot be obtained by experiments alone. In the case of the monohydride, the geometry was determined by cluster calculations and then used to calculate quantities accessible to the experiment. The angle made by the Si-H bond with the surface normal was calculated to be $\alpha=20°$. It was found that the back projections of $\mu_{//}$ and $\mu_{\perp}$ along the Si-H bond do not give the *same* value for the dynamic dipole moment. That is, the angle α cannot be obtained using $\tan\alpha=\mu_{//}/\mu_{\perp}$.

4.2 H-saturated Si(100) and Ge(100) surfaces

Despite its limitations in determining precise adsorbate geometries, surface IR spectroscopy is a powerful tool to unravel complex structures at surfaces. The H-saturated phases on Si(100) and Ge(100) are good examples because they both display a number of different phases which could not be detected by other probes such as EELS due to poor frequency resolution [27]. We refer the reader to refs. [28,29] for details in the summary presented here.

Let us consider the Si(100) nd Ge(100) saturated with H atoms at room temperature. The former displays a reasonably sharp (1x1) LEED pattern while the latter displays weak $\frac{1}{2}$-order beams, remnant of a (2x1) LEED pattern. The accepted view was that Si(100)-(1x1)H was made up of a uniform *dihydride* phase with all the dimers broken and that the Ge(100)-(2x1)H had only partial dihydride coverage. The initial key findings of SIRS were that at saturation (1) the Si(100)-(1x1)H surface was characterized by strong modes characteristic of *monohydride* with additional modes present and, (2) the Ge(100)-(2x1)H was characterized by a spectrum identical to that obtained for lower coverages for which a very sharp (2x1) pattern is present. Based on the above observations, it was apparent that the monohydride structure was an important part of the H-saturated Si(100) surface and the *dominant* part of the Ge(100) surface.

The next important observation was that a Si(100) surface saturated with H atoms while maintained at 380K displayed a sharp (3x1) LEED pattern, and was characterized by an IR spectrum *identical* to that of the Si(100)-(1x1)H surface prepared by H-saturation at room temperature. These two observations pointed to the fact that the Si(100)-(1x1)H surface was in fact a *disordered* phase with local (3x1) arrangement. The problem was then reduced to interpreting the SIR spectra associated with the (3x1) phase only. As for the Si(100)-(2x1)H case, single domain structures could be prepared by means of vicinal samples so that polarized SIRS could be profitably used. The data in Fig. 9 was obtained by cooling the sample to 80K so as to sharpen the modes (see section 4.3) in order to isolate different spectral components more easily.

As shown in Fig. 9(b), annealing the sample to 475K produced two very well defined changes: (1) the LEED pattern changed from a (3x1) to a (2x1), and (2) the IR spectrum displayed the loss of two lines only, labelled $DI_\perp$ and $DI_{//}$ in Fig. 9(a). Intermediate annealing showed that the intensity of these two lines varied simultaneously. The unambiguous assignment of these lines to the dihydride structure (Fig. 7b) relied again on cluster calculations [28] which accounted for the following characteristic experimental observations: (1) the mode perpendicular to the surface is now a lower frequency than that parallel to the surface, i.e. $\Delta_H \equiv \tilde{\nu}^o_{<100>} - \tilde{\nu}^o_{<0\bar{1}1>} = -12.5cm^{-1}$ for H; (2) the magnitude of the splitting between these two modes is larger for deuterium, i.e. $\Delta_D = -21cm^{-1}$; and (3) the intensity of $DI_{//}$ is substantially larger than that of $D_\perp$, $r \approx 3$ assuming $\epsilon_{\infty z} \simeq 1.4$. The theoretical predictions for the dihydride structure were $\Delta_H = -9cm^{-1}$, $\Delta_D = -22cm^{-1}$ and $r \approx 3$. In addition, the calculations showed that the "isolated" frequency of the dihydride structure should be $+7cm^{-1}$ *higher* than the isolated frequency of the monohydride structure, and $+5cm^{-1}$ for the corresponding isolated deuterium modes. Fig. 10 shows the results of isotopic mixture experiments, clearly showing that DI is $6.5cm^{-1}$ higher in frequency than M. For this dilute H concentration, no lines were observed at $DI_{//}$ and $DI_\perp$ or $M_{//}$ and $M_\perp$; only the isolated frequencies remained, beautifully confirming that both the dihydride and monohydride structures are simultaneously present in the (3x1) phase.

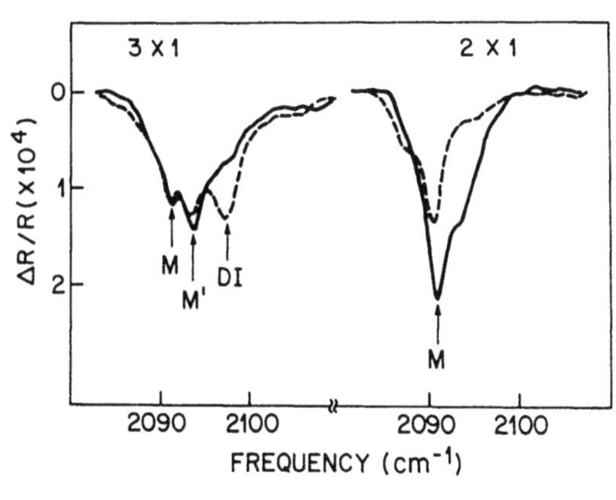

Fig. 10: Surface infrared spectra of the SiH stretch region obtained upon saturation exposure of an isotopic mixture at 375K [(3x1) phase] and after a subsequent 475K anneal [(2x1) phase]. Conditions are therefore identical to those yielding Fig. 9(a) and (b) except that hydrogen represents only 7.5% of the total saturation coverage (measured from IR intensities). The data are taken at room temperature on a Si(100) 5° sample with the s-polarization (dashed lines) and p-polarization (solid lines) defined as in Fig. 9. The resolution is $1cm^{-1}$. M, M′ and DI correspond to the isolated frequencies of pure monohydride, monohydride with a neighboring dihydride and dihydride, respectively.

It is worth noting that, due to excellent resolution ($1cm^{-1}$) and S/N, the spectra are good enough to show that the monohydride modes are shifted by $\sim 2cm^{-1}$ whenever the dihydride is observed. This can be seen in the lines M′ in Fig. 10 and $M'_{//}$ and $M'_{\perp}$ in Fig. 9(a). As discussed at the end of ref. [28], this small shift arises from small back-bond relaxation of the monohydride structure caused by a neighboring dihydride. This observation was taken as supporting evidence that a dihydride structure was always formed right next to a monohydride structure and the model shown in Fig. 1 of ref. [28] was proposed to account for the observations on the (100) terraces of H-saturated Si(100).

The other unlabelled features in the spectra of Fig. 9(a and b) were shown to arise from H (or D) at steps and defects. The presence of these features in Fig. 9 (a and b) indicates that H bonded at steps and defects is more strongly bound than H in a dihydride structure. However, the removal of these features upon annealing to 625K shows that they are less strongly bound than H in a monohydride structure [30].

Based on the above semi-quantitative knowledge of the various hydride phases on Si(100) and their associated IR spectra, we now summarize the qualitative information obtained from IR studies of the H/Ge(100) system for which no calculations were performed. Details can be obtained in ref. [29].

The main observation on this system is that the IR spectrum is dominated at all coverages (except for $\theta \lesssim 0.1ML$) by two modes at $1979cm^{-1}$ and $1991cm^{-1}$. The first is polarized parallel, the second perpendicular to the surface. This observation, along with the (2x1) LEED pattern and the absence of scissor mode in the $750-1000cm^{-1}$ region (characteristic of the dihydride structure) at all coverages, indicated that the main hydride phase is monohydride, even at saturation coverage. Dihydride is *not* formed on Ge(100) at any coverage. As expected, the faint (2x1) LEED pattern at saturation shows that the (2x1) unit cell have a poor *long range* arrangement due to etching of the surface by H atoms.

For the purpose of this lecture, it is worth noting that, although the splitting $\Delta_H = 12cm^{-1}$ is identical to that of the monohydride on Si(100), its origin is quite different. Isotopic mixture experiments show that the two modes *continuously* shift with increasing D concentration at saturation coverage to the value of $1980.5cm^{-1}$. The extended dynamical interaction (inter-unit cell) is therefore dominant for this system.

4.3 Dynamics of H on Si(100)

In this section, we outline the procedures by which dynamical information can be obtained from SIRS. In the favorable case of Si(100)-(2x1)H phase for which a *single* structure (the monohydride) is present, analysis of the line shape is warranted. Such an analysis for the monohydride lines was particularly interesting as variations in width and line positions were observed at a function of temperature. Since inhomogeneous broadening arising from variation in occupation of defect sites as a function of temperature could be ruled out, line broadening and shift must be due to interaction with phonons. Such anharmonic coupling to phonons can be calculated by molecular dynamics simulations [31].

The main drawback of such simulations is that, in general, they rely on fitted parameters, i.e. they are not first principle, and they are classical calculations. In the case of the Si(100)-(2x1) H system, ab initio calculations of the force fields. Both diagonal and off-diagonal stretch and bond force constants, including non-negligible cubic and quantic terms were calculated from first principles. Quantum effects were included by means of a novel stochastic method described in ref. [32].

The main results of such calculations were that (1) lifetime and dipole broadening was negligible ($\sim 10^{-3}cm^{-1}$ for H and $\sim 10^{-2}cm^{-1}$ for D), and (2) the dominant broadening mechanism was due to dephasing, i.e. anharmonic coupling between the Si-H bending mode and the Si-H stretch. We refer the reader to ref. [31] for details. Because the silicon substrate lattice was taken to be harmonic, the simulations made no predictions on line shifts ($4.1cm^{-1}$ and shift from 40K to 500K for H). The anharmonic extension of the calculations is important as shifts can come from several mechanisms, including dephasing, thermal expansion and energy transfer mechanisms.

In conclusion, it appears that dynamical information can be obtained from line shape analysis in favorable cases. As more data are obtained on a number of systems for which the *structure* is *well known* and for which inhomogeneous broadening may be ruled out, interesting dynamical information will be obtained. From these model calculations, it appears that the vibrational lifetime of adsorbate on semiconductors is long enough (10^{-7} 10^{-9}sec) that *direct* measurements of the lifetimes using tunable pulsed lasers may be possible.

4.4 H_2O on Si(100)

In this section, we summarize the findings of SIRS for H_2O adsorption on Si(100) to give an example of SIRS contribution to understanding simple chemistry at surfaces. The motivation for performing an IR study was to decide between conflicting interpretations: UPS studies indicated that water adsorbed molecularly [33] while EELS studies showed that water was dissociated into H and OH [34]. Because it was argued that the probe electrons in EELS could dissociate molecular water, studies using a non-destructive probe such as SIRS were necessary.

As shown in Fig. 11, the Si-H and O-H modes of dissociated water could be clearly observed upon exposure at room temperature [35,36]. Molecular water could only be observed by condensing water on an oxide layer or on a layer with dissociated water (water was found to dissociate on a *clean* Si(100) surface at 80K).

By using vicinal samples with steps along the $<0\bar{1}1>$ direction as was done for the H/Si(100) studies, it was possible to test whether dissociation takes place at steps by monitoring possible Si-H modes polarized along the $<011>$ direction, i.e. arising from H at steps. Although the data could not rule out the presence of some H at steps, they clearly showed that steps are not *saturated* with hydrogen [30]. Since the substrate temperature is too low for H to diffuse on the surface, it was concluded that water dissociation takes place on the terraces themselves.

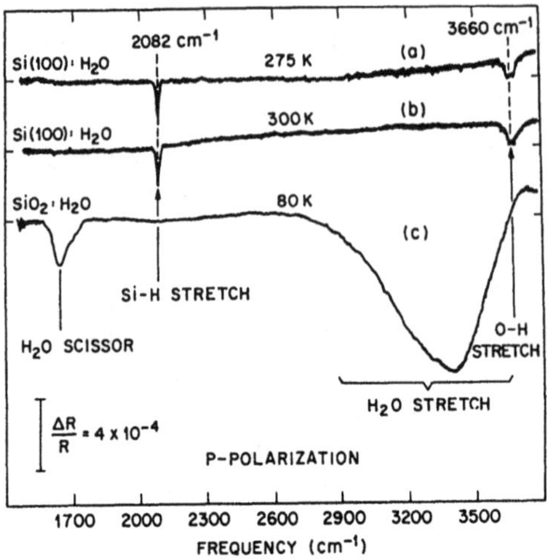

Fig. 11: Surface infrared spectra obtained upon exposure of clean Si(100)-(2x1) to (a) 0.5L water at T_s=275K, and (b) 10L water at T_s=300K. Curve (c) is obtained upon exposure of an oxidized Si(100) surface (native oxide) to 10L water at T_s=80K. Data were taken at the exposure temperatures as indicated on each spectrum.

5. PROBLEMS AND FUTURE DIRECTIONS:

While the previous section shows that chemical, structural and dynamical information of adsorbates at semiconductor surfaces can be obtained from SIRS and proper theoretical treatments, the technique faces severe limitations which must be addressed. In particular, the poor sensitivity to vibrational modes occurring in a frequency range where the substrate absorbs (usually the low frequency region) is very restrictive. Most of the adsorbate-substrate modes occur in this region. The substrate absorption rules out the use of multiple internal reflection geometry and photoacoustic detection. Emission spectroscopy [37] for which the sample acts as the IR source may be helpful if the largest solid angle compatible with the results of section 3.2 can be used. An interesting possibility regarding this problem is the use of synchrotron radiation in the range 100–1000cm^{-1}. For the geometrical requirements of surface studies (small spot size and low divergence), the synchrotron radiation is estimated to be one to two orders of magnitude brighter than a conventional source, depending on the frequency. Such an increase in intensity would make it possible to lose energy to the substrate in a total internal reflection configuration and yet have enough throughput for sensitive detection.

The second problem associated with the study of semiconductor surfaces is that of free carrier absorption when the substrate temperature is high. The ability to measure IR spectra on hot substrates is important for in situ studies of CVD processes. Emission spectroscopy is by far the most sensitive means of obtaining surface vibrational spectra for hot samples and might be performed without cooling the spectrometer if the samples are hot enough (i.e. T_{sample}>>300K).

The ability of SIRS to probe semiconductor surfaces under a variety of conditions is important because it makes it possible to use the technique for in situ measurements of technologically important processes. For instance, laser-induced photochemistry, plasma etching and thermal chemistry at semiconductor surfaces are processes which require the understanding of the microscopic mechanisms of surface reactions. SIRS has already been used to study photo-induced metal carbonyl decomposition of silicon surfaces [38] and several groups are setting up to do so.

Thanks to the improvement in fast scanning interferometers, SIRS appears to be a viable real-time probe for kinetic measurements requiring sub-second time resolution [2,3,39]. Resolutions of 25 msec are possible [40] using standard techniques. Better time resolution can be achieved for repetitive processes either by non-standard interferometric techniques or with grating spectrometers [41]. We feel that relatively large efforts will be made to tackle kinetic measurements in the next few years. Such measurements are particularly pertinent to the relevant technological processes at semiconductor interfaces.

6. CONCLUSIONS

In this lecture, the basic formalism necessary to design and analyze SIRS studies of semiconductor surfaces has been presented. The general conclusion is that the use of multiple internal reflection is more advantageous except in some cases for which external reflection or emission spectroscopy is better suited.

The various examples presented have emphasized the features of SIRS that make it a unique probe. These are:
(1) high resolution ($\sim$0.25cm^{-1}) necessary to resolve sharp modes ($\sim$1cm^{-1}) at semiconductor surfaces and to differentiate between various kinds of modes (e.g., H at steps or in monohydride structure, isolated monohydride and monohydride next to a dihydride);
(2) polarization properties, making it possible to probe and distinguish modes parallel to the surface. In contrast to MIR SIRS, specular EELS is *not* sensitive to the parallel modes. The combination of high resolution and polarization capabilities have made SIRS a much superior tool compared to EELS in the study of H on semiconductor surfaces. Finally,
(3) the compatibility of SIRS with ambient pressures and electric fields sets it apart as a powerful *in situ* probe. Other photon probes such as Raman scattering have not shown enough sensitivity so far to be used in a versatile way.

REFERENCES

[1] W. Ho, J. Vac. Sci. Technol. *A3,* 1432 (1985). T. H. Ellis, L. H. Dubois, S. D. Kevan, and M. J. Cardillo, Science *230,* 256 (1985).

[2] V. A. Burrows, S. Sundaresan, Y. J. Chabal and S. B. Christman, Surf. Sci. *160,* 122 (1985).

[3] V. A. Burrows, S. Sundaresan, Y. J. Chabal and S. B. Christman, Surf. Sci. *180,* 110 (1987).

[4] Y. J. Chabal, E. E. Chaban and S. B. Christman, J. Electron Spect. Rel. Phenom. *29,* 35 (1983).

[5] Y. J. Chabal, G. S. Higashi and S. B. Christman, Phys. Rev. *B28,* 4472 (1983).

[6] The proceedings of Fourth International Conference of Vibrations at Surfaces, published in J. Electron Spect. Rel. Phenom. (Volumes 38 & 39, 1986), is a good reference for other work done in the field.

[7] N. J. Harrick, "*Internal Reflection Spectroscopy*" (Wiley, New York, 1967). Second printing by Harrick Scientific Corporation, Ossining, N.Y. (1979).

[8] F. M. Mirabella and N. J. Harrick "*Internal Reflection Spectroscopy: Review and Supplement*" (Harrick Scientific Corporation, Ossining, N.Y. 1985). Also F. M. Mirabella, Appl. Spectrosc. Rev. *21*, 45 (1985).

[9] J. D. E. McIntyre and D. E. Aspnes, Surf. Sci. *24*, 417 (1971).

[10] D. E. Aspnes, private communication.

[11] L. D. Landau and E. M. Lifshitz, "*Electrodynamics of Continuous Media,*" (Pergamon Press, N.Y. 1960) Chapt. XI.

[12] P. Yeh, Surf. Sci. *96*, 41 (1980) and refs. therein.

[13] H. C. Chen, IEEE Trans. Micr. Theory Techn. *MTT-31*, 331 (1983).

[14] See p 27 of ref. 7.

[15] See pp 42-44 of ref. 7.

[16] E.g. F. Wooten, "*Optical Properties of Solids*" (Academic Press, N.Y., 1972).

[17] H. Ibach and D. L. Mills, "*Electron Energy Loss Spectroscopy and Surface Vibrations*" (Academic Press, N.Y. 1982) pp 93-98.

[18] G. D. Mahan and A. A. Lucas, J. Chem. Phys. *68*, 1344 (1978).

[19] Z. Schlesinger, L. H. Greene and A. J. Sievers, Phys. Rev. B *32*, 2721 (1985). See footnote 17.

[20] The value of C depends on the particular arrangement of the oscillators. For a planar triangular lattice, C=8.8904 as discussed in ref. 18.

[21] P. J. Feibelman, Progress Surf. Sci. *12*, 287 (1982).

[22] R. G. Greenler, J. Vac. Sci. Technol. *12*, 1410 (1975).

[23] P. Chiaradia "Optical Properties of Semiconductor Surfaces" published in the present volume.

[24] For example, some of the pioneering works were:
N. J. Harrick, Phys. Rev. Lett. *4*, 224 (1960).
G. E. Becker and G. W. Gobeli, J. Chem. Phys. *38*, 2942 (1963).
D. B. Novotny, J. Vac. Sci. Technol. *9*, 1447 (1972).
E. D. Palik, R. J. Holm, A. Stella and H. L. Hughes, J. Appl. Phys. *53*, 8454 (1982).
H. J. Stein and P. S. Peercy, Phys. Rev. *B22*, 6233 (1980).

[25] Y. J. Chabal and K. Raghavachari, Phys. Rev. Lett. *53*, 282 (1984).

[26] R. Kaplan, Surf. Sci. *93*, 145 (1980).

[27] F. Stucki, J. A. Schaefer, J. R. Anderson, G. J. Lapeyre and W. Göpel, Solid State Commun. *47*, 795 (1983).

[28] Y. J. Chabal and K. Raghavachari, Phys. Rev. Lett. *54*, 1055 (1985).

[29] Y. J. Chabal, Surf. Sci. *168,* 594 (1986).

[30] Y. J. Chabal, J. Vac. Sci. Technol. *A3,* 1448 (1985).

[31] J. C. Tully, Y. J. Chabal, K. Raghavachari, J. M. Bowman and R. R. Lucchese, Phys. Rev. *31,* 1184 (1985).

[32] A. Nitzan and J. C. Tully, J. Chem. Phys. *78,* 3959 (1983).

[33] D. Schmeisser, F. J. Himpsel and G. Hollinger, Phys. Rev. *B27,* 7813 (1983); D. Schmeisser, Surf. Sci. *137,* 197 (1984).

[34] H. Ibach, H. Wagner and D. Bruchmann, Solid State Commun. *42,* 457 (1982) and Appl. Phys. A*29,* 113 (1982).

[35] Y. J. Chabal, Phys. Rev. B *29,* 3677 (1984).

[36] Y. J. Chabal and S. B. Christman, Phys. Rev. B*29,* 6974 (1984).

[37] S. Chiang, R. G. Tobin, P. L. Richards and P. A. Thiel, Phys. Rev. Lett. *52,* 648 (1984).

[38] J. R. Swanson, C. M. Friend and Y. J. Chabal, Proceedings of the Materials Research Society Meeting (Boston, Dec. 1-6, 1986).

[39] M. A. Chesters, J. Electron Spect. Rel. Phenom. *38,* 123 (1986) and J. Vac. Sci. Technol. [May/June, 1987].

[40] J. E. Reutt and Y. J. Chabal, unpublished.

[41] B. A. Hayden, unpublished.

Raman Scattering from Interface Regions: Structure, Composition and Electronic Properties

J. Geurts[+] *and W. Richter*

1. Physikalisches Institut der RWTH Aachen,
D-5100 Aachen, Fed. Rep. of Germany

The abilities of Raman scattering from phonons for the investigation of interface regions are discussed. Starting with a general discussion of optical methods suitable for probing interfaces, the specific properties of phonon Raman scattering in relation to interfaces are presented. Information which can be obtained concerns the overlayer, the immediate interface region and the upmost substrate region. In contrast to electron spectroscopic methods, Raman scattering gives also information about deeply buried interfaces and it can be applied in a non-vacuum environment.

1. Optical Methods - General Considerations

The interaction of electromagnetic fields with matter is described by Maxwell's equations together with appropriate material equations. In general the material equation for the interaction with EM fields at optical frequencies is given by the polarization P, induced by the incident E-field E_i :

$$P = \varepsilon_0 . \chi . E_i \tag{1}$$

where the electric susceptibility χ describes the material response. In order to classify the different optical methods it is useful to expand χ in terms of the quantities by which it is affected:

$$\chi = \chi_0 + \frac{\delta\chi}{\delta Q} . Q + \frac{\delta^2\chi}{\delta Q \delta E_{st}} . Q . E_{st} + \chi_2 . E_i + \dots \tag{2}$$

The first term on the right hand side, χ_o is the normal first order susceptibility describing e.g. transmission, reflectivity or ellipsometry. The second and third term are so-called transition susceptibilities. In these interactions an elementary excitation of the solid is created or annihilated. As a consequence this causes a frequency shift of the emitted light with respect to the incident light.
Thus both terms describe inelastic light scattering. Here we deal with optical phonons as elementary excitations. The macroscopic variable Q in (2) may then be associated with the average lattice deformation. For the corresponding

---

[+] present address: IBM Forschungslaboratorium, Zürich (CH)

frequency shift region (10 to 1000 cm^{-1}) inelastic light scattering is commonly called Raman scattering. The second term describes the normal Raman scattering process, while the third term gives additional scattering in the presence of a static electric field E_{st}. The fourth term gives a polarization quadratic in the incident E-field. χ_2 is therefore the second order electric susceptibility and is responsible for second harmonic generation (SHG) of the incident light.
More terms may be written down in (2) as indicated by the dots. The criterion for the above selection was their usefulness for interface investigation. Ellipsometry (term 1) has been used already for a long time to characterize surfaces/interfaces /1/. Quite recently SHG (term 4) has been shown to provide an excellent tool for structural surface studies /2/. Phonon Raman scattering (term 2 and 3) is the topic of the present article and will be discussed at length in the following sections. We exclude here Raman scattering from free carriers (plasmons, single particle excitations) whose scattering intensity is normally far below those of phonons. However, they are useful for studies of the twodimensional electron gas at interfaces /3/.

Common to all optical methods in their application for interface investigation is that (i) deeply buried interfaces can be probed because of the large penetration and escape depth of the light and (ii) interfaces, which are already completed or under formation in non-vacuum conditions (reactive ion etching, gas phase epitaxy) can be studied. This gives additional abilities as compared to electron spectroscopy.

2. Raman Scattering

A typical experimental setup for Raman scattering at surfaces and interfaces under UHV conditions is shown in fig.1. The light source is usually a CW argon or

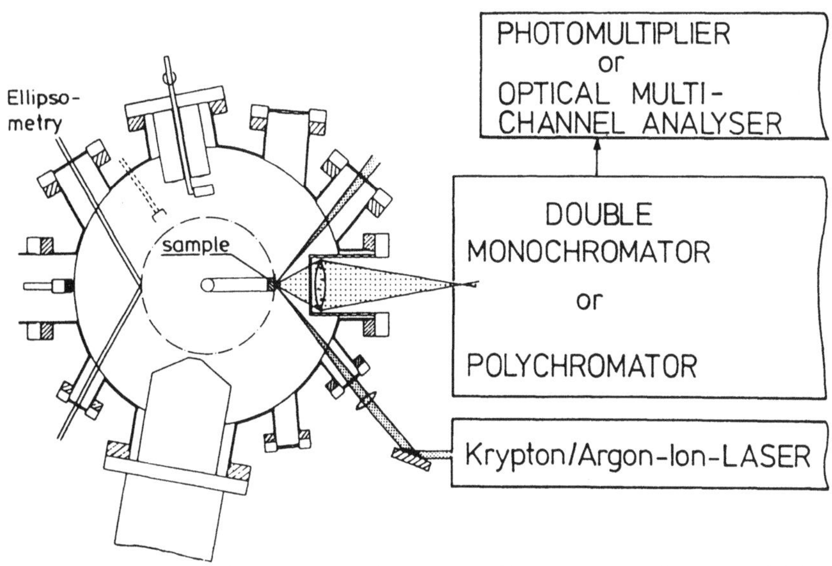

Fig.1: Experimental setup for Raman scattering at surfaces and interfaces under UHV conditions

krypton ion laser, sometimes combined with a dye laser. Besides having convenient properties for light scattering, such as well defined wavevector and high intensity, these lasers also offer a choice of many different light frequencies, quite often needed for resonance enhancement of the scattering cross section. The UHV chamber is equipped with a number of optical ports for easy access of the light beams. The scattered light is analyzed with a double monochromator or polychromator and detected by a photomultiplier or a multichannnel detector, respectively.

The information contained in a phonon Raman spectrum comes from (i) the frequency shift, (ii) the intensity and (iii) the linewidth. The frequency shifts, contained in a Raman spectrum are characteristic for the material and its structure and may serve as a fingerprint.

Besides, they may be changed under the influence of external forces. This effect is most prominent for mechanical stresses and has been studied quite extensively in bulk as well as in interface systems /4/. Changes of the strain in interface regions during various process stages can be monitored in this way.

The intensity is essentially given by the first order and second order derivatives in terms 2 and 3 of (2). They correspond in general to third and fourth rank tensors. Crystal symmetry reduces the number of possible nonzero components. These selection rules are quoted as 'Raman tensors' /5/. The magnitude of each tensor component can be expressed in terms of perturbation theory, based on the electron-radiation interaction and the electron-phonon interaction.

The most important mechanisms for the electron-phonon interaction are the deformation potential and the Fröhlich interaction /6/. While the former is always present for phonons, the latter occurs mainly for longitudinal phonons carrying a macroscopic electric field (IR-active phonons). Besides, the Fröhlich interaction, normally nearly zero for phonons with wavevectors close to zero, is strongly enhanced in the presence of a static electric field /7/ (see term 3 in (2)). This is the basis for probing space charge layers at interfaces.

Finally the phonon lifetime in the interface region and thus the corresponding spectral linewidth might differ from the bulk value. This can be interpreted in terms of geometrical effects such as crystallite size and/or layer thickness but also defects created during the processing stage of the interface (sputtering, ion implantation), which decrease the phonon lifetime.

3. Examples

In order to demonstrate the capabilities of Raman scattering we will discuss in the following some typical examples. They are grouped according to whether the information obtained concerns the overlayer, the immediate interface region or the substrate region.

3.1 Overlayer

Fig.2 gives phonon Raman spectra of an Sb layer on GaAs, recorded in situ during stepwise evaporation in UHV at substrate temperatures of 90K and 300K, respectively /8/. Sb Raman signals appear for coverages above approximately 3 ML. Its broad structure indicates amorphous growth of the Sb overlayer. At 90K

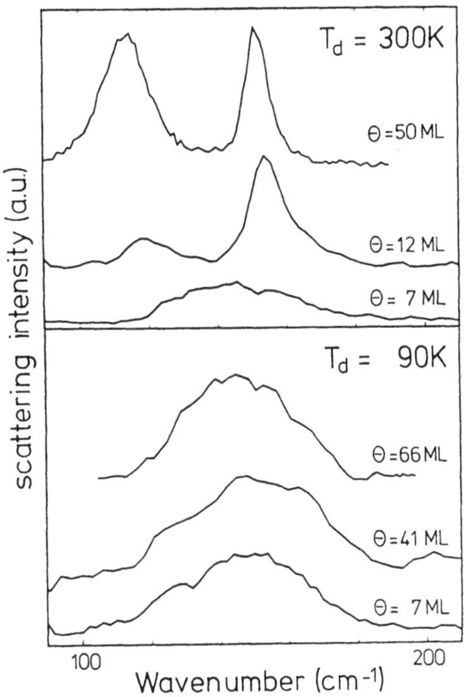

Fig.2:
Raman spectra of Sb on GaAs(110)
upper half: deposition at 300K substrate temperature
lower half: deposition at 90K substrate temperature
broad structure: amorphous Sb
peaks at 300K: crystalline Sb
(E_g and A_{1g} mode)

the Sb stays amorphous up to the highest investigated coverages (66 ML). At 300K, however, two distinct peaks appear for coverages above 12ML. They correspond to the phonon modes of crystalline rhombohedral Sb and show, that crystallization of the Sb layer occurs. Quantitatively the same behaviour is observed for Sb layers on InP substrates /9/.
However, the two Sb phonon peaks have a larger half width and are shifted in frequency as compared to the bulk values. From this we deduce a tensile stress of about 10 kbar immediately after the crystallization, while the crystallite size is estimated below 10nm. With increasing layer thickness the strain is reduced and the crystallite size grows /10/.
ABSTREITER et al. analyzed the influence of substrate temperature for Ge films on GaAs, materials lattice matched in the crystalline state. They observed amorphous growth for 100K, whereas epitaxial growth was achieved for 300K and 675K. The Ge phonon half width indicates an improved epitaxial growth at the higher temperature /11/.

The growth morphology can be analyzed from the attenuation of the substrate TO phonon intensity through the overlayer, as shown in fig.3 for Sb on GaAs /8/. At the start of the crystallization process (10ML) there is a dramatic change in the slope. Between 10 ML and 20 ML the substrate TO intensity is nearly constant. This is interpreted as an increase of surface roughness due to a partially 3-dimensional growth. After the completion of the crystallization, the slope increases again. This indicates a renewed homogeneous Sb growth.

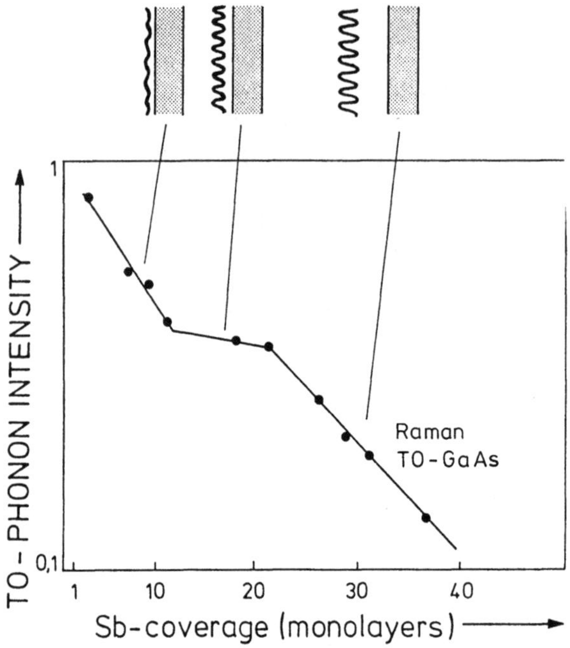

Fig.3:
integrated Raman peak intensity of the GaAs TO phonon as a function of Sb coverage.
Note the nearly constant intensity for 10-20 ML (due to Sb crystallization).

3.2 Interfacial Layers

Besides the analysis of layer growth, Raman spectroscopy can also yield information about chemical reactions between substrate and overlayer and reveal the composition of interlayers.

As an example, the growth of CdTe layers on InSb was analyzed for two different substrate temperatures: 300K and 500K /12/. For deposition at 300K a Te phonon peak is observed, indicating elemental Tellurium between substrate and overlayer. The results for 500K are shown in fig.4. In spectrum a the difference between the covered and the uncovered substrate is plotted to show the very weak interlayer signals, which prove that now an exchange reaction between overlayer and substrate occurs: we observe the A_{1g} mode of elemental Sb, as well as a series of peaks between 90 and 150 cm^{-1} whose position and relative intensities correspond to In_2Te_3 (cf. spectrum b: bulk In_2Te_3). The peak shifts of about 5 cm^{-1} in the interlayer with respect to the bulk indicate a strong strain of the interlayer. The existence of some interlayering In-Te compound was also confirmed by UPS results /13/. The structure at 190 cm^{-1} results mainly from changes in the InSb phonons.

A similar observation of interfacial reactions after annealing of Pt on Si was reported by Tsang et al. /14/, who conclude from Raman spectra the existence of different Pt-Si compounds.

Such reactions do not only occur after evaporation of overlayers, but can also take place after oxidation and annealing. Fig.5 shows the corresponding Raman spectra for the annealing of InSb at 510K in air. Increasing annealing time causes an attenuation of the InSb signal and the growth of the E_g and A_{1g} peaks of crystalline Sb. The Sb crystallites are strained, but they relax with increasing thickness, as it is seen from the shift of their phonon frequencies /15/.

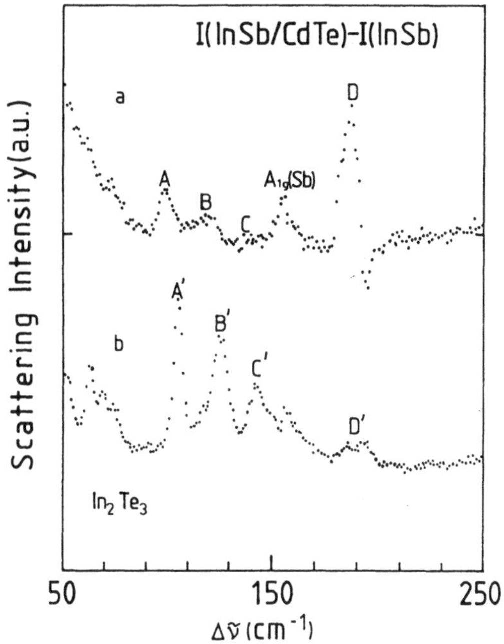

Fig.4:
Evaporation of CdTe on InSb (100) at 500K:
Spectrum a: Difference between Raman spectra for the covered and clean surface.
Peaks A,B, and C: In_2Te_3
Peak D: mainly changes in InSb phonon modes.
Peak at 155 cm^{-1} : elemental Sb
Spectrum b: bulk In_2Te_3

Sb phonon modes were also reported by Chang et al. after annealing InSb in O_2 atmosphere /16/. For GaAs similar results exist by Schwartz et al /17/. They observed crystalline and amorphous As after annealing an anodically oxidized GaAs substrate.

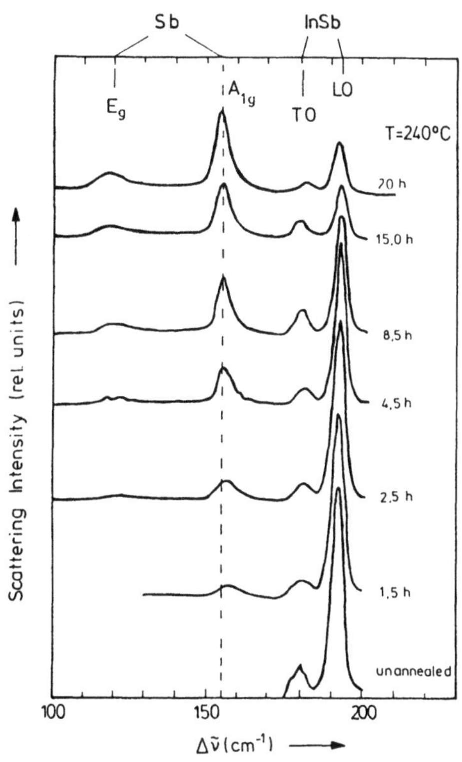

Fig.5:
Raman spectra of InSb (110) for different annealing times in air at 510K.
Spectra taken at RT. Note the increasing peaks of elemental Sb and the decreasing InSb signal.

3.3 Substrate Region

A unique ability of light scattering is the characterization of the substrate's interfacial region below a rather thick overlayer (up to about 100 ML). Here position and half width of the substrate phonon reveal eventual strain. Moreover, for III-V substrates, electronic band bending can be determined from the LO phonon intensity by electrical field induced Raman scattering.
Since for this scattering mechanism (term 3 in (2)) the transition susceptibility is proportional to the static electrical field E_{st}, its intensity is proportional to the square of this field. Therefore, in this method the band bending is detected through the corresponding electrical field in the space charge layer.
To derive the band bending from the LO intensity in the spectrum, the strongly inhomogeneous situation in the depletion layer has to be taken into account: (i) the static field $E_{st}(z)$ decreases with the distance z from the surface; in the Schottky approximation this decrease is linear, and the square of its value at the surface (or interface), $E_{st}^2(z=0)$ is proportional to the band bending V_s. (ii) the light absorption causes an exponential decrease of the incident electromagnetic field, as well as of the scattered (Lambert-Beer law).
The total LO phonon intensity is given by the integration over the whole space charge layer of thickness d_s. At suitable light frequencies the light penetration depth d is far below the space charge layer depth d_s and spatial variation of the field can be neglected. Then the integral is proportional to $E_{st}^2(z=0)$ i.e. proportional to the band bending V_s:

$$I_{LO} \propto \int E_{st}^2(z).\ \exp(-2z/d)dz \propto E_{st}^2(z=0) \propto V_s \ . \qquad (3)$$

Fig.6 shows Raman spectra, taken in situ during the stepwise evaporation of Sb onto GaAs in UHV /8/.The spectra show that the LO phonon increases strongly with increasing Sb coverage. The TO phonon, however, is constant after correction for the attenuation by the Sb layer. Therefore the TO intensity is convenient for normalization of the LO intensity, and the band bending will be discussed in terms of LO/TO intensity ratio, which is independent of overlayer attenuation or absolute laser intensity. This LO/TO ratio can be converted into a band bending value by calibration through Raman experiments on systems with a known band bending (e.g. from UPS or CPD), such as O_2 on GaAs. Details of these procedures can be found e.g. in /18/.
The converted LO/TO ratios for n-type GaAs/Sb are plotted in fig.7. Note the logarithmic coverage scale, going from .001 ML until about 100 ML. In the submonolayer region the band bending increases, and is in excellent agreement to UPS results from LÜTH et al /19/ (crosses). In addition, a strong variation in band bending is observed beyond 1ML. It shows a strong temperature dependence and is correlated to the Sb structure: for n-type GaAs amorphous Sb reduces the band bending whereas the crystallization at 300K near 12ML induces a reincrease. This reincrease might originate from additional surface defects by the crystallization, since the lattices of Sb and GaAs are very different. A detailed study of the influence of the substrate doping can be found in /8/.

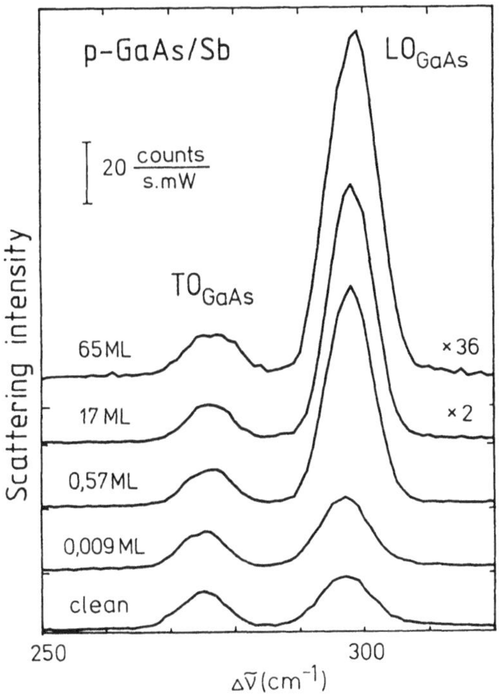

Fig.6:
Raman spectra of GaAs (110) for increasing Sb coverage, taken in situ under UHV conditions.
TO phonon: deformation potential scattering
LO phonon: electric field induced scattering
LO/TO intensity ratio is a measure of the band bending at the interface. (note the dynamics in coverage: from .01 ML up to 65 ML)

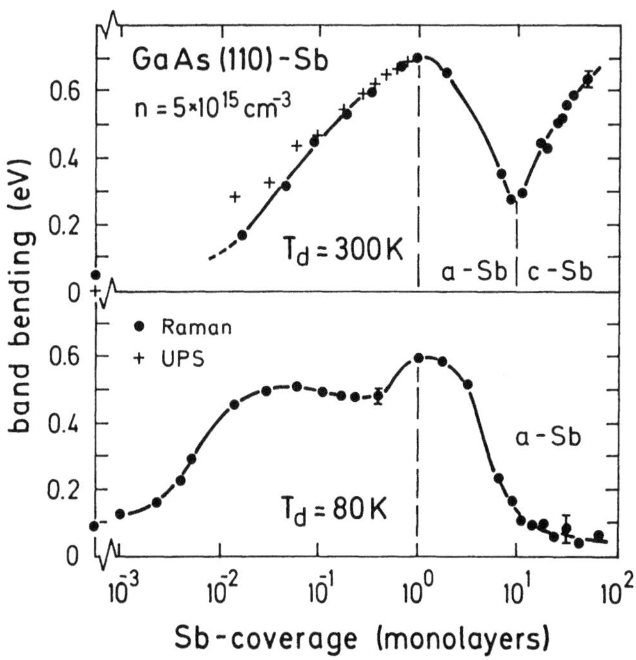

Fig.7: Band bending for the system n-GaAs(110)/Sb, determined by Raman scattering. Upper half: deposition at 300K; lower half: deposition at 80K
Note the strong dynamics in band bending above 1 ML, which is strongly influenced by the Sb structure (amorphous/crystalline)
Crosses: band bending values from UPS (from LüTH et al. /19/)

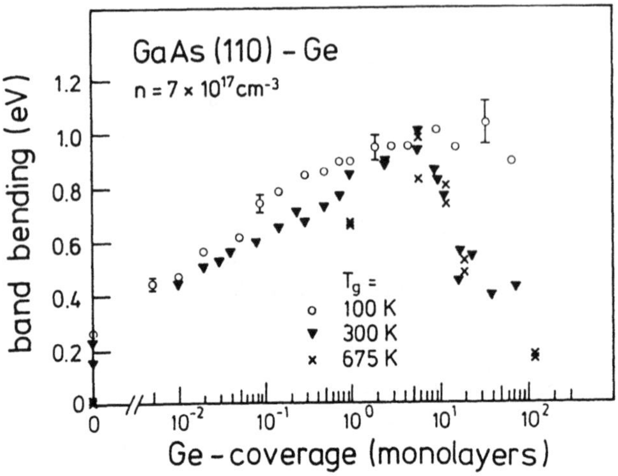

Fig.8: Raman-determined band bending for n-GaAs/Ge for different deposition temperatures. At 300K and 675K: recovering of flat band; at 100K: persistent band bending (from ABSTREITER et al. /11/)

For comparison, fig.8 gives results from Abstreiter et al. /11/ for the lattice matched system GaAs/Ge. Here the first ML of Ge also induces band bending. For amorphous growth at 100K this band bending is persistent, whereas crystalline epitaxial Ge (300K and 675K) does not induce surface defects, and flatband is recovered.

As an example for the determination of a substrate stress due to an overlayer, Fig.9 shows Raman spectra of the Si phonon at different positions on the same surface: one below a $TiSi_2$ electrode (thickness 100 nm) and the other beside

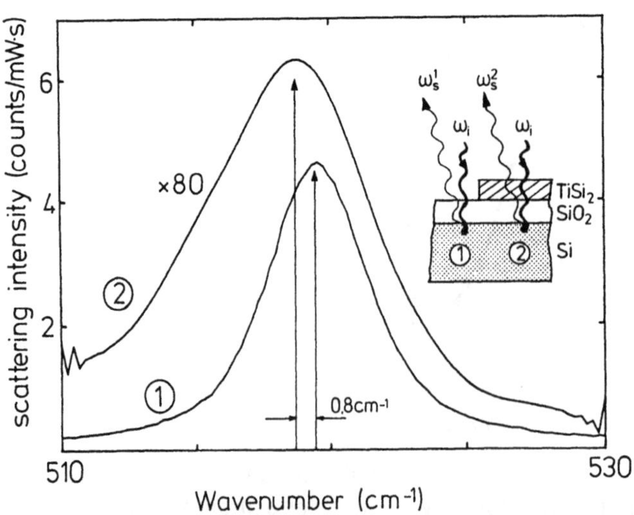

Fig.9: Determination of substrate stress under an overlayer by Raman scattering: Spectra of the Si phonon after a local $TiSi_2$ coverage of 100 nm.
(1) uncovered area; (2) below covering layer

/20/. A distinct phonon shift proves a tensile stress of 2kbar in the covered Si substrate region; the broadening is caused by the inhomogeneity of the stress, which decreases inside the substrate.

4. Conclusions

Raman spectroscopy has proved to be a valuable tool for the investigation of surfaces and interfaces. Its power is the ability to simultaneously analyse growth properties of overlayers and electronic properties of the interface up to high coverages. Investigations can be made in situ during growth as well as on completed systems under non-UHV conditions. Further, it has an excellent lateral resolution, which is important for the analysis of complex systems, such as contacted regions or devices.

References

1. D.E.Aspnes, Appl.of Surf.Sci. 22/23 (1985) 792
2. T.F.Heinz, M.M.T.Loy, and W.A.Thompson, Phys.Rev.Lett. 54 (1985) 63
3. For a review, see e.g.: G.Abstreiter, M.Cardona, and A.Pinczuk: In 'Light scattering in solids IV', ed. by M.Cardona and M.Güntherodt Topics in Applied Physics, Vol.54 (Springer, Berlin,Heidelberg, New York, Tokyo 1984)
4. S.R.J.Brueck, B.Tsaur, J.Fan, D.Murphy, T.Deutsch, and D.Silversmith, Appl.Phys.Lett. 40 (1982) 895
   E.Anastassakis: In Proc. 4th Int. School ISSPPME 1985, Varna (Bulgaria) Physical Problems in Microelectronics, p.128
5. R.Claus, L.Merten, and J.Brandmüller: In Springer Tracts in Modern Physics, vol.75, ed. G.Höhler (Springer, Berlin, Heidelberg, New York 1975)
6. W.Richter: In Springer Tracts in Modern Physics, vol.78, ed. G.Höhler (Springer, Berlin, Heidelberg, New York 1976)
7. M.L.Shand, W.Richter, E.Burstein, and J.G.Gay, J.Nonmetals 1 (1972) 53
8. W.Pletschen, N.Esser, H.Münder, D.Zahn, J.Geurts, and W.Richter Surf.Sci.178 (1986) 140
   W.Pletschen, PhD Thesis RWTH Aachen (FRG), 1986
9. D.Zahn, N.Esser, W.Pletschen, J.Geurts, and W.Richter, Surf.Sci.168 (1986) 823
   N.Esser, H.Münder, M.Hünermann, W.Pletschen, W.Richter, and D.Zahn J.Vac.Sci.Technol. (to be published)
10. W.Richter, U.Rettweiler, M.Hünermann, J.Geurts, W.Pletschen, and P.Lautenschlager (to be published)
11. H.Brugger, F.Schäffler, and G.Abstreiter, Phys.Rev.Lett. 52 (198 4) 141
    H.Brugger, Ph.D-Thesis, Technical University Munich (FRG), 1987
12. D.R.T.Zahn, K.J.Mackey, R.H.Williams, H.Münder, J.Geurts, and W.Richter, Appl.Phys.Lett.50 (1987) 742
13. K.J.Mackey, D.R.T.Zahn, P.M.G.Allen, R.H.Williams, W.Richter, and R.S.Williams, J.Vac.Sci.Technol.B (to be published)

14. J.C.Tsang, Y.Yokota, R.Matz and G.W.Rubloff, Appl.Phys.Lett. 44 (1984) 430
15. J.Geurts, Ph.D-Thesis, RWTH Aachen (FRG) 1984
16. R.L.Farrow, R.K.Chang, and S.Mroczkowski, Appl.Phys.Lett. 31 (1977) 768
17. G.P.Schwartz, B.Schwartz, D.Distefano, G.J.Gualtieri, and J.E.Griffiths, Appl.Phys.Lett. 34 (1979) 205
18. see e.g. ref. /11/, ref. /8/, and
F.Schäffler and G.Abstreiter, Phys.Rev.B 34 (1986) 4017
19. M.Mattern-Klosson, and H.Lüth, Sol.St.Comm. 56 (1985) 1001
20. J.Finders, J.Reuters, J.Geurts, W.Richter, and P.Balk, (to be published)

Part VI

Interfaces: Present Status and Perspectives

Role of Interfaces in Semiconductor Heterostructures

C. Delalande and G. Bastard

Groupe de Physique des Solides de l'Ecole Normale Supérieure,
24 rue Lhomond, F-75005 Paris, France

I - Introduction

The design of new microelectronics or optoelectronics devices often involves semiconductor heterolayers which are thin enough to allow pronounced quantum size effects. The latter are due to the confinement of electrons, holes or excitons within one layer with little penetration into the other layers. To achieve such a confinement it is necessary that potential energy steps exist between the heterolayers in order to prevent the leakage of the carrier wave function outside the slab where it is essentially localized. Thus, behind quantum confinement there is the idea that at the interfaces the conduction and valence band edges change rapidly over a small distance. One of the central questions of heterolayer physics is thus the magnitude of the fraction of the band-gap energy difference between two materials which is taken by, say, the valence band edge, i.e. the question of the band offset.

Although efforts are currently being made to provide a priori answers to this question, it is better solved experimentally, either through electrical measurements or through optical ones. To obtain accurate values of the band offsets, one should then design heterostructures of the two materials where band edge profiles lead to the existence of electronic states which are sensitive to the offsets. This is not a trivial matter. For instance, the most popular band edge profile, which is a rectangular quantum well, displays energy levels which are remarkably insensitive to the barrier heights.

In addition, much of the heterostructure modelling assumes that the interfaces which separate two consecutive layers are ideally sharp. This is only an approximation, which fortunately turns out to be realistic for the heterostructures whose growth is well controlled. The departure from ideally sharp interfaces, which will be termed interface grading, has in fact a modest effect on the energy levels of heterostructures.

Actual interfaces always display interface defects which scatter and trap the electrons, holes or excitons. These defects are better studied optically than electrically.

The purpose of the present paper is to present some theoretical and experimental results on the various effects that the interfaces may have on the electronic properties of semiconductor heterolayers. The paper will be organized as follows. In the first part we describe how the electronic states of the heterolayers are sensitive to the interfaces. In the second part we concentrate on optical probes of actual interfaces.

II - Effects of the interfaces on the energy levels of semiconductor heterolayers

There are various stages at which one may discuss the influence of interfaces on the heterostructure energy levels.

II - 1) Band offsets

Firstly, one may inquire how the one-electron potential energy spatially varies when going from one perfect layer to the next one. This requires fairly complex microscopic calculations of the band structure of the heterolayer, involving in particular the evaluation of the interface dipoles which are established at the interface. Such dipoles exist because there is no reason why the A-C bond at the interface should be the same as the C-B bond if one considers AC-BC heterolayers. Various theoretical schemes have been proposed to cope with this problem. Noticeably, Harrison's tight binding approach [1] or Flores-Tejedor-Tersoff's neutral gap [2-4] point rule aim to predict how the bandgap energy difference between two semiconductors is shared between the valence and conduction bands. These various theories provide conflicting results on the band offsets and are hardly more precise than a fraction of an electron volt. This is insufficient for most practical purposes. The knowledge of the band offsets between two semiconductors is however important in as much as the electronic properties of the heterolayers are largely controlled by this offset. This point is illustrated in Fig.1.

The best way to determine the band offsets is of course to measure them... Many measurements are conceivable : i) X ray photoemission at the interface [5]. A carefull measurement of the position of the core state of, say, As with respect to the last occupied level while growing GaAs and then AlAs provides a determination of the band alignment of AlAs with respect to GaAs. There are however difficulties, e.g. associated with the apparatus energy resolution (a fraction of an electron volt) which may cast doubt on the reliability of XPS measurements. ii) Electrical measurements on isotype heterostructures [6]. By measuring the current-voltage characteristics of a biased n-GaAs-n-Ga(Al)As heterojunction, it is possible to extract the conduction

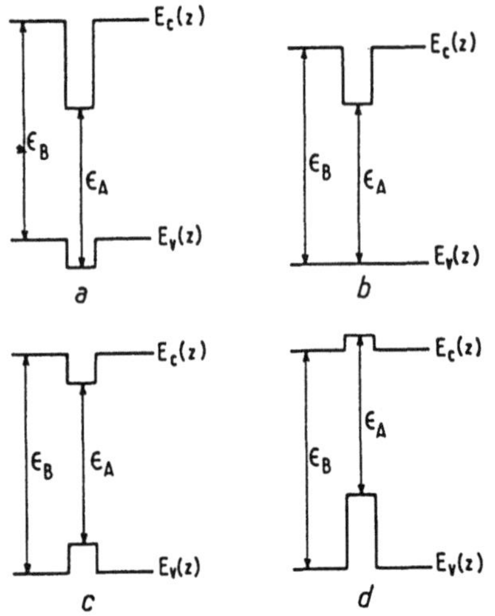

Fig. 1 Illustration of the part played by different apportionments between the valence and conduction bands of the bandgap energy difference $E_B - E_A$ of two semiconductors A and B on the electronic states of a BAB rectangular quantum well. The structure (c) which confines both electrons and holes in the A layers is the best candidate for low threshold lasing action.

band discontinuity between these two materials. The measurements are relatively easily done, but the interpretation relies on a knowledge of the actual doping of the heterostructure, which may be poorly known. iii) Optical determinations of the energy levels in quantum wells or superlattices built out of the two materials [7]. In high quality heterostructures the absorption or photoluminescence excitation lines are sharp enough and the experimental resolution is practically infinite. Thus, the measurements of the optical transition energies are very accurate. The determination of the band offset is, however, indirect since it relies on the sensitivity of the energy level differences to the conduction and valence band offsets. This sensitivity may be less than that to the structural parameters of the heterolayers : slab thicknesses, aluminium percentage in the Ga(Al)As barriers, etc. It is now possible to get rid of most of the experimental uncertainties on the structural parameters by using the RHEED oscillations [8], X ray diffraction, etc. Thus, one may believe that the measured transition energies in the optical experiments will allow the determination of the band offset. Yet, the band edge profile of the heterolayer has to be carefully optimized. In fact, the most popular band edge profile used so far is that of a rectangular quantum well. Unfortunately, this structure is very ill-suited to band offset determinations since the energy levels are very weakly dependent on the offsets. This is easily appreciated using the following argument. Suppose we wish to find the bound states supported by the one-dimensional potential (see Fig. 2)

$$V(z) = \begin{cases} V_b\,(z/L)^{\alpha} & 0 \leq z \leq L \\ V_b & z \geq L \end{cases} \qquad (1)$$

and to analyze their sensitivity to the conduction band offset V_b. We note that the limit $\alpha \to \infty$ corresponds to a rectangular quantum well, whereas $\alpha = 2$ corresponds to a finite parabolic well and $\alpha = 1$ to a v-shaped groove. Let m^* denotes the carrier

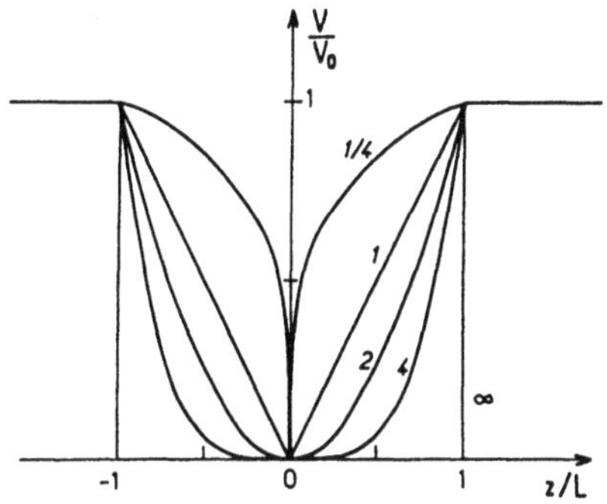

Fig. 2 : The potential energy V(z) defined in Eq.(1) is plotted versus z/L.

effective mass. Using the WKB approximation the bound energy levels ε_n ($0 \leqslant \varepsilon_n \leqslant V_b$) are such that

$$2\int_0^{L(\varepsilon_n/V_b)^{1/\alpha}} \sqrt{(2m^*/\hbar^2)\,[\varepsilon_n - V(z)]}\,dz = (n+\nu) \qquad (2)$$

where $0 \leqslant \nu \leqslant 1$. Equation (2) leads to

$$\varepsilon_n = V_b^{(2/\alpha+2)} \left(\frac{\hbar^2}{2m^*L^2}\right)^{(\alpha/\alpha+2)} \left(\frac{n+\nu}{2I_\alpha}\right)^{(2\alpha/\alpha+2)} , \qquad (3)$$

$$I_\alpha = \int_0^1 \sqrt{1-t^\alpha}\,dt \quad .$$

We see from (3) that the rectangular quantum wells ($\alpha \rightarrow \infty$) display energy levels which, at the WKB approximation, are V_b independent. This is $\alpha = 0$ which renders ε_n the most V_b dependent. Unfortunately $\alpha = 0$ means no well at all. Thus a compromise is necessary between the lower α, the feasibility of the structure and the existence of bound states which are not too shallow. Notice finally that the choice of an even potential is not necessary. We could have reached a conclusion similar to (3) by replacing the evenness condition on V(z) by $V(z) = V_b$ for negative z. In part II we shall present optical determinations of V_b in GaAs-Ga(Al)As heterolayers which have been obtained on structures (the separate confinement heterostructures) which are a rough approximation of v-shaped grooves ($\alpha = 1$).

II - 2) Interface grading

Another aspect of the effect of interfaces on the heterolayers' electronic states is the necessary allowance for a finite thickness of the interfaces [9]. Such an interface grading should extend, even in perfectly lattice-matched heterolayers, over one or two monolayers. This is to comply with the existence of mixed chemical bonds established between the interface atoms and their nearest neighbours in each layer. In GaAs-Ga(Al)As structures, it appears that under the condition of ideally controlled growth the interface grading retains its minimal value. In turn, the grading has a modest effect on the energy levels [10]. In other heterostructures, such as InP-Ga(In)As, the growth control appears more difficult and, as a result, the grading has more severe effects. This may induce significant changes in the energy levels and falsify the informations one could extract out of a comparison between the measurements of the energy level differences and models which assume the interfaces to be sharp. Quite generally however, one may show [11] that a quantum well with thickness L graded over distances b and b' on its two sides display bound states which, to lowest order, coincide with those of a perfectly rectangular one with thickness L + (b+b')/2. This result holds irrespective of the exact shape of the grading, provided it is smooth.

We have checked the accuracy of this prediction by numerically calculating the energy levels of a InP-Ga(In)As quantum well linearly graded on one side and by comparing these numerical results with those obtained for a rectangular quantum well with thickness L+(b/2). Such a comparison is presented in Fig. 3 where it is seen that over a wide range of grading and for several quantum well thicknesses the rule is well obeyed.

II - 3) Interface defects

The interface grading is not a defect since it is translationally invariant in the layer plane. What may scatter the plane waves and even trap the carriers are the local

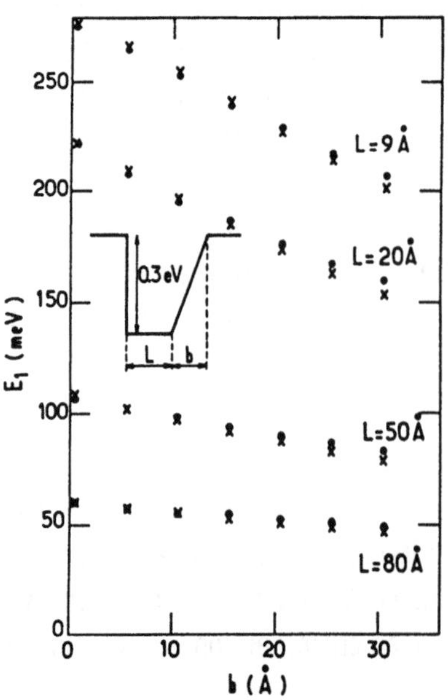

Fig. 3 : Comparison between the calculated energy levels of a graded quantum well (dots) and their approximation by those of a rectangular quantum well with thickness L + b/2 (crosses).

protrusions of the barrier into the well or vice versa. An interface defect is ,as shown below, and inefficient scatterer. This is because it is a short range scatterer (for the in-plane motion) and in addition a weak one since, in most instances, the probablility amplitude of finding the carrier integrated over the defect depth is small : the carrier wave function for the z motion peaks away from the interface and the probability of finding the carrier in the whole barrier seldom exceeds a few percent. Thus, on general grounds one expects the intrinsic defects to have little influence on the in-plane mobility of the carriers. Such is not the case if during the growth process there is an impurity segregation near the interface. This situation often occurs in MBE-grown GaAs-Ga(Al)As layers. If these impurities are coulombic (e.g. carbon in GaAs which behaves as a singly charged acceptor) their long range potentials drastically affect the carrier mobility. The impurity segregation, perhaps correlated with interface roughness, may explain the anomalously low mobility of inverted GaAs-Ga(Al)As heterojunctions, i.e. those corresponding to GaAs grown on top of Ga(Al)As. In fact, as will be reported below there is a clear correlation between the inverted Ga(Al)As-GaAs interface and the occurrence of an impurity-related photoluminescence of GaAs quantum wells or modified quantum wells.

Intrinsic interface defects, if attractive, give rise to a series of localized states below the edge of the extended states. This holds for single particles, the electrons and the holes, but also for two particle complexes : the excitons. In the latter case the internal motion may be assumed frozen, e.g. in a quasi 1S state, and exciton trapping corresponds to the formation of a bound state for the center of mass degree of freedom. In photoluminescence experiments the trapping time from a delocalized state to a bound state is often shorter than the radiative lifetime. Thus, the bound states associated with the interface defects may become populated and through a subsequent radiative recombination there is a possibility of optically detecting these bound states (see below).

Let us examine first the effect of interface defects on the extended states and then discuss the bound states. As often in semiconductor hetero-layers we concentrate on the effective Schrödinger equation fulfilled by the envelope function. In addition, we focus our attention on heterostructures which are characterized by a strong quantization of their z motion. By this we mean that the energy separation between the ground subband and the first excited one is large compared to kT, the level broadening and the effects of defects. Thus we write for the ground subband

$$\psi_{k_\perp}(\vec{r}) = \chi_1(z)\,\frac{\exp i\,\vec{k}_\perp \vec{r}_\perp}{\sqrt{S}} \qquad \textbf{(4)}$$

where S is the sample area and $\chi_1(z)$ is the ground bound state of

$$[T_z + V_b(z)]\,\chi_1(z) = E_1\,\chi_1(z) \quad ,$$

where T_z is the kinetic energy operator and $V_b(z)$ the band edge profile of the heterostructure. Let $V_{def}(\vec{r})$ be a defect potential. Neglecting the admixtures between subbands, the in-plane motion is the solution of

$$[T_\perp + \int \chi_1^2(z)\,V_{def}(\vec{r}_\perp,z)\,dz]\,\varphi(\vec{r}_\perp) = \varepsilon\,\varphi(\vec{r}_\perp) \quad , \tag{5}$$

where the energy zero has been taken at the edge of E_1 subband. In the absence of defects the two-dimensional plane waves of (10) are eigenstates of (5) with eigenenergies $\varepsilon = (\hbar^2 k_\perp^2/2m^*)$ where m^* is the in-plane effective mass.

We shall use very simplified models of interface defects :

$$V_{def}(\vec{r}_\perp,z) = \sum_{\vec{\rho}_i} V_{def}(\vec{r}_\perp - \vec{\rho}_i, z) \quad , \tag{6}$$

where

$$V_{def}(\vec{r}_\perp - \vec{\rho}_i, z) = -V_b\,Y(-z)\exp - \left[\frac{(\vec{\rho}-\vec{\rho}_i)^2}{2a_i^2} + \frac{z^2}{2b_i^2}\right] \tag{7}$$

or

$$V_{def}(\vec{r}_\perp - \vec{\rho}_i, z) = -V_b\,Y(-z)\,Y(z+b_i)\,Y\,[a_i - |\vec{r}_\perp - \vec{\rho}_i|] \tag{8}$$

for attractive interface defects. In (7,8) V_b is the conduction band offset between the two materials, $z = 0$ is the nominal interface and a_i, b_i are lengths which characterize the defect size in the layer plane or along the growth axis. The defect defined in (8) is a pill box defect while (7) will be termed a semi-Gaussian interface defect (see Fig. 4). Equations (7,8) correspond to protrusions of the weM acting material into the barrier acting material which are thus attractive. Repulsive interface defects correspond to the reverse situation and are described by (7,8) where V_b is changed into $-V_b$ and z into $-z$. The choice of the nominal interface at $z = 0$ means that the average of the potential created by all the defects, either attractive or repulsive, over the

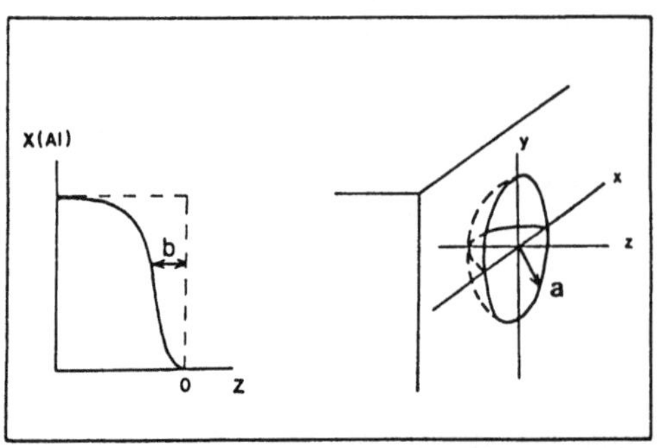

Fig. 4 Schematic representation of a semi-Gaussian interface defect.

unperturbed states $\Psi_{k_\perp}$ vanishes :

$$\frac{1}{S} \sum \int dz\ d^2r_\perp\ \chi_1{}^2(z)\ V_{def}(\vec{r}_\perp - \vec{\rho}_i, z) = 0 \quad , \tag{9}$$

Let N_a/S, N_r/S be the areal concentrations of attractive and repulsive defects respectively. Equation (9) can be rewritten

$$V_b < \int_{-\infty}^{0} dz\ \chi_1{}^2(z) \exp\frac{-z^2}{2b^2} \int_0^{\infty} 2\pi r_\perp\, dr_\perp \exp\frac{-r_\perp{}^2}{2a^2} >_{a,b} [-\frac{N_a}{S} + \frac{N_r}{S}] = 0, \tag{10}$$

where $< >_{a,b}$ means an average over the statistical distribution of the defect sizes a and b. Quite naturally the nominal interface z = 0 is such that there are as many attractive as repulsive defects. Such a result has been obtained because we have implicitly assumed that the a b statistics are the same for both attractive and repulsive defects, which may be true. An expression similar to (10) would be obtained if (8) were used instead of (7).

Let us now calculate the level lifetime $\tau_{k_\perp}$ of the state $\Psi_{k_\perp}$. In the Born approximation it is given by

$$\frac{1}{\tau_{k_\perp}} = \frac{2\pi}{\hbar} \sum_{\vec{q}_\perp} K \Psi_{k_\perp} |V_{def}| \Psi_{k_\perp + q_\perp} >|^2 \delta\, [\varepsilon_{k_\perp + q_\perp} - \varepsilon_{k_\perp}] \quad . \tag{11}$$

Using (7) we obtain

$$\frac{1}{\tau_{k_\perp}} = \frac{8\pi m^*}{\hbar^3} V_b{}^2 < a^4 g(b) \int_0^{\pi/2} d\varphi \exp[-2k_\perp{}^2 a^2 \sin^2\varphi] >_{a,b} \quad ,$$

$$g(b) = N_a \left| \int_{-\infty}^{0} \chi_1{}^2(z) e^{-(z^2/2b^2)} dz \right|^2 + N_r \left| \int_0^{\infty} \chi_1{}^2(z) e^{-(z^2/2b^2)} \right|^2 \quad . \tag{12}$$

It is interesting to see that the level lifetime is finite at the subband edge ($k_\perp = 0$). This feature reflects the two-dimensional density of states for the in plane motion which is step-like at the bottom of the subband. We also see that, in contrast to (10), attractive and repulsive defects add to shorten the level lifetime. This is a consequence of the Born approximation. Finally $\tau_{k_\perp}$ increases with increasing $\vec{k}_\perp$. Energetic carriers are less and less influenced by the interface defects since their de Broglie wavelength becomes much shorter than the defect in plane size (see Fig. 5).

To obtain an estimate of the interface defects contribution to the level lifetime we need to evaluate (12) explicitly. Under the assumption of a reasonably

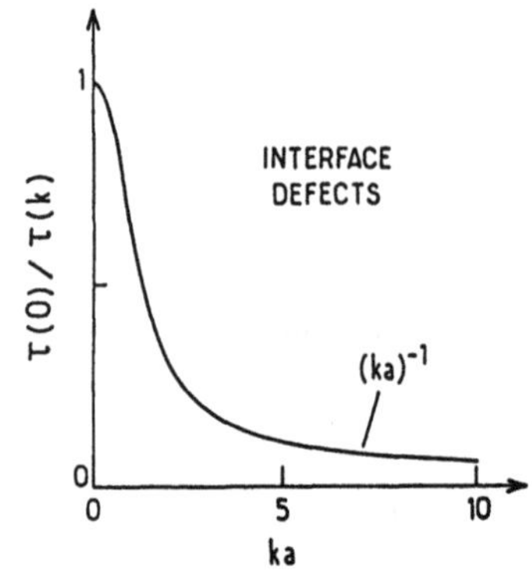

Fig. 5 : Relative variation of the level lifetime τ_k versus ka where a is the in plane extension of a semi-Gaussian defect.

good interface (say as currently obtained in GaAs-Ga(Al)As heterolayers) the defect depth b is equal to one or two monolayers, thus much smaller than the usual scale of variation of χ_1 near the interface. Let us calculate $\tau_{k_\perp = 0}$ to obtain lower bound of the level lifetime. Then

$$\frac{1}{\tau_0} \sim \frac{2\pi}{\hbar}\frac{m^*}{\hbar^2} V_b^2 \, 4\pi \left(\frac{N_a + N_r}{S}\right) \chi_1^4(0) < a^4 b^2 >_{a,b} \qquad (13)$$

We evaluate $\chi_1^4(0)$ for a rectangular quantum well and obtain

$$\chi_1^4(0) \sim \left(\frac{k_0}{{}_B}\right)^4 \left(\frac{2}{L}\right)^2 \sin^4 \frac{k_0 L}{2} \qquad (14)$$

where k_0 ($_B$) is the carrier wave vector in the well (in the barrier) and L the quantum well thickness

$$\frac{\hbar^2 k_0^2}{2m^*} = E_1 \qquad \frac{\hbar^2 \; {}_B^2}{2m^*} = V_b - E_1 \qquad (15)$$

Finally

$$\frac{\hbar}{2\pi\tau_0} \sim \frac{2m^*}{\hbar^2} \left(\frac{V_b}{V_b - E_1}\right)^2 \frac{E_1^2}{L^2} \left(\frac{N_a + N_r}{S}\right) < 8a^4 b^2 >_{a,b} \sin^4 \left(\frac{k_0 L}{2}\right) \qquad (16)$$

Taking L = 100 Å, V_b = 220 meV, E_1 = 30 meV, $m^+ = 0.07\, m_0$, $(N_a+N_r)/S = 10^{10}\ \mathrm{cm}^{-2}$, $k_0 L \sim \pi$, $< 8a^4 b^2 >_{a,b} = 25\ 10^{-40}\ \mathrm{cm}^{-6}$ as representative figures we get

$$\frac{\hbar}{2\pi\tau_0} \sim 0.17\ \mathrm{meV} \qquad (17)$$

If we were to convert the level lifetime into carrier mobility, we would obtain mobility figures in the 10^4 cm^2/Vs range, which is considerably less than obtained in

GaAs-Ga(Al)As heterojunctions. However, several factors add to raise the mobility relaxation time τ_t ($\tau_t = (m^*\mu)/e$) with respect to the level lifetime. Firstly, the measured mobility relaxation time involves the Fermi wave vector k_F and not $\vec{k}_\perp = 0$. With $k_F = (2\pi n_e)^{1/2}$, where n_e is the areal concentration of free carriers, the condition $k_F a \sim 1$ is easily fulfilled. As seen from (12) this increases τ_k. Secondly the free carriers screen the interface defect potential. This adds at least an extra factor $(\varepsilon_r)^2$ to τ where $\varepsilon_r \sim 12$ is the static dielectric constant of the heterostructure. Finally the transport relaxation time is always larger than the level lifetime. While the latter represents an unweighted average of the elemental transition probability $W_{k \to k'}$ over the angle θ between $\vec{k}_\perp$ and $\vec{k}'_\perp$, the former is weighted by $(1-\cos\theta)$ since it is a measure of the relative decrement of the carrier velocity. The $(1-\cos\theta)$ term effectively suppresses the small angle scattering events which, as may be directly checked from (12) provide the larger contribution to $1/\tau$.

In fact, when the scattering by interface defects is calculated for GaAs-Ga(Al)As heterojunctions it is found to be of negligible importance with respect to the coulombic scattering (see e.g. reference [11]). Notice that the evaluation of the level lifetime associated with coulombic scatterers requires the coulombic potential to be screened in order to suppress the $1/q$ divergence of $< \Psi_k |(1)/(r-r_i)|\Psi_{k+q} >$. Without screening (or an ad hoc cut off factor) we would find $\tau_k = 0$, a singularity which arises from the long range nature of the coulombic potential. Without entering into the details of the calculations, one may notice that the two-dimensional electron gas located near the inverted GaAs-Ga(Al)As interfaces that experience impurity segregation, display systematically lower (or much lower) mobilities than that located near the good interface. One may be tempted to blame the segregated impurities for such a behaviour since they constitute efficient, long range scatterers which are by (inadvertent !) construction placed near the two dimensional electron gas.

When the defect potential is attractive, bound states for electrons, holes or excitons are formed below the respective subband edges. Let us concentrate on excitonic bound states and assume that the interface defects are diluted enough to neglect multi defect effect (for a recent analysis of the latter effects on single particle states see reference (13)). Neglecting the valence subband intricacies the effective exciton-defect hamiltonian is

$$\left\{\frac{P_\perp^2}{2M_\perp} + \frac{P_\perp^2}{2\mu_\perp} - \text{—} < \frac{1}{r_{eh}} > - \tilde{V}_e \exp[-\rho_e^2/2a^2] - \tilde{V}_h \exp[-\rho_h^2/2a^2]\right\}$$

$$\times\; F(\vec{\rho}_e,\vec{\rho}_h) = [\varepsilon - \varepsilon_g - E_1 - HH_1]\, F(\vec{\rho}_e,\vec{\rho}_h) \quad , \qquad \textbf{(18)}$$

where

$$< \frac{1}{r_{eh}} > = \iint dz_e\, dz_h\, \chi_{1e}^2(z_e)\, \chi_{1h}^2(z_h)\, [\rho_{eh}^2 + (z_e - z_h)^2] \quad , \tag{19}$$

$$\tilde{V}_e = V_e \int \chi_{1e}^2(z_e) \exp[-z_e^2/2b^2]\, dz_e \quad , \tag{20}$$

$$\tilde{V}_h = V_h \int \chi_{1h}^2(z_h) \exp[-z_h^2/2b^2]\, dz_h \quad , \tag{21}$$

V_e and V_h are barrier heights for electrons and holes respectively, E_1 and HH_1 are the ground electron and heavy hole confinement energies and $\chi_{1e}(z_e)$, $\chi_{1h}(z_h)$ the ground bound state envelope functions for electrons and holes respectively.

The in-plane kinetic energy in (8) has been rewritten in terms of a part associated with i) the center of mass motion : $(P_\perp^2)/(2M_\perp)$ where $M_\perp = m_e + m_{h\perp}$ is the total exciton mass, and ii) the exciton internal degrees of freedom : $(p^2/2\mu_\perp)$ where $\mu_\perp^{-1} = m_e^{-1} + m_{h\perp}^{-1}$ is the reduced exciton mass.

In the absence of defect potential, the exciton eigenstates factorize into

$$F^{(0)}(\vec{\rho}_e, \vec{\rho}_h) = 1/\sqrt{S}\ \exp(i\vec{K}_\perp . \vec{R}_\perp)\, g_\nu(\vec{\rho}_e - \vec{\rho}_h) \quad , \tag{22}$$

which describes an exciton moving freely in the layer plane. Each eigenenergy ε_ν of the hamiltonian describing the relative electron-hole motion is the onset of a parabolic two-dimensional subband and one can write

$$\varepsilon_{\nu K_\perp} = \varepsilon_g + E_1 + HH_1 + \varepsilon_\nu + (\hbar^2 K_\perp^2)/(2M_\perp) \quad . \tag{23}$$

The envelope functions g, which describe the internal degrees of freedom may correspond either to a bound electron-hole pair ($\varepsilon_\nu < 0$) or to a dissociated one ($\varepsilon_\nu > 0$). A good variational estimate of the ground bound exciton state (quasi 1S) is

$$g_{1S}(\vec{\rho}_e - \vec{\rho}_h) = 1/\lambda \sqrt{2/\pi}\ \exp[(-1/\lambda)|\vec{\rho}_e - \vec{\rho}_h|] \quad . \tag{24}$$

We show in Fig. 6 the calculated free exciton binding energy versus the GaAs slab thickness L in GaAs-$Ga_{0.7}Al_{0.3}As$ single quantum wells using the variational wave function written in (24) [14]. One notices that the exciton binding energy first increases with decreasing L (increasing quasi bi-dimensionality) and then decreases either to zero (exciton attached to excited subbands) or to the bulk exciton Rydberg value corresponding either to the well or to the barrier. Notice that the correct behaviour at very small or very large L cannot be obtained using (24) since bulk exciton envelope functions are described by envelope functions which are not separable into $\rho_e - \rho_h$ and z_e, z_h.

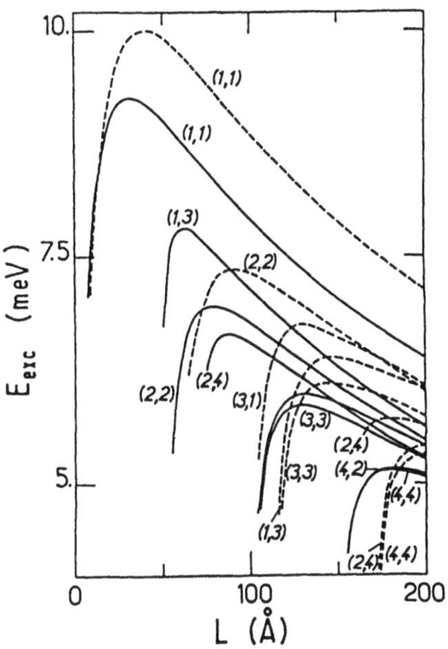

Fig. 6 : The free exciton binding energies in $GaAs-Al_{0.7}Al_{0.3}$ are plotted versus the GaAs slab thickness L. After reference [14]

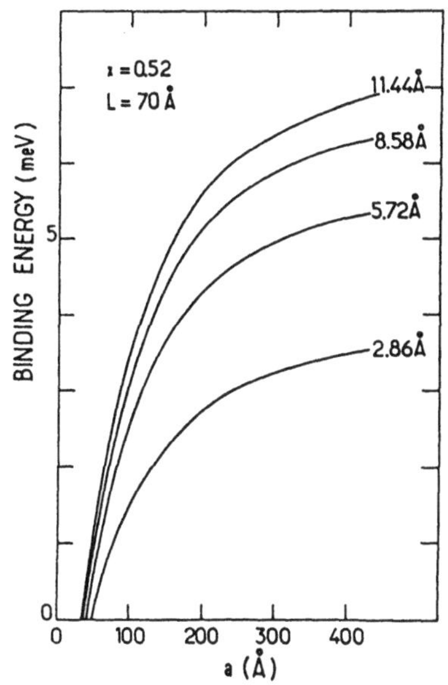

Fig. 7 : The calculated binding energy of E_1-HH_1 excitons bound to semi-Gaussian interface defects is plotted versus the defect in plane extension a for several defect depths b in a $GaAs-Ga_{0.48}Al_{0.52}As$ single quantum well. L = 70 Å. After reference [15]

For L ~ 100 Å the exciton binding energy is about twice as large as in bulk GaAs. This suggests that in the presence of interface defects one may consider the exciton internal degrees of freedom as being frozen and thus that the exciton binds to the interface defect through a localization of the center of mass. A simple trial wave function for the bound exciton is thus [15]

$$F(\vec{\rho}_e,\vec{\rho}_h) = 1/(\lambda_{1S})\sqrt{(2/\pi)}\ \exp(-|\rho_e-\rho_h|/\lambda_{1S})\,(1/\beta\sqrt{\pi})\exp(-R_\perp^2/2\beta^2)\,. \quad \textbf{(25)}$$

The bound exciton binding energy measured with respect to $\varepsilon_{1S}(K_\perp = 0)$ is plotted versus the characteristic in-plane extension of the defect a for several defect characteristic depths b in Fig. 7 for a $GaAs-Ga_{0.48}Al_{0.52}As$ single quantum well with L = 70 Å. In Fig. 8 we show the trapped exciton binding energy versus a for several quantum well thicknesses L and a fixed b : b = 2.83 Å in $GaAs-Ga_{0.7}Al_{0.3}As$ single quantum wells. One notices that the binding energy is significant only in narrow wells. This had to be expected since the effective well depths V_e, V_h are very small in thick wells due to negligible penetration of the $\chi_{1e}(z_e)$ and $\chi_{1h}(z_h)$ wave- functions in the barrier.

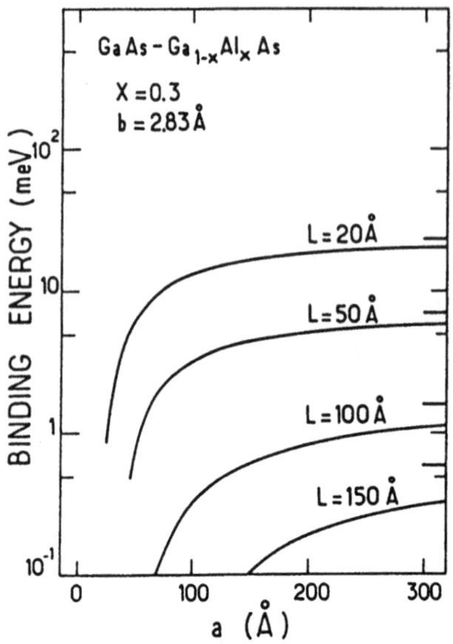

Fig. 8 : The calculated binding energy of E_1-HH_1 excitons bound to semi-Gaussian interface defects is plotted versus the defect in plane extension a for several quantum well thicknesses L in GaAs- $Ga_{0.7}Al_{0.3}As$ single quantum wells. b = 2.83 Å.

Owing to the magnitude of the trapped exciton binding energies, their observation requires low temperature photoluminescence experiments. As will be shown below it is possible to observed the thermal activation of the bound excitons at moderate temperatures.

III - Optical probes of quantum well interfaces

We briefly report in this section our contribution to the study of three different kinds of physical effects which are linked to the existences of interfaces separating GaAs from $Ga_{1-x}Al_xAs$ in GaAs-$Ga_{1-x}Al_xAs$ heterostructures.

III - 1) Optical determination of the band offset between GaAs and $Ga_{1-x}Al_xAs$

As pointed out in section II) there is little hope of obtaining accurate values of the conduction and valence band offsets between GaAs and $Ga_{1-x}Al_xAs$ by means of optical investigations of the intersubband optical absorption in rectangular quantum wells. Dingle's pioneering works [16] on multiple rectangular quantum wells were interpreted in terms of a Q_c value of 85% where $Q_c = \Delta E_c/\Delta E_g$ is the fraction of the band gap energy difference between GaAs and $Ga_{1-x}Al_xAs$ which is taken by the conduction band. Subsequently, Wang et al. [17] found that this value was incompatible with the measured concentration of transferred holes in modulation-doped p type GaAs-$Ga_{1-x}Al_xAs$ single heterojunctions. By fitting the transferred charge versus the space thickness Wang et al. [17] found Q_c = 62%. The optical investigations of

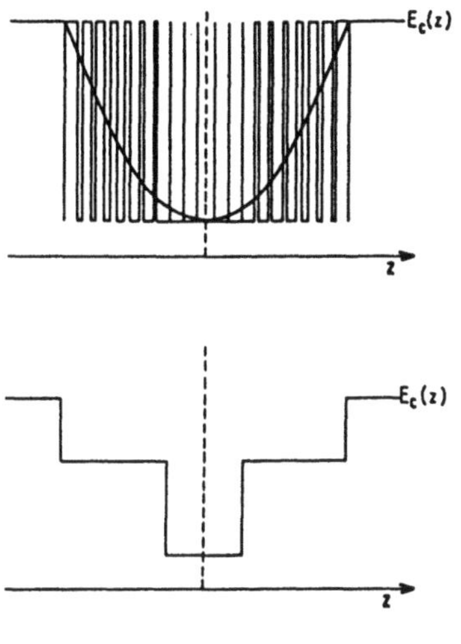

Fig. 9 Band edge profiles of a pseudo-parabolic quantum well and of a separate confinement heterostructure.

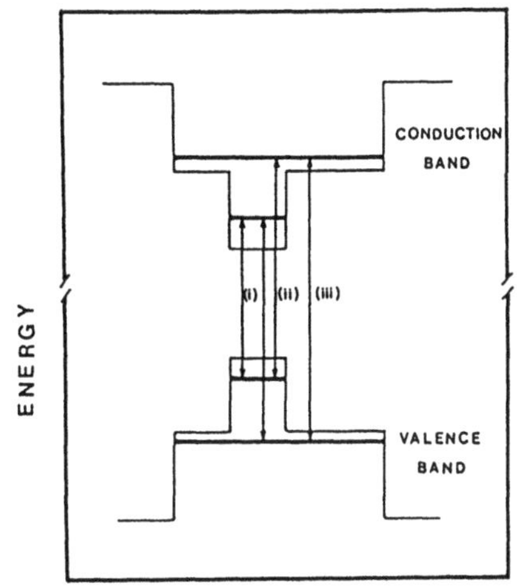

Fig. 10 ; Band edge profile and schematic representation of the three kinds of interband optical transition observed in a separate confinement heterostructure.

pseudo-parabolic GaAs-$Ga_{1-x}Al_xAs$ wells (see Fig. 9) undertaken by Miller et al.[18] led to a value Q_c = 57%.

The heterostructures we have investigated (19) consist of a narrow GaAs well embedded in a $Ga_{1-x_1}Al_{x_1}As$ one, the latter being sandwiched between thick $Ga_{1-x_2}Al_{x_2}As$ barriers (see Fig. 9). In these structures, it is possible to observe three kinds of optical transitions in the photoluminescence excitation spectra (see Fig. 10). These transitions involve quantized levels which are i) both essentially localized within the narrow GaAs well, ii) with one state (initial or final) confined within the GaAs quantum well and the other one (final or initial) in the intermediate $Ga_{1-x_1}Al_{x_1}As$ barrier and, iii) both in the intermediate barrier. It has been found that the transitions i) are very sensitive to the width of the GaAs QW, the transitions iii) to the percentage of aluminium x_1 and the transitions ii) to the fraction Q_c. The latter feature is evidenced in Fig. 11 which shows that the E_1-HH_3 transition, which involves one E_1 final state essentially localized within the GaAs well and a HH_3 initial state delocalized over the intermediate barrier, is very sensitive to Q_c. The investigations of several separate confinement GaAs-Ga(Al)As heterostructures with the GaAs well symmetrically or asymmetrically located in the intermediate barrier led in all cases to Q_c = 059 ± 0.03 with x_1 = 0.13. Such a Q_c value enables the interpretation of charge transfer in p-type[17] and also n-type[20] modulation-doped GaAs-$Ga_{1-x}Al_xAs$ heterojunctions (see Fig. 12).

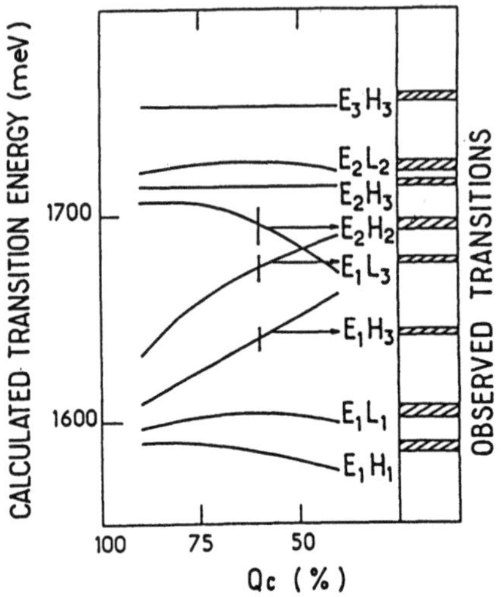

Fig. 11 : Calculated dependence of the transition energies upon Q_c for a 45 Å thick GaAs quantum well embedded in a 245 Å thick $Ga_{0.87}Al_{0.13}As$ one. The observed transitions, corrected for the binding energies of the excitons, are indicated on the right band side of the figure.

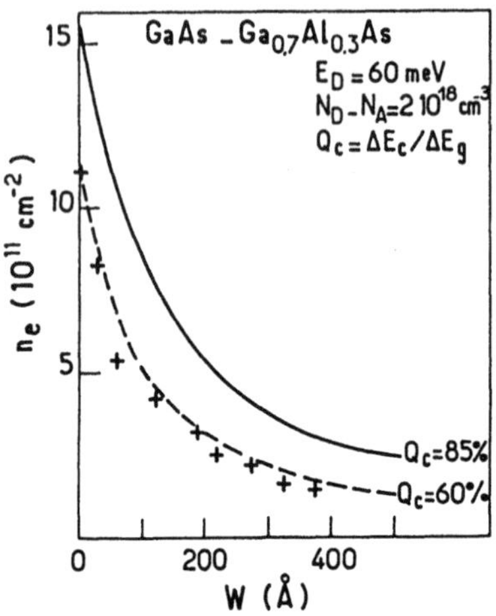

Fig. 12 : Calculated and measured electron transfer in modulation-doped n-type $GaAs-Ga_{0.7}Al_{0.3}As$ single heterojunction. Two Q_c values have been used in the calculations : Q_c = 85% and Q_c = 60%. Calculations after reference [20]. Experiments after reference [24].

III - 2) Low temperature excitation localization on interface defects

The low temperature photoluminescence of GaAs quantum wells is often associated with excitonic recombination (21). There also exist Stokes shifts between the photoluminescence line and the absorption for photoluminescence excitation spectra. This is evidenced in Fig.13 which shows the results obtained on a 70 Å thick $GaAs-Ga_{0.48}Al_{0.52}As$ single quantum well (15) grown by metal organic vapor deposition. The magnitude of the Stokes shift is sample dependent but in high quality materials it ranges from 0 to a few meV's. The interface defects with in-plane extensions larger than the exciton diameter are likely candidates to explain the existence of Stokes shifts. In fact, the absorption spectra probe the electron-hole states which have a large density of states, i.e. the free excitons, while the photoluminescence lines are associated with the electron-hole states of the lower possible energies. This is because the times which are associated with the relaxation effects (acoustical phonon emission) towards lower energies are considerably shorter than the radiative lifetimes. Hence, the electron hole pairs which recombine are the bound excitons, despite their low density of states. The exciton binding energies on attractive interface defects have already been discussed in section II). By comparing the trapped

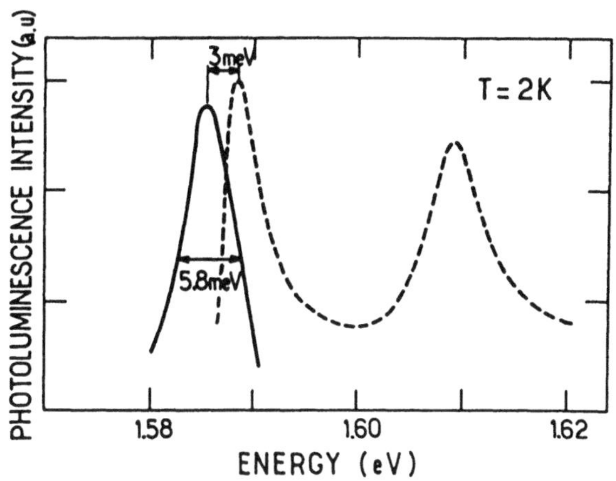

Fig. 13 : Low temperature photoluminescence and excitation spectra of a 70 Å thick GaAs- $Ga_{0.48}Al_{0.52}As$ single quantum well. Solid line : photoluminescence spectrum. Dashed line : photoluminescence excitation spectrum. After reference [7].

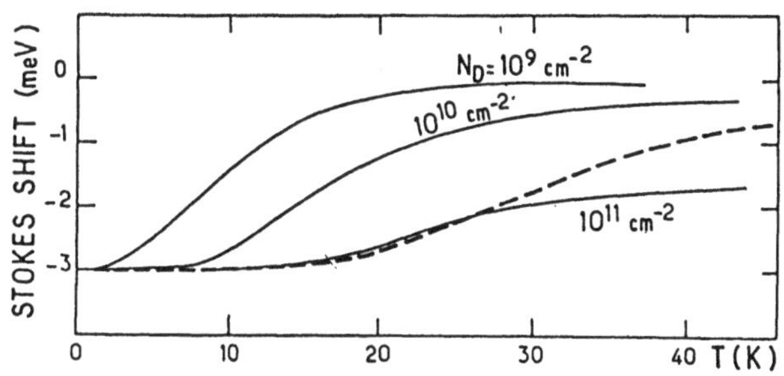

Fig. 14 : Temperature dependence of the Stokes shift between the energies of the E_1-HH_1 excitation line (dots) and of the related photoluminescence line Also shown (dashed lines) are the theoretical positions of the photoluminescence peaks for various densities of interface defects (in cm^{-2}). After reference [7].

exciton density of states to the photoluminescence results (which is sensible since the photoluminescence lines were shown to lack a high energy thermal tail) it was found that the interface defects had an average in-plane extension of 300 Å and were separated on the average by 1000 Å. Such values were found to be consistent with the temperature dependence of the Stokes shift (see Fig. 14). When the temperature is raised, the trapped excitons can detrap, and the luminescence line involves an increasing fraction of delocalized excitons. Thus the maximum of the photoluminescence line approaches that of the free exciton line seen in the excitation spectra.

III - 3) Optical investigation of the impurity segregation at the Ga(Al)As-GaAs inverted interface

The E_1-HH_1 heavy hole exciton photoluminescence line of nominally undoped GaAs-Ga(Al)As quantum wells grown by molecular beam epitaxy is often accompanied by an impurity-related feature on its low energy side [22]. The latter is attributed to the recombination of free electrons with holes bound to acceptors segregated near the inverted GaAs-Ga(Al)As interface. A possible explanation of such a segregation could be that carbon, which is a well-known contaminant of MBE chambers, is less soluble in

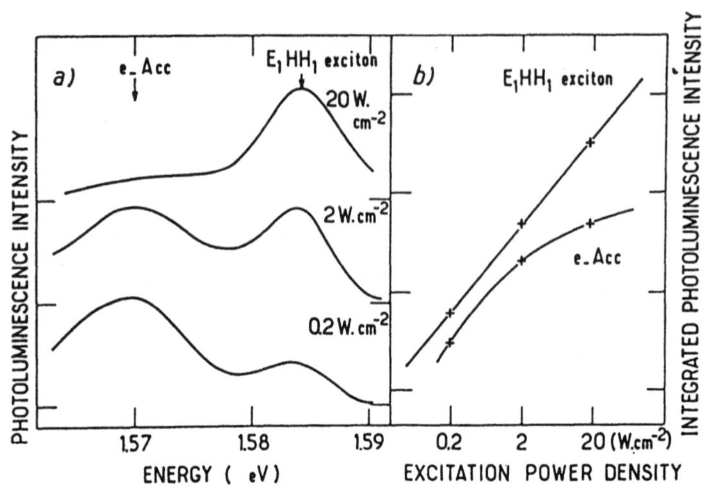

Fig. 15 : a) Photoluminescence spectrum of a 50 Å thick GaAs-$Ga_{0.85}Al_{0.15}As$ quantum well for three different excitation power densities. b) Integrated intensity of the n = 1 electron to heavy hole exciton and acceptor- related photoluminescence versus excitation power density (logarithmic scale).

Ga(Al)As than in GaAs. Consequently, it would keep floating on Ga(Al)As before being incorporated in the first few monolayers of GaAs.

We have determined [23] the spatial distribution of impurities near the inverted interface by fitting the impurity photoluminescence lineshape of e-Å recombinations at low temperatures and studied the effect of thin GaAs prelayers grown prior to the inverted GaAs-Ga(Al)As interface.

Figure 15 presents a typical photoluminescence spectrum of a nominally undoped GaAs-$Ga_{0.85}Al_{0.15}As$ quantum well. The spectrum displays a narrow E_1-HH_1 exciton peak followed at lower energy by an e-Å line which saturates at high laser power intensity. The energy difference between the two peaks shows that the it is possible to derive the binding energy of the acceptor (see Fig. 16). This binding energy is in fair agreement with the one calculated for on-edge acceptors.

The linewidth of the e-Å line suggests that the acceptors are spatially distributed near the edge of the well. We have therefore calculated the e-Å photoluminescence lineshape by accounting for a spread of the acceptor distribution. The results of such calculations is compared to experiments (see Fig. 17) for three different GaAs slab thicknesses.

Our numerical simulations have evidenced that the impurity profile must be assumed asymmetrical with respect to the interface. This is consistent with a model

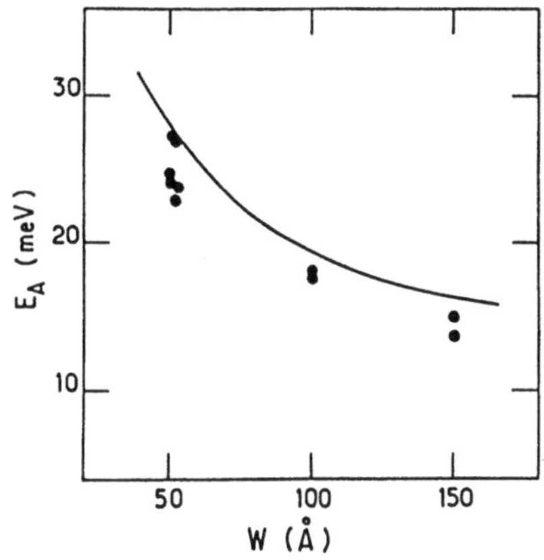

Fig. 16 : Calculated (solid line) and experimental (dots) dependence of the on-edge acceptor binding energy versus the well thickness for an aluminium concentration of 0.15

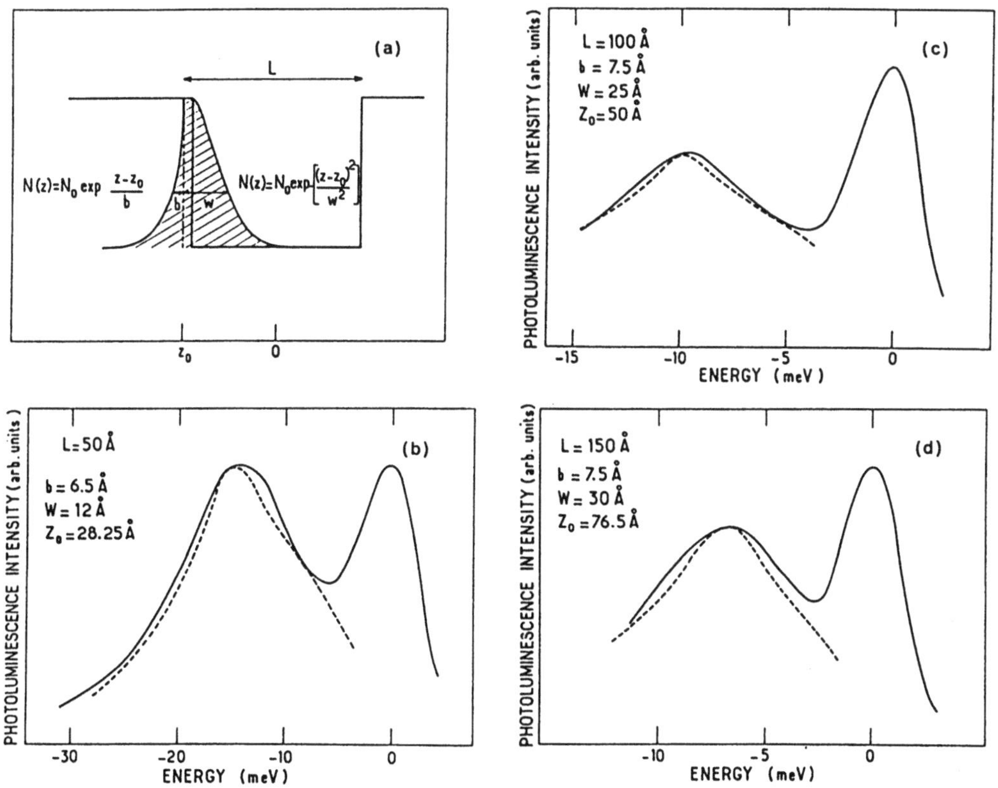

Fig. 17 . Calculated acceptor concentration profile at the well interface. Experimental (solid line) and calculated (dashed line) electron-to-acceptor photoluminescence lineshape for quantum wells of thicknesses 50 Å, 100 Å and 150 Å respectively. The fitting parameters are described in Fig. 1a.

acceptors involved in the recombination are mainly localized near the edge of the quantum well. By computing the energy position of the E_1-HH_1 line and of the E_1 level of impurity trapping occurring once the aluminium flux is terminated. A semi-Gaussian distribution provides a reasonably good fit of the low energy part of the impurity line. It is related to the acceptors of the higher binding energies, i.e. the in-well ones. The spreading of such a distribution varies from 12 Å to 30 Å (4 to 10 monolayers). As for the out-well acceptors' profile, a very good agreement with experiments is found for an exponential distribution of a nearly constant spreading : 6-8 Å (2 to 3 monolayers). These acceptors may be due to out diffusion processes of the in-well incorporated acceptors.

We have found that the impurity related luminescence is substantially reduced by growing thin GaAs prelayers prior to the main quantum well. These layers act as impurity trapping centers. The efficiency of this trapping points out the beneficial replacement of $Ga_{1-x}Al_xAs$ barriers by short period GaAs-AlAs superlattices.

Acknowledgements

We are much indebted to Drs M.H. Meynadier, J.A. Brum and M. Voos for their active participation in the work reported here. The generous supply of high quality GaAs-Ga(Al)As heterostructures by Drs P.M. Frijlink (LEP Limeil Brevannes) and F. Alexandre and J.L. Liévin (CNET Bagneux) has been much appreciated. The Groupe de Physique des Solides de l'Ecole Normale Supérieure is Laboratoire Associé au C.N.R.S. (LA17).

REFERENCES

(1) W. Harrison, J. Vac. Sci. Technol. 14, 1016 (1977) and B 3, 1231 (1985).
(2) C. Tejedor and F. Flores, J. Phys. C 11, L 19 (1978).
(3) F. Flores and C. Tejedor, J. Phys. C 12, 731 (1979).
(4) J. Tersoff, Phys. Rev. Lett. 52, 465 (1984) and 56, 2755 (1986).
(5) See e.g. R.S. Bauer and G. Magaritondo, Physics Today 40, 3 (1987).
(6) See e.g. H. Kroemer, Surf. Sci. 174, 299 (1986).
(7) See e.g. C. Delalande and M. Voos, Surf. Sci. 174, 111 (1986).
(8) See e.g. B.A. Joyce, P.J. Dobson, J.H. Neave and J. Zhang, Surf. Sci. 174, 1 (1986).
(9) P.J. Price and F. Stern, Surf. Sci. 132, 577 (1983).
(10) F. Stern and J.N. Schulman, Superl. and Microst. 1, 303 (1985).
(11) G. Bastard and M. Voos (1985) unpublished.
(12) T. Ando, J. Phys. Soc. Japan 51, 3900 (1982).
(13) D. Paquet, Superl. and Microst. (1987) in press.
(14) J.A. Brum and G. Bastard, J. Phys. C 18, L 789 (1985).

(15) G. Bastard, C. Delalande, M.H. Meynadier, P.M. Frijlink and M. Voos, Phys. Rev. B 29, 7042 (1984).
(16) R. Dingle in : Advances in Solid State Physics, Vol. 15. Festkörper-probleme, Ed. H.J. Queisser (Pergamon/Vieweg, London/Braunschweig, 1975), p.21.
(17) W.I. Wang, E.E. Mendez and F. Stern, Appl. Phys. Lett. 45, 639 (1984).
(18) R.C. Miller, A.C. Gossard, D.A. Kleiman and O. Munteanu, Phys. Rev. B 29, 3740 (1984).
(19) M.H. Meynadier, C. Delalande, G. Bastard, M. Voos, F. Alexandre and J.L. Liévin, Phys. Rev. B 31, 5539 (1985).
(20) G. Bastard, M. Voos, (1985) unpublished.
(21) C. Weisbuch, R.C. Miller, R. Dingle, A.C. Gossard and W. Wiegmann, Solid State Commun. 37, 219 (1981).
(22) R.C. Miller, W.T. Tsang and O. Munteanu, Appl. Phys. Lett. 41, 374 (1982).
(23) M.H. Meynadier, J.A. Brum, C. Delalande, M. Voos, F. Alexandre and J.L. Liévin, J. Appl. Phys. 58, 4307 (1985).
(24) J.C.M. Hwang, A. Kastalsky, H.L. Störmer and V.G. Keramidas, Appl. Phys. Lett. 44, 802 (1984).

The Physics of Metal Base Transistors

E. Rosencher, F. Arnaud d'Avitaya, P.A. Badoz, C. d'Anterroches, G. Glastre, G. Vincent, and J.C. Pfister

Centre National d'Etudes des Télécommunications, B.P. 98, Chemin du Vieux Chêne, F-38243 Meylan Cedex, France

Epitaxial Si/$CoSi_2$/Si structures can be grown under ultra-high vacuum conditions. The metallic $CoSi_2$ films can be extremely thin, typically between 1 nm and 20 nm. The electrical properties of these heterostructures are presented, mainly the transport of electrons in the metallic films parallel to the interfaces and the transfer of electrons through the metal film. The influence of pinholes in the $CoSi_2$ layers will be discussed.

I. Introduction

Thanks to advances in ultra-high vacuum technology, it has recently become possible to realize epitaxial Semiconductor / Metal/ Semiconductor (SMS) structures [1, 2] using a Si/$CoSi_2$/Si sandwich. The rather small lattice mismatch (~ 1.2 %) between Si and $CoSi_2$ crystals as well as their similar cubic structures allow the production of monocrystalline Si/$CoSi_2$ and Si/$CoSi_2$/Si heterostructures. These SMS structures open the way to promising ultra-low base resistance devices for millimeter wave applications.

In this paper, we intend to review the main results on the physics of metal base transistors (MBT). This term includes two different kinds of structures : the SMS-transistor where the $CoSi_2$ film is intended to be continuous and Permeable Base Transistors (PBT) where discontinuities in the metallic film are intentionally introduced by nanolithography techniques. Though this paper is mainly focused on the electrical properties of these structures, some information on the morphological quality of these sandwiches is given in Sect. II that is necessary for the understanding of the transport properties. In Sect. III, parallel transport in the ultra-thin metal films will be addressed while the perpendicular transport through the SMS structure will be developed in Sect. IV. We shall describe the current status of the Si/$CoSi_2$/Si permeable base transistor in Sect. V. A brief discussion is given in Sect. VI which tentatively describes the general trends in the future research on metal base transistors.

II. Morphology of Si/$CoSi_2$/Si heterostructures

This section is intended to describe the results necessary as a background for the understanding of transport properties rather than to provide an explanation of the growth mechanisms, which is still clearly lacking. The morphology observations are made using Scanning Electron Microscopy (SEM), Transmission Electron Microscopy (TEM) and High Resolution Transmission Electron Microscopy (HRTEM).

One of the key points for the growth mechanisms is the cleaning of the silicon surface before $CoSi_2$ formation. Up to now, we have used the Shiraki process, the details of which are described in Ref. [3] and [4]. This process is based on chemical oxidation-etching cycles, leaving a very thin protective SiO_x film which evaporates when heated above 750°C. The SMS structures are grown in a typical ultra-high vacuum system described in [4]. The surface crystallinity and cleanliness are checked in situ by Low Energy Electron Diffraction (LEED) and Auger Spectroscopy (AES). Let us note that no electronic grade materials can be grown when carbon contamination higher than 5 % of a monolayer is present at the silicon surface [5].

In our structures, $CoSi_2$ is grown by solid phase epitaxy on the <111> Si surface. Co layers are electron gun evaporated under a pressure less than 5 x 10^{-8} Torr with the sample kept at room temperature and then annealed at 650°C (± 25°C) for 10 minutes in order to obtain the $CoSi_2$ layers. Other growth techniques (molecular beam epitaxy, hot substrates, higher annealing temperatures, etc.) have led to lower quality structures [4-7]. Grazing angle SEM and TEM observations have shown that smooth $CoSi_2$ layers could thus be grown for a thickness up to 20 nm. Above this limit, the metal layers are rough and discontinuous, leading to a milky aspect of the wafers [4].

Fig.1 Cross-sectional TEM image of a 5 nm thick $CoSi_2$ layer on Si substrate. One notes the bowl-shaped $CoSi_2$ intrusion in Si bulk for an otherwise sharp $CoSi_2$/Si interface.

Cross-sectional TEM observations show that, for a thickness less than 20 nm, the interface between Si and $CoSi_2$ is rather smooth, with eventual presence of bowl-shaped $CoSi_2$ intrusions in Si, while the $CoSi_2$/vacuum interface is much smoother (Figure 1) [8]. Plane view TEM photographs of a $CoSi_2$/Si structure are shown in Figure 2a and b. The shape of the Moiré fringes indicates that strain fields are present in the $CoSi_2$ layers and reflects the 1.2 % lattice mismatch between Si and $CoSi_2$. Moreover, no pinholes are observable over a few square micrometers [9].

These pinholes, when they are present, appear as regions where Moiré fringes are absent, as described in [10]. Figure 2b is a plane view TEM observation which stresses the orientation of the $CoSi_2$ crystal relatively to the Si one in the <111> direction. It is clear that $CoSi_2$ is a mixture of two types of grains : type A grains where the Co planes are an extension of the Si bulk planes and type B grains where the $CoSi_2$ lattice is rotated 180° around the <111> direction relatively to the Si lattice [8]. We must stress that some other groups [7, 10] have found a great majority of type B grains in their $CoSi_2$ films (> 95 %) : the difference in the results of these different groups is still unexplained, though the influence of substrate cleaning most probably plays a major role.

Silicon layers are deposited by MBE on $CoSi_2$ films at 650°C and doped by Sb coevaporation under a pressure in the 10^{-9} Torr range. Figure 3 is a cross-sectional TEM photograph of a Si/$CoSi_2$/Si heterostructure. It is clear

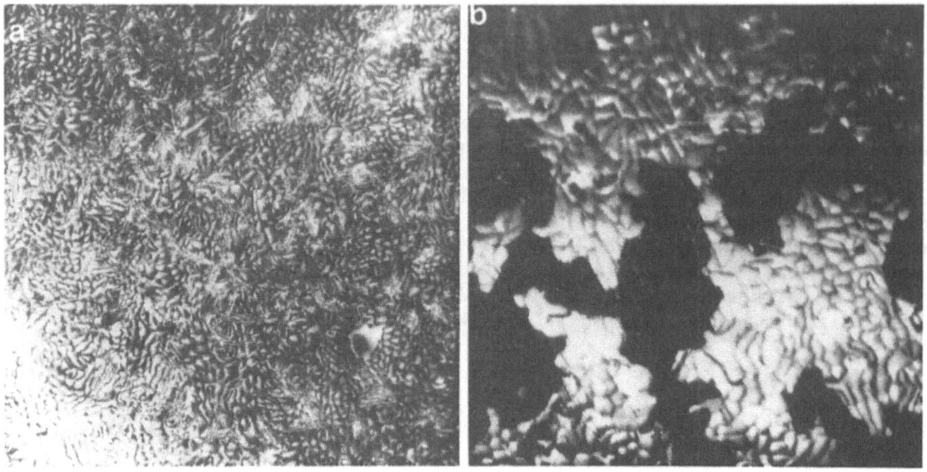

Fig. 2 TEM plane view observations of 5 nm thick $CoSi_2$ layers (a). The bright field image evidences the Moiré fringes (due to lattice mismatch and strains) and the absence of pinholes (b). The dark field image using a diffraction spot specific of type B grains indicates an equivalent density of both grains [9].

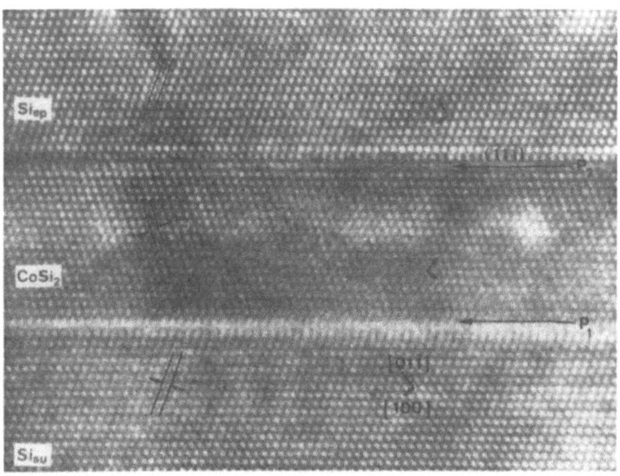

Fig. 3 Cross-section TEM image of $aSi/CoSi_2/Si$ heterostructure. Note the planarisation of interface relatively to Figure 1 [8].

that the roughness of the interfaces has dramatically decreased, the transition between $CoSi_2$ and Si material occurring within one monolayer over long distances. This effect (we call it" planarisation") is still unexplained. Moreover, grazing angle SEM observations indicate that the growth mechanism of Si over $CoSi_2$ is three-dimensional (Fig. 4). This results in a high density of sub-grain boundaries, twins... in the epitaxial Si layer. Finally, let us note that HRTEM observations indicate that, for such small $CoSi_2$ film thicknesses (< 20 nm), the $CoSi_2$ lattice is entirely strained in the $Si/CoSi_2/Si$ sandwich in order to fit the Si bulk lattice [8].

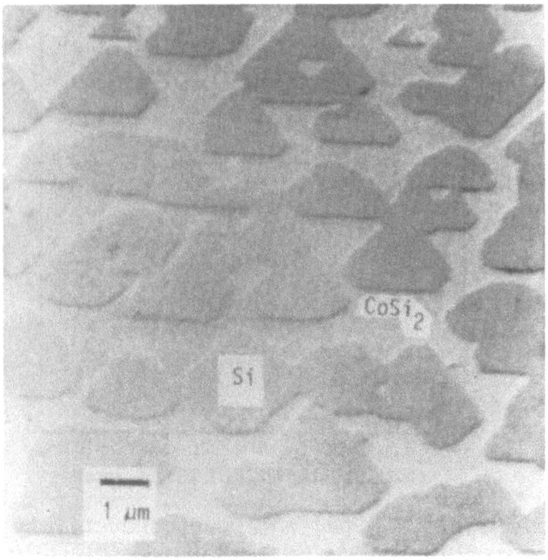

Fig. 4 SEM image of 23 nm thick Si epilayer on top of $CoSi_2$ (8 nm)/Si_{bulk} structure [11].

III. Parallel transport in Si/$CoSi_2$ heterostructures

Because of their outstanding crystalline quality, $CoSi_2$ films are ideal candidates for the study of electron transport in ultra-thin metallic films. Indeed, other possible systems (like Au on NaCl) are usually unstable against exposure to air and temperature cycles.

Hensel et al. were the first to show that down to very low thickness, i.e. 10 nm, the film resistivity exhibits little dependence on the $CoSi_2$ film thickness [12]. These thicknesses are much less than the bulk transport scattering length of ~ 100 nm as determined by magneto-resistance measurements, so that boundary scattering of the carriers is essentially specular in this system. In order to account for this phenomenon, these authors have used the well-known Fuchs-Sondheimer theory, introducing a specularity parameter p which is the fraction of electrons specularly reflected from the interfaces. The resistivity ρ of a film of thickness d is thus given by :

$$\rho = \rho_\infty \left[1 - 3/2k \int_0^1 \frac{(u - u^3)(1 - p)(1 - \exp(-k/u))}{1 - p\exp(-k/u)} du\right]^{-1} \quad (1)$$

where ρ_∞ is the bulk resistivity and k the ratio of the film thickness d to the mean free path λ_e, i.e. $k = d/\lambda_e$. Figure 5a shows the set of curves in $CoSi_2$ films for specularity parameter p ranging from 0 to 1 (i.e., from purely diffuse scattering to purely specular) compared to experimental data from Badoz et al. [13]. It is clear that for film thicknesses less than 10 nm, the variation of the film resistivity with thickness is far steeper than expected from Fuchs-Sondheimer theory. Figure 5b shows the values of the superconducting critical temperature T_c of $CoSi_2$ films as a function o thikness. Here again, there is an abrupt drop of T_c in $CoSi_2$ films thinner than 10 nm. It has to be noted that these results are different in nature from the usual experiments in thin metal films near the localisation regime ($R_\square \sim \hbar/e^2 \sim 4000\ \Omega_\square$). The sudden change in T_c and ρ occurs in films with resistances in the 30 $\Omega_\square$ range.

A phenomenological explanation of our results is the presence of a layer of about 5 nm at the Si/$CoSi_2$ interface in which the electronic transport properties are dramatically perturbed : in this region, the carrier mobility is extremely low and no superconductivity occurs. This layer is without doubt $CoSi_2$, as evidenced by LEED, AES, TEM and X-ray photoemission spectroscopy

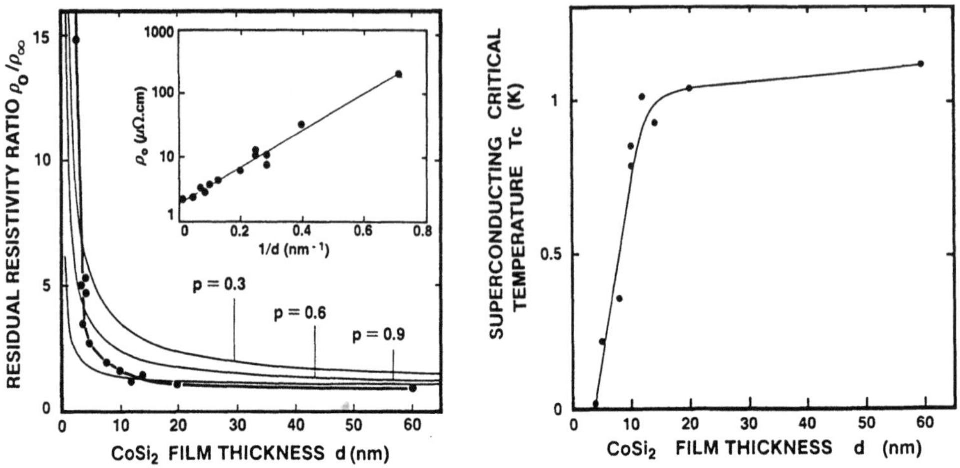

Fig. 5 $CoSi_2$ film resistivity ρ_0 measured at 4 K (a) and superconductivity critical temperature T_c (b) as a function of film thickness d [13].

measurements as well as by the Hall effect (constant density of carriers in this layer). A possible origin for this perturbed layer could be the presence of magnetic cobalt atoms : magnetic impurities are indeed known to be highly effective in quenching superconductivity as well as being efficient scattering centers for electronic transport. These magnetic Co atoms could stem from either i) a small departure from $CoSi_2$ stoichiometry [14] or from ii) ill-coordinated interface Co atoms as evidenced by careful observation of HRTEM photographs [8].

In an attempt to observe quantization of the metallic electron gas in ultra-thin $CoSi_2$ film, Tunneling Spectroscopy (TS) has been performed in degenerate Si/$CoSi_2$ Schottky diodes [15]. Sharp features have been observed in TS spectra up to high energies (600 meV). However, those peaks are weakly dependent on the film thickness (see Figure 6). Very recent experiments tend to prove that those features are due to phonon emission by the relaxation of hot electrons in the depletion layer of Si. The lack of observable quantization may be due to the high value of the electron energy at the Fermi level E_F, leading to a high value of the quantization energy uncertainty ΔE_F near the Fermi level given by :

$$\Delta E_F/E_F \sim 2\ \Delta d/d \qquad (2)$$

where Δd is the residual film roughness. For a ratio $\Delta d/d$ as low as 1 %, the energy spreading is already in the 100 meV range !

IV. Perpendicular transport in Si/$CoSi_2$/Si heterostructures

The first evidence of transistor effect in a monolithic SMS structure was reported by Rosencher et al. in 1984 [16]. Figure 7 shows the energy band diagram of a SMS transistor. This device consists basically of two back- to -back Schottky diodes. One of these diodes, the emitter, is forward biased, while the other one, the collector, is reverse biased. The carrier transport between the emitter and the collector is the subject of intensive investigations [16-19]. Two mechanisms may indeed be involved, with relative weight strongly dependent on technology.

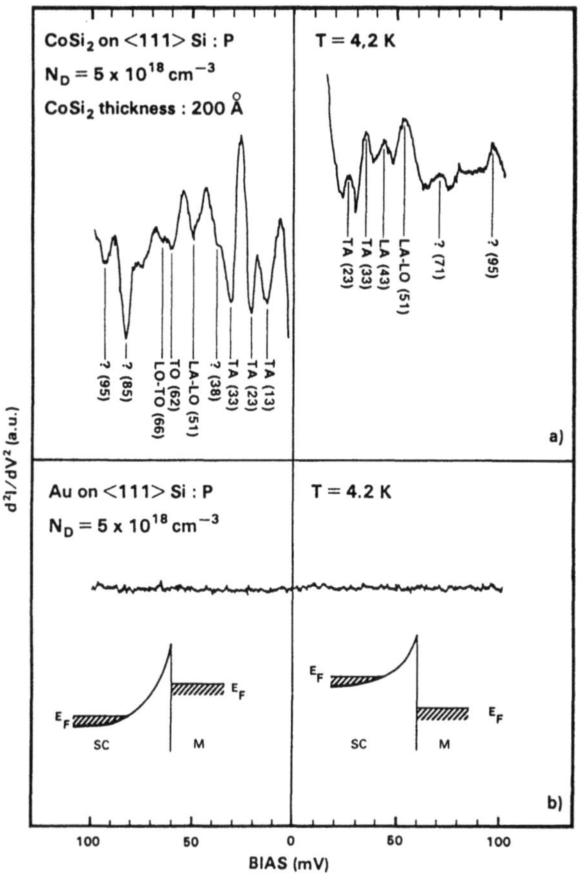

Fig. 6 Tunneling spectra of epitaxial $CoSi_2$/Si tunnel junction (a) and of nonepitaxial Au/Si junction (b). The main Si phonons are indicated [15].

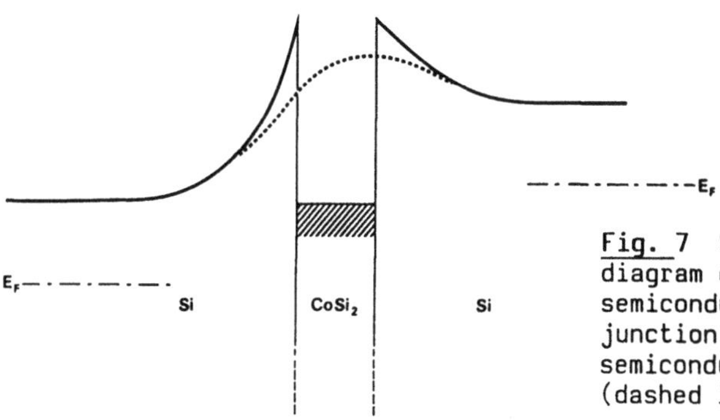

Fig. 7 Schematic energy-band diagram of a SMS transistor in a semiconductor-metal-semiconductor junction (solid line) and in a semiconductor pinhole channel (dashed line).

1. Electrons are emitted via thermionic emission from the forward biased emitter junction into the metal. A fraction of these carriers cross the metal film via ballistic transport and are collected by the reverse biased collector junction [16, 19]. This mechanism is described by the solid lines in Figure 7.
2. Pinholes are present in the metal film, in which silicon channels are imbedded. The electrons are transferred from the emitter to the collector via these semiconducting channels and the current flow is controlled by the barrier lowering in the pinholes (dashed line in Figure 7) [17, 18].

A theoretical model has been developed in [20] in order to evaluate the different weights of mechanisms 1 and 2. The conclusion is that, for usual doping levels, a single 150 nm radius pinhole in the metal base is enough to short circuit the whole ballistic transport in a 20 µm x 20 µm SMS transistor ! It is thus clear that TEM observations, which investigate only few square micrometers of a device, have no statistical significance in concluding the predominance of one mechanism over the other. We have thus developed an electrical measurement, a transconductance technique described in [17] and [19], which allows to measure the relative weights of mechanisms 1 and 2. This technique, based on the screening of the collector potential by the metallic $CoSi_2$ film when no pinholes are present, ensures that, in "pinhole-free" SMS, electron transport occurs almost entirely through the metal base. Independent evidence of the dominant role of hot electron transfer through the base in SMS-T is given in Figures 8a and 8b showing the common-base characteristics of a SMS structure (7 nm base thickness) using either the overgrown (Figure 8a) or bulk (Figure 8b) silicon as the emitter. The transfer ratios α are 15 % and less than 10^{-4}, respectively, for the same values of emitter current I_E (500 µA). Since the pinhole current must be of the same order of magnitude in both directions, the value 15 % is clearly due to the transfer through the metal base. The asymmetry of current gain is consistent with the already reported systematic difference in barrier heights $\emptyset_{ms}$ between the Si bulk/$CoSi_2$ ($\emptyset_{ms} \sim$ 0.63 eV) and $Si_{epi}/CoSi_2$ ($\emptyset_{ms} \sim$ 0.69 eV) junctions [19] : the injected electrons are well above the collector barrier when emitted from the overgrown Si and below when emitted from the bulk Si.

The ballistic transport [21] in "pinhole-free" SMS-T is described by a mean free-path $\lambda_B(T)$, so that the emitter to collector transfer ratio of electrons is expected to be :

$$\alpha = \alpha_0 (T) \cdot \exp(-d/\lambda_B(T)) \qquad (3)$$

where T is the measurement temperature and α_0 is the current gain extrapolated to zero metal base thickness. The departure of α_0 from unity is due to

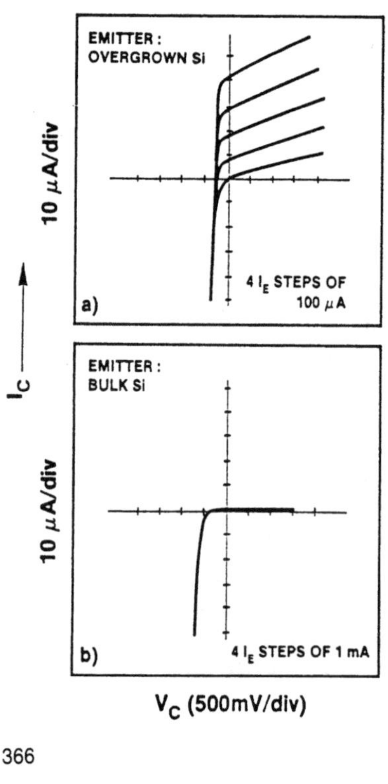

Fig. 8 Common base current-voltage characteristics of a SMS-T with a 7 nm thick base using either the regrown Si(a) or the bulk Si(b) as emitter. The measurements are performed at room temperature.

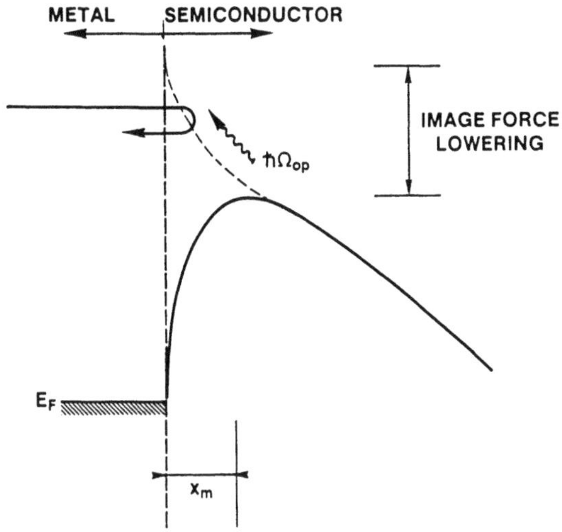

Fig. 9 Electron potential energy versus distance from the collector metal-semiconductor interface. An example of an electron backscattering event is symbolized.

collector as well as emitter losses :

$$\alpha_0 = \alpha_c \cdot \alpha_q \cdot \alpha_e \tag{4}$$

where α_c is the current gain upper limit associated with scattering in the Si collector, α_q is the quantum mechanical transmission of the base-collector potential barrier and α_e is the emitter efficiency coefficient [21]. The collector scattering contribution is expected to follow :

$$\alpha_c = \exp(- x_m/\lambda_{ph}) \tag{5}$$

where x_m is the position of the maximum of the collector barrier potential in the image force approximation and λ_{ph} is the mean free path in the Si collector (see Figure 9).

Figures 10 and 11 compare the above theory with experiment. Figure 10 shows the transfer ratios α obtained on samples with various values of $CoSi_2$ base thickness, at room temperature and at 77 K. Error bars correspond to the dispersion of values for different devices fabricated on the same wafer. The results clearly show that Eq. (4) is verified, with values of λ_b of 8 ± 1.5 nm at 300 K and 35 ± 5 nm at 77 K. This agreement strongly favors ballistic theory. Moreover, the λ_B values are close to the mean free path deduced from resistivity measurements [22]. This indicates that the same scattering mechanisms control both the electron transport close to the Fermi level and the hot electron relaxation for energies in the 0.7 eV range above E.

Figure 11 shows α versus $V_{BC}^{-1/4}$ curves, taken at different temperatures, where V_{BC} is the base-collector bias. Equation (5) clearly holds since $V_{BC}^{-1/4}$ is directly proportional to x_m [21]. Furthermore, a mean free path is extracted from the slope of the curves and its temperature variation is shown in the inset of Figure 10. The very low values obtained for λ_{ph} (< 2 nm) remain to be understood.

Another problem requiring explanation is the overall low values of α_0 (~ 0.3 at RT) corresponding to α/α_c ~ 0.6. The answer is most probably related to the quantum nature of the electron. Indeed, the electron energy E_1 in the metal is in the 5 eV range while its value E_2 in the semiconductor is in the 50 meV range, i.e. the Schottky barrier is lowered. Consequently, the abrupt change in the electron wavelength leads to a quantum reflection at the metal-semiconductor interface. If the crude model of the abrupt-step barrier is assumed, the quantum transmission coefficient α_q is [22] :

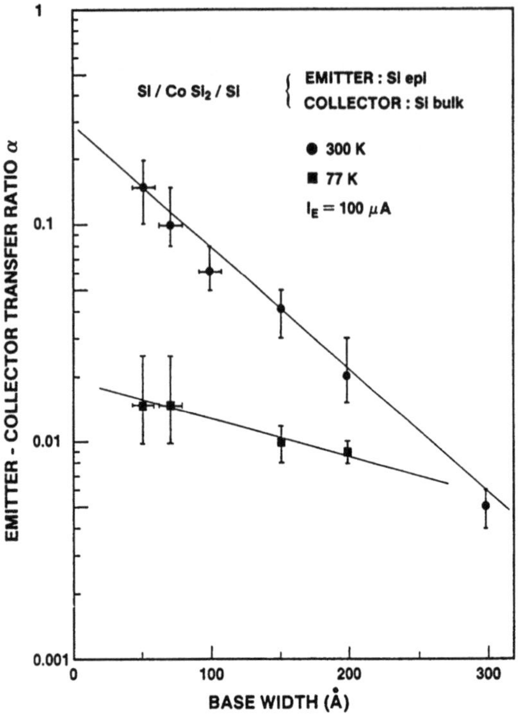

Fig. 10 Transfer ratio versus $CoSi_2$ base thickness measured at 77 K and 300 K in SMS transistors [9].

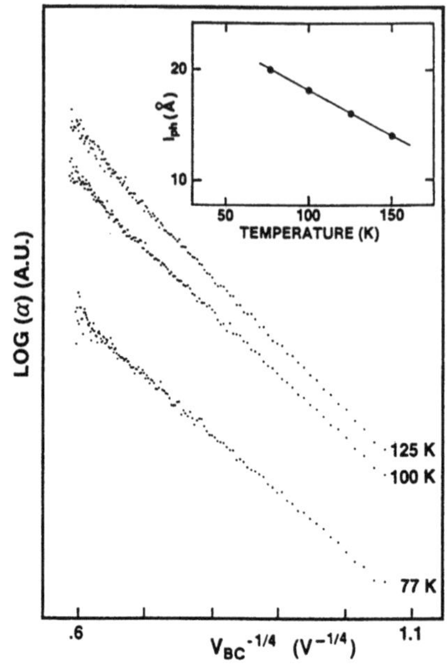

Fig. 11 Transfer ratio versus base-collector bias at different temperatures. The mean free path λ_{ph} deduced is shown in the inset as a function of temperature [22].

$$\alpha_q = 1 - [(m_2{}^*E_1)^{1/2} - (m_1{}^*E_2)^{1/2}/((m_2{}^*E_1)^{1/2} + (m_1{}^*E_2)^{1/2})]^2 \qquad (6)$$

where $m_i{}^*$ is the effective mass in the ith medium. Taking $m_1{}^* = 1$ in $CoSi_2$ and $m_2{}^* \sim 0.3$ in Si, one obtains $\alpha_q \sim 0.5$, which is in fair agreement with the experimental data of $\alpha_0 \sim 0.3$ taking into account the collector backscattering.

All these results, as well as those obtained by Sze and his co-workers [21], show that the quantum reflection is a severely limiting factor for the device interest of SMS transistors. These results suggest, in order to reduce this reflection, the use of highly asymmetrical SMS structures, for instance by use of two different semiconductors and/or metals.

V. Si/CoSi$_2$/Si Permeable Base Transistors

The idea of the Permeable Base Transistor (PBT) is to take advantage of mechanism 2 described in the preceding section. The main advantages of this device are :

- For small enough Si channels, PBTs behave like thermionic devices so that their transconductance is very high [23].
- There are no fundamental limitations such as quantum reflection in SMS transistors.
- It is easily shrinkable with no problems of punchthrough such as in bipolar transistors.

They are two possible ways to embed semiconductor channels in a metal grid.

1. The first one is to use the natural porosity of a metal layer on the surface of a semiconductor, leading to a natural permeable base transistor (also called metal grid transistor [24]). Discontinuous $CoSi_2$ films [17, 10] but also W layers deposited during Si CVD growth [24] have been used. However, the high input capacitance as well as the lack of control in the geometry of metal openings are not in favour of such a device.
2. The second way is to define the opening by lithographic techniques. Bozler and Alley [25] were the first to realize such a structure. They have used W grids of 320 nm periodicity and CVD deposited GaAs as the semiconductor, though the W/GaAs system is not epitactic.

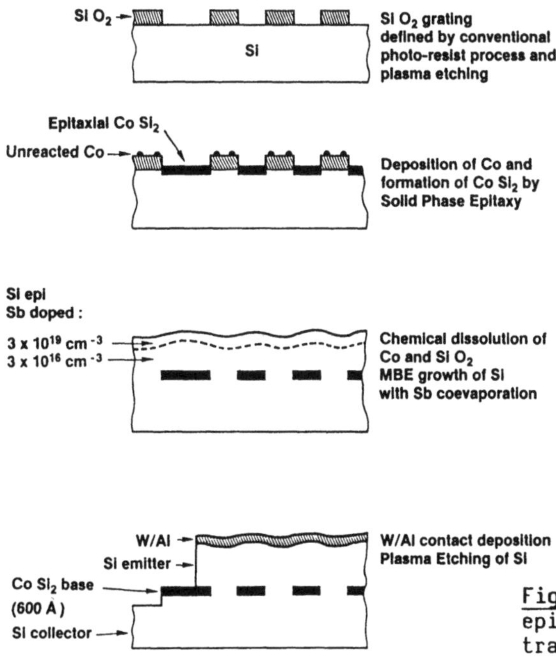

Fig. 12 Main fabrication sequence of epitaxial $Si/CoSi_2/Si$ permeable base transistor [27].

$Si/CoSi_2/Si$ permeable base transistor with micronic si channel dimensions have been independently obtained by Ishibashi et al.[26] and Rosencher et al [27]. Submicron $Si/CoSi_2/Si$ PBTs are currently under study in a few laboratories. Though the problems of materials associated with epitaxial growth in submicron structures are far beyond the scope of this paper, we show in Figure 12 the main fabrication sequence of epitaxial $Si/CoSi_2/Si$ permeable base transistors. Figure 13 shows the backscattered electron image in a scanning electron micrograph of a 300 nm $CoSi_2$ grid embedded in a Si lattice. Electrical behaviour of such PBTs is currently under study.

VI. Conclusion

The physics and technology of metal-semiconductor heterostructures are clearly a rapidly expanding new field of research. Many of the experimental results described in this review paper are still at an early stage. A non exhaustive list of unresolved problems and future development is :

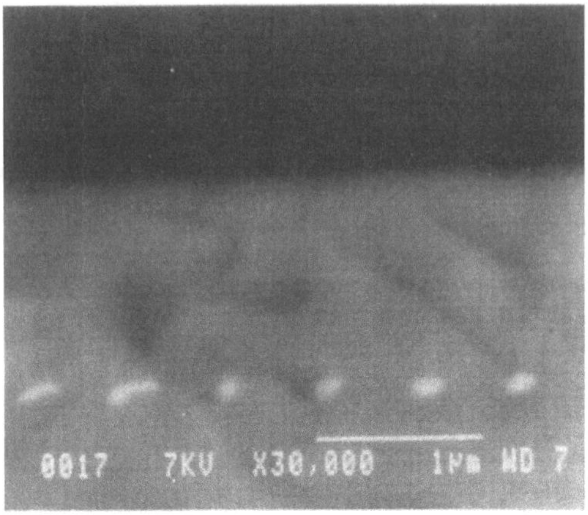

Fig. 13 Backscattered electron image in SEM of epitaxial $CoSi_2$ lines, 0.3 µm wide (bright areas) buried in Si lattice.

- Mechanism of $CoSi_2$ growth on top of Si (role of the early stage nucleation, roughness, planarisation, influence of strains...).
- Microscopic nature of the $CoSi_2$/Si interface with regard to parallel transport properties.
- 2D localisation effects.
- Scattering mechanisms in SMS transistors (hot electrons in the metal, Si collector backscattering...).
- Frequency limitation of permeable base transistors.
- Other epitaxial metals on Si such as ternary silicides.

References

[1] **S. Saitoh, H. Ishiwara, S. Furukawa** : Appl. Phys. Lett. 37, 203 (1980).
[2] **J.C. Bean, J.M. Poate** : Appl. Phys. Lett. 37, 643 (1980).
[3] **A. Ishizaka, K. Nakagawa, Y. Shiraki** : In Proceedings of the symposium on molecular beam epitaxy and clean surface techniques, (The Japan Society of Appl.Phys., Tokyo, 1982), p. 183.
[4] **F. Arnaud d'Avitaya, S. Delage, E. Rosencher, J. Derrien** : J. Vac. Sci. Technol. B3, 770 (1985).
[5] **E. Rosencher, S. Delage, F. Arnaud d'Avitaya** : J. Vac. Sci. Technol. B3, 762 (1985).
[6] **K. Ishibashi, S. Furukawa** : Appl. Phys. Lett. 43, 660 (1983).
[7] **B.D. Hunt, N. Lewis, E.L. Hall, L.G. Turner, L.S. Schowalter, M. Okamoto, S. Hashimoto** : In Proceedings of the MRS Conference, 37, "Layer Structures, Epitaxy and Interfaces", Ed. J.M. Gibson and L.R. Dawson, p. 131.
[8] **C. d'Anterroches, F. Arnaud d'Avitaya** : Thin Solid Films 137, 351 (1986).
[9] **E. Rosencher, P.A. Badoz, C. d'Anterroches, G. Glastre, G. Vincent, F. Arnaud d'Avitaya** (to be published).
[10] **R.T. Tung, A.F.J. Levi, J.M. Gibson** : Appl. Phys. Lett. 48, 635 (1986).
[11] **F. Arnaud d'Avitaya, J.A. Chroboczek, G. Glastre, Y. Campidelli, E. Rosencher** : J. of Crystal Growth (to be published).
[12] **J.C. Hensel, R.T. Tung, J.M. Poate, F.C. Unterwald** : Phys. Rev. Lett. 54, 1840 (1985).
[13] **P.A. Badoz, A. Briggs, E. Rosencher, F. Arnaud d'Avitaya** (to be published).
[14] **R. Madar, A. Briggs** (to be published).
[15] **E. Rosencher, P.A. Badoz, A. Briggs, Y. Campidelli, F. Arnaud d'Avitaya** : In Proceedings of the first international symposium of silicon molecular beam epitaxy, ed. J.C. Bean, The Electrochemical Society, Vol. 85-7, p. 268 (1985).

[16] E. Rosencher, S. Delage, Y. Campidelli, F. Arnaud d'Avitaya : Electron. Letters 20, 762 (1984).
[17] E. Rosencher, S. Delage, F. Arnaud d'Avitaya, C. d'Anterroches, K. Belhaddad, J.C. Pfister : Physica B134, 106 (1985).
[18] J.C. Hensel, A.F. Levi, R.T. Tung, J.M. Gibson : Appl. Phys. Lett. 47, 151 (1985). See also Ref. 10.
[19] E. Rosencher, P.A. Badoz, J.C. Pfister, F. Arnaud d'Avitaya, G. Vincent, S. Delage : Appl. Phys. Lett. 49, 271 (1986).
[20] J.C. Pfister, E. Rosencher, K. Belhaddad, A. Poncet : Solid State Electron. 29, 907 (1986).
[21] S.M. Sze : In Physics of Semiconductor Devices, (Wiley-Interscience, New York, 1969) Chap. 11.
[22] P.A. Badoz, E. Rosencher, S. Delage, G. Vincent, F. Arnaud d'Avitaya : In Proceedings of the 18th International Conference of Physics of Semiconductors (1986, Stockholm) (to be published).
[23] A. Marty, J. Clarac, J.P. Bailbe, G. Rey : IEE Proc. 130, 24 (1983).
[24] J. Lindmayer : Proc. IEEE, 1751 (1964).
[25] C.O. Bozler, G.D. Alley : IEEE Trans. Electron Devices 27, 1128 (1980).
[26] K.Ishibashi, S. Furukawa : IEEE Trans. on Electron Devices 33, 322 (1986).
[27] E. Rosencher, G. Glastre, G. Vincent, A. Vareille, F. Arnaud d'Avitaya Electron. Letters 22, 699 (1986).

Perspectives on Formation and Properties of Semiconductor Interfaces

R.S. Bauer[1], R.H. Miles[2], and T.C. McGill[2]

[1]Xerox Palo Alto Research Center, 3333 Coyote Hill Road, Palo Alto, CA 94304, USA

[2]Department of Applied Physics, California Institute of Technology, Pasadena, CA 91125, USA

Recent progress in experimentally and theoretically understanding interfaces at the atomic level suggest that ultimate electronic systems may one day be fabricated on a single integrated chip. If such elements as Si VLSI processors, GaAs/AlAs integrated optoelectronic IO devices, II-VI superlattice visible displays and high speed III-V processors are to be integrated, interface formation and *in situ* processing will be required at a level of sophistication well beyond what is available today. In this paper, we review recent developments in interface formation by both MOCVD and MBE. To illustrate the power of our diagnostic methods, the details of epitaxial interface formation on an atomic scale are reviewed for lattice matched systems (Ge/GaAs/AlAs) and epitaxial silicides (Ni/Si_2/Si) as well as oxidation of silicon to form Si/SiO_2 interfaces. New developments in using lattice mismatched superlattices with strained layers are discussed for CdTe/ZnTe. Additional complications of growing compound semiconductors on elemental substrates (e.g., anti-phase domains) are discussed for GaAs growth on Si(100).

The implications of these interface formation methods for understanding and controlling the physical properties of these material systems are also explored. A review of the current status of heterojunction band offsets is given. Emission and adsorption of light is discussed for lattice matched quantum well lasers and strained-layer superlattices for light sources and detectors. New developments in transport both parallel and perpendicular to the interface are explored with particular attention to new device structures employing layers of a few atomic dimensions for tunnel structures. It is shown that by understanding and controlling the formation of heterojunction and superlattice interfaces, novel device properties can be established to provide integrated device performance well beyond current capabilities. The opportunities for advances in this field are only just beginning to be realized.

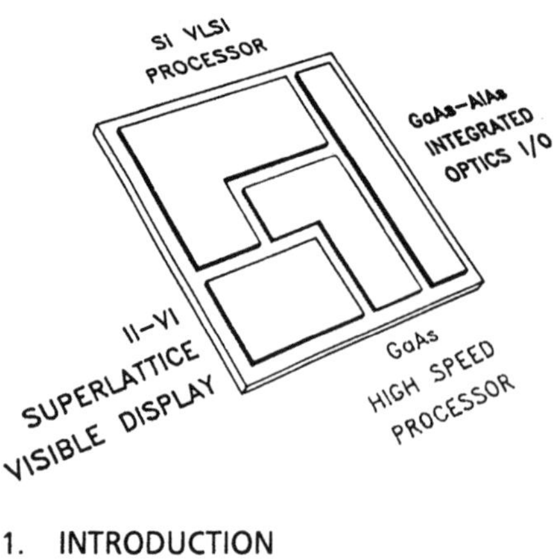

Fig. 1. Schematic representation of the ultimate electronic system. Combining interfaces from group IV, III-V, and II-VI interfaces on a single substrate requires advances in interface growth and *in situ* processing to achieve the advantageous performance of each of the device interfaces (e.g., visible light emission from II-VI's, high speed IO and processing from III-V's, and complex integrated electronics in Si).

1. INTRODUCTION

This paper is meant to take a broad overview of recent developments in understanding and controlling semiconductor interface formation. While presenting details at an atomic level, the intent is to provide insight into new ideas, trends, and opportunities for further research. References are given to work in the literature to provide entry to more

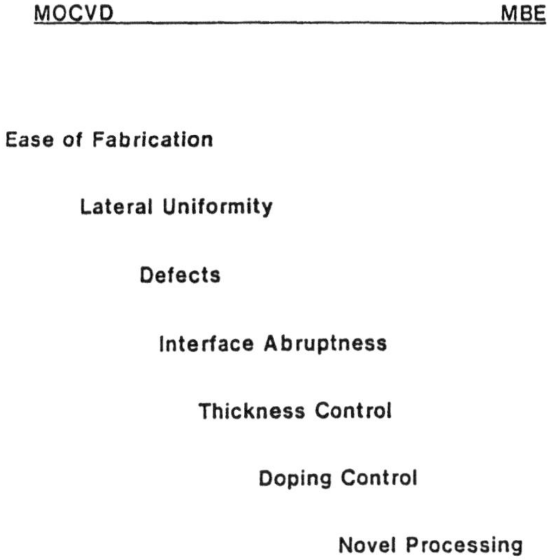

Fig. 2. Schematic comparison of advantages for MOCVD and MBE. A scale is depicted where a property such as ease of fabrication is an MOCVD advantage while at present novel *in situ* processing capabilities are a distinct advantage of MBE. Depending on the characteristic considered, either of the new epitaxial interface growth methods may be advantageous. For such areas as interface abruptness on an atomic scale, thickness control, defect formation, the two techniques are comparable. The challenge for future research is to improve the characteristics that are not well controlled (i.e., lateral film uniformity in MBE materials.

comprehensive treatment of the topics discussed. By presenting an overview of a broad range of recent developments, it is suggested that future advances will occur from extending knowledge in new areas such as strained-layer superlattices and combining them with the extraordinary atomic level tools developed in surface physics during the past 15 years.

2. FORMATION

Recent developments in the preparation of complex semiconductor interfaces are due to the development of deposition methods capable of atomic level control. The new methods of molecular beam epitaxy (MBE) and metal organic chemical vapor deposition (MOCVD) result in layers of extraordinarily high quality at thicknesses down to interatomic dimensions. This is opening up applications involving new levels of integration which exploit new materials and device structures.

2.A. Growth Methods

2.A.1. Metal Organic Chemical Vapor Deposition (MOCVD)

While scientific understanding of the interface formation process by MOCVD is at a more primitive level than MBE, the flexibility and potential for large-scale manufacturing makes this an extraordinarily exciting interface formation methodology. MOCVD systems are composed of a series of valves and stainless steel gas lines which are controlled by highly sophisticated mass flow controllers. The basic chemical reactions occurring at the semiconductor surface are relatively straightforward. For example, when trimethylaluminum and gallium are combined in arsine at between 600° and 800°C, an AlGaAs alloy of composition is formed in proportion to the amounts of aluminum and gallium compounds with methane as a byproduct:

$$(CH_3)_3Al + 4(CH_3)_3Ga + 5AsH_3 \xrightarrow[600^\circ-800^\circ C]{} 5Al_{0.2}Ga_{0.8}As + 15CH_4 \ . \qquad (1)$$

Because this technique involves the combination of organic compounds, there are a myriad of potential sources for controlling the growth of semiconductors and forming *in situ* interfaces with other semiconductors as well as metals and insulators. [1]

For group III elements there are at least a dozen potential sources involving combinations of methyl (CH_3), ethyl (C_2H_5), and isobutyl (C_4H_9) with the appropriate metallic or hydrogenated metallic species of aluminum, gallium, indium, etc. For group V sources, arsine (AsH_3) and phosphine (PH_3) have traditionally been used because of their very high toxicity. Recent attempts have been made to substitute organic compounds involving methyl, ethyl, and isobutyl groups with the arsenic, phosphorus or antimony sources. One area of particular promise is the use of adducts. These are organic compounds which are combinations of triethyl or trimethyl radicals with both a group III and group V element (e.g. $(CH_3)_3Ga{\cdot}(CH_3)_3As$). Literally dozens of sources exist for doping III-V compounds both n and p type. Se, Te, Si, S, Sn, and Ge are all available for n-type dopants. Zn, Mg, Be, Cd, Si, and Ge can be employed for p-type doping (though under normal vapor growth conditions Si and Ge are n-type dopants). To create complex interfaces with other materials, sources for semi-insulating layers are possible with such compounds as $VO(OC_2H_5)_3$ and $(C_5H_5)_2$ Fe. Metal contacts can also be deposited by MOCVD using such organic sources as $C_7H_{13}O_2Au$ for depositing gold. Diffusion masks can be deposited in such forms as Si_3N_4 by reacting SiH_4 with NH_3.

With such great potential and myriad of variables, MOCVD reactors have become increasingly complex. At Xerox PARC, R. D. Burnham [1] began building his first reactor in 1979 with two inject lines. In 1982 a next generation system was built with three lines in order to grow controlled quantum well structures and allow atomic level planar doping of III-V compounds. The present state of the art is exemplified by a new six inject line reactor which became operational in December, 1986. These are used for a multiple chamber configuration where choices of gases can be changed to include the types of materials discussed in the last paragraph plus selective etchants. Such reactors involve computer control of the switching manifold with the potential for extensive *in situ* monitoring. In addition, *in situ* lateral modification of the layers is possible using laser assisted growth techniques as discussed in section 3.B.1 below.

The resulting MOCVD layers exhibit interface abruptness of one interatomic spacing [2]. The uniformity of such layers are as good as or better than similar structures fabricated by MBE, as shown by quantum well laser structures. [3] The result is a technique for formation of interfaces which offer advantages and disadvantages comparable to MBE.

2.A.2. <u>Molecular Beam Epitaxy (MBE)</u>

Molecular Beam Epitaxy offers an interface formation technique which is compatible with sophisticated methods for characterizing surface properties. Present state of the art involves combining multiple ultra-high vacuum chambers with sample transport and interlock mechanisms. In this way interfaces can be grown by MBE in one special purpose chamber, transferred to another chamber for characterizing the properties of the surface and subsurface regions, and ultimately transferred to chambers which measure the performance of interfaces in producing physical phenomena of interest (e.g., light emission, transport). An example of such a sophisticated multi-chamber system is shown in Fig. 3, [4].

A system constructed by Ross Bringans and Lars-Erik Swartz of Xerox PARC combines three chambers for MBE preparation, characterization and measurement of synchrotron radiation photoelectron spectroscopies (i.e., angle resolved valence electrons using UV and high resolution core levels using soft x-rays). [5]

The current state of the art in molecular beam epitaxial systems has been reviewed in these proceedings by Klaus Ploog. [6] Here we will review some of the understanding of MBE interface formation that can be obtained through the combination of synchrotron radiation with MBE growth during the initial stages of heterojunction formation. Because the valence band offset evolves during the first few monolayers of growth, it is vital to probe the detailed interatomic interactions during the initial stages of deposition. Using the core-level emission from both the substrate and the overlayer adatoms, both stoichiometric and chemical bonding information can be directly obtained for the interface species involved. The intensity of the emission from the 3d core levels can be analyzed directly to follow the characteristics of the interface formation. Taking

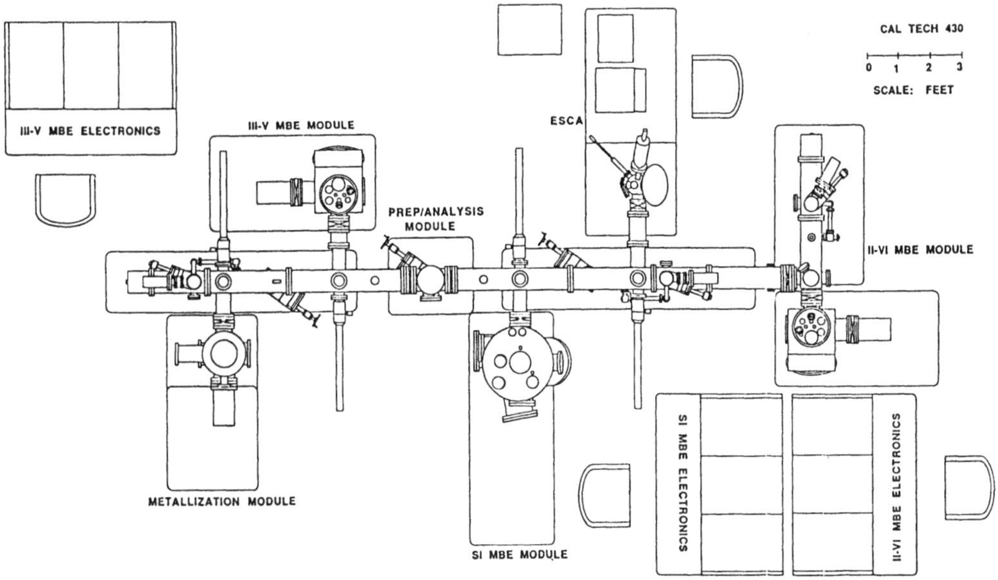

Fig. 3. Schematic diagram of the Caltech 430 MBE and analysis system manufactured by Physical Electronics, Inc. The system allows group IV, III-V, and II-VI MBE combined with a metalization module for sample contacts and silicide formation. Various *in situ* analysis techniques are possible including the ESCA facility shown in the diagram.

as an example the formation of the Ge interface with GaAs, we find that the thin film growth is dramatically different as a function of substrate temperature. [7] Normal photoemission analysis of core level intensity versus log of overlayer thickness, can distinguish abrupt epitaxial interface formation at 350°C from island formation at 22°C. [7] At lower temperatures much greater thicknesses are necessary to fully attenuate the substrate signal indicating island formation. At higher temperatures, interdiffusion occurs and the substrate constituents behave differently. Since in all cases the Ga substrate is known to be immobile, we have devised a novel method for analyzing core photoemission intensities. [8]

Figure 4 is a plot of the percentage of total 3d core photoemission intensity as a function of the percentage of the element which is stationary at the interface (i.e., Ga). In the ideal case shown by the dashed line, as the overlayer thickness increases (i.e., Ge), the substrate constituent signal would decrease to the origin (e.g., both Ga and As signals disappear at the same time). This is the case for 350°C Ge MBE growth on GaAs (110) [7]. In the case of GaAs (100) surfaces in Fig. 4, excess As evolves from the substrate to the top of the growing Ge film for all initial surface structures. This "stationary cation analysis" of core-level photoelectron spectra provides overviews for trends in the atomic level motion at interfaces during the initial stages of formation.

This result for Ge growth on (100) surfaces of GaAs demonstrates the importance of surface energy in determining the ultimate configuration of semiconductor interfaces. In another Ge MBE growth experiment involving (110) non-polar substrates, it was shown that even though AlAs is a more stable bulk semiconductor, the surface interactions were such that interdiffusion occurred even more abruptly than for GaAs. [9] Even though the enthalpy of AlAs is -28 kcal/mole compared to GaAs's enthalpy of -17 kcal/mole, deposition of Ge at between 325° and 350°C leads to out-diffusion of the substrate species from the non-polar (110) face <u>only</u> in the case of the AlAs. Thus bulk equilibrium thermodynamic quantities are not predictive of the surface and kinetic processes involved in MBE interface formation.

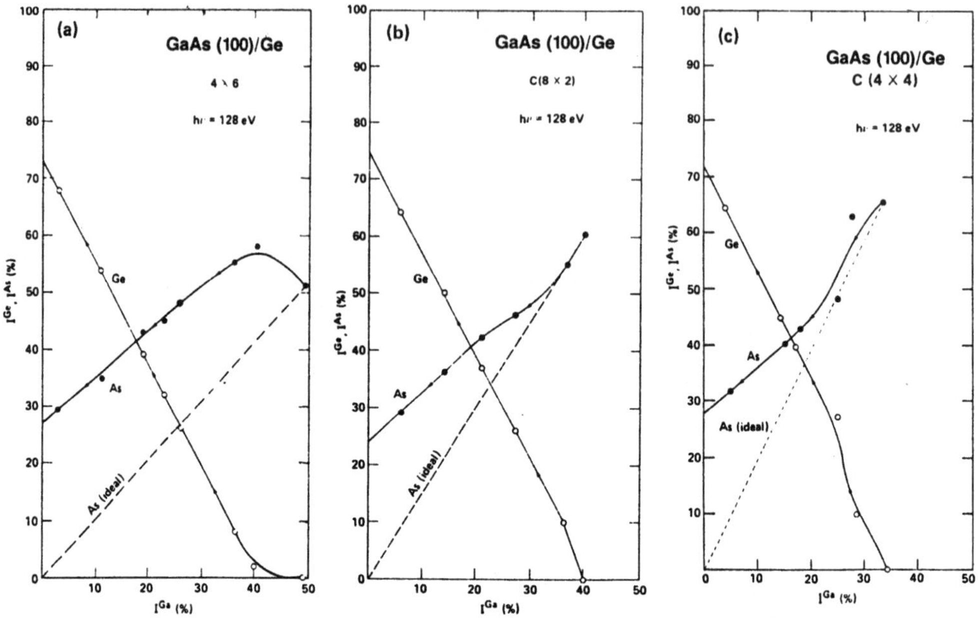

Fig. 4. Stationary cation analysis of core-level photoemission intensity to determine interface formation characteristics on an atomic scale for Ge deposited by MBE in *in situ* on GaAs (100) surfaces under epitaxial growth conditions [8]

2.A.3. Oxidation

No discussion of semiconductor interface formation and properties would be complete without consideration of the Si/SiO_2 interface. This gift of nature provides the physical properties allowing the marvels of the current electronics age. One of the striking features of this naturally occurring interface growth method is that the growth of the silicon dioxide/silicon interface proceeds in an atomically abrupt, planar manner. [10] Photoelectron spectroscopy has provided an extraordinarily valuable tool for characterizing this interface on an atomic scale. [11] Grunthaner et al. have used high resolution XPS techniques to identify the various bonding species of silicon with oxygen. Recently, they have extended this work using synchrotron radiation to vary the photon energy and obtain depth profiles of the oxide and substrate species. They have developed models for the bonding of silicon in its various charge states with oxygen across the (100) and (111) interfaces. By correlating core-level electron energy distributions at fixed photon energy with the photon energy variation of photoelectron yield at the silicon 2p threshold around 100 eV, we have been able to identify detailed formation mechanisms for SiO_2 and sub-oxide species of silicon at the interface. For the growth methods employed, we can develop the detailed atomic model for the interface which extends over one to two interatomic planes. [12,5]

Using such techniques, one can explore modifications in the growth procedures which may allow modification of interface defect formation, incorporation of foreign species to tailor electrical and transport properties, and ultimately achievement of multi-layer silicon/SiO_2 structures for advanced three-dimensional devices.

2.B. Epitaxy

Having explored methods for forming semiconductor interfaces, we discuss the process of epitaxy in detail for both lattice matched and greatly lattice mismatched semiconductor interfaces.

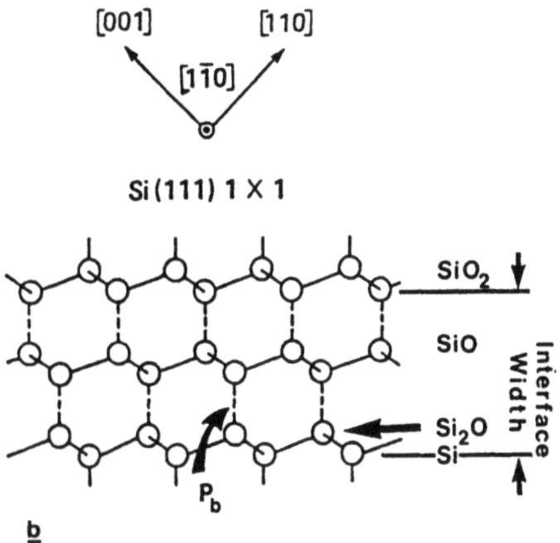

Fig. 5. Model for the interface width and species of silicon atoms surrounded by 4 silicons (Si),1 oxygen and 3 silicons (Si_2O), silicon with 2 oxygen and 2 silicon nearest neighbors (SiO), and fully oxidized SiO_2 (Silicon surrounded with four oxygen near neighbors). This result for (111) oxidation was determined using both the high resolution and tunable characteristics of soft X-ray synchrotron radiation in the 100 eV range for Si cleaved surfaces at elevated temperature. [12]

2.B.1. Lattice Matched Interfaces

2.B.1.a. GaAs/Ge/AlAs

There are many examples of heterojunction interfaces grown by MBE displaying pathologically different properties depending on such parameters as growth sequence. [9] However, device properties that are extremely sensitive to band offsets indicate symmetric barrier formation for reciprocal growth conditions. [7,13] It is important to understand the details of the chemical and physical structure on an atomic scale in order to understand the physical origin of such variations in interface properties as well as to evaluate possible methods of providing property control through growth variation.

We must recognize that the free surfaces of compound semiconductors are extremely complex. For a particular crystallographic orientation, different surface superstructures having varying stoichiometries can drastically affect the initial atomic configuration and chemical composition that incident adatoms bond with. Further, as discussed above, outdiffusions can occur under ideal conditions based on the stabilization of the growing free surface. Detailed decompositions of photoemission core levels into surface and bulk contributions can be individually studied as sub-monolayer coverages of foreign adatoms are deposited. Detailed kinetics for interface formation and models for the ultimate interface [10] can be obtained from 3d core line shapes and intensity variations for Ga, As, and Ge with growth. [14,7,5,8] An example of such a result is given in Fig. 6.

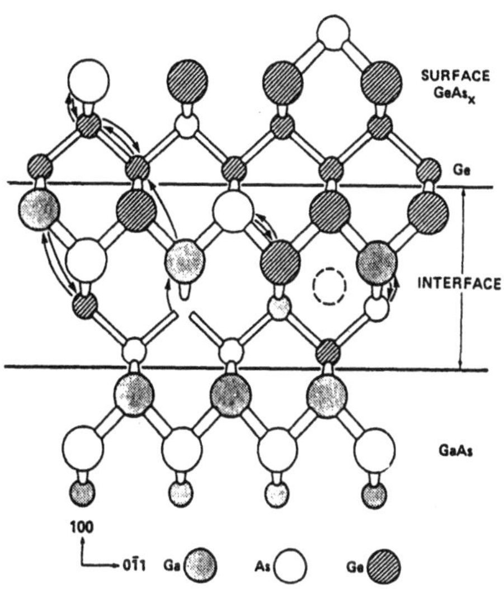

Fig. 6. Model for the GaAs (100)/Ge heterojunction grown by MBE at 340°C. The diagram shows some of the interatomic exchange processes that occur during MBE growth that lead to an atomically distributed (though otherwise abrupt) transition from one semiconductor to the other. This transition and the resulting complex structure at the interface will depend on the stoichiometry and structure of the initial GaAs surface because the growth process is dominated by physical and chemical forces which continually minimize the free surface energy. [10]

From such studies of epitaxial interface formation we conclude that metastable surface phases are driving forces that stabilize the ultimate interface atomic structure. Understanding the stable arsenic-terminated GeAs surface is as vital as having a model for the clean GaAs (100) starting surface. [15] This free energy driving force also means that the epitaxial overlayers are not necessarily characteristic of the bulk stoichiometric material. Detailed surface and interface physics and chemistry must be determined to understand formation of heterojunction interfaces.

2.B.1.b. $NiSi_2/Si$

Formation of lattice matched semiconductor interfaces is not limited to heterojunctions alone. One of the most exciting developments in this regard is the formation of insulator/semiconductor lattice-matched interfaces [16] and epitaxial metal/semiconductor interfaces [17,18]. Here we only briefly discuss the latter case as an example of exciting new directions in forming novel epitaxial interfaces.

The basic process of creating epitaxial Si/silicide structures involves (1) surface cleaning by argon ion bombardment and annealing, (2) deposition of a metal film onto the clean surface, and (3) formation of the silicide by solid phase reaction by annealing at elevated temperatures. Si MBE is much more difficult than GaAlAs MBE because of the higher substrate temperature required to achieve epitaxial growth. For example, *in situ* heating up to 1200°C while maintaining 10^{-9} torr UHV ambient is necessary for Si surface preparation. The result for Ni growth is the formation of atomically abrupt epitaxial $NiSi_2/Si$ interfaces. The $NiSi_2$ lattice matches Si and has two possible orientations relative to Si (111). (Type A interfaces have $NiSi_2$ aligned with the silicon lattice, while type B is the same structure rotated by 180°.) The important question is whether the properties of the abrupt interface depend on this different relative atomic orientation of the two lattice-matched materials.

To investigate this controversial question [17], photoresponse measurements using conventional Fowler analysis have been combined with analysis of the forward I-V characteristics of the same Schottky barrier samples by Hauenstein et al. [18] The rotated type B interface is found to be a compound barrier of both high and low barrier height. The Schottky barrier height for Type B differs by nearly 200 meV from the 0.62 ± .01 eV value obtained for the type A interface. Understanding the origin of such differences in properties and their relation to interface formation is crucial for tailoring device properties. For example, in this current $NiSi_2/Si$ controversy, understanding of the origin and role of planar defects along the interface must be improved; simple misfit dislocations or clustering of the two different types of interfaces appear to be absent for this system. Understanding the nature of the defects and their formation process is an issue for surface and interface physicists.

2.B.2. Strained Layer Superlattices

It is often desirable to tailor properties of semiconductor interfaces independent of whether the new material to be grown matches the lattice structure of the semiconductor substrate. In the past, quaternary alloys of column III and column V elements would have had to be combined in order to create an alloy with optimal lattice match across a heterojunction interface. Examples include InGaAsP and InGaAlP to achieve a band gap greater than that of the GaAs substrate. By being able to grow thin layers of semiconductors, one can create superlattice structures which accommodate the strain caused by the mismatch in lattice constant for materials with otherwise desirable properties (e.g., band gap). In this section we explore the cases of II-VI materials and of compound semiconductor growth on elemental semiconductor substrates. Both are examples of new methods of forming semiconductor structures to tailor specific physical properties.

2.B.2.a. CdTe/ZnTe

Because of their larger band gaps, II-VI materials offer the possibility for creating structures that absorb and emit light in the visible region of the spectrum. The promise of these materials has not been realized to date because of the difficulties in producing defect free structures and structures with controlled doping. With the advent of MOCVD and MBE, growth temperatures can be significantly lowered. For the zinc II-VI compounds, MOCVD growth temperatures are in the range of 350° to 400°C with MBE growths at 180°C compared to solution growth temperatures at 1600°C and even LPE at 800° to 850°C. This leads to better control of purity, layer structure, and dopant incorporation. By constructing structures of multiple layers of only tens of angstroms of thickness, one can in addition tune properties and utilize substrates compatible with other heterojunction combinations.

In a series of well-controlled MBE growths by Faurie and co-workers, Miles et al. have studied in detail the nature of the strain accommodation for II-VI growth on GaAs substrates and the resulting luminescent properties. [19] The superlattice structure studied is shown schematically in Fig. 7.

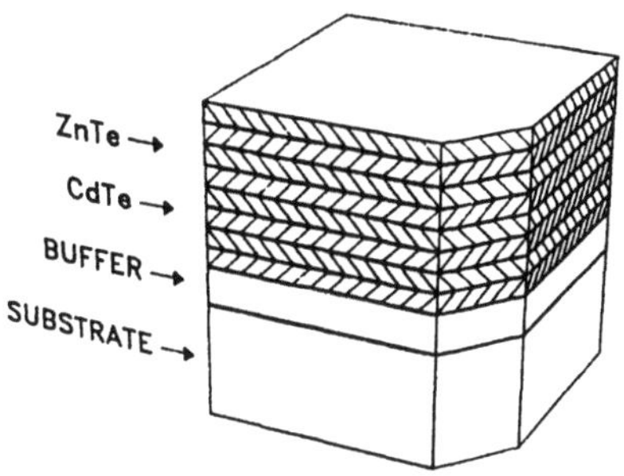

Fig. 7. Schematic diagram of ZnTe/CdTe superlattice structure grown on GaAs substrates with CdTe, ZnTe, and $Cd_{0.5}Zn_{0.5}Te$ buffer layers [19]

The bulk properties of CdTe and ZnTe show band gaps of 1.6 eV and 2.38 eV respectively and zinc blende lattice constants of 6.481 and 6.104 Å respectively. This corresponds to a lattice mismatch of 6.2%. The strained layer epitaxy for such highly mismatched pairs of materials can proceed in two ways. One, the overlayers can be unstrained with the lattice mismatch accommodated by dislocations at the interfaces between the materials (see Fig. 8a). Two, for layers below a critical thickness, the strain may be coherently accommodated by elastic deformation to a single lattice constant in the plane of growth. This results in alternate growth-direction expansion and contraction of consecutive layers, as dictated by a biaxial analogue of the Poisson ratio. The closer this in-plane lattice constant lies to the lattice constant which minimizes strain energy in the structure, the larger the "critical thickness" below which the structure is defect-free.

The best experimental way to resolve which of these strain accommodation mechanisms is operable for various layer thicknesses of CdTe/ZnTe, is through X-ray diffraction measurements. Through a measurement of superlattices of thicknesses in the range of 20 to 50 Å with approximately 200 periods each, Miles et al. studied the

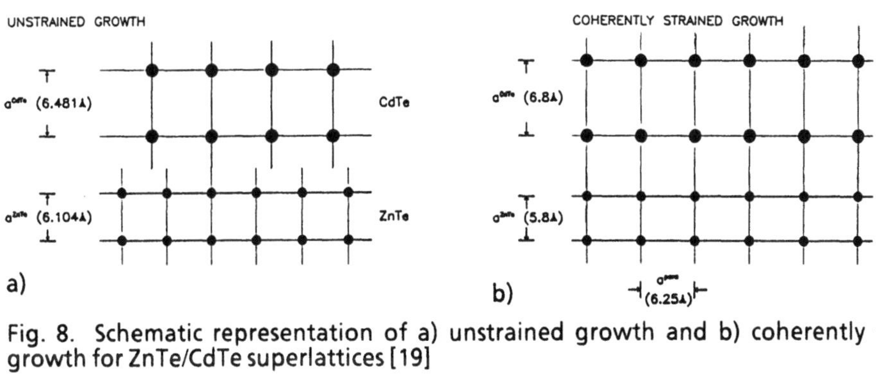

Fig. 8. Schematic representation of a) unstrained growth and b) coherently strained growth for ZnTe/CdTe superlattices [19]

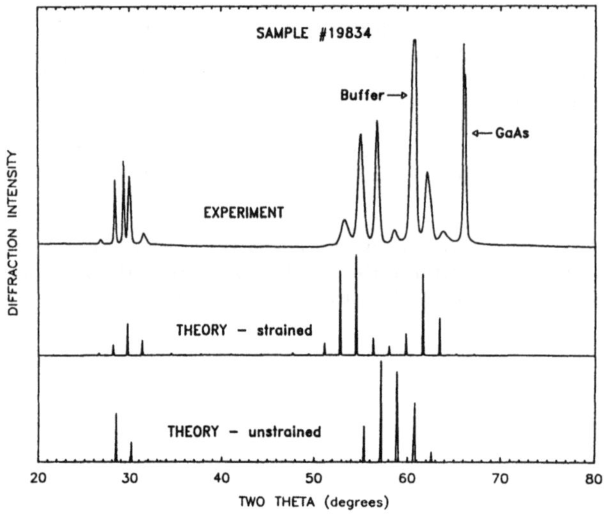

Fig. 9. X-ray diffraction measurements and kinematical model calculations for CeTe/ZnTe superlattices. [19]

contributions to structure from single superlattice cells and the effect of the superlattice periodicity. As summarized in Fig. 9, comparison of kinematical calculations with X-ray diffraction data suggests that the superlattices are highly strained with a structure close to that of a free-standing superlattice. Combining this data with TEM and in situ RHEED measurements has led to the conclusion that defects created at the buffer layer/superlattice interface do not propagate far into the superlattices, leaving high-quality structures away from this interface.

It is important to note also that such high quality superlattices exhibited strain accommodation independent of the composition of the buffer layer. It is believed that in these cases, creation of dislocations at the buffer layer to superlattice boundary accommodated the structural differences. As will be discussed in 3.B.2, the layers were quite uniform in thickness and exhibited intense luminescence compared to the equivalent II-VI alloy. Importantly, high quality II-VI superlattices could be grown for layer thicknesses exceeding the critical thickness predicted by standard theoretical approaches of Bean, Matthews and Van der Merwe [20]. This appears to be the result of a sudden jump of the in-plane lattice constant at the superlattice/buffer layer interface. Thus, although this lattice constant is not a parameter which can be adjusted at will in systems with this large a lattice mismatch, the choice of substrate appears to have little impact on the high quality of the superlattices. Understanding the detailed physical forces that define the critical thickness for such large mismatched superlattices is necessary if formation of such semiconductor interfaces is to be properly predicted and controlled.

2.B.2.b. GaAs/Si

Epitaxial interface formation of GaAs on Si offers major opportunities for merging the physical properties of silicon interfaces with those of III-V's, as well as providing cheap, durable, large area, and high thermal conductivity substrates. Problems in achieving this involve not only the large lattice mismatch of 4% but also the formation of antiphase domains within the compound semiconductor epitaxial layer. This latter problem occurs with growth on elemental substrates because of the absence of a preferential site for adsorption of the group III or group V element. One way suggested to avoid this situation is to use non-primitive crystallographic planes such as the Si (211) orientation. [21] The more technologically interesting result would be to employ the Si (100) surface because it is the orientation used for standard VLSI technologies.

Recently Aspnes and Ihm [22] have suggested a topological solution to the problem of GaAs growth on Si. Simply stated, if one can create steps that are two atoms in height, the Ga and As planes will interleave. This occurs because the first layer formed on Si will always be terminated by As due to the preferential energy of the arsenic terminated silicon surface. [15] Note that this is the same cause for As termination of the growing Ge surface in the III-V lattice-matched Ge/GaAs epitaxy discussed above. With steps in increments of two atomic heights, the terraces would be high enough to accommodate a layer of Ga plus As on top which is adjacent to As directly terminating Si on silicons on the neighboring plateau. Through a series of calculations of the energy of silicon atoms at step edges, Aspnes and Ihm have found that there is a net reduction of 0.04 eV per step atom through a π-bonded chain reconstruction. In addition, by having vicinal cuts in the (110) rather than (100) axis relative to the (100) surface normal, step terraces of retangular shape provide a favorable situation to accommodate low Ga or As surface mobility. Simple annealing for five minutes at 1000°C for such Si (100) substrates is found to eliminate all monoatomic steps. This provides a substrate for GaAs growth which should not produce antiphase domains. The sole origin of this effect appears to be thermodynamic driving forces caused by step atom reconstruction. [22] This new result for compound semiconductor growth on elemental semiconductor substrates deserves considerable effort by surface physicists to explore experimentally and theoretically the properties of steps.

3. PROPERTIES

Motivation for understanding the formation of semiconductor interfaces on an atomic scale is driven by the desire to tailor the properties of the resulting structures. In this section we explore both the parameters which determine properties of semiconductor interfaces as well as demonstrating new and exciting trends in properties of novel semiconductor interfaces.

3.A. Barrier Heights (Heterojunction)

The principal parameter determining semiconductor interface properties is the height of the electrostatic barrier at the interface. Whether these be epitaxial, amorphous, or interfaces created by doping variations in a single host material, changes in the potential within the semiconductor are responsible for the interesting physical properties.

3.A.1. Schottky Barriers and Ohmic Contacts

Much has been written about the controversies in understanding Schottky barrier heights. [23,24] Debate over the role of defects versus metal induced gap states (independent of what they are called!) seems to go on forever. The issues are important and perhaps some of the more subtle effects such as the differences in Type A and B interfaces of $NiSi_2$/Si will help to resolve these issues.

As will be shown for heterojunctions, Schottky barrier properties can be tailored. For example, Brillson has shown that metallic interlayers can cause metal/semiconductor interfaces to exhibit either rectifying or ohmic-current voltage characteristics depending on the thickness of the Al placed between Au and CdS. [26] See Fig. 10.

In discussing the formation of semiconductor interfaces, no consideration has been given to the problem of ohmic contacts. While major questions of the origin and theory for Schottky barriers remain (such as to explain the difference in type A versus type B $NiSi_2$/Si barrier heights), no comparable effort to understand and improve ohmic contacts exists. The metallurgy is quite complex [10] and epitaxial ohmic contact work is only just beginning [25].

3.A.2. Heterojunctions

There has been much recent discussion of the heterojunction band offset problem. [5,24] Here we provide a summary perspective on the current state of understanding.

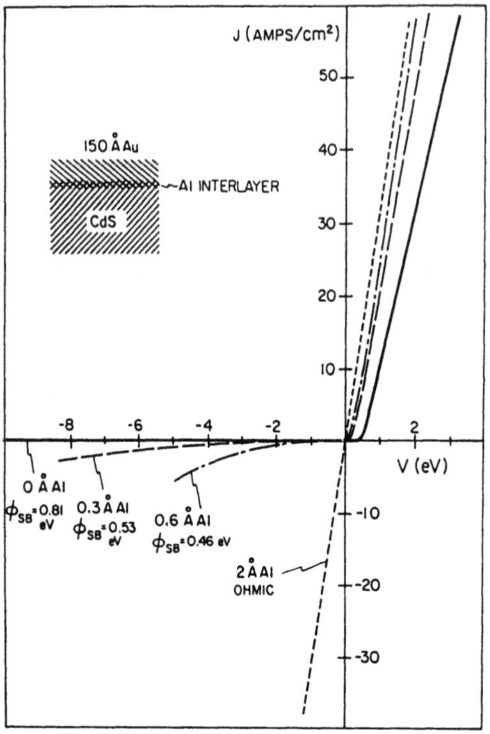

Fig. 10. Variation in Au/CdS interface properties by *in situ* deposition of submonolayer thicknesses of Al [27]

Recent work has firmly established that each heterojunction pair can have a single, unique band offset value independent of preparation details if the interfaces are formed in a controlled, ideal manner. For example, the crystallographic orientation, [8] the presence of large excess of a single constituent, [27] or the intentional deposition of a metal inner layer [28] do not change the value of the GaAs/Ge band offset. The controversey over possible violations of the commutative and transitive properties of band offsets has been resolved in favor of reciprocal interfaces which are clearly known to occur in real device structures. [29,30] Much of the discussion of tailoring intrinsic properties of interfaces is due to extrinsic effects in the formation of the heterojunction interface; for example, the use of cleaved surfaces and room temperature amorphous overlayers artificially causes variation as large as half the band gap compared to epitaxial crystalline layers on defect-free substrates. Even though charged levels of a single sign at the interface cannot pin the Fermi level [10,27], dipole layers created by intentional monolayer dopants of opposite sign intentionally deposited within at most a few atomic layers of each side of the interface can produce intentional potential changes. [31] It must be emphasized, however, that the properties of heterojunctions are controlled by band offsets with sensitivities down to meV. [32,5] While some of the most interesting theoretical advances have involved trying to find commonality between transition metal impurities, Schottky barrier heights, and heterojunction band offsets, [25] the general trends obtained by such theoretical approaches are only road maps for the experimentalist who must probe these interface properties to orders of magnitude better accuracy. The best "rule of thumb" for the band offset value of a new, unknown semiconductor/semiconductor pair is provided by the table of valence band edges of Margaritondo and co-workers. [33,15] However, caution must be used as inconsistencies are often found (e.g., ZnSe/SnTe), and the table values have evolved significantly over time (see for comparison ref. [33]).

Table 1. Positions of the band edges are given relative to the valence band edge of germanium. The difference between two terms gives a zeroth-order estimate of the discontinuity ΔE_V.

Energies of valence band edges

Semiconductor	E_V (eV)
Ge	-0.00
Si	-0.16
α-Sn	0.22
AlAs	-0.78
AlSb	-0.61
GaAs	-0.35
GaP	-0.89
GaSb	-0.21
InAs	-0.28
InP	-0.69
InSb	-0.09
CdS	-1.74
CdSe	-1.33
CdTe	-0.88
ZnSe	-1.40
ZnTe	-1.00
PbTe	-0.35
HgTe	-0.75
CuBr	-0.87
GaSe	-0.95
$CuInSe_2$	-0.33
$CuGaSe_2$	-0.62
$ZnSnP_2$	-0.48

3.B. Light Emission/Absorption

3.B.1. Quantum Well Lasers/Superlattice Luminescence

The ability to form heterojunctions having atomic scale interface abruptness has provided dramatic enhancements in the quality and properties of light-emitting devices. A number of excellent reviews have been published on the quantum size effect [34], quantum well lasers [35], and general reviews on new directions made possible by advances in MOCVD and MBE interface formation techniques [36]. The ability to control composition, interface abruptness, and complex structures of III-V alloys allows tailorability of opto-electronic properties.

One particular important recent advance involves impurity induced homogenation of superlattice regions at relatively low annealing temperatures. This promises to provide a new generation of novel opto-electronic properties. [37] Dramatic advances have been made in utilizing both n and p-type impurity dopants to cause homogenation of quantum well superlattice regions. [38] However, the detailed physical mechanism that causes impurities at lattice matched III-V heterojunction interfaces to produce massive interdiffusion some 400-^{o}500°C below normal pure superlattice interdiffusion temperatures is not understood and is little studied. [39]

As shown in Fig. 2, one of the distinct advantages of MBE is its ability to achieve *in situ* lateral modification of the atomic layer growth. One of the challenges for MOCVD is to develop ways to achieve such *in situ* processing of semiconductor interfaces. As shown schematically in Fig. 11, one method is to use focussed *ex situ* laser beams transmitted through the reactor so as to produce fine lines on the growing semiconductor interface. In this way, lasers can assist MOCVD, by local heating for device structure formation homogenizing alloys, by controlled photochemical etching, and by laterally defined deposition utilizing through photochemical and photothermal growth rate enhancements.

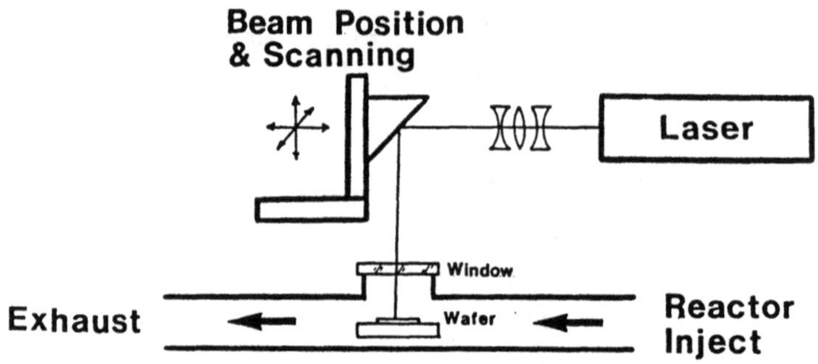

Fig. 11. Schematic diagram of a scanned laser system for *in situ* lateral structure processing of epitaxial layers grown by MOCVD.

The impurity-induced disordering effect smears the structure of superlattices with layers hundreds of angstroms in thickness. The region of homogenized alloy material has the composition of the average superlattice composition. The quality of the homogenized layer is equivalent in optical and electrical properties to that of the equivalent alloy layer grown directly by MOCVD. Thus, one can grow 2-D superlattice structures and then choose regions in which to locally induce homogenation of the lattice in order to create lateral regions of different optical and electrical characteristics. By choosing the proper impurity, one can also type convert to either n-type using silicon impurities or p-type by using zinc impurities. This provides a novel mechanism for creating regions of different indices of refraction laterally. In this way laser structures of enhanced capability can be produced.

The results described for impurity induced disordering have been nearly all the result of empirical studies. [38] Basic questions need to be resolved such as how the presence of the interfaces is important in promoting the impurity induced interaction in the superlattice. [39] Further questions of what range such forces may operate over (given that the effect has been seen for layers many hundreds of angstroms in thickness) present interesting challenges to the surface and interface scientist. Understanding and controlling this phenomenon promises to provide a new, powerful method for tailoring light emitting and detecting properties of semiconductor interfaces.

3.B.2. Strained-Layer Superlattices

The major advantage of creating strained-layer superlattices is to form interfaces among materials with desired properties where the structural parameters are incompatible. For the ZnTe/CdTe case described in 2.B.2.a, the band structure is shown in Fig. 12. Note that the valence bands include light and heavy hole contributions which are important in understanding the ultimate-superlattice optical properties. These superlattices exhibit intense luminescence compared to a rather weak and shifted broad luminescent intensity for the equivalent $Cd_xZn_{1-x}Te$ alloy. [19] The power of the strained layer superlattice idea is perhaps illustrated best by Fig. 13 where a k·p calculation for the contours of constant energy gap are shown as functions of the number of CdTe and ZnTe layers per superlattice period. A range of band gaps is achievable. Note also that because of the strain effects on the light hole valence states in ZnTe, the superlattice gap actually decreases as ZnTe layers are added for fixed thicknesses of CdTe layers. This "bowing" is an intriguing phenomenon that provides a region of relative stability for tailoring the optical properties of strained-layer superlattices.

For Ge/Si strained-layer superlattices, band offsets can be changed from mainly electron to mainly hole barriers depending on alloy composition and layer thickness as described by Bean [40]. The uniqueness of the optical properties of such Ge/Si superlattices has recently been demonstrated by Pearsall et al. [41] where new optical

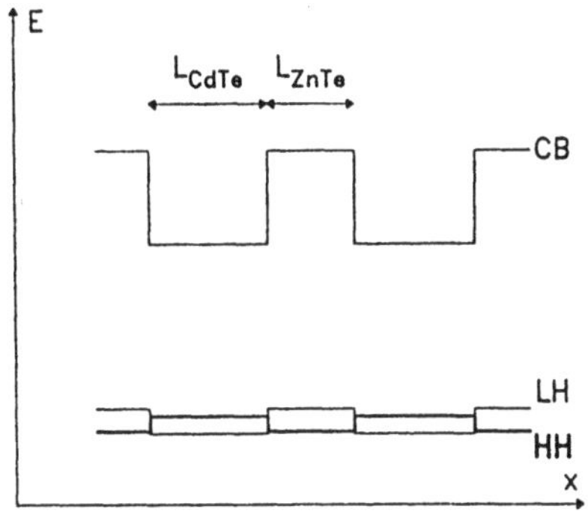

Fig. 12. Schematic band structure for the ZnTe/CdTe strained-layer superlattice

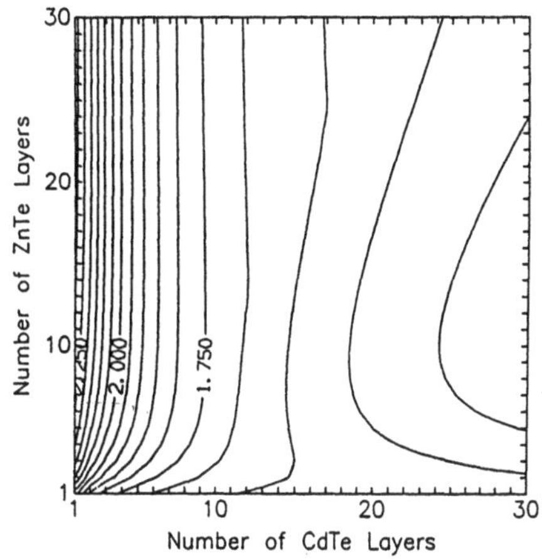

Fig. 13. Calculated superlattice band gaps with strain effects included. Contour interval is 50 meV. Calculations assumed the lattice constant of the free-standing superlattice. Note the shift to lower energies for high CdTe and low ZnTe content.

transitions were measured which are not present in Si, Ge, nor Si_xGe_{1-x} alloys. The tailoring of such superlattice structures offers the ability to vary optical properties of devices directly on Si substrates. With advances in these areas, the formation of opto-electronic circuits directly on Si substrates may be possible utilizing strictly group IV semiconductor interfaces.

3.C. Transport

With the ability to form interfaces with properties controlled to an atomic scale, one must understand the interactions of carriers being transported both parallel to such material transition regions as well as effects when they traverse across such abrupt boundaries. In this section, we note some of the considerations for the interactions of carriers near interfaces and suggest areas of great potential for novel interface properties.

3.C.1. Transport Parallel to Interfaces (High Electron Mobility Transistors)

The extraordinarily high mobilities for transport in the potential notch at heterojunction interface regions has received considerable attention. [42,33] Carrier interactions occur not only at the interface but across the interface for transport along interfaces. We must understand the extent of material transition regions, atomic scale interface composition, the formation of possible intrinsic clustering of atoms at interfaces, interface roughness, and long-range coulomb effects with dopants in layers near the interface. Note the fundamental role that the heterojunction band offset and doping play in determining the potential region in which carriers are transported.

3.C.2. Transport Perpendicular to Interfaces (Tunnel Structures)

With the ability to form interfaces of atomic abruptness, MBE and MOCVD are providing the ability to construct barriers of only 10-20 atomic layers in thickness. This creates the possibility for fabricating a series of heterojunction interfaces which can be used to control conduction via tunelling phenomena. Such structures can provide inherently high speed devices because of the rapid variation in current with voltage. There have been a number of major experimental programs begun to create negative differential resistance using a GaAs quantum well separated by two thin GaAlAs barriers. Many device configurations are under active investigation with exciting results beginning to appear. [43]

Detailed understanding of the interaction of carriers through these interfaces is crucial to making progress with these structures. Tailorability will depend on the affect of modifications to interface chemical and physical structure on transport and energy loss phenomenon. Reports of large, reproducible interface phonon structures have been reported for I-V measurements. [44] Perhaps these tunnel structures offer one of the most challenging problems for surface and interface studies because of their extreme sensitivity to the details of the entire interface environment. Both theoretical and experimental probes will be necessary to increase knowledge of these structures.

4. CONCLUSIONS

Of all semiconductor interfaces, heterojunctions and superlattices provide some of the most exciting new opportunities. These man-made materials provide a broad range of new device structure possibilties. Tailorability can be achieved in systems that previously had been believed to be unsuitable for applications due to such effects as lattice mismatch induced strain. The key to realizing such promise is to increase our understanding of surfaces and interfaces. The advances in MOCVD and MBE provide new atomic level control of interface formation. Advances in achieving lateral modification of these atomically controlled layers offer ways of directly tuning device structures. To realize this potential both thermodynamic and kinetic effects during the growth of semiconductor interfaces must be understood better. In this way semiconductor interface properties become controllable. Band offsets (barrier heights in general) can be controlled and tuned to provide novel light emission, optical absorption, and carrier adsorption (parallel and perpendicular to interfaces). The opportunities in this field are truly exploding!

5. ACKNOWLEDGEMENTS

The portions of this paper reporting results from the Xerox photoemission research would not have been possible without major collaborative contributions from H. W. Sang, Jr., A. Katnani, P. Zurcher, P. Chiaradia, and J. C. McMenamin, Jr. Perspectives on MOCVD are due to the insight and technical contributions of R. D. Burnham (Amaco). We acknowledge the exciting contributions and advice of R. L. Thornton (Xerox) on impurity-induced disordering of superlattices. The work at Caltech on strained-layer superlattices could not have progressed without the generous assistance of M. B. Johnson and C. Y. Wu. Ted Woodward (Caltech) provided stimulating discussions and shared results on resonant tunnelling device structures prior to publication. The support from the Office of Naval Research (L. R. Cooper) and DARPA (R. Reynolds, S. Roosild, and J. Murphy) is greatfully acknowledged for both its financial contribution as well as the technical perspective provided by the individuals. We are grateful to D. E. Aspnes of Bellcore for generously providing a preprint of his GaAs/Si work prior to publication.

REFERENCES

1. R. D. Burnham, W. Streifer, T. L. Paoli, and N. Holonyak, Jr., J. Crystal Growth **68**, 370 (1984); Y. Takahashi, T. Soga, S. Sakai, M. Umeno, and S. Hattori, Jpn. J. Appl. Phys. **23**, 709 (1984).

2. J. M. Brown, N. Holonyak, M. J. Ludowise, W. T. Deitze, and C. R. Lewis, Electron. Lett. **20**, 204 (1984).

3. R. D. Burnham, D. R. Scifres and W. Streifer, Appl. Phys. Lett. **40**, 118 (1982); D. R. Scifres, R. D. Burnham, M. Bernstein, H. Chung, F. Endicott, W. Mosby, J. Tramontana, J. Walker and R. D. Yingling, Appl Phys. Lett. **41**, 501 (1982); A. R. Bonnefoi, T. C. McGill and R. D. Burnham, Appl. Phys. Lett. **47**, 307 (1985).

4. Caltech 430 MBE system by Perkin-Elmer/Physical Electronics.

5. R. S. Bauer and G. Margaritondo, Physics Today **40**, Number 1, 26 (January, 1987).

6. K. Ploog, these proceedings, (1987).

7. R. S. Bauer and H. W. Sang, Jr., Surf. Sci. **132**, 479 (1983); R. S. Bauer and J. C. McMenamin, J. Vac. Sci. Techol. **15**, 1444 (1978).

8. A. D. Katnani, P. Chiaradia, H. W. Sang, Jr., P. Zurcher, and R. S. Bauer, Phys. Rev. B**31**, 2146 (1985).

9. R. S. Bauer, Thin Solid Films, **89**, 419 (1982).

10. R. S. Bauer and T. C. McGill, in: VSLI Electronics Microstructure Science, Eds., N. G. Einspruch and R. S. Bauer, **10**, "Interfaces and Devices," Chapt. 1, page 3 (1985).

II. F. J. Grunthaner and P. J. Grunthaner, Materials Science Reports, **1**, 65 (1986).

12. R. S. Bauer, SPIE Conf. Proc. **452**, 160 (1984).

13. A. D. Katnani and R. S. Bauer, Phys. Rev. B**33**, 1106 (1986).

14. R. S. Bauer and J. C. Mikelsen, Jr., J. Vac. Sci. Technol. **21**, 491 (1982).

15. F. Stucki, G. J. Lapeyre, R. S. Bauer, P. Zurcher and J. C. Mikkelsen, Jr. J. Vac. Sci. Technol. B**1**, 865 (1983); R. D. Bringans, R. I. G. Uhrberg and R. Z. Bachrach, Phys. Rev. Lett. **55**, 533 (1985); R. D. Bringans, R. I. G. Uhrberg, M. A. Olmstead, R. Z. Bachrach and J. E. Northrup, Physica Scripta **35** (1987).

16. C. M. Hanson and H. H. Wieder, J. Vac. Sci. Technol. B**5**, 971 (1987).

17. R. T. Tung, J. M. Gibson, and J. M. Poate, Phys. Rev. Lett. **50**, 429 (1983); M. Liehr, P. E. Schmid, F. K. LeGoues, and P. S. Ho, Phys. Rev. Lett. **54**, 2139 (1985).

18. R. J. Hauenstein, T. E. Schlesinger, T. C. McGill, B. D. Hunt, and L. J. Schowalter, J. Vac. Sci. Technol. A**4**, 860 (1986).

19. R. H. Miles, G. Y. Wu, M. B. Johnson, T. C. McGill, J. P. Faurie, and S. Sivananthan, Appl. Phys. Lett. **48**, 1383 (1986); R. H. Miles, T. C. McGill, S Sivananthan, X. Chu, and J. P. Faurie, J. Vac. Sci. Technol. B**5**, 1263 (1987).

20. J. H. Van der Merwe, J. Appl. Phys. **34**, 123 (1963); J. W. Matthews and A. E. Blakeslee, J. Cryst. Growth **27**, 118 (1974); **29**, 273 (1975); **32**, 265 (1976); R. People and J. C. Bean, Appl. Phys. Lett. **47**, 322 (1985).

21. S. L. Wright, M. Inada, and H. Kroemer, J. Vac. Sci. Technol. **21**, 534 (1982).

22. D. E. Aspnes and J. Ihm, Phys. Rev. Lett. **57**, 3054 (1986); J. Vac. Sci. Technol. **B5**, 939 (1987).

23. W. Monch, "On The Present Understanding of Schottky Contacts" in: Festkörperprobleme XXVI, (1986); J. Tersoff and W. A. Harrison, J. Vac. Sci. Technol. B**5**, 1221 (1987).

24. R. S. Bauer, Ed., "Proceedings of the 14th Annual Conference on the Physics and Chemistry of Semiconductor Interfaces," J. Vac. Sci. Technol. B**5**, 922-1311 (1987).

25. J. M. Woodall, J. Freeouf, G. D. Pettit; T. Jackson, and P. Kirchner, J. Vac. Sci. Technol. **19**, 626 (1981).

26. C. F. Brucker and L. J. Brillson, Appl Phys. Lett. **39**, 67 (1981).

27. P. Chiaradia, A. Katnani, H. W. Sang, Jr., and R. S. Bauer, Phys. Rev. Lett. **52**, 1246 (1984).

28. A. D. Katnani, P. Chiaradia, Y. Cho, P. Mahowald, P. Pianetta, and R. S. Bauer, Phys. Rev. B**32**, 4071 (1985).

29. J. R. Waldrop, R. W. Grant, and E. A. Kraut, J. Vac. Sci. Technol. B**5**, 1209 (1987).

30. A. D. Katnani and R. S. Bauer, Phys. Rev. B**33**, 1106 (1986).

31. F. Capasso, Surf. Sci. **132**, 527 (1983).

32. H. Kroemer, in: VSLI Electronics Microstructure Science, **10**, Eds. N. G. Einspruch and R. S. Bauer, Chapter 4, 121 (1985).

33. A. D. Katnani and G. Margaritondo, Phys. Rev. B**28**, 1944 (1983).

34. N. Holonyak, Jr., R. M. Kolbas, R. D. Dupuis, and P. D. Dapkus, IEEE J. Quantum Elect. **QE-16**, 170 (1986).

35. W. Streifer, A. Hardy, D. R. Scifres, and R. D. Burnham, IEEE J. Quantum Electron. **QE-19**, 991 (1983).

36. R. D. Burnham, W. Streifer, T. L. Paoli, and N. Holonyak, Jr., J. Crystal Growth **68**, 370 (1984).

37. W. D. Laidig, N. Holonyak, Jr., M. D. Camras, K. Hess, J. J. Coleman, P. D. Dapkus, and J. Bardeen, Appl. Phys. Lett. **38**, pp. 776-8 (1981); N. Holonyak, Jr., W. D. Laidig, M. D. Camras, J. J. Coleman, and P. D. Dapkus, Appl. Phys. Lett. **39**, pp. 102-4 (1981).

38. R. L. Thornton, R. D. Burnham, N. Holonyak, Jr., J. E. Epler, and T. L. Paoli, SPIE Conference Proceedings, **797** (1987).

39. J. A. Van Vechten, J. Vac. Sci. Technol. **B2**, 569 (1984).

40. J. C. Bean, "Silicon Based Semiconductor Heterostructures" in: Silicon Molecular Beam Epitaxy, Chapter 1, Eds., E. Kasper and J. C. Bean, CRC Press (1986).

41. T. Pearsall, J. Bevk, L. C. Feldman, J. M. Bonar, and A. Ourmazd, J. Vac. Sci. Technol. B**5**, 1274 (1987).

42. H. L. Stomer, Surf. Sci. **32**, 519 (1983).

43. T. K. Woodward, T. C. McGill, and R. D. Burnham, Appl. Phys. Lett. **50**(8) 451 (1985); N. Yokoyama, K. Imamura, S. Muto, S. Hiyamizu, and H. Nishi, Jpn. J. Appl. Phys. **24**, L853 (1985); F. Capasso and R. Kiehl, J. Appl. Phys. **58**, 1366 (1985); A. R. Bonnefoi, T. C. McGill, and R. D. Burnham, IEEE Electron Dev. Lett. **EDL-6**, 636 (1985); B. Ricco and M. Y. Azbel, Phys. Rev. B**29**, 1979 (1984).

44. A. R. Bonnefoi, D. H. Chow, T. C. McGill, R. D. Burnhan, and F. A. Ponce, J. Vac. Sci. Technol. B**4**, 988 (1986); R. T. Collins, J. Lambe, T. C. McGill, and R. D. Burnham, Appl. Phys. Lett. **44**, 532 (1984); A. R. Bonnefoi, T. C. McGill, R. D. Burnham, and G. B. Anderson, Appl. Phys. Lett. **50**(6), 344 (1987).

Index of Contributors

Arnaud d'Avitaya, F. 360

Badoz, P.A. 360
Bastard, G. 340
Bauer, R.S. 372
Binh, V.T. 126

Calandra, C. 162
Chabal, Y.J. 301
Chevalier, J.-P. 109
Chiaradia, P. 290
Cohen, C. 134

d'Anterroches, C. 360
Delalande, C. 340
Derrien, J. 102

Flores, F. 182

Geurts, J. 328
Glastre, G. 360

Himpsel, F.J. 196
Hollinger, G. 210

Lassabatère, L. 239
Le Lay, G. 48

Manghi, F. 162
McGill, T.C. 372
Miles, R.H. 372
Moison, J.M. 43,232
Morin, P. 114
Muret, P. 282

Pfister, J.C. 360
Ploog, K. 10

Richter, W. 328
Rosencher, E. 273,360
Rossi, G. 69,88

Salvan, F. 119
Sauvage-Simkin, M. 66
Sébenne, C.A. 2

Taglauer, E. 150

Vincent, G. 360

MIX
Papier aus verantwortungsvollen Quellen
Paper from responsible sources
FSC® C105338

If you have any concerns about our products,
you can contact us on
ProductSafety@springernature.com

In case Publisher is established outside the EU,
the EU authorized representative is:
Springer Nature Customer Service Center GmbH
Europaplatz 3, 69115 Heidelberg, Germany

Printed by Libri Plureos GmbH
in Hamburg, Germany